PEARSON EDEXCEL INTERNATIONAL AS/A LEVEL

CHEMISTRY

Student Book 1

Cliff Curtis
Jason Murgatroyd
with David Scott

Published by Pearson Education Limited, 80 Strand, London, WC2R 0RL.

www.pearsonglobalschools.com

Copies of official specifications for all Edexcel qualifications may be found on the website: https://qualifications.pearson.com

Text © Pearson Education Limited 2018
Designed by Tech-Set Ltd, Gateshead, UK
Typeset by Tech-Set Ltd, Gateshead, UK
Edited by Sarah Ryan and Katharine Godfrey Smith
Original illustrations © Pearson Education Limited 2018
Illustrated by © Tech-Set Ltd, Gateshead, UK
Cover design © Pearson Education Limited 2018

Cover images: Front: **Getty Images:** David Malin/Science Faction
Inside front cover: **Shutterstock.com**/Dmitry Lobanov

The rights of Cliff Curtis, Jason Murgatroyd and David Scott to be identified as authors of this work have been asserted by them in accordance with the Copyright, Designs and Patents Act 1988.

First published 2018

26 25 24
17 16 15

British Library Cataloguing in Publication Data
A catalogue record for this book is available from the British Library
ISBN 978 1 2922 4486 0

Printed in Slovakia by Neografia

Acknowledgements

(Key: b-bottom; c-centre; l-left; r-right; t-top)

Images:
123RF.com: Preecha bamrungrai 116b; **Alamy Stock Photo:** Annie Eagle 143c, Bernd Kroger/Insadco Photography 260-261, Brian North Gardens 202l, D. Hurst 121, David Taylor 206, Dbimages 35, E. R. Degginger 199r, GL Archive 42t, Jacek Nowak 112, James Harrison 142, Leslie Garland Picture Library 275, Linh Hassel/Age footstock 42c, MaximImages 90br, Nigel Cattlin 90cr, Paul Fleet 89, Peter D Noyce 270, Reinhard Dirscherl 143t, RGB Ventures/Superstock 42b, RooM the Agency 94, Sandy Young 120, Sara Richards 254, Science History Images 34, Sciencephotos 11, Shawn Hempel 10; **Royal Society of Chemistry:** 36; **Armin Kubelbeck:** 24; **Pearson Education:** Coleman Yuen 27b, Miguel Domínguez Muñoz 211, Trevor Clifford 32; **Corbis:** David Muench 234-235, Ann Johansson 286; **Getty Images:** A & F Michler/Photolibrary 90cl, Cybrain/iStock 64-65, Ishan Hassan/EyeEm Premium 168-169, Jeff Foot/Discovery Channe 230, Jonas Yaya/EyeEm ixt, 2-3, Jude Evans/Moment 149-150, Katsiaryna Kapusta/iStock 90bl, Mario Gutiérrez 98-99, Patrick Orton 184-185, Peepo/E+ 40-41, Photos.com/360 100, Ultra. F/DigitalVision 128-129; **NASA:** ESA/S. Beckwith (STScI) and The HUDF Team ixc, 61; **Omega Pharma:** 201b; **Science Photo Library:** 13, 198r, 201c, 202r, 205, Andrew Lambert Photography 105, 109, 135, 196, 199c, 210, 216, 221, 266, Carlos Goldin 119, Charles D. Winters 18, 199t, 199b, Hagley Museum And Archive 100b, Jannicke Wiik-Nielsen/Vetinst 124, Martyn F. Chillmaid 199l, 217, 219l, c, r, Sinclair Stammers 264; **Shutterstock:** Aleksey Stemmer 207, Christian Lagerek 21, Janprachal 227, Martin Nemec 124, Minerva Studio 116t, PhotoSGH 22, Pitsanu Kraichana 12, Sergio Ponomarev 229, wk1003mike 198l, Yellowj 123; **The Bridgeman Art Library Ltd:** De Agostini Picture Library/Chomon viii, 17; **Veer/Corbis:** Fabrizio 231

All other images © Pearson Education Limited 2018

We are grateful to the following for permission to reproduce copyright material:

Figures:
Figure on page 43 from 'Mass Number and Isotope', http://www.shimadzu.com, copyright © 2014 Shimadzu Corporation. All rights reserved; Figures on page 273 from Education Scotland © Crown copyright 2012; Figures on page 144 from 'Catalysts for a green industry' Education in Chemistry (Tony Hargreaves),July 2009, http://www.rsc.org/education/eic/issues/2009July /catalyst-green-chemistry-research-industry.asp, copyright © Royal Society of Chemistry; Figures on page 286 from 'Five rings good, four rings bad', Education in Chemistry (Dr Simon Cotton), March 2010, http://www.rsc.org/education/eic/issues/2010Mar/Five Rings Good Four Rings Bad.asp, copyright © Royal Society of Chemistry.

Text:
Extracts on page 36 from 'Ancient coins',http://www.rsc.org/Education/EiC/issues/2006Nov /AncientCoins.asp, copyright © Royal Society of Chemistry; Extract on page 60 from From Stars to Stalagmites: How Everything Connects by Paul Braterman, World Scientific Publishing Co.,2012, p.76, copyright © 2012 World Scientific Publishing Co.Pte Ltd; Extract on page 94 From 'Cooling chemical fuels snowy spat' by Victoria Gill copyright © 2007 Royal Society of Chemistry; Extract on page 124 From 'Alkanes: Natural Products' by Prof. Dr. Rainer Herges and Dr. Torsten Winkler translator Dr. Guenter Grethe © Wiley-VCH Verlag GmbH & Co. KGaA; Extracts on pages 144 and 286 from 'Catalysts for a green industry'http://www.rsc.org/ education/eic/issues /2009July/catalyst-greenchemistry-research-industry.asp, and 'Five rings good, four ringsbad', http://www.rsc.org /education/eic/issues/2010Mar/FiveRingsGoodFour RingsBad.asp, copyright ©Royal Society of Chemistry; Extract on page 164 From 'Organic Chemists Contribute to Renewable Energy'http://www.rsc. org/Membership / Networking/ InterestGroups/OrganicDivision/organic-chemistry-case-studies/organic-chemistry-biofuels.asp, copyright © Royal Society of Chemistry; Extract on page 180 from 'How fish keep one step ahead of ice', New Scientist Magazine, 02/10/2004, Issue 2467 (Katharine Davis) Copyright © 2004 New Scientist Ltd. All rights reserved. Distributed by Tribune Content Agency; Extracts on pages 230 from 'Some of our selenium is missing', New Scientist Magazine, 18/11/1995, Issue 2004 (Michelle Knott), copyright © 2002 Reed Business Information. All rights reserved. Distributed by Tribune Content Agency; Extracts on pages 254 from 'Something in the water…', New Scientist Magazine, 13/04/2002, Issue 2338 (Emma Young), copyright © 2007 Reed Business Information - UK. All rights reserved. Distributed by Tribune Content Agency; Extract on page 287 from 'How athletics is still scarred by the reign of the chemical sisters', The Daily Mail, 06/08/2012 (Matt Lawton), copyright © Solo Syndication, 2012.

Endorsement statement
In order to ensure that this resource offers high-quality support for the associated Pearson qualification, it has been through a review process by the awarding body. This process confirmed that this resource fully covers the teaching and learning content of the specification at which it is aimed. It also confirms that it demonstrates an appropriate balance betw3een the development of subject skills, knowledge and understanding, in addition to preparation for assessment.

Endorsement does not cover any guidance on assessment activities or processes (e.g. practice questions or advice on how to answer assessment questions) included in the resource, nor does it prescribe any particular approach to the teaching or delivery of a related course.

While the publishers have made every attempt to ensure that advice on the qualification and its assessment is accurate, the official specification and associated assessment guidance materials are the only authoritative source of information and should always be referred to for definitive guidance.

Pearson examiners have not contributed to any sections in this resource relevant to examination papers for which they have responsibility.

Examiners will not use endorsed resources as a source of material for any assessment set by Pearson. Endorsement of a resource does not mean that the resource is required to achieve this Pearson qualification, nor does it mean that it is the only suitable material available to support the qualification, and nay resource lists produced by the awarding body shall include this and other appropriate resources.

UNIT 1: STRUCTURE, BONDING AND INTRODUCTION TO ORGANIC CHEMISTRY

UNIT 2: ENERGETICS, GROUP CHEMISTRY, HALOGENOALKANES AND ALCOHOLS

ABOUT THIS BOOK

This book is written for students following the Pearson Edexcel International Advanced Subsidiary (IAS) Chemistry specification. This book covers the full IAS course and the first year of the International A Level (IAL) course.

The book contains full coverage of IAS units (or exam papers) 1 and 2. Students can prepare for the written Practical Paper by using the IAL Chemistry Lab Book (see page x of this book). Each unit has five Topic areas that match the titles and order of those in the specification. You can refer to the Assessment Overview on page xii for further information.

Each Topic is divided into chapters and sections to break the content down into manageable chunks. Each section features a mix of learning and activities.

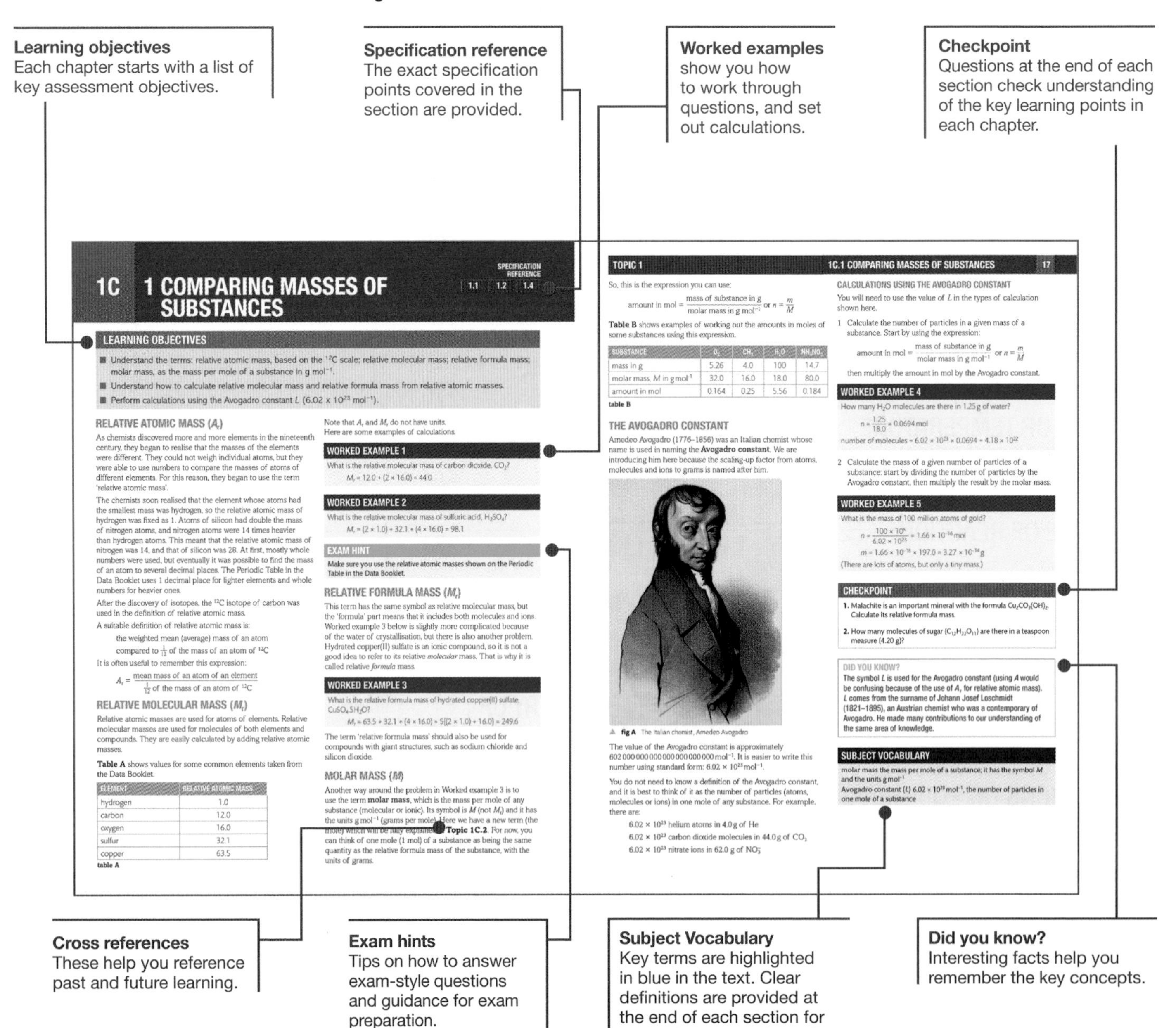

1C 1 COMPARING MASSES OF SUBSTANCES

SPECIFICATION REFERENCE 1.1 1.2 1.4

LEARNING OBJECTIVES

- Understand the terms: relative atomic mass, based on the ^{12}C scale; relative molecular mass; relative formula mass; molar mass, as the mass per mole of a substance in g mol^{-1}.
- Understand how to calculate relative molecular mass and relative formula mass from relative atomic masses.
- Perform calculations using the Avogadro constant L (6.02×10^{23} mol^{-1}).

RELATIVE ATOMIC MASS (A_r)

As chemists discovered more and more elements in the nineteenth century, they began to realise that the masses of the elements were different. They could not weigh individual atoms, but they were able to use numbers to compare the masses of atoms of different elements. For this reason, they began to use the term 'relative atomic mass'.

The chemists soon realised that the element whose atoms had the smallest mass was hydrogen, so the relative atomic mass of hydrogen was fixed as 1. Atoms of silicon had double the mass of nitrogen atoms, and nitrogen atoms were 14 times heavier than hydrogen atoms. This meant that the relative atomic mass of nitrogen was 14, and that of silicon was 28. At first, mostly whole numbers were used, but eventually it was possible to find the mass of an atom to several decimal places. The Periodic Table in the Data Booklet uses 1 decimal place for lighter elements and whole numbers for heavier ones.

After the discovery of isotopes, the ^{12}C isotope of carbon was used in the definition of relative atomic mass.

A suitable definition of relative atomic mass is:

the weighted mean (average) mass of an atom compared to $\frac{1}{12}$ of the mass of an atom of ^{12}C

It is often useful to remember this expression:

$$A_r = \frac{\text{mean mass of an atom of an element}}{\frac{1}{12}\text{ of the mass of an atom of }^{12}C}$$

RELATIVE MOLECULAR MASS (M_r)

Relative atomic masses are used for atoms of elements. Relative molecular masses are used for molecules of both elements and compounds. They are easily calculated by adding relative atomic masses.

Table A shows values for some common elements taken from the Data Booklet.

ELEMENT	RELATIVE ATOMIC MASS
hydrogen	1.0
carbon	12.0
oxygen	16.0
sulfur	32.1
copper	63.5

table A

Note that A_r and M_r do not have units.
Here are some examples of calculations.

WORKED EXAMPLE 1

What is the relative molecular mass of carbon dioxide, CO_2?

$M_r = 12.0 + (2 \times 16.0) = 44.0$

WORKED EXAMPLE 2

What is the relative molecular mass of sulfuric acid, H_2SO_4?

$M_r = (2 \times 1.0) + 32.1 + (4 \times 16.0) = 98.1$

EXAM HINT

Make sure you use the relative atomic masses shown on the Periodic Table in the Data Booklet.

RELATIVE FORMULA MASS (M_r)

This term has the same symbol as relative molecular mass, but the 'formula' part means that it includes both molecules and ions. Worked example 3 below is slightly more complicated because of the water of crystallisation, but there is also another problem. Hydrated copper(II) sulfate is an ionic compound, so it is not a good idea to refer to its relative *molecular* mass. That is why it is called relative *formula* mass.

WORKED EXAMPLE 3

What is the relative formula mass of hydrated copper(II) sulfate, $CuSO_4.5H_2O$?

$M_r = 63.5 + 32.1 + (4 \times 16.0) + 5[(2 \times 1.0) + 16.0] = 249.6$

The term 'relative formula mass' should also be used for compounds with giant structures, such as sodium chloride and silicon dioxide.

MOLAR MASS (M)

Another way around the problem in Worked example 3 is to use the term **molar mass**, which is the mass per mole of any substance (molecular or ionic). Its symbol is M (not M_r) and it has the units g mol^{-1} (grams per mole). Here we have a new term (the mole) which will be fully explained in **Topic 1C.2**. For now, you can think of one mole (1 mol) of a substance as being the same quantity as the relative formula mass of the substance, with the units of grams.

TOPIC 1

So, this is the expression you can use:

$$\text{amount in mol} = \frac{\text{mass of substance in g}}{\text{molar mass in g mol}^{-1}} \text{ or } n = \frac{m}{M}$$

Table B shows examples of working out the amounts in moles of some substances using this expression.

SUBSTANCE	O_2	CH_4	H_2O	NH_4NO_3
mass in g	5.26	4.0	100	14.7
molar mass, M in g mol^{-1}	32.0	16.0	18.0	80.0
amount in mol	0.164	0.25	5.56	0.184

table B

THE AVOGADRO CONSTANT

Amedeo Avogadro (1776–1856) was an Italian chemist whose name is used in naming the **Avogadro constant**. We are introducing him here because the scaling-up factor from atoms, molecules and ions to grams is named after him.

fig A The Italian chemist, Amedeo Avogadro

The value of the Avogadro constant is approximately 602 000 000 000 000 000 000 000 mol^{-1}. It is easier to write this number using standard form: 6.02×10^{23} mol^{-1}.

You do not need to know a definition of the Avogadro constant, and it is best to think of it as the number of particles (atoms, molecules or ions) in one mole of any substance. For example, there are:

6.02×10^{23} helium atoms in 4.0 g of He
6.02×10^{23} carbon dioxide molecules in 44.0 g of CO_2
6.02×10^{23} nitrate ions in 62.0 g of NO_3^-

1C.1 COMPARING MASSES OF SUBSTANCES 17

CALCULATIONS USING THE AVOGADRO CONSTANT

You will need to use the value of L in the types of calculation shown here.

1 Calculate the number of particles in a given mass of a substance. Start by using the expression:

$$\text{amount in mol} = \frac{\text{mass of substance in g}}{\text{molar mass in g mol}^{-1}} \text{ or } n = \frac{m}{M}$$

then multiply the amount in mol by the Avogadro constant.

WORKED EXAMPLE 4

How many H_2O molecules are there in 1.25 g of water?

$n = \frac{1.25}{18.0} = 0.0694$ mol

number of molecules $= 6.02 \times 10^{23} \times 0.0694 = 4.18 \times 10^{22}$

2 Calculate the mass of a given number of particles of a substance: start by dividing the number of particles by the Avogadro constant, then multiply the result by the molar mass.

WORKED EXAMPLE 5

What is the mass of 100 million atoms of gold?

$n = \frac{100 \times 10^6}{6.02 \times 10^{23}} = 1.66 \times 10^{-16}$ mol

$m = 1.66 \times 10^{-16} \times 197.0 = 3.27 \times 10^{-14}$ g

(There are lots of atoms, but only a tiny mass.)

CHECKPOINT

1. Malachite is an important mineral with the formula $Cu_2CO_3(OH)_2$. Calculate its relative formula mass.
2. How many molecules of sugar ($C_{12}H_{22}O_{11}$) are there in a teaspoon measure (4.20 g)?

DID YOU KNOW?

The symbol L is used for the Avogadro constant (using A would be confusing because of the use of A_r for relative atomic mass). L comes from the surname of Johann Josef Loschmidt (1821–1895), an Austrian chemist who was a contemporary of Avogadro. He made many contributions to our understanding of the same area of knowledge.

SUBJECT VOCABULARY

molar mass the mass per mole of a substance; it has the symbol M and the units g mol^{-1}
Avogadro constant (L) 6.02×10^{23} mol^{-1}, the number of particles in one mole of a substance

Your learning, Topic by Topic, is always put in context:

- Links to other areas of Chemistry include previous knowledge that is built on in the topic, and future learning that you will cover later in your course.
- A checklist details maths knowledge required. If you need to practise these skills, you can use the **Maths Skills** reference at the back of the book as a starting point.

TOPIC 10 ORGANIC CHEMISTRY: HALOGENOALKANES, ALCOHOLS AND SPECTRA

A GENERAL PRINCIPLES | B HALOGENOALKANES | C ALCOHOLS | D MASS SPECTRA AND IR

In Topics 4 and 5, you learned about the basics of organic chemistry and about two homologous series, alkanes and alkenes. In this topic, you will learn about two more homologous series, halogenoalkanes and alcohols. There is also a new reaction mechanism to consider, nucleophilic substitution.

Although you have already learned something about mass spectrometry, we will build on this knowledge to see how the technique was developed to become a useful tool to discover the structures of organic compounds. A new technique, infrared spectroscopy, is introduced to show another way to discover different information about the structures of organic compounds.

MATHS SKILLS FOR THIS TOPIC

- Use ratios to construct and balance equations
- Represent chemical structures using angles and shapes in 2D and 3D structures

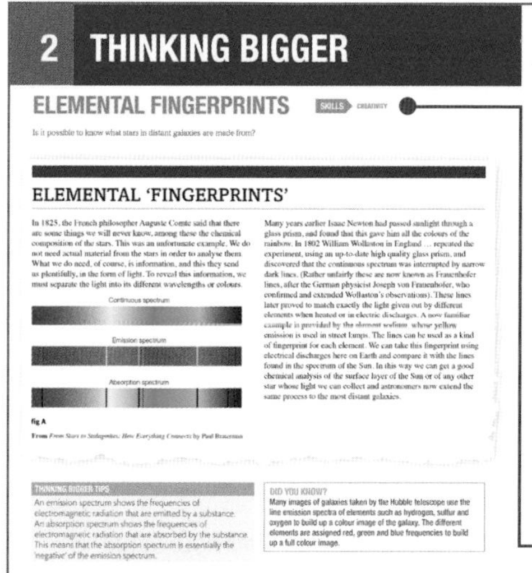
2 THINKING BIGGER

ELEMENTAL FINGERPRINTS

ELEMENTAL 'FINGERPRINTS'

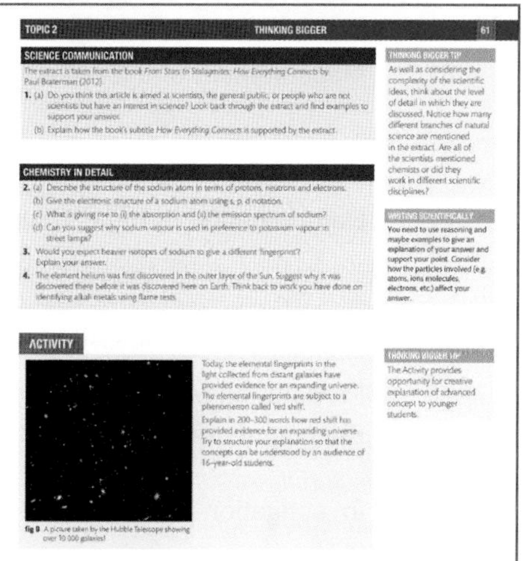

Thinking Bigger
At the end of each topic there is an opportunity to read and work with real-life research and writing about science. The activities help you to read real-life material that's relevant to your course, analyse how scientists write, think critically and consider how different aspects of your learning piece together.

Skills
These sections will help you develop transferable skills, which are highly valued in further study and the workplace.

10 EXAM PRACTICE

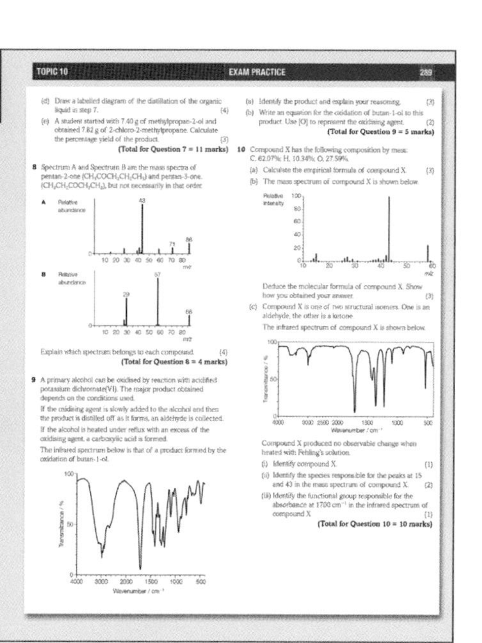

Exam Practice
Exam-style questions at the end of each chapter are tailored to the Pearson Edexcel specification to allow for practice and development of exam writing technique. They also allow for practice responding to the command words used in the exams (see the **command words glossary** at the back of this book).
You can also refer to the **Preparing for Your Exams** section in the back of the book, for sample exam answers with commentary.

PRACTICAL SKILLS

Practical work is central to the study of chemistry. The International Advanced Subsidiary (IAS) Chemistry specification includes eight Core Practicals that link theoretical knowledge and understanding to practical scenarios.

Your knowledge and understanding of practical skills and activities will be assessed in all examination papers for the IAS Level Chemistry qualification.

- Papers 1 and 2 will include questions based on practical activities, including novel scenarios.
- Paper 3 will test your ability to plan practical work, including risk management and selection of apparatus.

In order to develop practical skills, you should carry out a range of practical experiments related to the topics covered in your course. Further suggestions in addition to the Core Practicals are included in the specification which is available online.

STUDENT BOOK TOPIC	IAS CORE PRACTICALS	
TOPIC 1 **FORMULAE, EQUATIONS AND AMOUNT OF SUBSTANCE**	**CP1**	Measurement of the molar volume of a gas
TOPIC 6 **ENERGETICS**	**CP2**	Determination of the enthalpy change of a reaction using Hess's Law
TOPIC 8 **REDOX CHEMISTRY AND GROUPS 1, 2 AND 7**	**CP3**	Finding the concentration of a solution of hydrochloric acid
	CP4	Preparation of a standard solution from a solid acid and use it to find the concentration of a solution of sodium hydroxide
TOPIC 10 **ORGANIC CHEMISTRY: HALOGENOALKANES, ALCOHOLS AND SPECTRA**	**CP5**	Investigation of the rates of hydrolysis of some halogenoalkanes
	CP6	Chlorination of 2-methylpropan-2-ol with concentrated hydrochloric acid
	CP7	The oxidation of propan-1-ol to produce propanal and propanoic acid
	CP8	Analysis of some inorganic and organic unknowns

1E 1 MOLAR VOLUME CALCULATIONS

SPECIFICATION REFERENCE 1.8(ii) CP1

LEARNING OBJECTIVES

- Use chemical equations to calculate reacting volumes of gases and vice versa using the concept of molar volume of gases.

MOLAR VOLUME

The work of Avogadro and others led to the idea of **molar volume**, the volume of gas that contains one mole of that gas. The molar volume is approximately the same for all gases, but its value varies with temperature and pressure. The value most often used is for gases at room temperature and pressure (sometimes abbreviated as r.t.p.). Room temperature is 298 K (or 25 °C) and standard pressure varies, but is often quoted as 1.01×10^5 Pa. The value of molar volume is usually quoted as $24\,dm^3\,mol^{-1}$ (equivalent to $24\,000\,cm^3$). Its symbol is V_m.

$$V_m = 24\,dm^3\,mol^{-1} \text{ at r.t.p.}$$

and

$$V_m = 24\,000\,cm^3\,mol^{-1} \text{ at r.t.p.}$$

CALCULATIONS USING MOLAR VOLUME

CALCULATIONS INVOLVING A SINGLE GAS

If you are asked about a single gas, the calculation is straight forward. Assume that in these examples, all volumes are measured at r.t.p.

You need the expression $V_m = 24\,dm^3\,mol^{-1}$ which you might want to consider using in these alternative forms:

$$V_m = 24.0 = \frac{\text{volume in dm}^3}{\text{amount in mol}} \text{ or } V_m = 24\,000 = \frac{\text{volume in cm}^3}{\text{amount in mol}}$$

You will need to rearrange the expression depending on the actual question. Make sure that you use only 24 and dm^3, or only 24 000 and cm^3, in the calculation.

EXAMPLE 1

What is the amount, in moles, of CO in $3.8\,dm^3$ of carbon monoxide?

$$\text{Answer} = \frac{3.8}{24} = 0.16\,mol$$

EXAMPLE 2

What is the amount, in moles, of CO_2 in $500\,cm^3$ of carbon dioxide?

$$\text{Answer} = \frac{500}{24\,000} = 0.021\,mol$$

EXAMPLE 3

What is the volume of 0.36 mol of hydrogen?

$$\text{Answer} = 24 \times 0.36 = 8.64\,dm^3$$

CALCULATIONS INVOLVING GASES AND SOLIDS OR LIQUIDS

You may be given a chemical equation that involves one or more gases and a solid or a liquid. If you use information about the amount (in mol) of a solid or liquid, you can combine two of the calculation methods that we have already used.

The basis of this type of calculation is that for gases you can interconvert between amount and volume. For solids and liquids you can interconvert between amount and mass.

Step 1: Calculate the amount in moles from either the mass or the volume, depending on which one is given.

Step 2: Use the relevant reaction ratio in the equation to calculate the amount of the other substance.

Step 3: Convert this amount to a mass or a volume, depending on what the question asks.

EXAM HINT

When attempting multistep calculations involving different units, it is important to show the units for each step of the reaction. It will force you to think about what you have just calculated rather than leaving a number floating.

EXAMPLE 4

A piece of magnesium with a mass of 1.00 g is added to an excess of dilute hydrochloric acid. What volume of hydrogen gas is formed?

The equation for the reaction is:

$$Mg(s) + 2HCl(aq) \rightarrow MgCl_2(aq) + H_2(g)$$

You are not given any information about the hydrochloric acid, and you are not asked anything about magnesium chloride. You can use the mole expression to calculate the amount of magnesium:

$$n(Mg) = \frac{1.00}{24.3} = 0.0412\,mol$$

You can see that the $Mg:H_2$ ratio in the equation is 1 : 1, which means that 0.0412 mol of hydrogen is formed.

Convert this amount to a volume and you have the answer:

$$\text{volume} = 24 \times 0.0412 = 0.99\,dm^3$$

EXAMPLE 5

Calcium carbonate reacts with nitric acid to form calcium nitrate, water and carbon dioxide, as shown in the equation:

$$CaCO_3(s) + 2HNO_3(aq) \rightarrow Ca(NO_3)_2(aq) + H_2O(l) + CO_2(g)$$

In a reaction, $100\,cm^3$ of carbon dioxide is formed. What mass of calcium carbonate is needed for this?

You are not told anything about nitric acid, or asked anything about calcium nitrate or water. You can use the molar volume expression to calculate the amount of carbon dioxide:

$$\text{amount} = \frac{100}{24000} = 0.00417\,mol$$

You can see that the $CaCO_3:CO_2$ ratio in the equation is 1 : 1, which means that 0.00417 mol of calcium carbonate is needed.

Convert this amount to a mass and you have the answer:

$$m = n \times M = 0.00417 \times 100.1 = 0.42\,g$$

In the **Student Book**, the Core Practical specifications are supplied in the relevant sections.

This Student Book is accompanied by a **Lab Book**, which includes instructions and writing frames for the Core Practicals for you to record your results and reflect on your work. Practical skills practice questions and answers are also provided. The Lab Book records can be used as preparation for the Practical Skills Paper.

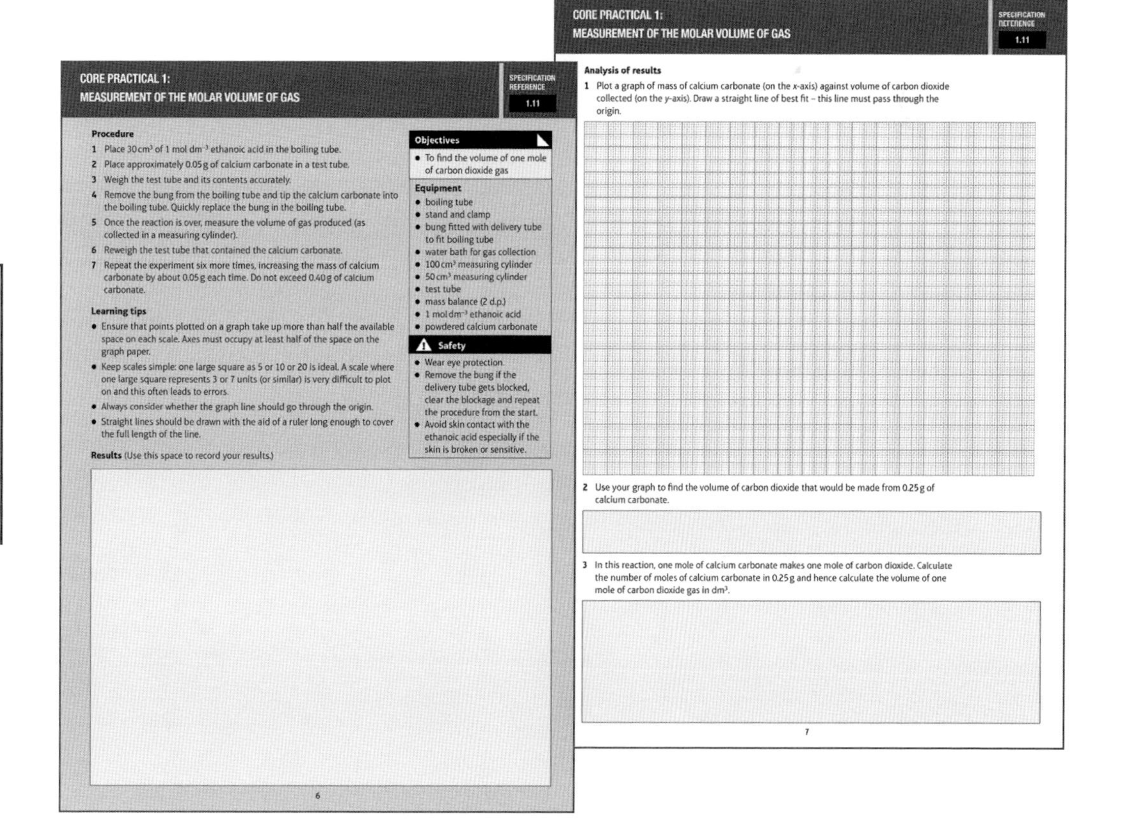

CORE PRACTICAL 1:
MEASUREMENT OF THE MOLAR VOLUME OF GAS

SPECIFICATION REFERENCE 1.11

Procedure

1 Place $30\,cm^3$ of $1\,mol\,dm^{-3}$ ethanoic acid in the boiling tube.
2 Place approximately 0.05 g of calcium carbonate in a test tube.
3 Weigh the test tube and its contents accurately.
4 Remove the bung from the boiling tube and tip the calcium carbonate into the boiling tube. Quickly replace the bung in the boiling tube.
5 Once the reaction is over, measure the volume of gas produced (as collected in a measuring cylinder).
6 Reweigh the test tube that contained the calcium carbonate.
7 Repeat the experiment six more times, increasing the mass of calcium carbonate by about 0.05 g each time. Do not exceed 0.40 g of calcium carbonate.

Learning tips

- Ensure that points plotted on a graph take up more than half the available space on each scale. Axes must occupy at least half of the space on the graph paper.
- Keep scales simple: one large square as 5 or 10 or 20 is ideal. A scale where one large square represents 3 or 7 units (or similar) is very difficult to plot on and this often leads to errors.
- Always consider whether the graph line should go through the origin.
- Straight lines should be drawn with the aid of a ruler long enough to cover the full length of the line.

Results (Use this space to record your results.)

Objectives

- To find the volume of one mole of carbon dioxide gas

Equipment

- boiling tube
- stand and clamp
- bung fitted with delivery tube to fit boiling tube
- water bath for gas collection
- $100\,cm^3$ measuring cylinder
- $50\,cm^3$ measuring cylinder
- test tube
- mass balance (2 d.p.)
- $1\,mol\,dm^{-3}$ ethanoic acid
- powdered calcium carbonate

Safety

- Wear eye protection.
- Remove the bung if the delivery tube gets blocked, clear the blockage and repeat the procedure from the start.
- Avoid skin contact with the ethanoic acid especially if the skin is broken or sensitive.

6

CORE PRACTICAL 1:
MEASUREMENT OF THE MOLAR VOLUME OF GAS

SPECIFICATION REFERENCE 1.11

Analysis of results

1 Plot a graph of mass of calcium carbonate (on the *x*-axis) against volume of carbon dioxide collected (on the *y*-axis). Draw a straight line of best fit – this line must pass through the origin.

2 Use your graph to find the volume of carbon dioxide that would be made from 0.25 g of calcium carbonate.

3 In this reaction, one mole of calcium carbonate makes one mole of carbon dioxide. Calculate the number of moles of calcium carbonate in 0.25 g and hence calculate the volume of one mole of carbon dioxide gas in dm^3.

7

ASSESSMENT OVERVIEW

The following tables give an overview of the assessment for Pearson Edexcel International Advanced Subsidiary (IAS) Level course in Chemistry. You should study this information closely to help ensure that you are fully prepared for this course and know exactly what to expect in each part of the examinations. More information about this qualification, and about the question types in the different papers, can be found in *Preparing for your exams* on page 296 of this book.

PAPER / UNIT 1	PERCENTAGE OF IAS	PERCENTAGE OF IAL	MARK	TIME	AVAILABILITY
STRUCTURE, BONDING AND INTRODUCTION TO ORGANIC CHEMISTRY Written exam paper Paper code WCH11/01 Externally set and marked by Pearson Edexcel Single tier of entry	40%	20%	80	1 hour 30 minutes	January, June and October First assessment: January 2019

PAPER / UNIT 2	PERCENTAGE OF IAS	PERCENTAGE OF IAL	MARK	TIME	AVAILABILITY
ENERGETICS, GROUP CHEMISTRY, HALOGENOALKANES AND ALCOHOLS Written exam paper Paper code WCH12/01 Externally set and marked by Pearson Edexcel Single tier of entry	40%	20%	80	1 hour 30 minutes	January, June and October First assessment: June 2019

PAPER / UNIT 3	PERCENTAGE OF IAS	PERCENTAGE OF IAL	MARK	TIME	AVAILABILITY
PRACTICAL SKILLS IN CHEMISTRY 1 Written exam paper Paper / Unit code WCH13/01 Externally set and marked by Pearson Edexcel Single tier of entry	20%	10%	50	1 hour 20 minutes	January, June and October First assessment: June 2019

ASSESSMENT OBJECTIVES AND WEIGHTINGS

		% IN IAS	% IN IA2	% IN IAL
A01	Demonstrate knowledge and understanding of science.	34–36	29–31	32–34
A02	(a) Application of knowledge and understanding of science in familiar and unfamiliar contexts.	34–36	33–36	33–36
	(b) Analysis and evaluation of scientific information to make judgements and reach conclusions.	9–11	14–16	11–14
A03	Experimental skills in science, including analysis and evaluation of data and methods.	20	20	20

RELATIONSHIP OF ASSESSMENT OBJECTIVES TO UNITS

UNIT NUMBER	ASSESSMENT OBJECTIVE			
	A01	A02 (a)	A02 (b)	A04
UNIT 1	17–18	17–18	4.5–5.5	0.0
UNIT 2	17–18	17–18	4.5–5.5	0.0
UNIT 3	0.0	0.0	0.0	20
TOTAL FOR INTERNATIONAL ADVANCED SUBSIDIARY	**34–36**	**34–36**	**9–11**	**20**

TOPIC 1 FORMULAE, EQUATIONS AND AMOUNT OF SUBSTANCE

A ATOMS, ELEMENTS AND MOLECULES | B EQUATIONS AND REACTION TYPES | C MOLE CALCULATIONS | D EMPIRICAL AND MOLECULAR FORMULAE | E CALCULATIONS WITH SOLUTIONS AND GASES

Studying chemistry at this level involves many skills. These skills include describing observations, giving explanations, planning and interpreting experiments. Although all of these skills are discussed in this topic, the most important feature of this part of the course involves using numbers. Some students have a natural ability for numbers, while others find them less attractive. Whatever your mathematical ability, it is important to spend as much time as possible focusing on this topic in order to improve.

Using numbers accurately is an important skill in so many occupations – here are just a few examples:

- in medicine, calculating the dosage of a drug to maximise benefit to the patient and minimise unwanted side-effects
- in business, predicting whether capital investment will produce a sufficient income to justify the investment
- in industry, considering whether the costs and polluting effects of a new method of making a molecule justify making the change
- in shipping operations (for example, the Costa Concordia in 2012), working out the volume of air needed to float a ship after an accident.

As far as your chemistry course is concerned, this topic is the foundation of success in other topics, such as energetics and kinetics.

MATHS SKILLS FOR THIS TOPIC

- Use appropriate units and conversions, including for masses, gas and solution volumes
- Use standard and ordinary form, decimal places, and significant figures
- Use ratios, fractions and percentages, including in yield and atom economy calculations
- Rearrange expressions and substitute values

What prior knowledge do I need?

- Using appropriate apparatus to measure masses and volumes, and recording values to the appropriate precision
- Knowing how to convert between different units of mass and volume
- Writing and balancing chemical equations, including the use of state symbols
- Using the mole as the unit for amount of substance
- Calculating the relative formula mass of a compound from relative atomic masses of elements

What will I study in this topic?

- Using the mole in new situations, to calculate masses, volumes, concentrations and formulae
- Using experiments to obtain evidence for chemical formulae and equations
- Calculating percentage yields and atom economies from equations
- Interpreting observations from test-tube reactions, including displacement, neutralisation and precipitation

What will I study later?

Topic 11 (Book 2: IAL)

- Kinetics calculations, including those based on measurements of mass and volume changes in experiments

Topic 12B (Book 2: IAL)

- Further calculations of enthalpy changes, including those relating to lattice energies

Topic 13 (Book 2: IAL)

- Calculation of equilibrium constants from concentrations

Topic 14 (Book 2: IAL)

- Calculations of pH, K_a, K_w and other terms related to acids and bases

1A ATOMS, ELEMENTS AND MOLECULES

SPECIFICATION REFERENCE 1.1

LEARNING OBJECTIVES

- Know the terms atom, element, ion, molecule, compound.

THE PURPOSE OF THIS SECTION

In your previous study of chemistry, you have learned about several terms at an introductory level. In this book, you will study them in greater depth.

This topic is designed to revise some of the terms that you have met before, and will meet again in more detail.

WHAT IS AN ELEMENT?

We can start by looking at a simplified version of the Periodic Table of Elements. You have already learned something about the Periodic Table and will learn more about it in **Topic 2**.

			H				
							He
Li	Be	B	C	N	O	F	Ne
Na	Mg	Al	Si	P	S	Cl	Ar

Each box in the Periodic Table contains a capital letter, or more often a pair of letters. The first one is always a capital letter and the second one is never a capital letter. Each letter or pair of letters represents an element. You will know some of them from your everyday life. For example, oxygen is an element you breathe, and iron is an element used to make bodies of cars and bicycles.

You may have been taught that an element is a substance that contains atoms of only one type. Elements are chemically the simplest substances so they cannot be broken down using chemical reactions. For example, neon is an element because it contains only neon atoms, which have the symbol Ne. It cannot be broken down into atoms of any other element. Water is not an element because it can be broken down into the elements hydrogen and oxygen.

This way of describing an element is sufficient for this stage of your studies, but more details will be required as you progress through this course. For example, some elements contain isotopes. You will learn about isotopes in **Topic 2**. Neon contains three stable isotopes, ^{20}Ne, ^{21}Ne, and ^{22}Ne. All isotopes of the same element have the same number of protons and electrons, but different numbers of neutrons.

WHAT IS AN ATOM?

You already know what an atom is, because we have used the term to help you understand what an element is. Atoms are far too small to be seen with the human eye. You can see a tiny grain of sand and this contains many billions of atoms of silicon and oxygen. An atom can be described as the smallest part of an element that has the properties of that element. This meaning is fine when you work through **Topic 1**. However, in **Topic 2** you will need to recall that atoms contain even smaller particles (protons, neutrons and electrons).

WHAT IS A MOLECULE?

This is easier now that we know the meanings of element and atom.

You could describe a molecule as a particle made of two or more atoms bonded together.

If a molecule contains atoms of the same element, then the result is a molecule of an element. For example, a molecule that contains two atoms of hydrogen joined together can be represented by the formula H_2. This formula contains only one symbol (H), so it is the formula of an element.

If a molecule contains atoms of two or more different elements, then the result is a molecule of a compound. For example, a molecule that contains two atoms of hydrogen and one atom of oxygen joined together can be represented by the formula H_2O. This formula contains two different symbols, so it is the formula of a compound (water).

WHAT IS A COMPOUND?

Now we can describe a compound. This is a substance containing atoms of different elements combined together.

Note that some compounds contain large numbers of atoms bonded together, but other compounds contain molecules with only two atoms. Some compounds contain oppositely charged ions. We will explore these differences in later sections of this book.

WHAT IS AN ION?

One way to describe an ion is as a species consisting of one or more atoms joined together and having a positive or negative charge.

Note that this is not a description of how an ion is formed. **Topic 3A.1** will cover the formation of ions.

An ion with a positive charge is called a cation. An ion with a negative charge is called an anion.

Table A shows diagrams to illustrate the terms referred to. Each atom is shown as a circle containing the symbol of an element. The lines show bonds between atoms.

TERM	DIAGRAM	NAME	SYMBOL OR FORMULA	NOTE
element	Cu Cu Cu Cu Cu	copper	Cu	This is an element. All the atoms are the same.
atom	He	helium	He	This is an atom of an element.
molecule	Br—Br	bromine	Br_2	This is a molecule of an element. The atoms are the same.
compound	H—Br	hydrogen bromide	HBr	This is a molecule of a compound. The atoms are different.
ion	$^-$O, $^-$O, C=O	carbonate	CO_3^{2-}	This is an ion. There are two negative charges shown.

table A Illustrations of terms used in this topic.

OTHER TERMS

Elements that are made up of single atoms are described as monatomic. One example is helium, the gas used in weather balloons. The symbol for helium is He.

Elements and compounds made up of two atoms joined together are described as diatomic. The two main diatomic gases in the atmosphere are nitrogen (N_2) and oxygen (O_2).

Elements and compounds with molecules made up of several atoms joined together are described as polyatomic. Examples of polyatomic molecules are phosphorus (P_4) and methane (CH_4).

The same terms can be used for ions. Chloride (Cl^-) is an example of a monatomic ion. Hydroxide (HO^- or OH^-) is a diatomic ion. A sulfate ion (SO_4^{2-}) is polyatomic as it contains five atoms.

LEARNING TIP

It is often helpful to refer to a substance by its symbol or formula as well as its name so that your meaning is clear. For example, the word hydrogen could refer to a hydrogen atom (H), a hydrogen molecule (H_2) or a hydrogen ion (H^+).

CHECKPOINT

1. Classify each of these symbols and formulae as atoms, molecules or ions.
   Ne CO_2 H^+ S_8 Al^{3+}
2. Which of these formulae represent elements, compounds, or neither elements nor compounds? Explain your answer.
   Br_2 H_2O_2 NO_3 O_3 CaO

DID YOU KNOW?

The names and symbols of many elements come from the Greek and Latin languages. For example, hydrogen comes from the Greek words for *water* and *producer*. This is because when hydrogen gas is burned, it forms (or 'produces') water. Another example is copper (Cu) – the symbol Cu comes from the Latin word *cuprum*, which means 'metal from Cyprus'. Cyprus is the island where the Romans obtained much of their copper.

1B 1 WRITING CHEMICAL EQUATIONS

SPECIFICATION REFERENCE 1.3

LEARNING OBJECTIVES

- Write balanced full and ionic equations, including state symbols, for chemical reactions.

WRITING EQUATIONS: WHAT TO REMEMBER

This topic is a useful summary of what you are expected to do when you write equations.

WRITING FORMULAE FOR NAMES

You may be given the formulae of unfamiliar compounds in a question. You need to work out the formulae of many familiar compounds from their names, and to remember other formulae. You also need to show elements with the correct formulae. It is not possible to provide a complete list, but here are some examples.

You need to remember that:

- oxygen is O_2 and not O
- hydrogen is H_2 and not H
- nitrogen is N_2 and not N
- water is H_2O
- sodium hydroxide is NaOH
- nitric acid is HNO_3.

You should be able to work out that:

- iron(II) sulfate is $FeSO_4$
- iron(III) oxide is Fe_2O_3
- calcium carbonate is $CaCO_3$.

EXAM HINT

Remember to use your Periodic Table in the exam. If you know the formula of *magnesium* sulfate then you can assume *strontium* sulfate has the same formula because Mg and Sr are in the same group.

WRITING AN EQUATION FROM A DESCRIPTION

You will need to convert words into formulae and decide which ones are reactants and which ones are products.

Consider this description: when carbon dioxide reacts with calcium hydroxide, calcium carbonate and water are formed. The wording of the description makes it clear that carbon dioxide and calcium hydroxide are the reactants, and that calcium carbonate and water are the products.

Now you have to write the formulae in the correct places:

$$CO_2 + Ca(OH)_2 \rightarrow CaCO_3 + H_2O$$

The next step is balancing the equation. You need to add up the numbers of all the atoms to make sure that, for each element, the totals are the same on both the left and the right side of the equation. In this example, there is one carbon, one calcium, two hydrogen and four oxygen atoms on each side, so the equation is already balanced.

Here is an example where the first equation you write is not balanced.

The description is:

hydrogen peroxide decomposes to water and oxygen.

The formulae are:

$$H_2O_2 \rightarrow H_2O + O_2$$

This is already balanced for hydrogen, but not for oxygen. With careful practice, which sometimes involves guessing until you get it right, you should be able to write the **coefficients** needed to balance the equation. In this case, the balanced equation is:

$$2H_2O_2 \rightarrow 2H_2O + O_2$$

Most equations are balanced using whole-number coefficients, but using fractions or decimals is usually acceptable. This is especially the case in organic chemistry.

Consider this unbalanced equation for the complete combustion of butane:

$$C_4H_{10} + O_2 \rightarrow CO_2 + H_2O$$

The balanced equation can be either:

$$2C_4H_{10} + 13O_2 \rightarrow 8CO_2 + 10H_2O$$

or

$$C_4H_{10} + 6\tfrac{1}{2}O_2 \rightarrow 4CO_2 + 5H_2O$$

Using 6.5 instead of $6\frac{1}{2}$ is also acceptable.

USING STATE SYMBOLS

Many chemical equations include state symbols. The symbols are:

- (s) = solid
- (l) = liquid
- (g) = gas
- (aq) = aqueous (dissolved in water).

It is important to distinguish between (l) and (aq). A common error is to write $H_2O(aq)$ instead of $H_2O(l)$.

Although it is good practice to include state symbols in all equations, in some cases they are essential, while in other cases they are often omitted. For example, when writing equations to represent ionisation energies in **Topic 2A.4**, it is important to include (g) after each atom and ion. However, in **Topics 4 and 5** you will write equations for organic reactions, where state symbols are often not included.

Here is another description:
when aqueous solutions of silver nitrate and calcium chloride are mixed, a white precipitate of silver chloride forms. As this precipitate settles, a solution of calcium nitrate becomes visible.

After writing the correct formulae, balancing the equation and including state symbols, the equation is:

$$2AgNO_3(aq) + CaCl_2(aq) \rightarrow 2AgCl(s) + Ca(NO_3)_2(aq)$$

ARROWS IN EQUATIONS

Most equations are shown with a conventional (left to right) arrow $\rightarrow$. However, some important reactions are reversible. This means that the reaction can go both in the forward and backward (reverse) directions. The symbol $\rightleftharpoons$ is used in equations for these reactions. You can find guidance later in this book (**Topic 9B**) about when this reversible arrow should be used.

Sometimes a conventional arrow is made longer to allow information about the reaction to be shown above the arrow (and sometimes below it). This information might be about reaction conditions, such as temperature, pressure and the use of a catalyst. In organic chemistry, where reaction schemes are important, a label indicating, for example, Step 1, may be placed on the arrow in a sequence of reactions.

IONIC EQUATIONS

SIMPLIFYING FULL EQUATIONS

Ionic equations show any atoms and molecules involved, but only the ions that react together, and not the **spectator ions**.

This is the easiest method to follow for simplifying equations:

1 Start with the full equation for the reaction.
2 Replace the formulae of ionic compounds by their separate ions.
3 Delete any ions that appear identically on both sides.

WORKED EXAMPLE 1

What is the simplest ionic equation for the neutralisation of sodium hydroxide solution by dilute nitric acid?

The full equation is:

$$NaOH(aq) + HNO_3(aq) \rightarrow NaNO_3(aq) + H_2O(l)$$

You should now consider which of these species are ionic and replace them with ions. In this example, the first three compounds are ionic:

$$Na^+(aq) + OH^-(aq) + H^+(aq) + NO_3^-(aq) \rightarrow Na^+(aq) + NO_3^-(aq) + H_2O(l)$$

After deleting the identical ions, the equation becomes:

$$H^+(aq) + OH^-(aq) \rightarrow H_2O(l)$$

WORKED EXAMPLE 2

What is the simplest ionic equation for the reaction that occurs when solutions of lead(II) nitrate and sodium sulfate react together to form a precipitate of lead(II) sulfate and a solution of sodium nitrate?

The full equation is:

$$Pb(NO_3)_2(aq) + Na_2SO_4(aq) \rightarrow PbSO_4(s) + 2NaNO_3(aq)$$

Replacing the appropriate species by ions gives:

$$Pb^{2+}(aq) + 2NO_3^-(aq) + 2Na^+(aq) + SO_4^{2-}(aq) \rightarrow PbSO_4(s) + 2Na^+(aq) + 2NO_3^-(aq)$$

After deleting the identical ions, the equation becomes:

$$Pb^{2+}(aq) + SO_4^{2-}(aq) \rightarrow PbSO_4(s)$$

These remaining ions are not deleted because they are not shown identically. Before the reaction they were free-moving ions in two separate solutions (Pb^{2+} and SO_4^{2-}). After the reaction they are joined together in a solid precipitate ($PbSO_4$).

WORKED EXAMPLE 3

Carbon dioxide reacts with calcium hydroxide solution to form water and a precipitate of calcium carbonate.

The full equation is:

$$CO_2(g) + Ca(OH)_2(aq) \rightarrow CaCO_3(s) + H_2O(l)$$

Replacing the appropriate species by ions gives:

$$CO_2(g) + Ca^{2+}(aq) + 2OH^-(aq) \rightarrow CaCO_3(s) + H_2O(l)$$

Note that carbon dioxide and water are molecules, so their formulae are not changed. In this example, no ions are shown identically on both sides, so this is the simplest ionic equation.

LEARNING TIP

Consider carefully what the correct state symbol for a species should be. It may be different in different reactions. For example, water is never $H_2O(aq)$, but it may be $H_2O(s)$, $H_2O(l)$ or $H_2O(g)$, depending on the temperature.

IONIC HALF-EQUATIONS

We write ionic half-equations for reactions involving oxidation and reduction, and they usually show what happens to only one reactant. A simple example is the reaction that occurs at the negative electrode during the electrolysis of aqueous sulfuric acid:

$$2H^+(aq) + 2e^- \rightarrow H_2(g)$$

You will learn much more about ionic half-equations in **Topic 8**.

CHECKPOINT

1. Sodium thiosulfate ($Na_2S_2O_3$) solution reacts with dilute hydrochloric acid to form a precipitate of sulfur, gaseous sulfur dioxide and a solution of sodium chloride. Write an equation, including state symbols, for this reaction.
2. Solutions of ammonium sulfate and sodium hydroxide are warmed together to form sodium sulfate solution, water and ammonia gas. Write the simplest ionic equation for this reaction.

SUBJECT VOCABULARY

coefficient the technical term for the number written in front of species when balancing an equation

spectator ion an ion that is there both before and after the reaction but is not involved in the reaction

1B 2 TYPICAL REACTIONS OF ACIDS

SPECIFICATION REFERENCE 1.12(ii)

LEARNING OBJECTIVES

- Relate ionic and full equations, with state symbols, to observations from simple test-tube reactions, for reactions of acids.

INTRODUCTION

Acids are common reagents in chemistry. In this topic, we summarise some of their typical reactions, using hydrochloric, nitric, sulfuric and phosphoric acids.

In each of these reactions, a salt is formed. The reactions can be used to prepare samples of salts.

ACIDS WITH METALS

A general equation for these reactions is:

metal + acid → salt + hydrogen

Bubbles of hydrogen gas form, and if the salt formed is soluble, then a solution forms.

The metal must be sufficiently reactive to react in this way. For example, magnesium reacts but copper does not.

Typical equations for magnesium and hydrochloric acid are:

$$Mg + 2HCl \rightarrow MgCl_2 + H_2$$

$$Mg(s) + 2H^+(aq) \rightarrow Mg^{2+}(aq) + H_2(g)$$

These reactions may appear to be examples of neutralisation reactions because the H^+ ions are removed from the solution when they react with the metal. However, as the H^+ ions gain electrons from the metal and are converted to $H_2(g)$, it means that the H^+ ions are reduced, not neutralised.

ACIDS WITH METAL OXIDES AND INSOLUBLE METAL HYDROXIDES

A general equation for these reactions is:

metal oxide + acid → salt + water

metal hydroxide + acid → salt + water

The reactivity of the metal does not matter in these reactions because in the reactant it is present as metal ions, not metal atoms. The only observation is likely to be the formation of a solution.

Typical equations for copper(II) oxide and zinc hydroxide reacting with sulfuric acid are:

$$CuO + H_2SO_4 \rightarrow CuSO_4 + H_2O$$

$$CuO(s) + 2H^+(aq) \rightarrow Cu^{2+}(aq) + H_2O(l)$$

$$Zn(OH)_2 + H_2SO_4 \rightarrow ZnSO_4 + 2H_2O$$

$$Zn(OH)_2(s) + 2H^+(aq) \rightarrow Zn^{2+}(aq) + 2H_2O(l)$$

These reactions can be classified as neutralisation reactions because the H^+ ions react with O^{2-} or OH^- ions. They are not redox reactions because there is no change in the oxidation number of any of the species.

ACIDS WITH ALKALIS

Metal hydroxides that dissolve in water are called alkalis. A general equation for these reactions is:

alkali + acid → salt + water

There are no visible changes during these reactions, although if a thermometer is used, a temperature rise can be noted.

Typical equations for sodium hydroxide reacting with phosphoric acid are:

$$NaOH + H_3PO_4 \rightarrow NaH_2PO_4 + H_2O$$

$$2NaOH + H_3PO_4 \rightarrow Na_2HPO_4 + 2H_2O$$

$$3NaOH + H_3PO_4 \rightarrow Na_3PO_4 + 3H_2O$$

There are three replaceable hydrogens in phosphoric acid. The salt formed depends on the relative amounts of acid and alkali used. The ionic equation for all these reactions is:

$$H^+(aq) + OH^-(aq) \rightarrow H_2O(l)$$

These reactions can be classified as neutralisation reactions because the H^+ ions react with OH^- ions. They are not redox reactions because there is no change in the oxidation number of any of the species.

ACIDS WITH CARBONATES

A general equation for these reactions is:

metal carbonate + acid → salt + water + carbon dioxide

Bubbles of carbon dioxide gas form. If the salt formed is soluble, then a solution forms.

Typical equations for lithium carbonate reacting with hydrochloric acid are:

$$Li_2CO_3 + 2HCl \rightarrow 2LiCl + H_2O + CO_2$$

$$CO_3^{2-}(aq) + 2H^+(aq) \rightarrow H_2O(l) + CO_2(g)$$

These reactions can be classified as neutralisation reactions because the H^+ ions react with CO_3^{2-} ions. They are not redox reactions because there is no change in the oxidation number of any of the species.

▲ **fig A** Many old buildings are made from carbonates such as limestone. Centuries of reaction between limestone and acids in the atmosphere have caused damage to the walls of Kolossi Castle in Cyprus.

ACIDS WITH HYDROGENCARBONATES

Hydrogencarbonates are compounds containing the hydrogencarbonate ion (HCO_3^-), and they react with acids in the same way as carbonates. The best-known example is sodium hydrogencarbonate ($NaHCO_3$), commonly known as bicarbonate of soda or baking soda. Baking soda is used in cooking at home and in the food industry. The 'lightness' of baked food such as cakes is due to the formation of bubbles of carbon dioxide in the cake mixture, which cause the cake to rise.

A word equation for the reaction between baking soda and the acid in lemon juice is:

sodium hydrogencarbonate + citric acid → sodium citrate + water + carbon dioxide

A suitable test for the presence of carbonate or hydrogencarbonate ions in a solid or solution is to add an aqueous acid and to test the gas produced with limewater (see **Topic 8B.4**).

LEARNING TIP

Practise writing full and ionic equations for different reactions of acids.

CHECKPOINT

1. Write full equations for the reactions between:

(a) zinc and sulfuric acid

(b) aluminium oxide and hydrochloric acid.

2. Write the simplest ionic equations for the reactions between:

(a) zinc and hydrochloric acid

(b) magnesium carbonate and nitric acid.

3 DISPLACEMENT REACTIONS

LEARNING OBJECTIVES

- Relate ionic and full equations, with state symbols, to observations from simple test-tube reactions, for displacement reactions.

WHAT IS A DISPLACEMENT REACTION?

As you learn about more chemical reactions, you will know that they are often classified into different types of reaction. You will recognise reaction types such as addition, neutralisation, combustion, oxidation and several others.

In this topic, we look at a reaction type called displacement. In simple terms, it is a reaction in which one element replaces another element in a compound.

DISPLACEMENT REACTIONS INVOLVING METALS

Here are the equations for two **displacement reactions** of metals:

Reaction 1 $Mg(s) + CuSO_4(aq) \rightarrow Cu(s) + MgSO_4(aq)$

Reaction 2 $2Al(s) + Fe_2O_3(s) \rightarrow 2Fe(s) + Al_2O_3(s)$

What do these reactions have in common?

- Both involve one metal reacting with the compound of a different metal.
- Both produce a metal and a different metal compound.
- Both are redox reactions.
- You can see that the metal element on the reactants side has taken the place of the metal in the metal compound on the reactants side.

What are the differences between the reactions?

- Reaction 1 takes place in aqueous solution, but Reaction 2 involves only solids.
- Reaction 1 occurs without the need for energy to be supplied, but Reaction 2 requires a very high temperature to start it.
- Reaction 1 is likely to be done in the laboratory, but Reaction 2 is done for a specific purpose in industry.

METAL DISPLACEMENT REACTIONS IN AQUEOUS SOLUTION

Take a closer look at Reaction 1, shown above. When magnesium metal is added to copper(II) sulfate solution, the blue colour of the solution becomes paler. If an excess of magnesium is added, the solution becomes colourless, as magnesium sulfate forms. The magnesium changes in appearance from silvery to brown as copper forms on it.

The equation can be rewritten as an ionic equation:

$$Mg(s) + Cu^{2+}(aq) + SO_4^{2-}(aq) \rightarrow Cu(s) + Mg^{2+}(aq) + SO_4^{2-}(aq)$$

Cancelling the ions that appear identically on both sides gives:

$$Mg(s) + Cu^{2+}(aq) \rightarrow Cu(s) + Mg^{2+}(aq)$$

Now you can see that this is a redox reaction. Electrons are transferred from magnesium atoms to copper(II) ions, so magnesium atoms are oxidised (loss of electrons) and copper(II) ions are reduced (gain of electrons).

This reaction is just one example of many similar reactions in which a more reactive metal displaces a less reactive metal from one of its salts. You may come across this and other similar reactions as examples used in the measurement of temperature changes.

▲ **fig A** The photo shows what happens when a copper wire is placed in silver nitrate solution for some time. You can see the results of the displacement reaction, the 'growth' on the wire is silver metal and the blue solution contains the copper(II) nitrate that is formed.

METAL DISPLACEMENT REACTIONS IN THE SOLID STATE

Reaction 2 is used in the railway industry to join rails together. You might imagine that a good way to join rails together would be by welding, but the metal rails are good conductors of heat and it is very difficult to get the ends of two rails hot enough for them to melt and join together.

For this reason, the thermite method is used. A mixture of aluminium and iron(III) oxide is positioned just above the place where the two rails are to be joined. A magnesium fuse is lit, and Reaction 2 occurs. It is so exothermic that the iron is formed as a molten metal, which flows into the gap between the two rails. The molten iron cools, joining the rails together.

As in Reaction 1, the equation can be rewritten ionically and simplified:

$$2Al(l) + 2Fe^{3+}(l) + 3O^{2-}(l) \rightarrow 2Fe(l) + 2Al^{3+}(l) + 3O^{2-}(l)$$

then:

$$2Al(l) + 2Fe^{3+}(l) \rightarrow 2Fe(l) + 2Al^{3+}(l)$$

As with Reaction 1, you can see that this is s a redox reaction. Electrons are transferred from aluminium atoms to iron(III) ions, so aluminium atoms are oxidised (loss of electrons) and iron(III) ions are reduced (gain of electrons).

▲ **fig** B The thermite reaction. The flame comes from the highly exothermic reaction forming molten iron. The molten iron will be used to fill the gap between two rails and form a strong join between them.

DISPLACEMENT REACTIONS INVOLVING HALOGENS

In **Topic 8C.2**, we will look at how more reactive halogens can displace less reactive halogens from their compounds. For example, chlorine will displace bromine from a potassium bromide solution. The full, ionic and simplified ionic equations for this reaction are:

$$Cl_2(aq) + 2KBr(aq) \rightarrow Br_2(aq) + 2KCl(aq)$$

$$Cl_2(aq) + 2K^+(aq) + 2Br^-(aq) \rightarrow Br_2(aq) + 2K^+(aq) + 2Cl^-(aq)$$

$$Cl_2(aq) + 2Br^-(aq) \rightarrow Br_2(aq) + 2Cl^-(aq)$$

As with the metal displacement reactions, this is a redox reaction. Electrons are transferred from bromide ions to chlorine, so bromide ions are oxidised (loss of electrons) and chlorine is reduced (gain of electrons).

LEARNING TIP

When describing displacement reactions, be careful to refer to the correct species. For example, in the reaction between magnesium and copper(II) sulfate, magnesium *atoms* and copper(II) *ions* are involved.

CHECKPOINT

1. Iron metal reacts with silver nitrate in a displacement reaction to form silver and iron(II) nitrate. Write a full equation, an ionic equation and a simplified ionic equation for this reaction. Include state symbols in all your equations.
2. A mixture of zinc metal and copper(II) oxide is ignited, causing an exothermic reaction to occur. Write a full equation, an ionic equation and a simplified ionic equation for this reaction. Do not include state symbols in your equations.

SUBJECT VOCABULARY

displacement reaction a reaction in which one element replaces another, less reactive, element in a compound

1B 4 PRECIPITATION REACTIONS

SPECIFICATION REFERENCE 1.12(iii)

LEARNING OBJECTIVES

- Relate ionic and full equations, with state symbols, to observations from simple test-tube reactions, for precipitation reactions.

In this topic, we focus on two aspects of **precipitation reactions**:

- their use in chemical tests
- their use in working out chemical equations.

CHEMICAL TESTS

CARBON DIOXIDE

This may be your earliest memory of a precipitation reaction. When carbon dioxide gas is bubbled through calcium hydroxide solution (often called limewater), a white precipitate of calcium carbonate forms. The relevant equation is:

$$Ca(OH)_2(aq) + CO_2(g) \rightarrow CaCO_3(s) + H_2O(l)$$

The formation of the white precipitate was probably described as the limewater going milky or cloudy.

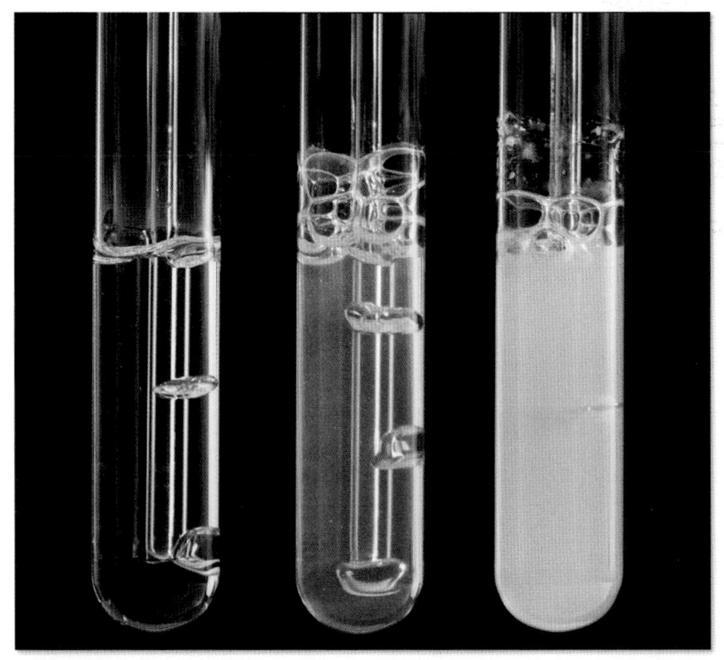

▲ **fig A** Limewater is a colourless solution. As more carbon dioxide is bubbled through it, the amount of white precipitate increases.

SULFATES

The presence of sulfate ions in solution can be shown by the addition of barium ions (usually from solutions of barium chloride or barium nitrate). The white precipitate that forms is barium sulfate.

For example, when barium chloride solution is added to sodium sulfate solution, the relevant equations are:

$$Na_2SO_4(aq) + BaCl_2(aq) \rightarrow BaSO_4(s) + 2NaCl(aq)$$

$$SO_4^{2-}(aq) + Ba^{2+}(aq) \rightarrow BaSO_4(s)$$

This test is covered in more detail in **Topic 8B.4.**

HALIDES

The presence of halide ions in solution can be shown by the addition of silver ions (from silver nitrate solution). The precipitates that form are silver halides.

For example, when silver nitrate solution is added to sodium chloride solution, the relevant equations are:

$$NaCl(aq) + AgNO_3(aq) \rightarrow AgCl(s) + NaNO_3(aq)$$

$$Cl^-(aq) + Ag^+(aq) \rightarrow AgCl(s)$$

This test is covered in more detail in **Topic 8C.4**.

WORKING OUT EQUATIONS

A good example of using a precipitation reaction to work out an equation is the reaction between aqueous solutions of lead nitrate and potassium iodide. Both reactants are colourless solutions. When they are mixed, a yellow precipitate of lead iodide forms.

The word equation for this reaction is:

lead nitrate + potassium iodide → lead iodide + potassium nitrate

Here is an outline of the experiment.

- Place the same volume of a potassium iodide solution in a series of test tubes.
- Add different volumes of a lead nitrate solution to the tubes.
- Place each tube in a centrifuge and spin the tubes for the same length of time.
- Measure the depth of precipitate in each tube.

Table A shows the results of one experiment.

The concentration of both solutions is 1.0 mol dm^{-3}.

The depth of each precipitate indicates the mass of precipitate formed.

TUBE	1	2	3	4	5	6	7
volume of potassium iodide solution / cm^3	5.0	5.0	5.0	5.0	5.0	5.0	5.0
volume of lead nitrate solution / cm^3	0.5	1.0	1.5	2.0	2.5	3.0	3.5
depth of precipitate / cm	2.5	3	4	5	6	6	6

table A Results of the reaction between aqueous solutions of lead nitrate and potassium iodide in one experiment.

The diagram shows the tubes at the end of the experiment.

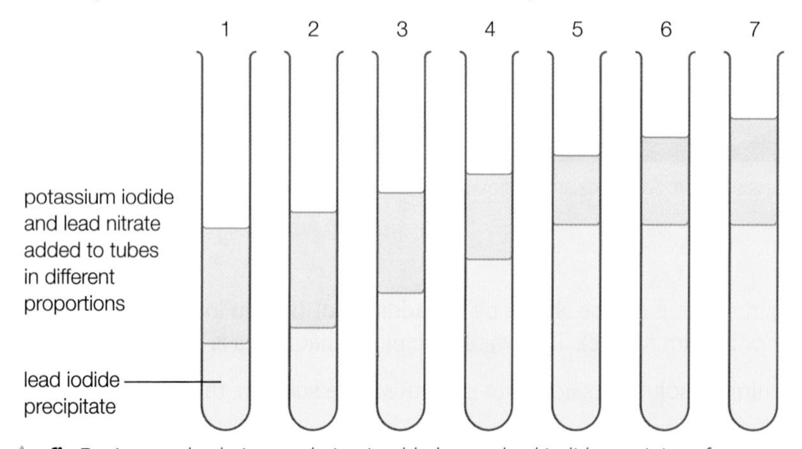

▲ **fig B** As more lead nitrate solution is added, more lead iodide precipitate forms.

You can see that there is no increase in the amount of precipitate from tube 5 to tube 6. This shows that the reaction is incomplete in tubes 1, 2, 3 and 4, but is complete in tube 5. The amounts, in moles, of reactants used in tube 5 are calculated as follows:

n(potassium iodide) = 0.005 × 1.0 = 0.005 mol

n(lead nitrate) = 0.0025 × 1.0 = 0.0025 mol

This shows that lead nitrate reacts with potassium iodide in the ratio 1 : 2.

The equations for the reaction are:

$Pb(NO_3)_2(aq) + 2KI(aq) \rightarrow PbI_2(s) + 2KNO_3(aq)$

$Pb^{2+}(aq) + 2I^-(aq) \rightarrow PbI_2(s)$

LEARNING TIP

Practise calculating the amounts of reactants and products in tubes 1–4.

CHECKPOINT

1. Write the simplest ionic equation, including state symbols, for:
   (a) the test for a sulfate
   (b) the test for a chloride.
2. Calculate the amounts, in moles, of each reactant and product in tube 7 in **fig B**.

SKILLS PROBLEM SOLVING

SUBJECT VOCABULARY

precipitation reaction reaction in which an insoluble solid is formed when two solutions are mixed

1C 1 COMPARING MASSES OF SUBSTANCES

SPECIFICATION REFERENCE 1.1 1.2 1.4

LEARNING OBJECTIVES

- Understand the terms: relative atomic mass, based on the ^{12}C scale; relative molecular mass; relative formula mass; molar mass, as the mass per mole of a substance in g mol^{-1}.
- Understand how to calculate relative molecular mass and relative formula mass from relative atomic masses.
- Perform calculations using the Avogadro constant L (6.02×10^{23} mol^{-1}).

RELATIVE ATOMIC MASS (A_r)

As chemists discovered more and more elements in the nineteenth century, they began to realise that the masses of the elements were different. They could not weigh individual atoms, but they were able to use numbers to compare the masses of atoms of different elements. For this reason, they began to use the term 'relative atomic mass'.

The chemists soon realised that the element whose atoms had the smallest mass was hydrogen, so the relative atomic mass of hydrogen was fixed as 1. Atoms of silicon had double the mass of nitrogen atoms, and nitrogen atoms were 14 times heavier than hydrogen atoms. This meant that the relative atomic mass of nitrogen was 14, and that of silicon was 28. At first, mostly whole numbers were used, but eventually it was possible to find the mass of an atom to several decimal places. The Periodic Table in the Data Booklet uses 1 decimal place for lighter elements and whole numbers for heavier ones.

After the discovery of isotopes, the ^{12}C isotope of carbon was used in the definition of relative atomic mass.

A suitable definition of relative atomic mass is:

the weighted mean (average) mass of an atom compared to $\frac{1}{12}$ of the mass of an atom of ^{12}C

It is often useful to remember this expression:

$$A_r = \frac{\text{mean mass of an atom of an element}}{\frac{1}{12}\text{ of the mass of an atom of }^{12}\text{C}}$$

RELATIVE MOLECULAR MASS (M_r)

Relative atomic masses are used for atoms of elements. Relative molecular masses are used for molecules of both elements and compounds. They are easily calculated by adding relative atomic masses.

Table A shows values for some common elements taken from the Data Booklet.

ELEMENT	RELATIVE ATOMIC MASS
hydrogen	1.0
carbon	12.0
oxygen	16.0
sulfur	32.1
copper	63.5

table A

Note that A_r and M_r do not have units.

Here are some examples of calculations.

WORKED EXAMPLE 1

What is the relative molecular mass of carbon dioxide, CO_2?

$M_r = 12.0 + (2 \times 16.0) = 44.0$

WORKED EXAMPLE 2

What is the relative molecular mass of sulfuric acid, H_2SO_4?

$M_r = (2 \times 1.0) + 32.1 + (4 \times 16.0) = 98.1$

EXAM HINT

Make sure you use the relative atomic masses shown on the Periodic Table in the Data Booklet.

RELATIVE FORMULA MASS (M_r)

This term has the same symbol as relative molecular mass, but the 'formula' part means that it includes both molecules and ions. Worked example 3 below is slightly more complicated because of the water of crystallisation, but there is also another problem. Hydrated copper(II) sulfate is an ionic compound, so it is not a good idea to refer to its relative *molecular* mass. That is why it is called relative *formula* mass.

WORKED EXAMPLE 3

What is the relative formula mass of hydrated copper(II) sulfate, $CuSO_4.5H_2O$?

$M_r = 63.5 + 32.1 + (4 \times 16.0) + 5\{(2 \times 1.0) + 16.0\} = 249.6$

The term 'relative formula mass' should also be used for compounds with giant structures, such as sodium chloride and silicon dioxide.

MOLAR MASS (M)

Another way around the problem in Worked example 3 is to use the term **molar mass**, which is the mass per mole of any substance (molecular or ionic). Its symbol is M (not M_r) and it has the units g mol^{-1} (grams per mole). Here we have a new term (the mole) which will be fully explained in **Topic 1C.2**. For now, you can think of one mole (1 mol) of a substance as being the same quantity as the relative formula mass of the substance, with the units of grams.

So, this is the expression you can use:

$$\text{amount in mol} = \frac{\text{mass of substance in g}}{\text{molar mass in g mol}^{-1}} \text{ or } n = \frac{m}{M}$$

Table B shows examples of working out the amounts in moles of some substances using this expression.

SUBSTANCE	O_2	CH_4	H_2O	NH_4NO_3
mass in g	5.26	4.0	100	14.7
molar mass, M in g mol^{-1}	32.0	16.0	18.0	80.0
amount in mol	0.164	0.25	5.56	0.184

table B

THE AVOGADRO CONSTANT

Amedeo Avogadro (1776–1856) was an Italian chemist whose name is used in naming the **Avogadro constant**. We are introducing him here because the scaling-up factor from atoms, molecules and ions to grams is named after him.

▲ **fig A** The Italian chemist, Amedeo Avogadro

The value of the Avogadro constant is approximately 602 000 000 000 000 000 000 000 mol^{-1}. It is easier to write this number using standard form: 6.02×10^{23} mol^{-1}.

You do not need to know a definition of the Avogadro constant, and it is best to think of it as the number of particles (atoms, molecules or ions) in one mole of any substance. For example, there are:

6.02×10^{23} helium atoms in 4.0 g of He

6.02×10^{23} carbon dioxide molecules in 44.0 g of CO_2

6.02×10^{23} nitrate ions in 62.0 g of NO_3^-

CALCULATIONS USING THE AVOGADRO CONSTANT

You will need to use the value of L in the types of calculation shown here.

1 Calculate the number of particles in a given mass of a substance. Start by using the expression:

$$\text{amount in mol} = \frac{\text{mass of substance in g}}{\text{molar mass in g mol}^{-1}} \text{ or } n = \frac{m}{M}$$

then multiply the amount in mol by the Avogadro constant.

WORKED EXAMPLE 4

How many H_2O molecules are there in 1.25 g of water?

$$n = \frac{1.25}{18.0} = 0.0694 \text{ mol}$$

number of molecules = $6.02 \times 10^{23} \times 0.0694 = 4.18 \times 10^{22}$

2 Calculate the mass of a given number of particles of a substance: start by dividing the number of particles by the Avogadro constant, then multiply the result by the molar mass.

WORKED EXAMPLE 5

What is the mass of 100 million atoms of gold?

$$n = \frac{100 \times 10^6}{6.02 \times 10^{23}} = 1.66 \times 10^{-16} \text{ mol}$$

$$m = 1.66 \times 10^{-16} \times 197.0 = 3.27 \times 10^{-14} \text{ g}$$

(There are lots of atoms, but only a tiny mass.)

CHECKPOINT

1. Malachite is an important mineral with the formula $Cu_2CO_3(OH)_2$. Calculate its relative formula mass.

2. How many molecules of sugar ($C_{12}H_{22}O_{11}$) are there in a teaspoon measure (4.20 g)?

DID YOU KNOW?

The symbol L is used for the Avogadro constant (using A would be confusing because of the use of A_r for relative atomic mass). L comes from the surname of Johann Josef Loschmidt (1821–1895), an Austrian chemist who was a contemporary of Avogadro. He made many contributions to our understanding of the same area of knowledge.

SUBJECT VOCABULARY

molar mass the mass per mole of a substance; it has the symbol M and the units g mol^{-1}

Avogadro constant (L) 6.02×10^{23} mol^{-1}, the number of particles in one mole of a substance

1C 2 CALCULATIONS INVOLVING MOLES

LEARNING OBJECTIVES

- Know that the mole (mol) is the unit for the amount of a substance.

WHAT IS A MOLE?

So far, we have referred to the mole and have used it in a simple form of calculation, but we have not properly explained what it is.

THE DEFINITION OF A MOLE

A **mole** is the amount of substance that contains the same number of particles as the number of carbon atoms in exactly 12 g of the ^{12}C isotope.

This definition is not easy to understand, but will be explained as you read on.

▲ **fig A** From left to right: tin (Sn), magnesium (Mg), iodine (I) and copper (Cu). Each sample contains 6.02×10^{23} atoms.

COUNTING ATOMS

As you know, atoms are very tiny particles that cannot be seen by the human eye. When we look at the sand on a beach, we are looking at billions and billions of atoms of, mostly, silicon and oxygen in the compound silicon dioxide (SiO_2). Quoting the actual numbers of atoms involved in a reaction, whether in a test tube or on an industrial scale, would involve extremely large numbers that would be very difficult to handle.

You are already familiar with the use of relative atomic masses to compare the relative masses of atoms. In a water molecule, the oxygen atom has a mass that is 16 times greater than the mass of a hydrogen atom. You know this because in the Periodic Table the relative atomic masses are H = 1.0 and O = 16.0. Note the word 'relative'. These values do not tell us the actual mass in grams of an oxygen atom or a hydrogen atom. They only tell us that an oxygen atom has a mass 16 times greater than that of a hydrogen atom.

Now consider using these numbers (16.0 and 1.0) with the familiar unit g (grams). You can more easily visualise 16.0 g of oxygen than a single atom of oxygen. Doing this is effectively scaling up on a very large scale. The number of oxygen atoms in 16.0 g of oxygen is the same as the number of hydrogen atoms in 1.0 g of hydrogen.

CALCULATIONS USING MOLES

WHAT TO REMEMBER WHEN DOING CALCULATIONS

You can use the mole to count atoms, molecules, ions, electrons and other species. So, it is important to include an exact description of the species being referred to. Consider the examples of hydrogen, oxygen and water.

One mole of water has a mass of 18.0 g, but what is the mass of one mole of hydrogen or oxygen? It depends on whether you are referring to atoms or molecules, so you need to make this clear.

Remember that the symbol n is used for the amount of substance in mol.

Consider the substances in **table A**. The masses are the same, but the amounts are different.

	1g OF OXYGEN ATOMS (O)	1g OF OXYGEN MOLECULES (O_2)	1g OF OZONE MOLECULES (O_3)
n (amount in mol)	1 ÷ 16.0 = 0.0625 mol	1 ÷ 32.0 = 0.0313 mol	1 ÷ 48.0 = 0.0208 mol

table A

You should always clearly identify the species that you are referring to if there is any possibility that there could be more than one meaning. This is not usually necessary for compounds.

It is good practice to refer to both the formula and the name. Examples include:

- the amount, in moles, of O in 9.4 g of oxygen atoms (0.59 mol)
- the amount, in moles, of O_2 in 9.4 g of oxygen molecules (0.29 mol)
- the amount, in moles, of O_3 in 9.4 g of ozone molecules (0.20 mol)
- the amount, in moles, of CO_2 in 9.4 g of carbon dioxide (0.21 mol)
- the amount, in moles, of SO_4^{2-} in 9.4 g of sulfate ions (0.098 mol).

THE EQUATION FOR CALCULATING MOLES

The equation for calculating moles is:

$$\text{amount of substance in moles} = \frac{\text{mass in grams}}{\text{molar mass}} \text{ or } n = \frac{m}{M}$$

You will use this expression in many calculations during your study of chemistry. It is often rearranged as:

$$M = \frac{m}{n} \text{ or } m = n \times M$$

WORKED EXAMPLE 1

What is the amount of substance in 6.51 g of sodium chloride?

$$n = \frac{m}{M} = \frac{6.51}{58.5} = 0.111 \text{ mol}$$

WORKED EXAMPLE 2

What is the mass of 0.263 mol of hydrogen iodide?

$$m = n \times M = 0.263 \times 127.9 = 33.6 \text{ g}$$

WORKED EXAMPLE 3

A sample of 0.284 mol of a substance has a mass of 17.8 g. What is the molar mass of the substance?

$$M = \frac{m}{n} = \frac{17.8}{0.284} = 62.7\,\text{g mol}^{-1}$$

LEARNING TIP

When using moles, always make clear what particles you are referring to: atoms, molecules, ions or electrons. It is also a good idea to state the formula.

CHECKPOINT

1. What is the amount of substance in each of the following?

(a) 8.00 g of sulfur, S

(b) 8.00 g of sulfur dioxide, SO_2

(c) 8.00 g of sulfate ions, SO_4^{2-}

2. How many particles are there of the specified substance?

(a) atoms in 2.00 g of sulfur, S

(b) molecules in 4.00 g of sulfur dioxide, SO_2

(c) ions in 8.00 g of sulfate ions, SO_4^{2-}

SUBJECT VOCABULARY

mole the amount of substance that contains the same number of particles as the number of carbon atoms in exactly 12 g of ^{12}C

1C 3 CALCULATIONS USING REACTING MASSES

SPECIFICATION REFERENCE
1.7 1.10

LEARNING OBJECTIVES

- Use chemical equations to calculate reacting masses and vice versa, using the concepts of amount of substance and molar mass.
- Determine a formula or confirm an equation by experiment, including evaluation of the data.

INTRODUCTION TO REACTING MASSES

You can use the ideas from previous topics about amounts of substance and equations to do calculations involving the masses of reactants and products in equations.

A balanced equation for a reaction shows the same number of each species (atoms, molecules, ions or electrons) on both sides of the equation. It is also balanced for the masses of each species. This means that we can make predictions about the masses of reactants, which are needed to form a specified mass or amount of a product, or the other way round.

Consider this equation used in the manufacture of ammonia:

$$N_2 + 3H_2 \rightarrow 2NH_3$$

This shows that one molecule of nitrogen reacts with three molecules of hydrogen to form two molecules of ammonia.

This statement can be made about the amounts involved:

1 mol of N_2 reacts with 3 mol of H_2 to form 2 mol of NH_3.

This statement can be made about the masses involved:

28.0 g of N_2 reacts with 6.0 g of H_2 to form 34.0 g of NH_3.

These amounts and masses can have many other values, as long as the ratio does not change.

CALCULATING REACTING MASSES FROM EQUATIONS

Using a balanced equation, predictions can be made about reacting masses.

WORKED EXAMPLE 1

The equation for a reaction is:

$$SO_3 + H_2O \rightarrow H_2SO_4$$

What mass of sulfur trioxide is needed to form 75.0 g of sulfuric acid?

Step 1: calculate the molar masses of all substances you are told about and asked about, in this case, sulfur trioxide and sulfuric acid

$M(SO_3) = 80.1\ \text{g mol}^{-1}$ and $M(H_2SO_4) = 98.1\ \text{g mol}^{-1}$

Step 2: calculate the amount of sulfuric acid

$$n = \frac{m}{M} = \frac{75.0}{98.1} = 0.765\ \text{mol}$$

Step 3: use the reaction ratio in the equation to work out the amount of sulfur trioxide needed

As the ratio is 1 : 1, the amount is the same, so $n(SO_3) = 0.765$ mol

Step 4: calculate the mass of sulfur trioxide

$$m = n \times M = 0.765 \times 80.1 = 61.2\ \text{g}$$

WORKED EXAMPLE 2

The equation for a reaction is:

$2NH_3 + H_2SO_4 \rightarrow (NH_4)_2SO_4$

What mass of ammonia is needed to form 100 g of ammonium sulfate?

Step 1:

$M(NH_3) = 17.0\,g\,mol^{-1}$ and $M((NH_4)_2SO_4) = 132.1\,g\,mol^{-1}$

Step 2:

$n((NH_4)_2SO_4) = \frac{100}{132} = 0.757\,mol$

Step 3:

$n(NH_3) = 2 \times 0.757 = 1.51\,mol$
(note the 2:1 ratio in the equation)

Step 4:

$m(NH_3) = n \times M = 1.51 \times 17.0 = 25.7\,g$

WORKING OUT FORMULAE AND EQUATIONS FROM REACTING MASSES

You might assume that the formulae and equations for all reactions are already known. However, there are sometimes two or more possible formulae for a substance. There can also be more than one reaction for the same reactants. Reacting masses can be used to identify the correct formula, or which of the reactions is occurring.

WORKED EXAMPLE 3

Sodium carbonate exists as the pure (anhydrous) compound but also as three **hydrates**. Careful heating can decompose these hydrates to one of the other hydrates or to the anhydrous compound. The measurement of reacting masses can allow you to determine the correct equation for the decomposition.

Question

A 16.7 g sample of a hydrate of sodium carbonate ($Na_2CO_3.10H_2O$) is heated at a constant temperature for a specified time until the reaction is complete. A mass of 3.15 g of water is obtained. What is the equation for the reaction occurring?

Method

Step 1: calculate the molar masses of the relevant substances

$M(Na_2CO_3.10H_2O) = 286.1\,g\,mol^{-1}$
and $M(H_2O) = 18.0\,g\,mol^{-1}$

Step 2: calculate the amounts of these substances

$Na_2CO_3.10H_2O$: $n = \frac{m}{M} = \frac{16.7}{286.1} = 0.0584\,mol$

water: $n = \frac{m}{M} = \frac{3.15}{18.0} = 0.175\,mol$

Step 3: use these amounts to calculate the simplest whole-number ratio for these substances

$Na_2CO_3.10H_2O$ and H_2O are in the ratio 0.0584 : 0.175 or 1 : 3

Step 4: use this ratio to work out the equation for the reaction

$Na_2CO_3.10H_2O \rightarrow Na_2CO_3.7H_2O + 3H_2O$

We have worked out the $Na_2CO_3.7H_2O$ formula by considering the ratio of the other two formulae.

WORKED EXAMPLE 4

Copper forms two oxides. Both oxides can be converted to copper by heating with hydrogen.

Question

An oxide of copper is heated in a stream of hydrogen to constant mass. The masses of copper and water formed are Cu = 17.6 g and H_2O = 2.56 g. What is the equation for the reaction occurring?

Method

Step 1: $M(Cu) = 63.5\,g\,mol^{-1}$ and $M(H_2O) = 18.0\,g\,mol^{-1}$

Step 2: $n(Cu) = \frac{17.6}{63.5} = 0.277\,mol$ and $n(H_2O) = \frac{2.56}{18.0} = 0.142\,mol$

Step 3: ratio is 0.277 : 0.142 = 2 : 1

Step 4: the equation has 2 mol of Cu and 1 mol of H_2O, so the products must be $2Cu + H_2O$

So the equation is:

$Cu_2O + H_2 \rightarrow 2Cu + H_2O$ and not $CuO + H_2 \rightarrow Cu + H_2O$

LEARNING TIP

One important part of both these calculation methods is the use of the relevant ratio from the equation. Practise deciding which substances should be used for the ratio and which way round to use the ratio.

CHECKPOINT

SKILLS ADAPTIVE LEARNING

1. A fertiliser manufacturer makes a batch of 20 kg of ammonium nitrate. What mass of ammonia, in kg, does the manufacturer need to start with?
2. A sample of an oxide of iron was reduced to iron by heating with hydrogen. The mass of iron obtained was 4.35 g and the mass of water was 1.86 g. Deduce the equation for the reaction that occurred.

fig A Chemistry on this scale needs careful calculations so that no reactants are wasted.

SUBJECT VOCABULARY

hydrate compound containing water of crystallisation, represented by formulae such as $CuSO_4.5H_2O$.

1C 4 THE YIELD OF A REACTION

SPECIFICATION REFERENCE 1.9

LEARNING OBJECTIVES

- Calculate percentage yields in laboratory and industrial processes using chemical equations and experimental results.

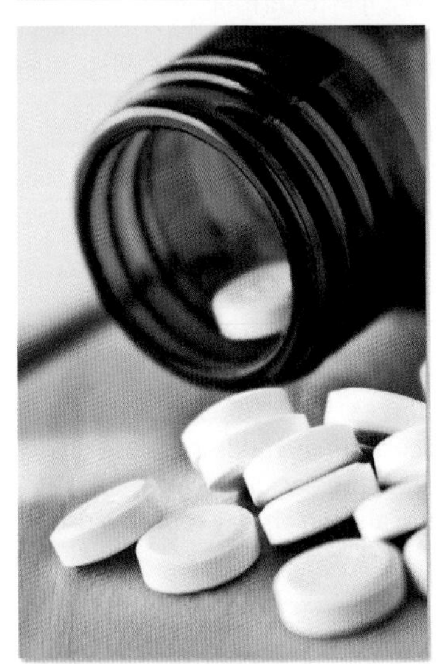

▲ **fig A** Pharmaceutical companies are always looking for ways to increase the percentage yield when manufacturing a drug.

THEORETICAL YIELD, ACTUAL YIELD AND PERCENTAGE YIELD

In the laboratory, when you are making a product, you naturally want to obtain as much of it as possible from the reactants you start with. In industry, where reactions occur on a much larger scale, and there is economic competition between manufacturers, it is even more important to maximise the product of a reaction.

There are some reasons why the mass of a reaction product may be less than the maximum possible.

- The reaction is reversible and so may not be complete.
- There are side-reactions that lead to other products that are not wanted.
- The product may need to be purified, which may result in loss of product.

TERMINOLOGY RELATING TO 'YIELD'

We normally use the term 'yield' with other words, such as:

- **theoretical yield**
- **actual yield**
- **percentage yield**.

In the laboratory, theoretical yield and actual yield may be measured in grams, but in industry, kilograms and tonnes are more likely to be used. Percentage yield is the term most often used, but you need to understand the other two terms first.

THEORETICAL YIELD

We calculate theoretical yield using the equation for the reaction, and we use a method you are familiar with from previous topics. It is always assumed that the reaction goes to completion, with no losses.

WORKED EXAMPLE 1

Copper(II) carbonate is decomposed to obtain copper(II) oxide. The equation for the reaction is:

$$CuCO_3 \rightarrow CuO + CO_2$$

What is the theoretical yield of copper(II) oxide obtainable from 5.78 g of copper(II) carbonate?

Step 1: calculate the amount of starting material

$$n(CuCO_3) = \frac{5.78}{123.5} = 0.0468\ \text{mol}$$

Step 2: use the reacting ratio to calculate the amount of desired product

$$n(CuO) = 0.0468\ \text{mol}$$

Step 3: calculate the mass of desired product

$$m = 0.0468 \times 79.5 = 3.72\ \text{g}$$

WORKED EXAMPLE 2

Magnesium phosphate can be prepared from magnesium by reacting it with phosphoric acid. The equation for the reaction is:

$$3Mg + 2H_3PO_4 \rightarrow Mg_3(PO_4)_2 + 3H_2$$

What is the theoretical yield of magnesium phosphate obtainable from 5.62 g of magnesium?

Step 1: $n(Mg) = \frac{5.62}{24.3} = 0.231\ \text{mol}$

Step 2: $n(Mg_3(PO_4)_2) = \frac{0.231}{3} = 0.0770\ \text{mol}$

Step 3: $m = 0.0770 \times 262.9 = 20.2\ \text{g}$

ACTUAL YIELD

This is the actual mass obtained by weighing the product obtained, not by calculation.

PERCENTAGE YIELD

Percentage yield is calculated using the equation:

$$\frac{\text{actual yield} \times 100}{\text{theoretical yield}} = \text{percentage yield}$$

This calculation may be done independently or in conjunction with the calculation of theoretical yield.

EXAM HINT

You may be asked in a question to suggest why you have a low yield. If so, you should give a specific example in the reaction method where you may have lost product such as, 'some solution was left on the filter paper during filtration'.

WORKED EXAMPLE 3

The theoretical yield in a reaction is 26.7 tonnes. The actual yield is 18.5 tonnes. What is the percentage yield?

$$\text{percentage yield} = \frac{18.5 \times 100}{26.7} = 69.3\%$$

WORKED EXAMPLE 4

A manufacturer uses this reaction to obtain methanol from carbon monoxide and hydrogen:

$$CO + 2H_2 \rightarrow CH_3OH$$

The manufacturer obtains 4.07 tonnes of methanol starting from 4.32 tonnes of carbon monoxide. What is the percentage yield?

First, calculate the theoretical yield.

Step 1: $n(CO) = \frac{4.32 \times 10^6}{28.0} = 1.54 \times 10^5$ mol

Step 2: $n(CH_3OH) = 1.54 \times 10^5$ mol (because of 1:1 ratio)

Step 3: $m = 1.54 \times 10^5 \times 32.0 = 4.94 \times 10^6$ mol

Then use that answer to calculate the percentage yield.

$$\text{Percentage yield} = \frac{4.07 \times 10^6 \times 100}{4.94 \times 10^6} = 82.4\%$$

LEARNING TIP

For calculations using kilograms and tonnes, remember that:

1 kg = 1×10^3 g

1 tonne = 1×10^6 g

CHECKPOINT

SKILLS ADAPTIVE LEARNING

1. A student prepares a sample of copper(II) sulfate crystals, $CuSO_4.5H_2O$, weighing 7.85 g. She started with 4.68 g of copper(II) oxide. What is the percentage yield?
2. A manufacturer makes some ethanoic acid using this reaction:

   $$CH_3OH + CO \rightarrow CH_3COOH$$

   Starting with 50.0 kg of methanol, the manufacturer obtains 89.2 kg of ethanoic acid. What is the percentage yield?

SUBJECT VOCABULARY

theoretical yield the maximum possible mass of a product in a reaction, assuming complete reaction and no losses

actual yield the actual mass obtained in a reaction

percentage yield the actual yield divided by the theoretical yield, expressed as a percentage

1C 5 ATOM ECONOMY

SPECIFICATION REFERENCE 1.9

LEARNING OBJECTIVES

- Calculate percentage atom economies using chemical equations and experimental results.

BACKGROUND TO ATOM ECONOMY

In **Topic 1C.4**, we looked at the percentage yield of a reaction. The closer the value is to 100%, the better. A higher percentage means that less of the starting materials are lost or end up as unwanted products.

Percentage yield is an important factor to take into consideration when assessing the suitability of an industrial process. However, it is not the only one. Other factors include:

- the availability or scarcity of non-renewable raw materials
- the cost of raw materials
- the quantity of energy needed.

HOW ATOM ECONOMY WORKS

Here is an example of **atom economy** in action.

There are two main processes in the manufacture of phosphoric acid. To make the comparison easier to follow, a single summary equation is shown for each process.

Process 1 $Ca_3(PO_4)_2 + 3H_2SO_4 \rightarrow 2H_3PO_4 + 3CaSO_4$

Process 2 $P_4 + 5O_2 + 6H_2O \rightarrow 4H_3PO_4$

There are advantages and disadvantages of both processes. However, what you can see from these equations is that all of the atoms in the starting materials for Process 2 end up in the desired product. In Process 1, many of the atoms end up in a second, unwanted product, calcium sulfate. Process 1 has a lower atom economy than Process 2.

THE CONTRIBUTION OF BARRY TROST

A chemist from the USA, Barry Trost, developed the idea of atom economy as an alternative way of assessing chemical reactions, especially in industrial processes. He believed that it was important to consider how many atoms from the reactants end up in the desired product.

▲ **fig A** Barry Trost is a pioneer of the concept of atom economy.

The expression we use to calculate atom economy (usually described as percentage atom economy) is:

$$\text{atom economy} = \frac{\text{molar mass of the desired product}}{\text{sum of the molar masses of all products}} \times 100$$

You can see that you do not need a calculator to work out the atom economy of Process 2. There is only one product, so it must be 100%.

$$\text{For Process 1, atom economy} = \frac{(98.0 \times 2) \times 100}{(98.0 \times 2) + (136.2 \times 3)} = 32.4\%$$

So, you can see that less than one-third of the mass of the starting materials ends up in the desired product, which does not look good if the $CaSO_4$ is a waste product that has to be disposed of. If the other product has a use, then the manufacturer can sell it, which would partly balance the low atom economy of the process. Even if the percentage yield of Process 1 was as high as 100%, the atom economy is still only 32.4%.

REACTION TYPES AND ATOM ECONOMY

We can make some generalisations about certain types of reaction.

- Addition reactions have 100% atom economy.
- Elimination and substitution reactions have lower atom economies.
- Multistep reactions may have even lower atom economies.

EXAMPLES OF CALCULATIONS

WORKED EXAMPLE 1

Sodium carbonate is an important industrial chemical manufactured by the Solvay process. The overall equation for the process is:

$CaCO_3 + 2NaCl \rightarrow Na_2CO_3 + CaCl_2$

A manufacturer starts with 75.0 kg of calcium carbonate and obtains 76.5 kg of sodium carbonate. Calculate the percentage yield and atom economy for this reaction.

M_r values are 100.1 for $CaCO_3$ and 106.0 for Na_2CO_3.

$$\text{Theoretical yield} = \frac{75.0 \times 106.0}{100.1} = 79.4\,\text{kg}$$

$$\text{Percentage yield} = \frac{76.5 \times 100}{79.4} = 96.3\%$$

$$\text{Atom economy} = \frac{106.0 \times 100}{106.0 + 111.1} = 48.8\%$$

LEARNING TIP

Remember that percentage yield indicates how efficient a reaction is at converting the reactants to the products.
Atom economy indicates the percentage of atoms from the starting materials that end up in the desired product.

EXAM HINT

Also note that if you have more than one product, you will need time and energy to separate the desired product from the mixture.

WORKED EXAMPLE 2

Hydrazine (N_2H_4) can be used as a rocket fuel and is manufactured using this reaction:

$2NH_3 + NaOCl \rightarrow N_2H_4 + NaCl + H_2O$

What is the atom economy for this reaction?

You first need to work out the molar masses of the products. These are 32.0, 58.5 and 18.0.

$$\text{Atom economy} = \frac{32.0 \times 100}{32.0 + 58.5 + 18.0} = 29.5\%$$

WORKED EXAMPLE 3

A manufacturer of ethene wants to convert some ethene into 1,2-dichloroethane. He considers two possible reactions:

Reaction 1 $H_2C{=}CH_2 + Cl_2 \rightarrow ClCH_2CH_2Cl$

Reaction 2 $2H_2C{=}CH_2 + 4HCl + O_2 \rightarrow 2ClCH_2CH_2Cl + 2H_2O$

Explain, without doing a calculation, which reaction would be a good choice on the basis of atom economy.

The answer is Reaction 1, because there is only one product, so all the atoms in the reactants end up in the desired product and the atom economy is 100%.

Reaction 2 has a lower atom economy because some of the atoms in the reactants form water, which has no value as a product.

CHECKPOINT

1. Ethanol can be manufactured by the hydration of ethene:

$C_2H_4 + H_2O \rightarrow C_2H_5OH$

What is the atom economy of this process?

2. Ethene can be manufactured by the dehydration of ethanol:

$C_2H_5OH \rightarrow C_2H_4 + H_2O$

What is the atom economy of this process?

SUBJECT VOCABULARY

atom economy the molar mass of the desired product divided by the sum of the molar masses of all the products, expressed as a percentage

1D 1 EMPIRICAL FORMULAE

SPECIFICATION REFERENCE 1.1 1.6

LEARNING OBJECTIVES

- Know the term empirical formula.
- Use experimental data to calculate empirical formulae.

INTRODUCTION TO EMPIRICAL FORMULAE

The three letters 'mol' appear in several words used in chemistry. You will know some of these from your previous studies, especially 'molecule', which is a group of two or more atoms joined together by covalent bonds. You know that a molecule of water can be represented by the formula H_2O, so this is the molecular formula of water. You have already come across the term 'relative molecular mass', which you will probably remember is 18 (or more accurately, 18.0) for water.

The term 'empirical' indicates that some information has been found by experiment. An **empirical formula** shows the smallest whole number ratio of the atoms of each element in a compound.

ONE EXAMPLE OF AN EXPERIMENTAL METHOD

A simple example involves determining the formula of an oxide of copper. This oxide can be converted to copper by removing the oxygen.

Here are the steps in the experiment.

- Place a known mass of the oxide of copper in the tube.
- Heat the oxide in a stream of hydrogen gas (or natural gas, which is mostly methane).
- The gas reacts with the oxygen in the copper oxide and forms steam.
- The colour of the solid gradually changes to orange-brown, which is the colour of copper.
- The excess gas is burned off at the end of the tube for safety reasons.
- After cooling, remove and weigh the solid copper.
- It is good practice to heat the solid again in the stream of the gas to check whether its mass changes. Heating to a constant mass suggests that the conversion to copper is complete.

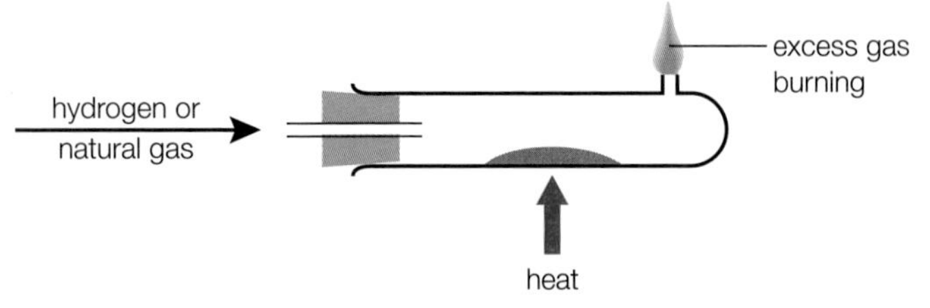

fig A This apparatus can be used to convert copper oxide to copper by removing the oxygen.

CALCULATING EMPIRICAL FORMULAE

The calculation method involves these steps.

- Divide the mass, or percentage composition by mass, of each element by its relative atomic mass.
- If necessary, divide the answers from this step by the smallest of the numbers.
- This gives numbers that should be in an obvious whole number ratio, such as 1 : 2 or 3 : 2.
- These whole numbers are used to write the empirical formula.

The numbers may not be in an exact ratio because of experimental error, but you should be able to decide what the nearest whole-number ratio is.

Use at least two significant figures in the calculation (preferably three), and beware of inappropriate rounding. For example, you cannot convert a 1.2 : 1.2 : 1.9 ratio to 1 : 1 : 2, because that would give you the wrong answers.

It may help you to organise the calculation using a table, although this is not essential.

CALCULATION USING MASSES

Assume that these are the results of the experiment outlined on the left.

mass of copper oxide = 4.28 g

mass of copper = 3.43 g

mass of oxygen removed is 4.28 − 3.43 = 0.85 g

Table A is the calculation table for these results.

	Cu	O
mass of element / g	3.43	0.85
relative atomic mass	63.5	16.0
division by A_r	0.0540	0.0531
ratio	1	1

table A

Here, the ratio is obviously 1 : 1, so the empirical formula is CuO.

CALCULATION USING PERCENTAGE COMPOSITION BY MASS

You do the calculation in the same way, except that you divide percentages instead of masses by the relative atomic masses.

The calculation table, **table B**, refers to a compound containing three elements. The compound has the percentage composition by mass C = 38.4%, H = 4.8%, Cl = 56.8%.

	C	H	Cl
% of element	38.4	4.8	56.8
relative atomic mass	12.0	1.0	35.5
division by A_r	3.2	4.8	1.6
ratio	2	3	1

table B

You can see that the empirical formula is C_2H_3Cl.

CALCULATION WHEN THE OXYGEN VALUE IS NOT PROVIDED

The results for some compounds do not include values for oxygen because it is often difficult to obtain an experimental value for the mass of oxygen. Sometimes you will need to remember to calculate the percentage of oxygen by subtraction. Here is an example.

A compound has the percentage composition by mass Na = 29.1%, S = 40.5%, with the remainder being oxygen.

The percentage of oxygen = 100 − (29.1 + 40.5) = 30.4%.

Table C is the calculation table.

	Na	S	O
% of element	29.1	40.5	30.4
relative atomic mass	23.0	32.1	16.0
division by A_r	1.27	1.26	1.90
division by the smallest	1	1	1.5
ratio	2	2	3

table C

You can now see that the empirical formula is $Na_2S_2O_3$.

EXAM HINT

Take care to look out for decimals that indicate obvious fractions, e.g., simplest ratio 1 : 1.33 indicates an empirical formula of 3 : 4. How would a ratio of 1 : 1.25 convert into an empirical formula?

CALCULATION USING COMBUSTION ANALYSIS

Many organic compounds contain carbon and hydrogen, or carbon, hydrogen and oxygen. When a known mass of an organic compound is completely burned, it is possible to collect and measure the masses of carbon dioxide and water formed. The calculation is more complex because there are extra steps. Here is an example.

A 1.87 g sample of an organic compound was completely burned, forming 2.65 g of carbon dioxide and 1.63 g of water.

In this type of calculation, the first steps are to calculate the masses of carbon and hydrogen in the carbon dioxide and water.

- The relative molecular mass of carbon dioxide is 44.0 but, because the relative atomic mass of carbon is 12.0, the proportion of carbon in carbon dioxide is always 12.0 ÷ 44.0.
- Similarly, the proportion of hydrogen in water is always (2 × 1.0) ÷ 18.0.

All of the carbon in the carbon dioxide comes from the carbon in the organic compound. Similarly, all of the hydrogen in the water comes from the hydrogen in the organic compound. In this example:

$$\text{mass of carbon} = \frac{2.65 \times 12.0}{44.0} = 0.723\,\text{g}$$

$$\text{mass of hydrogen} = \frac{1.63 \times 2.0}{18.0} = 0.181\,\text{g}$$

These two masses add up to 0.904 g.

The original mass of the organic compound was 1.87 g, so the difference must be the mass of oxygen present in the organic compound. The mass of oxygen = 1.87 − 0.904 = 0.966 g.

Table D is the calculation table. You can see that the empirical formula of the sample compound is CH_3O.

	C	H	O
mass of element / g	0.723	0.181	0.966
relative atomic mass	12.0	1.0	16.0
division by A_r	0.0603	0.181	0.0604
ratio	1	3	1

table D

Fig B shows a model of the glucose molecule. You can count the numbers of the three different atoms in the molecule.

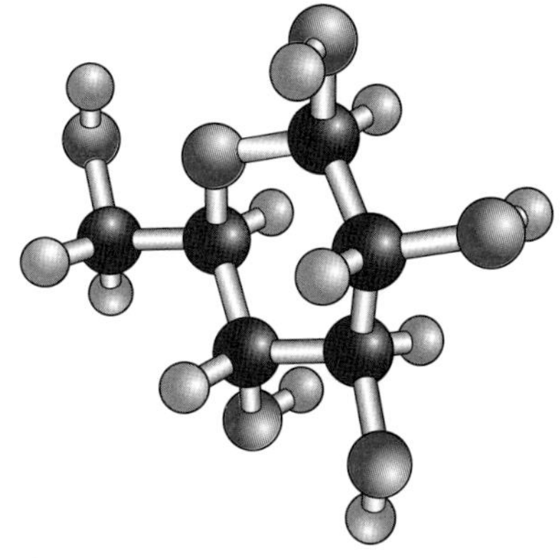

▲ **fig B** A ball-and-stick model of glucose.

LEARNING TIP

Divide by the relative atomic mass, not the atomic number or the relative molecular mass.

For oxygen, only divide by 16.0, not by 8 or 32.

CHECKPOINT

1. A compound has the percentage composition by mass Ca = 24.4%, N = 17.1% and O = 58.5%. What is its empirical formula?
2. Combustion analysis of 2.16 g of an organic compound produced 4.33 g of carbon dioxide and 1.77 g of water. What is its empirical formula?
3. Can you work out the empirical formula of this molecule? Black represents carbon and white represents hydrogen.

SUBJECT VOCABULARY

empirical formula the smallest whole-number ratio of atoms of each element in a compound

1D 2 MOLECULAR FORMULAE

SPECIFICATION REFERENCE 1.1 1.6 1.8(iii)

LEARNING OBJECTIVES

- Know the term molecular formula.
- Use experimental data to calculate molecular formulae.
- Use the expression $pV = nRT$ for gases and volatile liquids.

INTRODUCTION TO MOLECULAR FORMULAE

In **Topic 1D.1**, you learned how to calculate empirical formulae. Sometimes the empirical formula of a compound is the same as its molecular formula. Carbon dioxide (CO_2) and water (H_2O) are common examples you will know. The **molecular formula** of a compound shows the actual numbers of the atoms of each element in the compound.

Examples of compounds with different empirical and molecular formulae are:

hydrogen peroxide	empirical formula is HO	molecular formula is H_2O_2
butane	empirical formula is C_2H_5	molecular formula is C_4H_{10}

To determine the molecular formula of a compound, you need to already know or to calculate:
- the empirical formula
- the relative formula mass.

For example, if you had already found that the empirical formula of a compound was HO and you then found that its relative formula mass was 34, you could compare the relative mass of the empirical formula (17) with 34 and work out that the molecular formula was double the empirical formula.

CALCULATING MOLECULAR FORMULAE

You have already practised calculating an empirical formula from experimental data. Now look at how to calculate a molecular formula from an empirical formula.

WORKED EXAMPLE 1

In this example, the empirical formula is given.
A compound has the empirical formula CH and a relative formula mass of 104.
The 'formula mass' of the empirical formula is 13.0.
104 ÷ 13.0 = 8, so the molecular formula of the compound is C_8H_8.

WORKED EXAMPLE 2

In this example, you first have to work out the empirical formula. A compound contains the percentage composition by mass Na = 34.3%, C = 17.9%, O = 47.8%, and has a **molar mass** of 134 g mol^{-1}.
The calculations are shown below.

	Na	C	O
% of element	34.3	17.9	47.8
relative atomic mass	23.0	12.0	16.0
division by A_r	1.49	1.49	2.99
ratio	1	1	2

The empirical formula is $NaCO_2$.
The 'formula mass' of the empirical formula is 23.0 + 12.0 + (2 × 16.0) = 67.0.
The molar mass, 134, is 2 × 67.0, so the molecular formula of the compound is $Na_2C_2O_4$.

THE IDEAL GAS EQUATION $pV = nRT$

$pV = nRT$ is the ideal gas equation, and can be used for gases (or volatile liquids above their boiling temperatures) to find the amount of a substance in moles. If the mass of the substance is also known, then the molar mass of the substance can be calculated. This gives the extra information needed to work out a molecular formula from an empirical formula.

The expression can also be rearranged to calculate a value of p, or V or T.

SI UNITS

When using this equation, you need to be careful that the units are the correct ones. It is always safest to work in SI units.

The SI units you should use are:

- p = pressure in pascals (Pa)
- V = volume in cubic metres (m^3)
- T = temperature in kelvin (K)
- n = amount of substance in moles (mol)
- R = the gas constant – this appears in the Data Booklet provided for use in the examinations and has the value $8.31\,J\,mol^{-1}K^{-1}$.

Sometimes in a question you may find that the units quoted are not SI units. If this is the case, then you will need to convert them to SI units. **Table A** shows the conversions you are likely to need.

CONVERSION	HOW TO DO IT
kPa → Pa	multiply by 10^3
$cm^3 \rightarrow m^3$	divide by 10^6 or multiply by 10^{-6}
$dm^3 \rightarrow m^3$	divide by 10^3 or multiply by 10^{-3}
°C → K	add 273

table A

WORKED EXAMPLE 3

In this example, you will calculate the molar mass of the gas. It may help you (at least until you have had more practice) to write a list of the values with any necessary conversions.

A 0.280 g sample of a gas has a volume of $58.5\,cm^3$, measured at a pressure of 120 kPa and a temperature of 70 °C. Calculate the molar mass of the gas.

$p = 120\,kPa = 120 \times 10^3\,Pa$

$V = 58.5\,cm^3 = 58.5 \times 10^{-6}\,m^3$

$T = 70\,°C = 343\,K$

$R = 8.31\,J\,mol^{-1}K^{-1}$

So, $$n = \frac{pV}{RT} = \frac{120 \times 10^3 \times 58.5 \times 10^{-6}}{8.31 \times 343} = 0.00246\,mol$$

$$M = \frac{m}{n} = \frac{0.280}{0.00246} = 114\,g\,mol^{-1}$$

WORKED EXAMPLE 4

In this example, you will calculate the empirical formula, then the amount in moles. After that the molar mass, then the molecular formula.

A compound has the percentage composition by mass C = 52.2%, H = 13.0%, O = 34.8%. A sample containing 0.173 g of the compound had a volume of $95.0\,cm^3$ when measured at 105 kPa and 45 °C. What is the molecular formula of this compound?

Step 1: calculate the empirical formula

	C	H	O
% by mass of element	52.2	13.0	34.8
relative atomic mass	12.0	1.0	16.0
division by A_r	4.35	13.0	2.175
ratio	2	6	1

The empirical formula is C_2H_6O.

Step 2: calculate the amount in moles

$$n = \frac{pV}{RT} = \frac{105 \times 10^3 \times 95.0 \times 10^{-6}}{8.31 \times 318} = 0.00377\,mol$$

Step 3: calculate the molar mass

$$M = \frac{m}{n} = \frac{0.173}{0.00377} = 45.9\,g\,mol^{-1}$$

Step 4: calculate the molecular formula

The 'formula mass' of the empirical formula is

$(2 \times 12.0) + (6 \times 1.0) + 16.0 = 46.0$

As 46.0 is the same as the molar mass, then the empirical and molecular formulae are the same.

The molecular formula is C_2H_6O.

LEARNING TIP

Be careful when using the word 'amount'. It should only be used for the amount, in moles, of a substance. For example, 'The amount of magnesium used was 0.150 mol' is correct.

You should not use it instead of quantities with other units. For example, 'The amount of magnesium used was 3.6 g' should be 'The mass of magnesium used was 3.6 g'.

Another example:

'The amount of water used was $25.0\,cm^3$' should be 'The volume of water used was $25.0\,cm^3$'.

CHECKPOINT

SKILLS PROBLEM SOLVING

1. A 2.82 g sample of a gas has a volume of $1.26\,dm^3$, measured at a pressure of 103 kPa and a temperature of 55 °C. Calculate the molar mass of the gas.
2. A compound has the percentage composition by mass C = 40.0%, H = 6.7%, O = 53.3%. A sample containing 0.146 g of the compound had a volume of $69.5\,cm^3$ when measured at 98 kPa and 63 °C. What is the molecular formula of this compound?

SUBJECT VOCABULARY

molecular formula the actual number of atoms of each element in a molecule

molar mass the mass per mole of a substance; it has the symbol M and the units $g\,mol^{-1}$

1E 1 MOLAR VOLUME CALCULATIONS

SPECIFICATION REFERENCE 1.8(ii) CP1

LEARNING OBJECTIVES

- Use chemical equations to calculate reacting volumes of gases and vice versa using the concept of molar volume of gases.

MOLAR VOLUME

The work of Avogadro and others led to the idea of **molar volume**, the volume of gas that contains one mole of that gas. The molar volume is approximately the same for all gases, but its value varies with temperature and pressure. The value most often used is for gases at room temperature and pressure (sometimes abbreviated as r.t.p.). Room temperature is 298 K (or 25 °C) and standard pressure varies, but is often quoted as 1.01×10^5 Pa. The value of molar volume is usually quoted as $24\,dm^3\,mol^{-1}$ (equivalent to $24\,000\,cm^3$). Its symbol is V_m.

$V_m = 24\,dm^3\,mol^{-1}$ at r.t.p.

and

$V_m = 24\,000\,cm^3\,mol^{-1}$ at r.t.p.

CALCULATIONS USING MOLAR VOLUME

CALCULATIONS INVOLVING A SINGLE GAS

If you are asked about a single gas, the calculation is straight forward. Assume that in these examples, all volumes are measured at r.t.p.

You need the expression $V_m = 24\,dm^3\,mol^{-1}$ which you might want to consider using in these alternative forms:

$$V_m = 24.0 = \frac{\text{volume in dm}^3}{\text{amount in mol}} \text{ or } V_m = 24\,000 = \frac{\text{volume in cm}^3}{\text{amount in mol}}$$

You will need to rearrange the expression depending on the actual question. Make sure that you use only 24 and dm^3, or only 24 000 and cm^3, in the calculation.

EXAMPLE 1

What is the amount, in moles, of CO in $3.8\,dm^3$ of carbon monoxide?

$$\text{Answer} = \frac{3.8}{24} = 0.16\,\text{mol}$$

EXAMPLE 2

What is the amount, in moles, of CO_2 in $500\,cm^3$ of carbon dioxide?

$$\text{Answer} = \frac{500}{24\,000} = 0.021\,\text{mol}$$

EXAMPLE 3

What is the volume of 0.36 mol of hydrogen?

$$\text{Answer} = 24 \times 0.36 = 8.64\,dm^3$$

CALCULATIONS INVOLVING GASES AND SOLIDS OR LIQUIDS

You may be given a chemical equation that involves one or more gases and a solid or a liquid. If you use information about the amount (in mol) of a solid or liquid, you can combine two of the calculation methods that we have already used.

The basis of this type of calculation is that for gases you can interconvert between amount and volume. For solids and liquids you can interconvert between amount and mass.

Step 1: Calculate the amount in moles from either the mass or the volume, depending on which one is given.

Step 2: Use the relevant reaction ratio in the equation to calculate the amount of the other substance.

Step 3: Convert this amount to a mass or a volume, depending on what the question asks.

EXAM HINT

When attempting multistep calculations involving different units, it is important to show the units for each step of the reaction. It will force you to think about what you have just calculated rather than leaving a number floating.

EXAMPLE 4

A piece of magnesium with a mass of 1.00 g is added to an excess of dilute hydrochloric acid. What volume of hydrogen gas is formed?

The equation for the reaction is:

$$Mg(s) + 2HCl(aq) \rightarrow MgCl_2(aq) + H_2(g)$$

You are not given any information about the hydrochloric acid, and you are not asked anything about magnesium chloride. You can use the mole expression to calculate the amount of magnesium:

$$n(\text{Mg}) = \frac{1.00}{24.3} = 0.0412\,\text{mol}$$

You can see that the $Mg : H_2$ ratio in the equation is 1 : 1, which means that 0.0412 mol of hydrogen is formed.

Convert this amount to a volume and you have the answer:

$$\text{volume} = 24 \times 0.0412 = 0.99\,dm^3$$

EXAMPLE 5

Calcium carbonate reacts with nitric acid to form calcium nitrate, water and carbon dioxide, as shown in the equation:

$$CaCO_3(s) + 2HNO_3(aq) \rightarrow Ca(NO_3)_2(aq) + H_2O(l) + CO_2(g)$$

In a reaction, $100\,cm^3$ of carbon dioxide is formed. What mass of calcium carbonate is needed for this?

You are not told anything about nitric acid, or asked anything about calcium nitrate or water. You can use the molar volume expression to calculate the amount of carbon dioxide:

$$\text{amount} = \frac{100}{24000} = 0.00417\,\text{mol}$$

You can see that the $CaCO_3 : CO_2$ ratio in the equation is 1 : 1, which means that 0.00417 mol of calcium carbonate is needed.

Convert this amount to a mass and you have the answer:

$$m = n \times M = 0.00417 \times 100.1 = 0.42\,\text{g}$$

EXAMPLE 6

Ammonium sulfate reacts with sodium hydroxide solution to form sodium sulfate, water and ammonia, as shown in the equation:

$$(NH_4)_2SO_4(s) + 2NaOH(aq) \rightarrow Na_2SO_4(aq) + 2H_2O(l) + 2NH_3(g)$$

What volume of ammonia is formed by reacting 2.16 g of ammonium sulfate with excess sodium hydroxide solution?

You are not given any information about the sodium hydroxide, and you are not asked anything about sodium sulfate or water. You can use the mole expression to calculate the amount of ammonium sulfate:

$$n((NH_4)_2SO_4) = \frac{2.16}{132.1} = 0.01635\,\text{mol}$$

You can see that the $(NH_4)_2SO_4 : NH_3$ ratio in the equation is 1 : 2, which means that $0.01635 \times 2 = 0.0327$ mol of ammonia is formed.

Convert this amount to a volume and you have the answer:

$$\text{volume} = 24\,000 \times 0.0327 = 785\,\text{cm}^3$$

LEARNING TIP

Practise using the three-step method for calculating masses from volumes, and volumes from masses, in some reactions.

CHECKPOINT

SKILLS PROBLEM SOLVING

In these questions, assume that all volumes are measured at r.t.p.

1. A flask contains 2 dm^3 of butane. What is the amount, in moles, of gas in the flask?

2. 10.0 g of copper(II) oxide is heated with hydrogen according to this equation:

$$CuO(s) + H_2(g) \rightarrow Cu(s) + H_2O(l)$$

What volume of hydrogen gas is needed to react with the copper(II) oxide, and what mass of copper is formed?

SUBJECT VOCABULARY

molar volume the volume occupied by 1 mol of any gas; this is normally 24 dm^3 or 24 000 cm^3 at r.t.p.

1E 2 CONCENTRATIONS OF SOLUTIONS

SPECIFICATION REFERENCE 1.5

LEARNING OBJECTIVES

- Calculate the concentration of a solution, in $g\,dm^{-3}$ and $mol\,dm^{-3}$.

CALCULATIONS USING MASS CONCENTRATION ($g\,dm^{-3}$)

If you know the mass of a **solute** that you dissolve in a **solvent** (usually water), and the volume of the **solution** formed, then it is straightforward to calculate the **mass concentration**.

You use the expression:

$$\text{mass concentration in g dm}^{-3} = \frac{\text{mass of solute in g}}{\text{volume of solution in dm}^3}$$

In this topic, we only use values based on g and dm^3. You may sometimes see other units, such as $g\,cm^{-3}$ and $kg\,m^{-3}$.

As with other similar expressions, you will need to rearrange it, depending on the wording of the question. You also need to remember to convert cm^3 to dm^3 (by dividing by 1000).

EXAMPLE 1

200 cm^3 of a solution contains 5.68 g of sodium bromide. What is its mass concentration?

$$\text{mass concentration} = \frac{m}{V} = \frac{5.68}{0.200} = 28.4\text{ g dm}^{-3}$$

EXAMPLE 2

The concentration of a solution is 15.7 $g\,dm^{-3}$. What mass of solute is there in 750 cm^3 of solution?

$$m = \text{mass concentration} \times V = 15.7 \times 0.750 = 11.8\text{ g}$$

EXAMPLE 3

A chemist uses 280 g of a solute to make a solution of concentration 28.4 $g\,dm^{-3}$. What volume of solution does he make?

$$V = \frac{m}{\text{mass concentration}} = \frac{280}{28.4} = 9.86\text{ dm}^3$$

▲ **fig A** These containers indicate increasing relative solute concentrations by increasing colour intensity. Unfortunately, most solutions we use are not coloured, so we cannot rely on different colour intensities to indicate different concentrations.

CALCULATIONS USING MOLAR CONCENTRATION ($mol\,dm^{-3}$)

Molar concentration (it used to be called molarity) is used more often than mass concentration. If only the term 'concentration' is mentioned, then you should assume that it refers to molar concentration.

The units of molar concentration are $mol\,dm^{-3}$, and this is often denoted by using square brackets. If a solution of hydrochloric acid has a concentration of 0.150 $mol\,dm^{-3}$, this can be shown as [HCl] = 0.150 $mol\,dm^{-3}$. The symbol c is sometimes used to represent molar concentration.

You need to be able to use these two expressions together:

$$\text{amount} = \frac{\text{mass}}{\text{molar mass}} \text{ or } n = \frac{m}{M}$$

and

$$\text{(molar) concentration} = \frac{\text{amount}}{\text{volume}} \text{ or } c = \frac{n}{V}$$

As previously, you may need to rearrange these expressions. Look at the question wording to decide which one to use first.

WORKED EXAMPLE 1

A chemist makes 500 cm^3 of a solution of nitric acid of concentration 0.800 $mol\,dm^{-3}$. What mass of HNO_3 does she need?

Step 1: You are given values of V and c, so you can use the second expression to calculate a value for n.

$$n = c \times V = 0.800 \times 0.500 = 0.400\text{ mol}$$

Step 2: You can now use the first expression to calculate the mass of nitric acid.

$$m = n \times M = 0.400 \times 63.0 = 25.2\text{ g}$$

WORKED EXAMPLE 2

A student has 50.0 g of sodium chloride. What volume of a 0.450 $mol\,dm^{-3}$ solution can he make?

Step 1: You are given the value of m and can work out M from the Periodic Table, so you can calculate n.

$$n = \frac{m}{M} = \frac{50.0}{58.5} = 0.855\text{ mol}$$

Step 2: You can now use the second expression to calculate the volume of solution.

$$V = \frac{n}{c} = \frac{0.855}{0.450} = 1.90\text{ dm}^3$$

CALCULATIONS FROM EQUATIONS USING CONCENTRATION AND MASS

In this type of calculation, you can use an equation to calculate the mass of a reactant or product if you are given the volume and molar concentration of another substance, and vice versa.

The expressions you need are the same as those you have just used, but you also need the equation for the reaction so that you can see the reacting ratio.

WORKED EXAMPLE 3

An excess of magnesium is added to 100 cm^3 of 1.50 mol dm^{-3} hydrochloric acid. The equation for the reaction is:

$Mg + 2HCl \rightarrow MgCl_2 + H_2$

What mass of hydrogen is formed?

Step 1: You are given the values of *V* and *c*, so you can use the second expression to calculate the value of *n* for hydrochloric acid.

$n = 0.100 \times 1.50 = 0.150$ mol

Step 2: The ratio for HCl : H_2 is 2 : 1, so $n(H_2) = 0.150 \div 2 = 0.0750$ mol

Step 3: For hydrogen, $m = n \times M = 0.0750 \times 2.0 = 0.15$ g

WORKED EXAMPLE 4

A mass of 47.8 g of magnesium carbonate reacts with 2.50 mol dm^{-3} hydrochloric acid. The equation for the reaction is:

$MgCO_3 + 2HCl \rightarrow MgCl_2 + H_2O + CO_2$

What volume of acid is needed?

Step 1: You are given the value of *m* and can work out *M* from the Periodic Table, so you can calculate *n*.

$n = \frac{47.8}{84.3} = 0.567$ mol

Step 2: The ratio for $MgCO_3$: HCl is 1 : 2, so $n(HCl) = 2 \times 0.567 = 1.134$ mol

Step 3: for HCl, $V = \frac{1.134}{2.50} = 0.454$ dm^3

LEARNING TIP

Remember that the volumes referred to in this topic are of solutions, not solvents. If you dissolve a solute in 100 cm^3 of a solvent, the volume of the solution is not exactly 100 cm^3.

CHECKPOINT

SKILLS PROBLEM SOLVING

1. 50.0 g of sodium hydroxide is dissolved in water to make 1.50 dm^3 of solution. What is the molar concentration of the solution?

2. 150 cm^3 of 0.125 mol dm^{-3} lead(II) nitrate solution is mixed with an excess of potassium iodide solution. The equation for the reaction that occurs is:

$Pb(NO_3)_2(aq) + 2KI(aq) \rightarrow PbI_2(s) + 2KNO_3(aq)$

What mass of lead(II) iodide is formed?

SUBJECT VOCABULARY

solute a substance that is dissolved
solvent a substance that dissolves a solute
solution a solute dissolved in a solution
mass concentration (of a solution) the mass (in g) of the solute divided by the volume of the solution
molar concentration (of a solution) the amount (in mol) of the solute divided by the volume of the solution

1E 3 CONCENTRATIONS IN PPM

SPECIFICATION REFERENCE 1.4(iv)

LEARNING OBJECTIVES

- Understand the term parts per million (ppm), e.g. gases in the atmosphere.

CONCENTRATIONS OF SOLUTIONS

Concentrations of solutions can also be compared using the term **parts per million**, or **ppm**. It is often used for pollutants in water.

Think of the word 'percent' – the 'cent' part refers to 100, so 50% means 50 out of 100, and parts per million means out of a million. 50 ppm means 50 parts out of 1 000 000 parts. The 'parts' usually refer to mass. The parts can be measured in any unit of mass, but grams or milligrams are the most commonly used.

CALCULATIONS FOR SOLUTIONS IN PPM

A concentration of 1 ppm means 1 g in 1 000 000 g, or 1 mg in 1 000 000 mg.

This expression can be used to calculate concentrations in ppm:

$$\text{concentration in ppm} = \frac{\text{mass of solute} \times 1\,000\,000}{\text{mass of solvent}}$$

The masses can be in any units, but they must be the same units. If different units are given, then one of them must be converted.

WORKED EXAMPLE 1

A solution contains 0.176 g of solute dissolved in 750 g of solvent. What is the concentration in ppm?

As the units of solute and solvent are the same, the values can be directly inserted into the expression.

$$\text{concentration in ppm} = \frac{\text{mass of solute} \times 1\,000\,000}{\text{mass of solvent}}$$

$$= \frac{0.176 \times 1\,000\,000}{750} = 235\text{ ppm}$$

WORKED EXAMPLE 2

A mass of 23 mg of sodium chloride is dissolved in 900 g of water. What is the concentration of sodium chloride in the solution in ppm?

As the units of solute and solvent are different, either the mass of solute or the mass of solvent must be converted so they are the same. Then the values can be directly inserted into the expression.

First, convert the mass of sodium chloride from mg to g (divide by 1000):

$$\text{mass of sodium chloride} = 23 \div 1000 = 0.023\text{ g}$$

$$\text{concentration in ppm} = \frac{\text{mass of solute} \times 1\,000\,000}{\text{mass of solvent}}$$

$$= \frac{0.023 \times 1\,000\,000}{900} = 26\text{ ppm}$$

WORKED EXAMPLE 3

A sample of river water contains phosphate ions with a concentration of 17 ppm. What is the mass of phosphate ions in 500 g of the river water?

In this example, the expression needs to be rearranged:

$$\text{mass of solute} = \frac{\text{concentration in ppm} \times \text{mass of solvent}}{1\,000\,000}$$

$$= \frac{17 \times 500}{1\,000\,000} = 0.0085\text{ g}$$

GASES IN THE ATMOSPHERE

Mauna Loa is the name of a volcano in Hawaii. It is also the location of a weather observatory that has recorded the levels of carbon dioxide in the atmosphere over a long period of time.

Fig A shows how the levels of carbon dioxide at Mauna Loa have changed over a period of 50 years. There are small variations during each year. In 1960, the level was about 316 ppm. By 2010, the level had reached nearly 390 ppm. Since 2015 it has been over 400 ppm, and many scientists believe that it will always remain above this value.

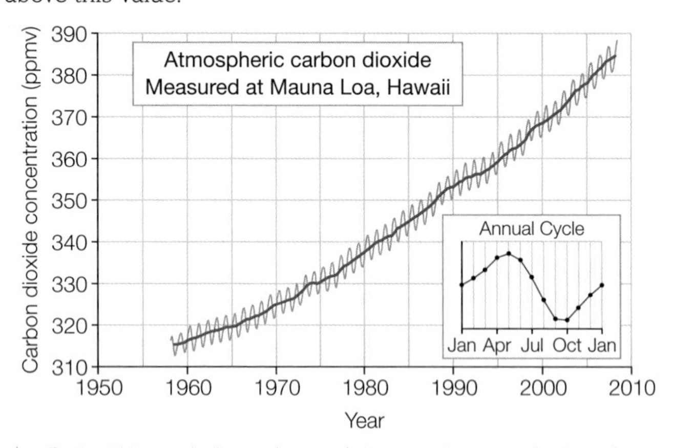

▲ **fig A** This graph shows the steady increase in atmospheric carbon dioxide over five decades.

The concentrations of pollutant gases in the atmosphere are often given as values in ppm. Instead of using masses, the comparison is usually by volume, so calculations are done in a different way. Sometimes the values are quoted in ppmv – the v shows that the value refers to concentration by volume.

A concentration of 1 ppmv means 1 cm^3 in 1 000 000 cm^3, or 1 cm^3 in 1 000 dm^3.

Residents of Dhaka, the capital city of Bangladesh, are concerned about air quality. As in many cities, road traffic leads to high levels of carbon monoxide, especially during rush hours, when many people are travelling to and from work. Levels of CO in the air can be as high as 100 ppm. Outside the city there are many factories producing bricks for constructing buildings. These factories are responsible for many pollutants, including sulfur dioxide and particulate matter.

▲ **fig B** This brick factory releases pollutant gases into the atmosphere.

CALCULATIONS FOR GASES IN PPM

This expression can be used to calculate concentrations in ppm.

$$\text{concentration in ppm} = \frac{\text{volume of gas} \times 1\,000\,000}{\text{volume of air}}$$

The volumes can be in any units, but they must be the same units. If different units are given, then one of them must be converted.

WORKED EXAMPLE 4

Some nitrogen dioxide gas, with a volume of 1.5 dm^3, mixes with 10 000 dm^3 of air. What is the concentration of nitrogen dioxide, in ppm, in the air?

As the volume units of both gases are the same, then the values can be directly inserted into the expression.

$$\text{concentration in ppm} = \frac{\text{volume of gas} \times 1\,000\,000}{\text{volume of air}}$$

$$= \frac{1.5 \times 1\,000\,000}{10\,000} = 150 \text{ ppm}$$

WORKED EXAMPLE 5

5000 dm^3 of air is found to contain ozone with a concentration of 87 ppm. What volume of ozone is in this sample of air?

In this example, the expression needs to be rearranged:

$$\text{volume of gas} = \frac{\text{concentration in ppm} \times \text{volume of air}}{1\,000\,000}$$

$$= \frac{87 \times 5000}{1\,000\,000} = 0.435 \text{ dm}^3$$

WORKED EXAMPLE 6

Two samples of air containing sulfur dioxide were analysed. The results for Sample 1 showed that 500 dm^3 of air contained 37 cm^3 of sulfur dioxide. The results for Sample 2 showed that there were 1.4 dm^3 of sulfur dioxide in 4000 dm^3 of air. Show, by calculation, which sample has the higher concentration, in ppm, of sulfur dioxide.

Sample 1

$$\text{concentration in ppm} = \frac{\text{volume of gas} \times 1\,000\,000}{\text{volume of air}}$$

$$= \frac{37 \div 1000 \times 1\,000\,000}{500} = 74 \text{ ppm}$$

Sample 2

$$\text{concentration in ppm} = \frac{\text{volume of gas} \times 1\,000\,000}{\text{volume of air}}$$

$$= \frac{1.4 \times 1\,000\,000}{4000} = 350 \text{ ppm}$$

Sample 2 has the higher concentration.

LEARNING TIP

For ppm calculations with solutions, use the expression involving masses. For calculations with gases, use the expression involving volumes.

CHECKPOINT

1. 0.2 g of potassium sulfate is dissolved in water to make 800 g of solution. What is the concentration of the salt in ppm?
2. 200 dm^3 of air contains 58 cm^3 of chlorine. What is the concentration of chlorine in ppm?

SUBJECT VOCABULARY

parts per million (ppm) the number of parts of one substance in one million parts of another substance; a measure used to describe chemical concentration; usually, 'parts' refers to masses of both substances, or to volumes of both substances

1 THINKING BIGGER

COINS IN HISTORY

SKILLS PROBLEM SOLVING

The chemical analysis of coinage from earlier periods in history can offer insights into the technology of metal extraction available at the time and give key insights into contemporary geopolitical questions. Note that these Thinking Bigger sections are synoptic and use knowledge from the different parts of the course. In this case, you will need to understand redox reactions (Topic 8) before you attempt the 'Chemistry in detail' section.

'ALL THAT GLISTENS IS NOT GOLD

fig A Trajan Decius, 249–251 BCE. Denomination: Silver Antoninianus. Mint: Rome.

Over the centuries, the base metals Fe, Cu, Ni, Zn, Al, Sn and Pb have been used as minor alloy constituents or as principal components in coins. Generally, a metal must be reasonably hard-wearing to ensure the economic lifetime of a coin, and must retain an acceptable appearance on exposure to the atmosphere. And it must not be too expensive. As a soft and expensive metal, gold is now usually alloyed with copper to give a hard-wearing alloy. British gold sovereigns, which are still minted, are made from 2 carats of alloy and 22 carats of gold. They comprise 91% Au and, 8.3% Cu. (A carat is a measure of the purity of gold, with pure gold being 24 carat.) Bronze (Cu–Sn alloy) farthings issued between 1897 and 1917 were darkened using $Na_2S_2O_3$, resulting in a surface layer of copper sulfide. This was done to avoid confusion between a newly minted farthing and a gold half-sovereign. The half-sovereign being worth 480 times as much as the farthing. Bronze pennies issued between 1944 and 1946 were similarly treated to deter the hoarding of new pennies at the end of the Second World War.

Platinum made only a brief appearance as a coinage metal. A few high-value coins in Russia were made in the mid-nineteenth century, at a time when platinum had recently been found in the Ural Mountains. For a while, the value of platinum fell below that of gold, and this gave rise to the emergence of counterfeit sovereigns in the 1870s. These were made from platinum and had a thin layer of electrochemically deposited gold.

Silver coinage was the universal medium of trading for centuries. The silver content of the coins made in city mints was a reflection of the status of the city as a trading centre. Traders of good repute only dealt in high quality silver coins, as evidenced by the high silver content of those coins found along the old established trading routes, e.g. from the Mediterranean, through the Middle East and Central Asia, to China.

If military expenditure increased, or during times of declining prosperity, a government would put only enough silver in its coins for them to remain acceptable to the public. Sometimes, however, there was a more subtle reason for the decline in silver content. Up to circa 100 BCE, Roman silver coins contained more than 90% Ag. Thereafter, silver supplies grew scarcer, and Roman technology was unable to extract silver from low-grade ores, which by the third century BCE were the only primary sources of silver available. However, at around this time, deposits of AgCl (chlorargynite) were found in Cornwall, Brittany and Alsace. The Romans observed that if copper or bronze coins were dipped in molten AgCl (mp 455 °C), they became coated with silver:

$$Cu + 2AgCl \rightarrow 2Ag + CuCl_2$$

From an article in *Education in Chemistry* magazine, published by the Royal Society of Chemistry

DID YOU KNOW?

In April 1940 during the Nazi occupation of Denmark, the Hungarian-born chemist George de Hevesy decided to hide the Nobel Prize medals of colleagues Max von Laue and James Franck by dissolving them in *aqua regia* (a mixture of concentrated nitric and hydrochloric acid). The resulting solution sat on a shelf for the next 5 years attracting no particular attention. At the end of the war, the gold was recovered from the solution and the two medals recast.

SCIENCE COMMUNICATION

1 The article is about the development of a coin-based currency from the earliest times. Answer the following three questions and explain your answers with reasons or examples. In what ways is an article of this type different from a research paper? Which audience is this article aimed at? Is there any bias present in the report?

CHEMISTRY IN DETAIL

2. (a) Modern gold sovereigns weigh 7.99 g (to two decimal places). Using the information in the article, calculate the number of moles of gold in a sovereign.

(b) Calculate the percentage of gold in an 18 carat ring.

3. (a) According to the reaction between copper and silver chloride in the article how many moles of silver are produced from 1 mole of copper?

(b) Calculate the number of moles of copper in 2.17 g (Ar Cu = 63.5)

(c) When 2.17 g of copper reacts with excess silver chloride according to the equation a total of 5.40 g of silver is recovered. Assuming that all of the copper reacts calculate the percentage yield of the silver recovered. (Ar Ag = 107.9)

4. Reread the third paragraph of the article. Using the information given, write half-equations for

(i) the reduction of thiosulfate ions to sulfide in acidic solution, and

(ii) the oxidation of copper metal to copper(II) ions.

5. A gold sovereign is analysed using the following chemical procedure.

1 The sovereign is reacted with excess concentrated nitric acid in a fume cupboard.

2 The resultant mixture is filtered to remove the unreacted gold.

3 The resultant solution is made up in distilled water to a volume of 250 cm^3.

4 A volume of 25 cm^3 of this solution is pipetted into a conical flask and 50 cm^3 of 0.1 mol dm^{-3} KI(aq) is added (this is an excess). The resultant solution immediately turns an orange-brown colour.

5 This solution is then titrated with 0.050 mol dm^{-3} $Na_2S_2O_3(aq)$ solution until the solution fades to a straw yellow colour.

6 Starch indicator is then added to give a blue-black colour.

7 Further $Na_2S_2O_3(aq)$ solution is added until the solution becomes colourless.

8 The first titration volume is recorded.

9 The titration is repeated 3 times and the results recorded to the nearest 0.05 cm^3.

Titre number	1	2	3	4
Final volume / cm^3	26.00	26.35	25.90	25.85
Initial volume / cm^3	0.00	0.55	0.00	0.00
Titre / cm^3	26.00			

Results table

(a) Suggest why step 1 must be done in a fume cupboard.

(b) Complete the results table and calculate a suitable mean titre.

(c) Calculate the percentage uncertainty in the reading for titre number 3.

(d) Match the equations below with the reactions taking place in the method described.

$Cu(s) + 4H^+(aq) + 2NO_3^-(aq) \rightarrow Cu^{2+}(aq) + 2NO_2(g) + 2H_2O(l)$

$2Cu^{2+}(aq) + 4I^-(aq) \rightarrow 2CuI(s) + I_2(aq)$

$2S_2O_3^{2-}(aq) + I_2(aq) \rightarrow 2I^-(aq) + S_4O_6^{2-}(aq)$

(e) Using appropriate calculations, decide whether the sovereign is genuine or not.

(f) What assumptions have been made in the procedure?

THINKING BIGGER TIP

In your previous study of chemistry, you will have covered metals and their methods of extraction. This would be a good chance for you to review some of that work. Consider how metal reactivity relates to methods of extraction.

INTERPRETATION NOTE

See if you can find a similar article on the development of pigments and dyes over the ages or the development of textiles. Consider how the the chemical ideas are represented.

ACTIVITY

The history of human technology is closely associated with metals.

Draw a timeline identifying the metals used and the cultures responsible for developing the technology.

- You should consider the following metals: gold, silver, copper, iron, aluminium, titanium and uranium.
- Consider also the methods of extraction and the key usages of the metals.

1 EXAM PRACTICE

1 How many molecules of oxygen are there in a 1.00 g sample of O_2?

(Avogadro constant, $L = 6.02 \times 10^{23}$ mol^{-1})

A 1.88×10^{22}

B 3.76×10^{22}

C 9.63×10^{24}

D 1.93×10^{25} [1]

(Total for Question 1 = 1 mark)

2 What is the concentration, in mol dm^{-3}, of a solution of sodium chloride containing 4.27 g of NaCl in 300 cm^3 of solution?

A 0.0219

B 0.243

C 4.11

D 45.7 [1]

(Total for Question 2 = 1 mark)

3 The overall equation for one method of manufacturing hydrogen from methane is

$$CH_4 + 2H_2O \rightarrow 4H_2 + CO_2$$

In one batch, 5.12 tonnes of hydrogen were obtained from 13.6 tonnes of methane.

What is the percentage yield of hydrogen for this batch?

A 37.6

B 42.5

C 54.4

D 75.3 [1]

(Total for Question 3 = 1 mark)

4 The equation for a reaction that can be used to manufacture lithium carbonate is

$$2Li_2O_2 + 2CO_2 \rightarrow 2Li_2CO_3 + O_2$$

What is the atom economy for this reaction?

A 58.9

B 62.1

C 69.8

D 82.2 [1]

(Total for Question 4 = 1 mark)

5 The equation for a displacement reaction is

$$Zn + FeSO_4 \rightarrow Fe + ZnSO_4$$

Which species is displaced in this reaction?

A zinc

B iron

C sulfur

D oxygen [1]

(Total for Question 5 = 1 mark)

6 The equation for a precipitation reaction is

$$AgNO_3(aq) + NaCl(aq) \rightarrow AgCl(s) + NaNO_3(aq)$$

A solution containing 0.04 mol silver nitrate is added to a solution containing 0.06 mol sodium chloride.

What mass of precipitate forms?

A 5.74 g

B 7.17 g

C 8.60 g

D 14.3 g [1]

(Total for Question 6 = 1 mark)

7 Magnesium chloride may be prepared by reacting magnesium with dilute hydrochloric acid.

A sample of magnesium of mass 1.215 g was added to 60.0 cm^3 dilute hydrochloric acid of concentration 2.00 mol dm^{-3}.

The equation for the reaction is

$$Mg(s) + 2HCl(aq) \rightarrow MgCl_2(aq) + H_2(g)$$

(a) State two observations that are made when magnesium reacts with dilute hydrochloric acid. [2]

(b) (i) Calculate the amount, in moles, of magnesium used. [1]

(ii) Calculate the amount, in moles, of hydrochloric acid used. [1]

(iii) Use your answers to (b)(i) and (b)(ii) to show that the hydrochloric acid is in excess. [2]

(c) Calculate the volume, measured at r.t.p. of hydrogen produced in this reaction.

(The molar volume of hydrogen at r.t.p. is 24 dm^3 mol^{-1}) [2]

(Total for Question 7 = 8 marks)

8 Ammonium nitrate (NH_4NO_3) can be prepared by the reaction of ammonia with nitric acid.

The equation for the reaction is

$$NH_3(aq) + HNO_3(aq) \rightarrow NH_4NO_3(aq)$$

(a) (i) Calculate the minimum volume of nitric acid, of concentration 0.500 mol dm^{-3}, that is required to react completely with 25.0 cm^3 of aqueous ammonia of concentration 2.00 mol dm^{-3}. [2]

(ii) How could a solid sample of ammonium nitrate be obtained from the reaction mixture? [1]

(b) When ammonium nitrate is heated, it decomposes according to the following equation:

$$NH_4NO_3(s) \rightarrow N_2O(g) + 2H_2O(l)$$

A 4.00 g sample of ammonium nitrate was carefully heated to produce only N_2O and water.

(i) Calculate the amount, in moles, of ammonium nitrate used. [2]

(ii) Calculate the volume of N_2O that was formed. [1]

(Assume that the molar volume of N_2O is $24\,dm^3\,mol^{-1}$ under the experimental conditions.)

(iii) N_2O can be decomposed into its elements by further heating. Write an equation for this reaction. Include state symbols. [2]

(Total for Question 8 = 8 marks)

9 Phosphorus forms three chlorides of molecular formulae PCl_3, PCl_5 and P_2Cl_4.

(a) State what is meant by molecular formula. [1]

(b) PCl_5 can be prepared by reacting white phosphorus (P_4) with chlorine gas.

Write an equation for this reaction. State symbols are not required. [1]

(c) Solid PCl_5 reacts vigorously with water to form a solution containing a mixture of two acids, H_3PO_4 and HCl.

Write an equation for this reaction. Include state symbols. [2]

(d) A compound was found to have the following percentage composition by mass:

P = 30.39%; Cl = 69.61%

(i) Use this information to identify the chloride. [3]

(ii) Give the systematic name for this chloride. [1]

(Total for Question 9 = 8 marks)

10 The diagram shows the apparatus used to collect and measure the volume of hydrogen given off when a sample of a Group 2 metal reacts with dilute hydrochloric acid.

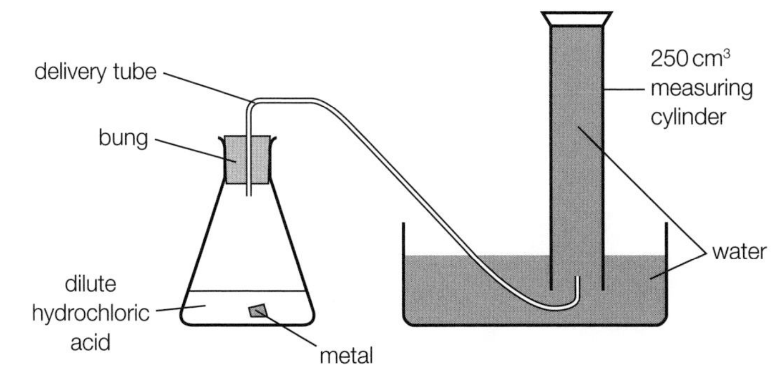

The equation for the reaction is

$$M(s) + 2HCl(aq) \rightarrow MCl_2(aq) + H_2(g)$$

where M is the symbol for the Group 2 metal.

The apparatus was set up as shown in the diagram.

A sample of the metal was weighed and then placed into the conical flask.

The bung was removed and an excess of acid was then added to the metal. The bung was replaced.

When the reaction was complete, and the gas collected had cooled to room temperature, the volume of gas collected in the measuring cylinder was measured.

The results are shown in the table.

Mass of metal	0.24 g
Volume of hydrogen collected	230 cm³

(a) (i) Use the results to calculate the molar mass of the metal.

Assume the molar volume of hydrogen is $24.0\,dm^3\,mol^{-1}$ under the experimental conditions. [3]

(ii) Give the most likely identity of the metal. Justify your answer. [2]

(b) (i) Identity the major procedural error in the experiment. [1]

(ii) State a modification that would reduce this procedural error. [1]

(Total for Question 10 = 7 marks)

11 Azides are compounds of metals with nitrogen. Some azides are used as detonators in explosives. However, sodium azide (NaN_3) is used in air bags in cars.

(a) Sodium azide decomposes into its elements when heated.

$$2NaN_3(s) \rightarrow 2Na(l) + 3N_2(g)$$

What is the volume of gas produced, measured at r.t.p., when one mole of sodium azide is decomposed?

A $24\,dm^3$
B $36\,dm^3$
C $48\,dm^3$
D $72\,dm^3$

(Assume one mole of gas occupies $24\,dm^3$ at r.t.p.) [1]

(b) (i) A student completely decomposed 3.25 g of sodium azide. Calculate the mass of sodium she obtained. [2]

(ii) She then carefully reacted the sodium obtained with water to form $25.0\,cm^3$ of aqueous sodium hydroxide.

$$2Na(s) + 2H_2O(l) \rightarrow 2NaOH(aq) + H_2(g)$$

Calculate the concentration, in $mol\,dm^{-3}$, of the aqueous sodium hydroxide. [2]

(Total for Question 11 = 5 marks)

TOPIC 2 ATOMIC STRUCTURE AND THE PERIODIC TABLE

A ATOMIC STRUCTURE | B THE PERIODIC TABLE

Chemistry is the study of matter and the changes that can be made to matter. Matter is anything that has mass and takes up space. This includes all solids, liquids and gases. The ancient Greeks thought that all matter was made from four elements: Air, Earth, Fire and Water. This idea was gradually abandoned as scientists began to experiment and apply logical scientific thinking to their observations.

In 1803, John Dalton presented his atomic theory. He developed this theory from the idea that the atoms of different elements could be distinguished by differences in their weights. Some people still consider that there is truth in some parts of his theory. However, scientists have gradually refined the theory into one that more precisely explains the observations we can make today. Science works by scientists continuing to experiment and make new observations. If scientists cannot explain those observations satisfactorily using the existing theory, then they have to change, or even possibly replace, the theory.

Atoms, and the particles they are made from, govern how everything works in the world around us. The continuing study and understanding of atoms has allowed us to progress with developing: new medicines to treat illnesses; the materials and sophisticated technology we use in communications and computing; and newer and better materials.
In fact, there is probably not a single area of modern life that is not affected by the study of atoms.

MATHS SKILLS FOR THIS TOPIC

- Recognise and use expressions in decimal and ordinary form
- Use ratios, fractions and percentages
- Use an appropriate number of significant figures
- Find arithmetic means

What prior knowledge do I need?

- The relative mass and relative charge of a proton, neutron and electron
- Where protons, neutrons and electrons exist within atoms
- What is meant by atomic number, mass number, isotopes and relative atomic mass (A_r)
- Calculations of relative atomic mass from relative abundances of isotopes
- The Periodic Table as an arrangement of elements in order of atomic number
- The electronic configurations of the first 20 elements in the Periodic Table
- The number of outer electrons in an atom of a main group element relates to its position in the Periodic Table

What will I study in this topic?

- How our understanding of the structure of the atom has developed over time
- How the masses of atoms and molecules can be determined
- The evidence for the existence of quantum shells, subshells and electron orbitals
- The electronic configurations of the first 36 elements in the Periodic Table
- Periodicity of properties in Periods 2 and 3 of the Periodic Table

What will I study later?

Topic 3

- The different types of bonding that can exist in elements and compounds
- The different types of structure that exist in elements and compounds
- How the type of structure and bonding affects the physical properties of elements and compounds

Topic 10

- Using mass spectrometry to determine masses and structures of organic molecules

2A 1 STRUCTURE OF THE ATOM AND ISOTOPES

SPECIFICATION REFERENCE 2.1 2.2 2.3 2.4 2.5

LEARNING OBJECTIVES

- Know the structure of an atom in terms of electrons, protons and neutrons.
- Know the relative mass and charge of protons, neutrons and electrons.
- Know the meaning of the terms atomic (proton) number and mass number.
- Be able to use the atomic number and the mass number to determine the number of each type of subatomic particle in an atom or ion.
- Understand the term isotope.

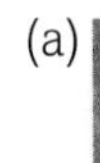

(a)

(b)

(c)

▲ **fig A**
(a) J.J. Thomson,
(b) Ernest Rutherford and
(c) James Chadwick.

WHO DISCOVERED ELECTRONS, PROTONS AND NEUTRONS?

Our current understanding of the structure of atoms is influenced by the theories put forward by scientists such as J.J. Thomson, Ernest Rutherford and James Chadwick.

J.J. Thomson discovered the electron in 1897. Ernest Rutherford discovered the proton in 1917. James Chadwick discovered the neutron in 1932.

STRUCTURE OF AN ATOM

Although scientists have discovered many other subatomic particles, chemistry is only concerned with electrons, protons and neutrons.

We can summarise the structure of the atom in terms of these three subatomic particles, as shown in **table A**.

PARTICLE	SYMBOL	RELATIVE MASS	RELATIVE CHARGE	POSITION IN THE ATOM
proton	p	1	+1	nucleus
neutron	n	1	0	nucleus
electron	e^-	$\frac{1}{1840}$	−1	energy levels surrounding the nucleus

table A The structure of the atom in terms of protons, neutrons and electrons.

Rutherford discovered the proton and he is also credited for first suggesting that the atom has a very small core containing the bulk of the mass of the atom. This core is called the nucleus and contains all of the protons and neutrons in that atom.

Electrons are also present in atoms. The electrons exist in energy levels surrounding the nuclei. The energy levels are called quantum shells. We will develop this concept throughout **Topic 2A**.

LEARNING TIP

Never state that the mass of an electron is zero. This is not true, an electron does have a mass.
You do not need to know the exact masses of subatomic particles in grams, or the exact charges in coulombs.
You need to know the relative values.

DID YOU KNOW?

All atoms of any element have the same atomic number, which is different from the atomic number of any other element.
The number of electrons in a neutral atom of an element is equal to the number of protons in the nucleus of that atom.
The electrons surround the nucleus in well-defined energy levels called quantum shells.

ATOMIC NUMBER, MASS NUMBER AND ISOTOPES

Fig B shows the **atomic number**, **mass number** and **isotopes** of carbon.

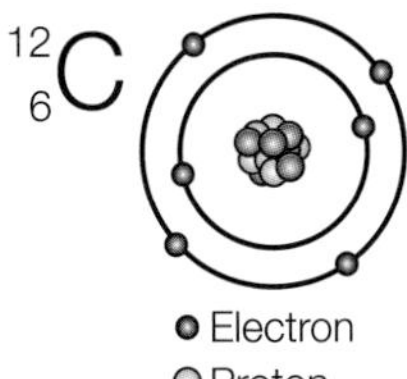

Carbon atoms

Mass number ⟶ 12, 13
Atomic number ⟶ 6

$${}^{12}_{6}C,\ {}^{13}_{6}C$$

Atomic number = number of protons
(= number of electrons)

Mass number = number of protons + number of neutrons

▲ **fig B** Atomic number, mass number and isotopes of carbon.

Two isotopes of carbon are ^{12}C and ^{13}C. They have the same atomic number, 6, and the same electronic configuration, 2.4, (see **Topic 2A.3**), so samples of carbon containing either isotope will have the same chemical properties. However, the mass of ^{13}C is larger than ^{12}C. This is because ^{13}C has one more neutron.

ISOTOPES

In 1919, Francis Aston invented an instrument called a mass spectrometer. He used it to discover that not all the atoms of an element have the same mass. It was eventually realised that the different masses were caused by the atoms having different numbers of neutrons.

These different atoms are called isotopes.

We can illustrate this point by considering two isotopes of chlorine: chlorine-35 and chlorine-37.

Table B shows the number of protons, neutrons and electrons in an atom of each isotope.

ISOTOPE	SYMBOL FOR ISOTOPE	NUMBER OF PROTONS	NUMBER OF NEUTRONS	NUMBER OF ELECTRONS
chlorine-35	${}^{35}_{17}Cl$	17	18	17
chlorine-37	${}^{37}_{17}Cl$	17	20	17

table B Isotopes of chlorine.

DID YOU KNOW?

Isotopes of the same element have identical chemical properties because they have identical electronic configurations.

CHECKPOINT

1. Use the information in the table below to answer the following questions about particles **A** to **F**.
   (a) Which two particles are isotopes of the same element?
   (b) Which two particles are positive ions?
   (c) Which two particles are negative ions?
   (d) Which two particles have the same mass number?

PARTICLE	NUMBER OF PROTONS	NUMBER OF NEUTRONS	NUMBER OF ELECTRONS
A	12	13	12
B	17	18	18
C	11	14	10
D	12	12	12
E	35	44	36
F	19	21	18

2. Complete the table below to show the numbers of protons, neutrons and electrons in these atoms, molecules and ions.

SYMBOL FOR ATOM OR ION	NUMBER OF PROTONS	NUMBER OF NEUTRONS	NUMBER OF ELECTRONS
${}^{3}_{1}H$			
${}^{18}_{8}O^{2-}$			
${}^{24}_{12}Mg^{2+}$			
${}^{14}_{7}N{}^{1}_{1}H_3$			
${}^{14}_{7}N{}^{2}_{1}H_4^{+}$			

SUBJECT VOCABULARY

atomic number (*Z*) the number of protons in the nucleus of an atom
mass number the sum of the number of protons and the number of neutrons in the nucleus of an atom
isotopes atoms of the same element that have the same atomic number but different mass number

2A 2 MASS SPECTROMETRY AND RELATIVE MASSES OF ATOMS, ISOTOPES AND MOLECULES

SPECIFICATION REFERENCE 2.6 2.7

LEARNING OBJECTIVES

- Understand the basic principles of a mass spectrometer and be able to analyse and interpret spectra to: deduce the isotopic composition of an element; calculate relative atomic mass from the relative abundance of isotopes and vice versa; determine the relative molecular mass of a molecule, and identify molecules in a sample.
- Understand that ions in a mass spectrometer may have a 2+ charge.
- Be able to predict the mass spectra, including relative peak heights, for diatomic molecules including chlorine, given the isotopic abundances.

RELATIVE ATOMIC MASS AND RELATIVE ISOTOPIC MASS

RELATIVE ATOMIC MASS

Chemists working in the 19th century established the chemical formulae of some common compounds. They also calculated the masses of atoms from which these compounds are often made. In fact, they hadn't measured the actual masses of the atoms but how the masses of atoms compared with one another, in other words **relative atomic masses**.

The sulfur atom was twice as heavy as the oxygen atom, and the oxygen atom was sixteen times heavier than the hydrogen atom. The hydrogen atom was the lightest atom, so its mass was given a value of 1. The masses of the atoms of the other elements were stated relative to this value. So the mass of an atom of oxygen on this scale was 16, and sulfur was 32.

The discovery of isotopes complicated the situation. In 1961, scientists decided to adopt an isotope of carbon, carbon-12, as the standard.

RELATIVE ISOTOPIC MASS

Relative isotopic masses and their percentage abundance in a sample of an element are used to calculate relative atomic masses.

WORKED EXAMPLE

Lithium, in its naturally occurring compounds, has two isotopes of relative isotopic masses 6.015 and 7.016. The percentage abundance of each isotope is 7.59 and 92.41 respectively.

This is how you can calculate the relative atomic mass of lithium.

There are two stages.

Stage 1: $(6.015 \times 7.59) + (7.016 \times 92.41) = 694.00241$

Stage 2: $\frac{694.00241}{100} = 6.9400241$

The relative atomic mass of lithium, to three significant figures, is 6.94.

EXAM HINT

The relative atomic mass will always be a number somewhere between the relative isotopic masses. If your calculator gives you a different number, you have made a mistake and should go back and check your workings.

LEARNING TIP

Remember that:

- the Z number is not the same as the relative isotopic mass; the mass number is always a whole number because it is the sum of two whole numbers (the number of protons and the number of neutrons in the nucleus of an atom)
- relative isotopic mass is relative to the mass of a carbon-12 atom; it is not likely to be a whole number
- relative isotopic masses should be used to calculate relative atomic masses, but sometimes mass numbers are used to make the arithmetic easier.

USING DATA OBTAINED FROM A MASS SPECTROMETER

WHAT IS A MASS SPECTROMETER?

A mass spectrometer measures the masses of atoms and molecules.

It produces positive ions that are deflected by a magnetic field according to their mass-to-charge ratio (m/z). It also calculates the relative abundance of each positive ion and displays this as a percentage.

The positive ions could be positively charged atoms, positively charged molecules or positively charged fragments of molecules.

DID YOU KNOW?

The relative atomic mass of an element is the weighted mean mass of an atom of the element compared to $\frac{1}{12}$ the mass of an atom of carbon-12, which has a mass of 12.

The relative isotopic mass is the mass of an atom of an isotope of the element compared to $\frac{1}{12}$ the mass of an atom of carbon-12, which has a mass of 12.

Dealing with fragments of molecules is particularly useful when you are analysing organic compounds. We will discuss this in more detail in **Topic 10**.

HOW DOES A MASS SPECTROMETER WORK?

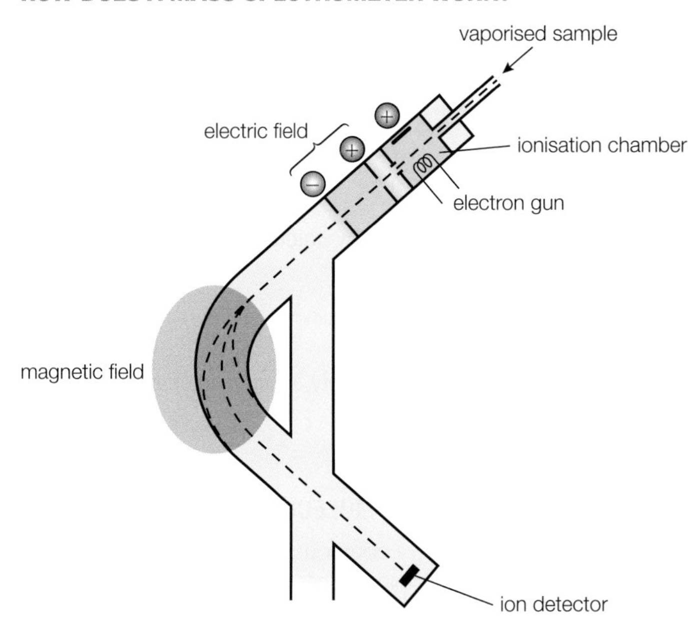

fig A A mass spectrometer.

1. The sample being analysed must be in the gaseous state so that its particles can move through the machine. The sample is injected into the mass spectrometer, where it is vaporised.
2. The vapour is bombarded with high energy electrons. These electrons collide with the atoms or molecules of the sample. One or more electrons are removed from the atoms or molecules to form positive ions.
3. An electric field causes the positive ions to accelerate.
4. The positive ions then enter, and are deflected by, a uniform magnetic field. The amount of deflection depends on the mass-to-charge (m/z) ratio of the ions. Ions with a large mass-to-charge ratio are deflected less than ions with a small mass-to-charge ratio. If all the ions have the same charge, usually 1+, the extent of the deflection is inversely proportional to their mass.
5. The deflected ions pass through a narrow slit and are collected on a metallic plate connected to an amplifier. For a given strength of magnetic field, only ions with a certain m/z ratio pass through the slit and are detected. For example, an ion with a mass of 28 and a charge of 1+ will be detected at the same time as an ion with a mass of 56 and a charge of 2+. Both ions have a m/z ratio of 28.

   The strength of the magnetic field is then changed to detect positive ions with other m/z ratios.

DETERMINING RELATIVE ISOTOPIC AND ATOMIC MASSES

You can determine the *exact* values of the relative masses of isotopes from a mass spectrum of an element, together with the percentage abundance of each isotope.

You can use this information to calculate the relative atomic mass of the element.

DETERMINING RELATIVE MOLECULAR MASS (M_r) OF DIATOMIC MOLECULES

Some elements and compounds contain two or more atoms covalently bonded together. If these substances are analysed by mass spectrometry, you can obtain the relative molecular mass of the element or compound by observing the peaks with the largest m/z ratios (assuming a value of $z = 1$).

Fig B shows the mass spectrum of chlorine, which exists as diatomic molecules: Cl_2.

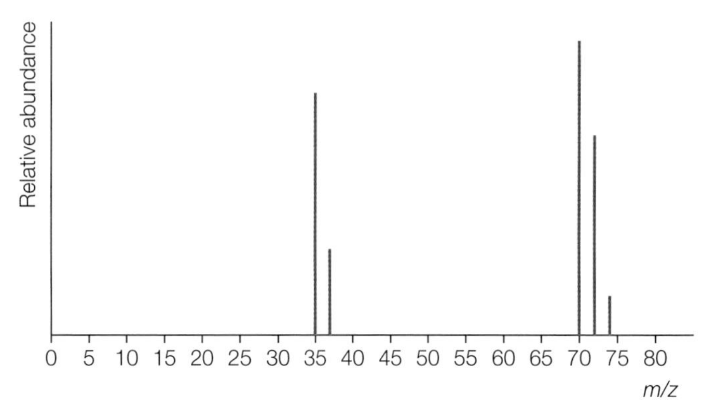

fig B Mass spectrum of chlorine.

To make calculations easier, in this example the relative molecular and isotopic masses are quoted as whole numbers.

There are two peaks corresponding to isotopic masses of 35 and 37. The approximate relative peak heights for each isotope are 3 : 1 for chlorine-35 : chlorine-37.

This gives an approximate relative atomic mass of this sample of chlorine of $((3 \times 35) + (1 \times 37)) \div 4 = 35.5$.

You will notice in the diagram that there are three peaks, which correspond to molecular masses of 70, 72 and 74. How can we explain the relative heights of these peaks?

Table A shows the atomic composition of each molecule of chlorine.

MASS OF MOLECULE	FORMULA OF MOLECULE
70	$^{35}Cl^{35}Cl$
72	$^{35}Cl^{37}Cl$
74	$^{37}Cl^{37}Cl$

table A Atomic composition of each molecule of chlorine.

The ratio of ^{35}Cl to ^{37}Cl in this sample is 3 : 1. The molecular composition is shown in **table B**.

FORMULAE OF MOLECULE	RATIO OF MOLECULE
$^{35}Cl^{35}Cl$	9
$^{35}Cl^{37}Cl$	6
$^{37}Cl^{37}Cl$	1

table B Molecular composition of sample.

LEARNING TIP

The peak heights will not be exactly 9 : 6 : 1 because the relative isotopic masses are not whole numbers.

LEARNING TIP

Make sure you always use the relative atomic masses you are given in any question. For example, the value for magnesium given in the Periodic Table on page 306 at the end of this book is 24.3. However, you may occasionally be asked to use a more accurate value, such as 24.305.

Let's explain this.

- There is a 3 in 4 chance of selecting a ^{35}Cl atom from a sample of chlorine atoms. This means that the total chance of two ^{35}Cl atoms combining together is $\frac{3}{4} \times \frac{3}{4} = \frac{9}{16}$.
- There is a 1 in 4 chance of selecting a ^{37}Cl atom. This means that the chance of ^{37}Cl combining with a ^{35}Cl is $\frac{1}{4} \times \frac{3}{4} = \frac{3}{16}$. The chance of a ^{35}Cl atom combining with a ^{37}Cl atom is also $\frac{3}{16}$. This means that the total chance of ^{35}Cl and ^{37}Cl combining together in any order is $2 \times \frac{3}{16}$ or $\frac{6}{16}$.
- The chance of two ^{37}Cl atoms combining together is $\frac{1}{4} \times \frac{1}{4} = \frac{1}{16}$

  $\frac{9}{16} : \frac{6}{16} : \frac{1}{16}$ in whole number ratios is 9 : 6 : 1.

This corresponds approximately to the peak heights of 70, 72 and 74.

You can then determine the relative molecular mass (M_r) of chlorine by calculating the weighted mean of the various molecules present:

$$\frac{M_r(Cl_2) = (9 \times 70) + (6 \times 72) + (1 \times 74)}{16} = 71$$

You will usually be asked to calculate relative molecular masses by simply adding together the relative atomic masses of the elements in the molecule.

If we use this method, the relative molecular mass of chlorine is $2 \times 35.5 = 71$.

DETERMINING RELATIVE MOLECULAR MASS OF A POLYATOMIC MOLECULE

It is quite difficult to use the data supplied by a mass spectrum of a compound to determine its exact relative molecular mass. You need to know the exact relative isotopic masses of all the atoms present, and the relative composition of all the different molecules.

You are more likely to be asked to work out the relative molecular mass of a compound by considering what is called the **molecular ion peak**.

You should be careful when analysing organic compounds. This is because there is always a small percentage of the carbon-13 isotope present in the compound, which can lead to what is referred to as an M+1 peak. This peak can often be seen in molecules with large masses, where the percentage of carbon-13 becomes significant. The peak is often missing, or insignificant, in molecules of small mass.

The following examples illustrate this point.

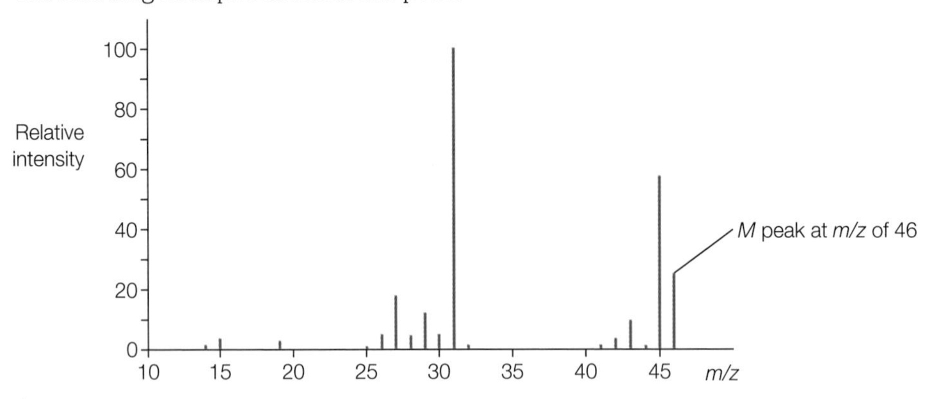

fig C Mass spectrum of ethanol (M_r = 46).

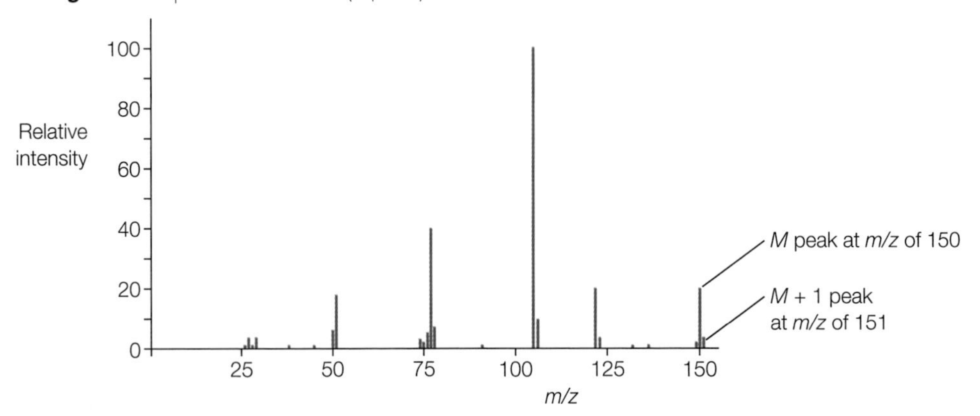

fig D Mass spectrum of ethyl benzoate (M_r = 150).

CHECKPOINT

1. Describe the difference between relative atomic mass and relative isotopic mass.
2. Explain why the relative atomic masses of many elements are not exact whole numbers.
3. Calculate the relative atomic mass of a sample of magnesium that has the following isotopic composition:
   - magnesium-24: 78.6%
   - magnesium-25: 10.1%
   - magnesium-26: 11.3%

   Give your answer to three significant figures.
4. A sample of copper contains two isotopes of relative isotopic mass 63.0 and 65.0. If the relative atomic mass of copper is 63.5, calculate the relative abundance of each isotope.
5. The mass spectrum of bromine vapour, Br_2, is shown in **fig E** below.

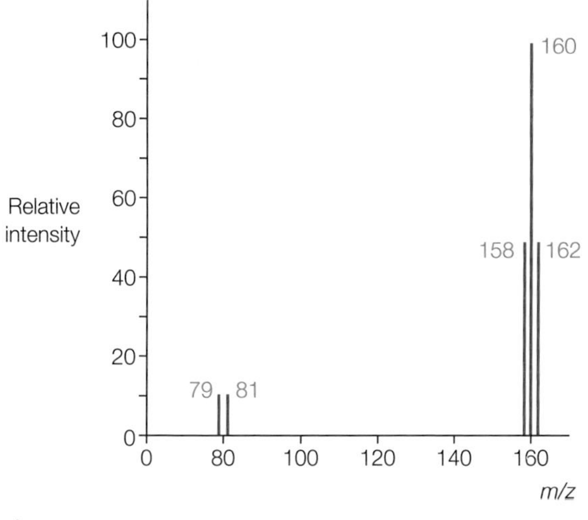

▲ **fig E**

(a) What are the relative isotopic masses of the two isotopes present in bromine?

(b) Identify the particles responsible for the peaks at m/z 158, 160 and 162.

(c) Deduce the relative abundance of the two isotopes and explain the relative heights of the three peaks at m/z 158, 160 and 162.

EXAM HINT

In exam questions, always give your answer to the number of significant figures asked for, otherwise you will probably lose a mark.

SKILLS REASONING

SUBJECT VOCABULARY

relative atomic mass (A_r) (of an element) the weighted mean (average) mass of an atom of the element compared to $\frac{1}{12}$ of the mass of an atom of carbon-12

relative isotopic mass the mass of an individual atom of a particular isotope relative to $\frac{1}{12}$ of the mass of an atom of carbon-12

molecular ion peak the peak with the highest m/z ratio in the mass spectrum, the M peak

2A 3 ATOMIC ORBITALS AND ELECTRONIC CONFIGURATIONS

SPECIFICATION REFERENCE 2.9 2.12 2.13 2.14 2.15

LEARNING OBJECTIVES

- Know that an orbital is a region within an atom that can hold up to two electrons with opposite spins.
- Be able to describe the shapes of s and p orbitals.
- Know that orbitals in sub-shells: each take a single electron before pairing up; pair up with two electrons of opposite spin.
- Be able to predict the electronic configuration of the atoms of the elements from hydrogen to krypton inclusive and their ions, using s, p, d notation and electron-in-boxes notation.
- Understand that electronic configuration determines the chemical properties of an element.
- Know the number of electrons that can fill the first four quantum shells.
- State the number of electrons that occupy s, p and d sub-shells.

QUANTUM SHELLS

Max Planck first presented the quantum theory in 1900. We can use this theory to describe the arrangement of electrons around the nuclei of atoms.

According to this theory, electrons can only exist in certain well-defined energy levels called **quantum shells**.

All electrons in a quantum shell have similar, but not identical, energies.

ELECTRONS IN THE FIRST FOUR QUANTUM SHELLS

Electrons in the first quantum shell have the lowest energy for that element. The first quantum shell is found in the region closest to the nucleus. The second quantum shell is in a region outside the first; the third is outside the second, and so on.

Each quantum shell, apart from the first, is further divided into sub-shells of slightly different energy levels.

- There is only one sub-shell in the first quantum shell. This is labelled the 1s sub-shell.
- The second quantum shell has two sub-shells, labelled 2s and 2p. Electrons in the 2p sub-shell have a slightly higher energy level than those in the 2s sub-shell.
- The third quantum shell is divided into three sub-shells, labelled 3s, 3p and 3d. Electrons in the 3p sub-shell have slightly higher energy than those in 3s, and those in the 3d sub-shell have slightly higher energy than those in 3p.
- The fourth quantum shell is divided into four sub-shells, labelled 4s, 4p, 4d and 4f. Again, the electron energies increase in the order 4s < 4p < 4d < 4f.

s ORBITALS

Each sub-shell contains **orbitals**. If you could see a single electron in a hydrogen atom and map its position at regular intervals, then you could construct a three-dimensional map of places where the electron is likely to be found. Most of the time, the electron is located within a fairly easily-defined region of space close to the nucleus.

Fig A shows a cross-section through a spherical s orbital.

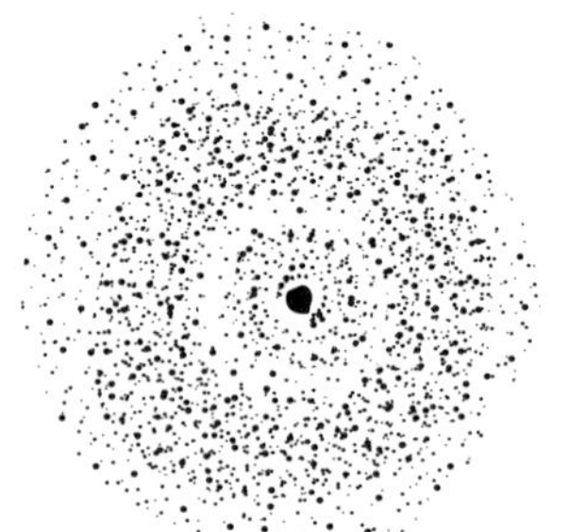

▲ **fig A** Cross-section through an s orbital.

Because the electron of the hydrogen atom is in an s sub-shell, the orbital is described as a *1s orbital*.

Electrons in the 2s sub-shell also exist in an s orbital, called the 2s orbital. The 2s orbital has the same shape as the 1s orbital but is larger. **Fig B** shows the shape of typical 1s and 2s orbitals. Note that *x*, *y* and *z* are 3D Cartesian axes (i.e. axes at mutual right angles).

The sizes and shapes of the orbitals mean that there is a 90% probability of finding the electron within their boundaries.

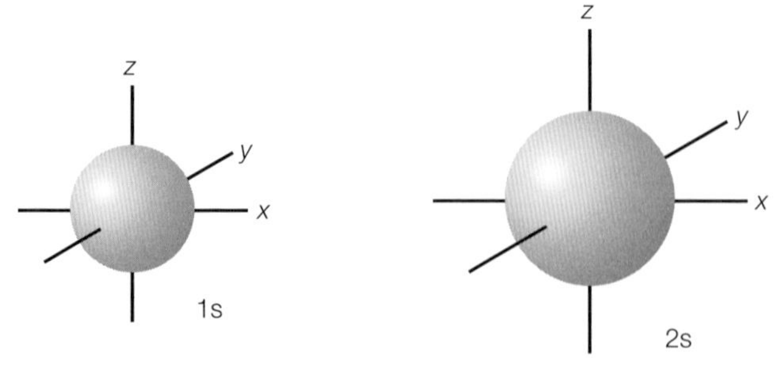

▲ **fig B** The shape of typical 1s and 2s orbitals.

p ORBITALS

The 2p sub-shell contains three separate p orbitals. These orbitals are more complex and harder to visualise. They have an elongated dumbbell shape and a variable charge density. The only difference between the three orbitals is their orientation in space, as shown in **fig C**.

Again, the size and shape of the orbitals means that there is a 90% probability of finding the electron within their boundaries. The three orbitals are arranged at mutual right angles.

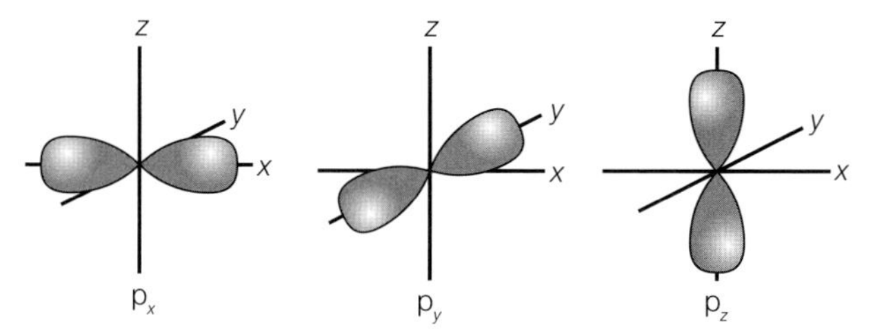

▲ **fig C** Orientation of the three p orbitals.

d ORBITALS

The d orbitals are very complicated. You do not need to know their shapes for IAS or IAL studies. There are five d orbitals in the d sub-shell.

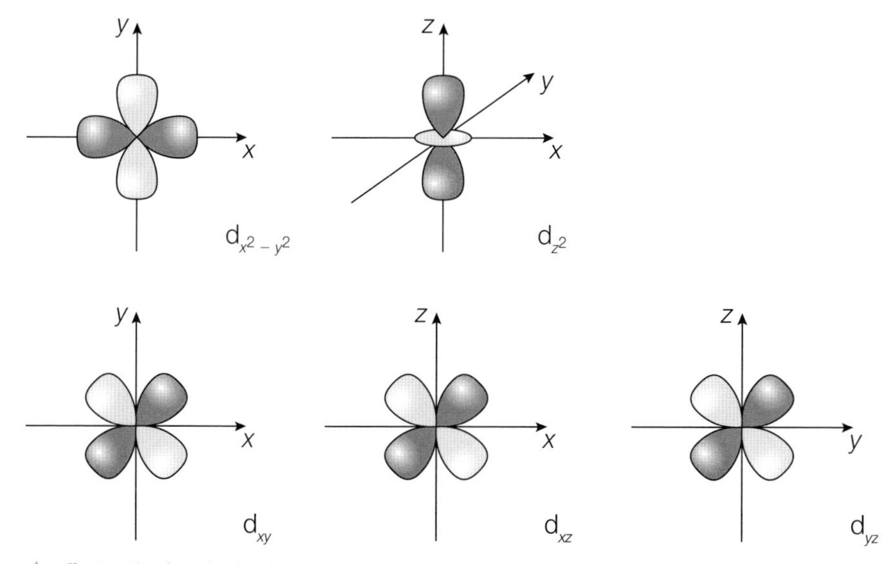

▲ **fig D** The five d orbitals.

There can be up to two electrons in each orbital. Using this information, the number of electrons that can exist in each sub-shell and each quantum shell is shown in **table A**.

	NUMBER OF ELECTRONS
s sub-shell	2 (1 × 2)
p sub-shell	6 (3 × 2)
d sub-shell	10 (5 × 2)
f sub-shell	14 (7 × 2)

	NUMBER OF ELECTRONS
first quantum shell	2
second quantum shell	8
third quantum shell	18
fourth quantum shell	32

table A The number of electrons in each sub-shell and each quantum shell.

ELECTRONIC CONFIGURATIONS

The **electronic configuration** of an atom of an element is the distribution of electrons among atomic orbitals.

Table B shows the electronic configurations for the first 36 elements in the Periodic Table.

ATOMIC NUMBER	SYMBOL	1s	2s	2p	3s	3p	3d	4s	4p
1	H	1							
2	He	2							
3	Li	2	1						
4	Be	2	2						
5	B	2	2	1					
6	C	2	2	2					
7	N	2	2	3					
8	O	2	2	4					
9	F	2	2	5					
10	Ne	2	2	6					
11	Na	2	2	6	1				
12	Mg	2	2	6	2				
13	Al	2	2	6	2	1			
14	Si	2	2	6	2	2			
15	P	2	2	6	2	3			
16	S	2	2	6	2	4			
17	Cl	2	2	6	2	5			
18	Ar	2	2	6	2	6			
19	K	2	2	6	2	6		1	
20	Ca	2	2	6	2	6		2	
21	Sc	2	2	6	2	6	1	2	
22	Ti	2	2	6	2	6	2	2	
23	V	2	2	6	2	6	3	2	
24	Cr	2	2	6	2	6	5	1	
25	Mn	2	2	6	2	6	5	2	
26	Fe	2	2	6	2	6	6	2	
27	Co	2	2	6	2	6	7	2	
28	Ni	2	2	6	2	6	8	2	
29	Cu	2	2	6	2	6	10	1	
30	Zn	2	2	6	2	6	10	2	
31	Ga	2	2	6	2	6	10	2	1
32	Ge	2	2	6	2	6	10	2	2
33	As	2	2	6	2	6	10	2	3
34	Se	2	2	6	2	6	10	2	4
35	Br	2	2	6	2	6	10	2	5
36	Kr	2	2	6	2	6	10	2	6

table B

As you can see, in general the lowest energy orbitals are occupied and filled first.

For example, the electronic configuration of helium (He) is $1s^2$ not $1s^1 2s^1$. This is because the 1s orbital can accommodate two electrons, and electrons in the 1s orbital have a lower energy than those in the 2s orbital.

PREDICTING ELECTRONIC CONFIGURATIONS

However, as you can see from **table B**, there are some exceptions to this general rule.

For example, the electronic configuration of potassium (K) is $1s^22s^22p^63s^23p^64s^1$ and not, as you might expect, $1s^22s^22p^63s^23p^63d^1$.

This is because the exact energies of the electrons in the orbitals are determined by the number of protons in the nucleus of the atom and the repulsion between all of the electrons present in the atom. For both the potassium atom and the calcium atom, the energy of the 4s orbital is lower than that of the 3d orbitals.

This means that the electronic configuration of calcium (Ca) is $1s^22s^22p^63s^23p^64s^2$ and *not* $1s^22s^22p^63s^23p^63d^2$.

DID YOU KNOW?

Interestingly, for all the elements after calcium in the Periodic Table, the energy of the 3d orbitals is *less* than that of the 4s orbitals.

This means that you would expect the electronic configuration of scandium (Sc) to be $1s^22s^22p^63s^23p^63d^3$.

However, it is $1s^22s^22p^63s^23p^63d^14s^2$.

As Professor Eric Scerri of the University of California indicated, the 3d orbitals accept electrons before the 4s orbitals: so why are two electrons pushed into the higher-energy 4s orbital?

The reason is that 3d orbitals are more compact than the 4s orbitals and, as a result, any electrons entering the 3d orbital will experience greater mutual repulsion than they would in the 4s orbital. When scandium ionises, it is the 4s electrons, i.e. those with the greater energy, that are lost first. This is what happens to the elements from scandium to zinc. We will discuss this in more detail in Topic 17 (Book 2: IAL).

However, other atoms do not always act in this way. The nuclear charge increases as we move from scandium to zinc. This complicates the interactions between the electrons and the nucleus, and between the electrons themselves. This is what ultimately produces an electronic configuration. Unfortunately, there is not one simple rule we can use in this complicated situation.

For example, you might think, incorrectly, that the most stable electronic configuration for chromium (Cr) is $3d^44s^2$ and for copper (Cu) is $3d^94s^2$. This is not the case, as we have shown in the table of electronic configurations (table B).

LEARNING TIP

You may have heard of a principle called the 'Aufbau principle'. Chemists often quote this principle as a method for working out the electronic configuration of the atoms. According to the Aufbau principle, as protons are added to the nucleus, the electrons are successively added to orbitals of increasing energy. This begins with the lowest energy orbitals, and continues until all electrons are accommodated. It can be a useful general rule for working out the configurations of the first 36 elements, with the exceptions of chromium and copper. However, as you study more elements you will discover more exceptions. You can use the Aufbau principle to work out electronic configurations, but it does not offer an explanation for them.

SPIN-SPIN PAIRING AND ELECTRONIC CONFIGURATIONS

HUND'S RULE AND THE PAULI EXCLUSION PRINCIPLE

Hund's rule states that electrons will occupy the orbitals singly before pairing takes place.

The **Pauli Exclusion Principle** states that two electrons cannot occupy the same orbital unless they have opposite spins.

Hund's rule and the Pauli Exclusion Principle are responsible for the way in which electrons fill the p and the d sub-shells.

If we apply Hund's rule to boron (B), carbon (C) and nitrogen (N), we obtain the following electronic configurations:

B $1s^22s^22p_x^{\,1}$

C $1s^22s^22p_x^{\,1}2p_y^{\,1}$

N $1s^22s^22p_x^{\,1}2p_y^{\,1}2p_z^{\,1}$

If we apply the Pauli Exclusion Principle to oxygen (O), fluorine (F) and neon (Ne), the electronic configurations are displayed as boxes.

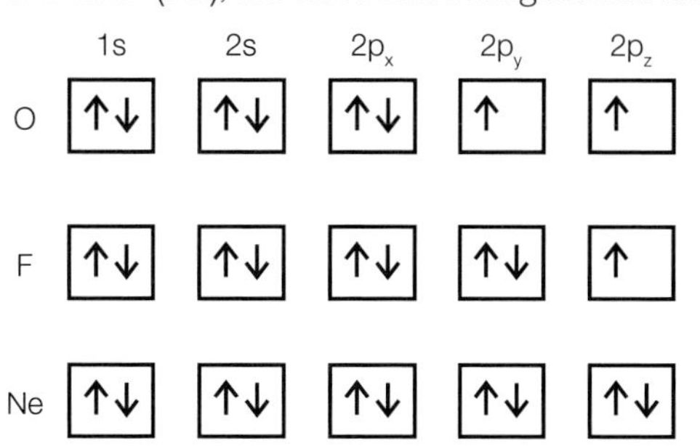

fig E Electronic configurations for oxygen, fluorine and neon displayed in boxes.

CHECKPOINT

1. The way that electrons of an atom are located in atomic orbitals is subject to Hund's rule and the Pauli Exclusion Principle.
   (a) State what you understand by Hund's rule and the Pauli Exclusion Principle.
   (b)

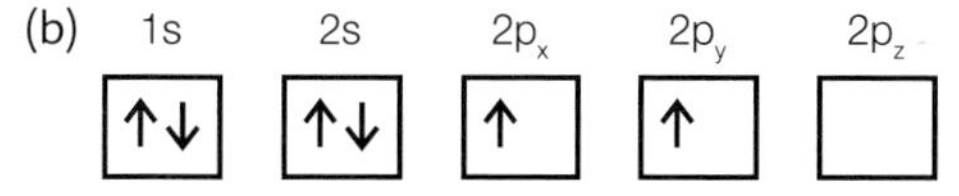

fig F Box notation diagram for Question 1 (b) and (c).

   (i) State the significance of the arrows and of the letters *x*, *y* and *z* in **fig F**.
   (ii) Suggest why the electron in the $2p_y$ orbital is not located in the $2p_x$ orbital.
   (iii) Describe the shapes of s and p orbitals.

   (c) Using the box notation shown above, give the electronic configuration of:
   (i) sulfur
   (ii) phosphorus
   (iii) chromium.

SUBJECT VOCABULARY

quantum shell the energy level of an electron

orbital a region within an atom that can hold up to two electrons with opposite spins

electronic configuration (of an atom) the number of electrons in each sub-shell in each energy level of the atom

Hund's rule electrons will occupy the orbitals singly before pairing takes place

Pauli Exclusion Principle two electrons cannot occupy the same orbital unless they have opposite spins; electron spin is usually shown by using upward and downward arrows: ↑ and ↓

4 IONISATION ENERGIES

LEARNING OBJECTIVES

- Be able to define first, second, and third ionisation energies and understand that all ionisation energies are endothermic.
- Understand how ionisation energies are influenced by the number of protons in the nucleus, the electron shielding and the sub-shell from which the electron is removed.
- Know that ideas about electronic configurations developed from an understanding that: successive ionisation energies provide evidence for the existence of quantum shells and the group to which the element belongs; the first ionisation energy of successive elements provides evidence for electron sub-shells.
- Be able to represent ionisation energy data in graphical form.
- Explain the general increase in first ionisation energy across a period.
- Explain the decrease in first ionisation energy down a group.

When you studied chemistry previously, you probably learned that electrons exist in shells surrounding the nucleus. You may not have been told how we know this.

There is lots of experimental evidence to support this theory, including data from considering the ionisation energies of elements.

IONISATION ENERGY

Ionisation energy is a measure of the energy required to completely remove an electron from an atom of an element.

We can represent the **first ionisation energy** of an element, A, by the equation:

$$A(g) \rightarrow A^{+}(g) + e^{-}$$

We can represent the **second ionisation energy** of A by the equation:

$$A^{+}(g) \rightarrow A^{2+}(g) + e^{-}$$

We can represent the **third ionisation energy** of A by the equation:

$$A^{2+}(g) \rightarrow A^{3+}(g) + e^{-}$$

SUCCESSIVE IONISATION ENERGIES

When successive energies of an element are listed, there are steady increases, and big jumps occur at defined places. This is one piece of evidence for the existence of quantum shells.

A sodium (Na) atom has 11 electrons. Its successive ionisation energies are shown in **table A**. The places where the ionisation energy has jumped significantly are shown in red.

1st	2nd	3rd	4th	5th	6th
496	4563	6913	9544	13352	16611

7th	8th	9th	10th	11th
20115	25491	28934	141367	159079

table A Successive ionisation energies, in kJ mol^{-1}, of sodium.

The first electron is considerably easier to remove than the second. There is a steady rise in ionisation energy for the next eight electrons. Then there is a big jump in energy from the ninth electron to the tenth. The last two electrons are much more difficult to remove than the previous eight.

The explanation for this trend is that the last two electrons to be removed are in the first quantum shell, the one that has the lowest energy. The next eight electrons are in the second quantum shell, which is of lower energy than the third. The first electron to be removed is in the third quantum shell of highest energy. This means that the basic electronic configuration of a sodium atom is 2,8,1.

The trend in successive ionisation energies can be seen when the logarithm of the ionisation energy is plotted against the order of electron removal. **Fig A** shows the graph for a sodium atom.

It is easy to see from the graph that there are three distinct quantum shells containing electrons. The first quantum shell contains two electrons, the second quantum shell contains eight electrons and the third quantum shell contains one electron.

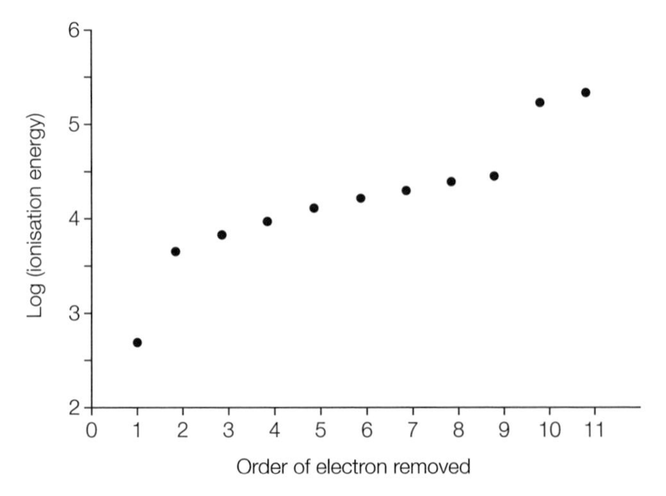

▲ **fig A** Graph showing the trend in successive ionisation energies for sodium.

EXAM HINT

In an exam, it is worth quickly sketching a graph of successive ionisation energies, so that you can clearly see the pattern.

WHY DO SUCCESSIVE IONISATION ENERGIES INCREASE IN MAGNITUDE?

We need to understand what ionisation energy represents so that we can answer this question.

The electron lost during ionisation is so far away from the influence of the nucleus that it no longer experiences an attractive force from the nucleus. We can say it is at an infinite distance from the nucleus. The energy of the electron has to be increased to a particular value for it to be removed. For any given atom, the energy value that the electron has when it reaches this position will always be the same, regardless of where in the atom the electron has come from.

If an electron already has a high energy, then it will not need to gain much energy to be removed. If, however, the electron is in an orbital of a low-energy quantum shell, for example the 1s orbital, then it will need to gain a lot more energy to be removed. The *difference* in energy between the electron when it has been removed and the energy it has when it is in its original orbital in the quantum shell is known as the ionisation energy. We can represent this by the equation:

Ionisation energy (IE) = energy of electron when removed − energy of electron when in the orbital

The ionisation energy for a particular electron in a given atom depends solely on the energy it has when it is in its orbital within the atom.

WHY DO THE SUCCESSIVE IONISATION ENERGIES OF SODIUM INCREASE?

Look at **table A** showing the successive ionisation energies of sodium.

The first electron to be removed has the highest energy of any of the electrons in the sodium atom. It is in the third quantum shell in a 3s orbital. This means that the amount of energy it has to gain in order to be removed is the lowest for any of the electrons. The first ionisation energy is therefore the lowest of all the ionisation energies.

There is a large jump from the first to the second ionisation energy. This is because the second electron to be removed is in a quantum shell of considerably lower energy, the second quantum shell. There is a steady rise in ionisation energy from the second to the ninth electron. This indicates that the eight electrons all exist within the same quantum shell. As each successive electron is removed from this shell, the electron–electron repulsion within the shell decreases. This results in a decrease in the energies of the remaining electrons, and therefore a steady increase in ionisation energy from the second to the ninth electron.

The large jump from the ninth to the tenth electron indicates another significant change in energy of the electron. This corresponds to the removal of an electron from the first quantum shell: the 1s orbital. The tenth ionisation energy is larger than the ninth; the reason for this is that as one electron is removed from the 1s orbital, the remaining electron now experiences zero repulsion, so its energy decreases.

LEARNING TIP

Remember that electron-electron repulsion:

- exists between two electrons that are in the same orbital
- exists between electrons in different orbitals within a given quantum shell
- is sometimes most significant between electrons in adjacent quantum shells
- is sometimes called shielding or screening.

DID YOU KNOW?

The terms 'shielding' and 'screening' are in some ways an unfortunate choice of words for the effect that electron–electron repulsion has on the energies of the electrons in their respective orbitals. The terms seem to suggest that, for example, the electrons in an inner quantum shell will set up a barrier to the attractive force of the nucleus to those electrons in an outer quantum shell. This is not necessarily the case. The effect of electron–electron repulsion is to raise the energy of the electrons involved above the value they would have if there was no repulsion between them. This then affects the amount of energy required to remove the electron from the atom or ion. That is, it affects the magnitude of the ionisation energy.

WHAT DETERMINES THE ENERGY OF AN ELECTRON?

You might think the answer to this question is very straightforward. You might think it is determined by the orbital in which the electron has been placed. As you will often find in chemistry, there are a number of different factors that affect the answer.

HYDROGEN AND HELIUM

First, consider the atoms of hydrogen (H) and helium (He).

Their electronic configurations are below.

H $1s^1$

He $1s^2$

The outer electrons of both atoms are in a 1s orbital, so you might think that they will have the same energy. However, the 1s orbital of helium contains two electrons. This increases the electron–electron repulsion within the orbital: each electron shields the other from the effect of the nuclear charge. The effect of this factor alone would increase the energy of the electrons.

However, the nuclear charge of helium is double that of hydrogen. This is because it contains two protons as opposed to hydrogen's one proton. The effect of this increased nuclear charge is to decrease the energy of the 1s electrons since they are attracted more strongly.

In this case, the effect of the increased nuclear charge is greater than that of the increased shielding. This means that the first ionisation energy of helium (2370 kJ mol^{-1}) is larger than that of hydrogen (1310 kJ mol^{-1}).

HELIUM AND LITHIUM

Now consider helium (He) and lithium (Li).

The electronic configuration of each is:

He $1s^2$

Li $1s^2 2s^1$

The nuclear charge of lithium is +3 (three protons), which is larger than the nuclear charge of helium (+2).

However, the outer electron of lithium is in the second quantum shell in a 2s orbital. The second quantum shell is at a higher energy level than the first quantum shell. Added to this, the electron in the 2s orbital experiences repulsion from the two inner 1s electrons, i.e. it experiences shielding from the nuclear charge.

The last two effects are more significant than the increased nuclear charge. This means that the first ionisation energy of lithium (519 kJ mol^{-1}) is smaller than the first ionisation energy of helium (2370 kJ mol^{-1}).

SUMMARY

The factors that affect the energy of an electron are:

- the orbital in which the electron exists
- the nuclear charge of the atom (i.e. the number of protons in the nucleus)
- the repulsion (shielding) experienced by the electron from all the other electrons present.

TRENDS IN IONISATION ENERGIES

The two major trends in ionisation energies in the Periodic Table are found:

- across a period
- down a group.

ACROSS A PERIOD

To illustrate this trend, consider Period 2; and the elements lithium (Li) to neon (Ne).

As we move from Li to Ne across Period 2, the nuclear charge increases as the number of protons increases. This, on its own, would lead to an increased attraction between the nucleus and the electron, and therefore a decrease in the energy of the outermost electron, and an increase in first ionisation energy.

But, counteracting this, one more electron is added to the same quantum shell on each occasion and this increases the electron–electron repulsion within the quantum shell. This, on its own, would cause an increase in energy of the outermost electron, and would lead to a decrease in first ionisation energy.

The increase in nuclear charge is more significant than the increase in electron–electron repulsion. So there is a general increase in first ionisation energy across Period 2.

We can say the same of the Period 3 elements, sodium (Na) to argon (Ar), excluding the d-block elements, scandium (Sc) to zinc (Zn).

LEARNING TIP

This trend is not perfect and there are two exceptions in each of Periods 2 and 3: in Period 2, beryllium and boron (Be and B); and in Period 3, magnesium and aluminium (Mg and Al).

DOWN A GROUP

To illustrate this trend, consider the Group 1 elements, lithium (Li) to caesium (Cs).

As we descend through Group 1 from Li to Cs, the nuclear charge increases as the number of protons increases. This would lead to an increased attraction between the nucleus and the electron, and therefore a decrease in energy of the outer electron. This, in turn, would lead to an increase in the first ionisation energy.

However, one new quantum shell is added on each occasion. This increases the energy of the outermost electron for two reasons. Firstly, the third quantum shell has a higher energy value than the second; the fourth quantum shell is higher in energy than the third.

This continues down the group. Secondly, as each new quantum shell is added, the outer electron experiences increased repulsion (i.e. increased shielding) from the inner electrons.

On this occasion, the combined effect of adding an extra shell and increasing the shielding is more significant than the increase in nuclear charge. So the first ionisation energy decreases down Group 1 from Li to Cs.

This trend is repeated in:

- Group 2: beryllium to barium (Be to Ba)
- Group 5: nitrogen to bismuth (N to Bi)
- Group 6: oxygen to polonium (O to Po)
- Group 7: fluorine to astatine (F to At)
- Group 8: neon to radon (Ne to Rn).

In Group 4, lead (Pb) is an anomaly (does not fit in with the others) because it has a first ionisation energy that is higher than that of tin (Sn), the element immediately above it.

There is no general trend in first ionisation energy in Group 3 (boron (B) to thallium (Tl)).

The explanations for the anomalies in Groups 3 and 4 are not included in this book, but they illustrate that, in chemistry, you can rarely apply a simple pattern or trend to all situations.

CHECKPOINT

1. State the three factors that determine the magnitude of the first ionisation energy of an element.

2. Write equations to represent:

(a) the first ionisation energy of sodium

(b) the second ionisation energy of calcium

(c) the third ionisation energy of carbon.

3. Draw a sketch graph for the logarithm to base 10 for the successive ionisation energies of phosphorus.

4. The table below shows the first four ionisation energies, in kJ mol^{-1}, of five elements: A, B, C, D and E.

ELEMENT	FIRST IONISATION ENERGY	SECOND IONISATION ENERGY	THIRD IONISATION ENERGY	FOURTH IONISATION ENERGY
A	496	4563	6913	9544
B	738	1451	7733	10 541
C	578	1817	2745	11 578
D	900	1757	14 849	21 007
E	631	1235	2389	7089

(a) Which two elements are in the same group of the Periodic Table? Explain your answer.

(b) In which group of the Periodic Table is element C likely to occur? Explain your answer.

(c) Which element requires the least amount of energy to form a 2+ ion? Explain your answer.

5. The first four ionisation energies, in kJ mol^{-1}, of calcium are 590, 1145, 4912 and 6474.

(a) Explain why the second ionisation energy of calcium is larger than the first.

(b) Explain why the third ionisation energy is much larger than the second.

SUBJECT VOCABULARY

first ionisation energy (of an element) the energy required to remove an electron from each atom in one mole of atoms in the gaseous state

second ionisation energy (of an element) the energy required to remove an electron from each singly charged positive ion in one mole of positive ions in the gaseous state

third ionisation energy (of an element) the energy required to remove an electron from each doubly charged positive ion in one mole of positive ions in the gaseous state

2B 1 THE PERIODIC TABLE

SPECIFICATION REFERENCE 2.15 2.16 PART

LEARNING OBJECTIVES

- Understand that electronic configuration determines the chemical properties of an element.
- Know that the Periodic Table is divided into blocks, such as s, p, and d.

GROUPS, PERIODS AND BLOCKS

GROUPS

We call the vertical columns in the Periodic Table **groups**. All the elements in a main group (i.e. Groups 1 to 8 inclusive) contain the same outer electronic configuration. In **table A**, *n* represents the quantum shell. For lithium to neon (Li to Ne), n = 2; for sodium to argon (Na to Ar), n = 3, etc.

	GROUP 1	GROUP 2	GROUP 3	GROUP 4
outer electronic configuration	ns^1	ns^2	ns^2np^1	ns^2np^2
	GROUP 5	**GROUP 6**	**GROUP 7**	**GROUP 8**
outer electronic configuration	ns^2np^3	ns^2np^4	ns^2np^5	ns^2np^6

table A Outer electronic configuration of Groups 1–8 in the Periodic Table.

The elements within a group have similar chemical properties because they have the same outer electronic configuration.

PERIODS

We call the horizontal rows in the Periodic Table **periods**. All elements in a period have the same number of quantum shells containing electrons. For example, all elements in Period 2, lithium to neon (Li to Ne), have electrons in both the first and second quantum shells.

BLOCKS

The Periodic Table is also divided into blocks.

The s-block consists of the elements in Groups 1 and 2. An s-block element has its highest energy electron in an s orbital.

s-block ⟷ (Groups 1–2); d-block ⟷; p-block ⟷ (Groups 3–8)

1	2											3	4	5	6	7	8
Li	Be											B	C	N	O	F	Ne
Na	Mg	d-block										Al	Si	P	S	Cl	Ar
K	Ca	Sc	Ti	V	Cr	Mn	Fe	Co	Ni	Cu	Zn	Ga	Ge	As	Se	Br	Kr
Rb	Sr	Y	Zr	Nb	Mo	Tc	Ru	Rh	Pd	Ag	Cd	In	Sn	Sb	Te	I	Xe

▲ **fig A** The s-, d- and p-blocks in the Periodic Table.

DID YOU KNOW?

We cannot define the d-block in the same way as the s- and p-blocks, because the electrons in the d orbitals are of lower energy than the electrons in the outer s orbital. For example, 3d has a lower energy than 4s for the d-block elements in the first row.

The p-block consists of the elements in Groups 3 to 8 inclusive. A p-block element has its highest energy electron in a p-orbital.

The d-block consists of the elements scandium to zinc (Sc to Zn) in Period 4 and yttrium to cadmium (Y to Cd) in Period 5. The number of electrons in the d orbitals gradually increases from left to right across the table. Therefore, a d-block element can be defined as one in which the d sub-shell is being filled.

CHECKPOINT

1. What is meant by the terms
   (a) s-block element and
   (b) p-block element?
2. Explain why the chemical properties of the elements in Group 1 are very similar.
3. The outer electronic configuration of carbon is $2s^22p^2$.
   Work out the outer electronic configuration of arsenic (As). Explain how you arrived at your answer.

SUBJECT VOCABULARY

groups the vertical columns in the Periodic Table
periods the horizontal rows in the Periodic Table

LEARNING OBJECTIVES

Be able to explain:

- the trends in melting and boiling temperatures of the elements of Periods 2 and 3 of the Periodic Table in terms of the structure of the element and the bonding between its atoms or molecules;
- the specific trends in ionisation energy of the elements across Periods 2 and 3 of the Periodic Table.

WHAT ARE PERIODIC PROPERTIES?

The elements in a period exhibit **periodic properties** (also sometimes called **periodicity**).

We can illustrate the idea of periodic properties by looking at the elements in Periods 2 and 3. We have already seen one example of periodic properties in Periods 2 and 3 – the regular repeating pattern of electronic configurations from ns^1 through to ns^2np^6. Other examples are the trends in atomic radii, melting and boiling temperatures, and first ionisation energies.

ATOMIC RADII

The atomic radius of an element is a measurement of the size of its atoms. It is the distance from the centre of the nucleus to the boundary of the electron cloud. Since the atom does not have a well-defined boundary, we can find the atomic radius by determining the distance between the two nuclei and dividing it by two.

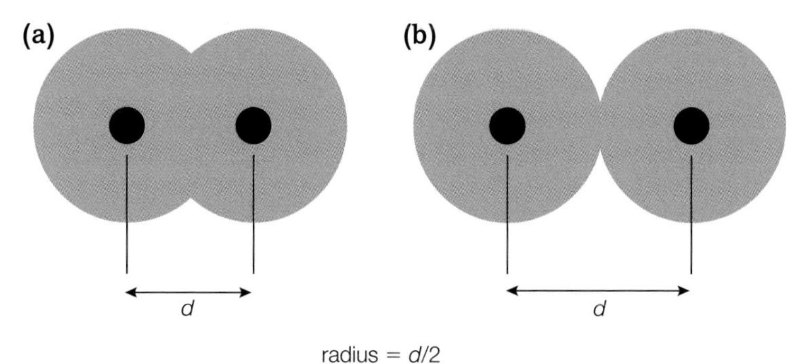

▲ **fig A** Measuring (a) the covalent radius and (b) the van der Waals radius.

Diagram (a) in **fig A** shows two bonded atoms. The atoms are closer together than they are in diagram (b) (where they are only just touching).

In diagram (a), we are measuring the covalent radius.

The radius we are measuring in diagram (b) is called the van der Waals radius. This is the only radius that we can determine for neon and argon, because they do not bond with other elements.

There is a third radius that is used for metals. It is called the metallic radius.

Fig B shows the trend in covalent radii across the second period (lithium to fluorine, or Li to F) and third period (sodium to chlorine, or Na to Cl) measured in nanometres (10^{-9} m).

DID YOU KNOW?
Different types of radii give different measurements for the same element. The van der Waals radius is always larger. So always compare like with like when you are looking at the trends in atomic radii.

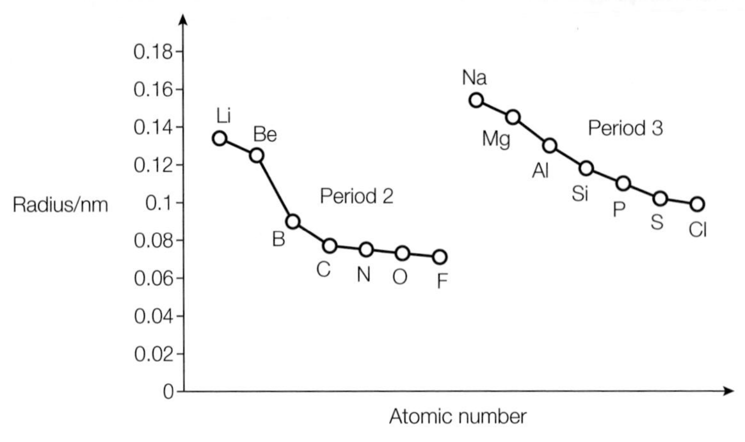

▲ **fig B** Trend in covalent radii across the second and third periods.

You can see that the radius decreases across each period. This is because as the number of protons in the nucleus increases, so does the nuclear charge. This results in an increase in the attractive force between the nucleus and the outer electrons. This increase in attractive force offsets (counterbalances) the increase in electron–electron repulsion as the number of electrons in the outer quantum shell increases.

MELTING AND BOILING TEMPERATURES

Table A below illustrates the changes in melting and boiling temperatures for the elements in Periods 2 and 3.

PERIOD 2	Li	Be	B	C (DIAMOND)	N	O	F
melting temperature / °C	181	1278	2300	3550	−210	−218	−220
boiling temperature / °C	1342	2970	3927	4827	−196	−183	−188
type of bonding	metallic	metallic	covalent	covalent	covalent	covalent	covalent
structure	giant lattice	giant lattice	giant lattice	giant lattice	simple molecular	simple molecular	simple molecular

PERIOD 3	Na	Mg	Al	Si	P	S	Cl
melting temperature / °C	98	649	660	1440	44	113	−101
boiling temperature / °C	883	1107	2467	2355	280	445	−35
type of bonding	metallic	metallic	metallic	covalent	covalent	covalent	covalent
structure	giant lattice	giant lattice	giant lattice	giant lattice	simple molecular	simple molecular	simple molecular

table A Changes in melting and boiling temperatures for Period 2 and 3 elements.

EXAM HINT

If you are asked to explain trends in melting points, you do not need to be able to remember specific melting points of elements. You only need to be able to explain the trends in terms of bond type.

You may have noticed that the elements with giant lattice structures have high melting and boiling temperatures, and those with simple molecular structures have low melting and boiling temperatures. We will explain the reason for this in **Topic 3**.

FIRST IONISATION ENERGIES

Fig C is a plot of first ionisation energy for the first three Periods.

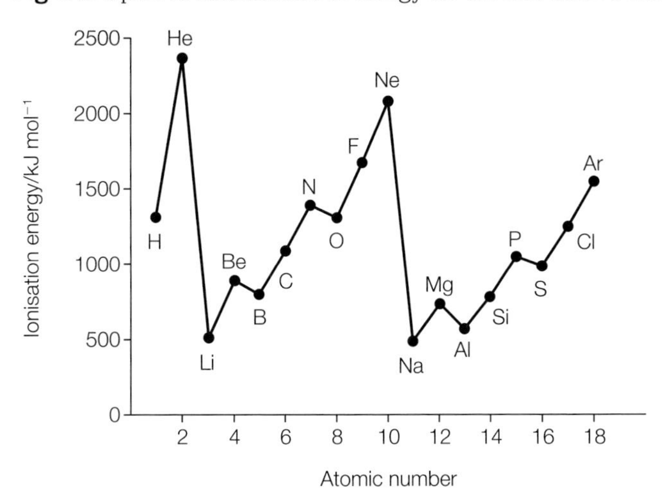

▲ **fig C** First ionisation energy for Periods 1–3.

HYDROGEN AND HELIUM

The electronic configurations of hydrogen and helium are $1s^1$ and $1s^2$, respectively.

We can explain the increase in first ionisation energy from hydrogen to helium by the increase in nuclear charge from 1 to 2 as an extra proton is added. This increase in nuclear charge more than offsets the increase in electron–electron repulsion in the 1s orbital as a second electron is added.

LEARNING TIP

You may hear or read about a suggestion that the first ionisation energy of boron is less than that of beryllium, because by losing an electron the boron atom will acquire a full 2s orbital and so become more stable. In fact, because energy has to be supplied to the boron atom in order to remove an electron, the ion formed is energetically less stable than the atom because it has a higher energy value.

ANOMALIES

We have already explained the general increase in first ionisation energy across the period from lithium to neon (Li to Ne) and from sodium to argon (Na to Ar), in terms of the increasing nuclear charge.

However, you will notice that there are two anomalies in each case. The first ionisation energy of the Group 3 element is less than that of the Group 2 element, and the first ionisation energy of the Group 6 element is less than that of the Group 5 element.

First, consider the case of beryllium (Be) and boron (B).

The electronic configurations are:

Be: $1s^2\ 2s^2$

B: $1s^2\ 2s^2\ 2p^1$

Although the nuclear charge of the boron atom is greater than that of the beryllium atom, the outer electron of boron has more energy, since it is in a 2p orbital as opposed to the 2s orbital for beryllium. For this reason, the energy required to remove 2p electron in boron is less than the energy required to remove a 2s electron from a beryllium atom. In addition, the 2p electron in boron experiences greater electron–electron repulsion (i.e. greater shielding) because there are two inner electron sub-shells as opposed to only one in the beryllium atom.

A similar argument applies to magnesium and aluminium (Mg and Al), except that in this case it involves the 3s and 3p electrons.

LEARNING TIP

You may hear or read about an explanation stating that the outer electron in the boron atom is further from the nucleus than the outer electron in the beryllium atom. This is not correct. The boron atom is smaller than the beryllium atom as shown in **fig B**.

NITROGEN AND OXYGEN

Now consider nitrogen (N) and oxygen (O).

The electronic configurations are:

N: $1s^2\ 2s^2\ 2p_x^{\,1}2p_y^{\,1}2p_z^{\,1}$

O: $1s^2\ 2s^2\ 2p_x^{\,2}2p_y^{\,1}2p_z^{\,1}$

The first electron removed from the oxygen atom is one of the two paired electrons in the $2p_x$ orbital.

The presence of two electrons in a single orbital increases the electron–electron repulsion in this orbital. So less energy is required to remove one of these electrons than is required to remove a 2p electron from a nitrogen atom, despite the larger nuclear charge of the oxygen atom.

LEARNING TIP

You may see an explanation stating that the first ionisation energy of oxygen is less than that of nitrogen, saying this is because when the oxygen atom loses an electron, it acquires a stable half-full 2p sub-shell. However, there is no special stability associated with a half-full sub-shell. Also, as with the boron atom, energy must be supplied to the oxygen atom in order to remove an electron, so the ion formed is energetically less stable than the atom.

You may also hear or read about an explanation stating that the outer electron in the oxygen atom is further from the nucleus than the outer electron in the nitrogen atom. This is not correct. The oxygen atom is smaller than the nitrogen atom as shown in **fig B**.

CHECKPOINT

SKILLS REASONING

1. An element has very high melting and boiling temperatures.
   (a) What type of structure is it likely to have?
   (b) Which physical property would help you determine whether the bonding was metallic or covalent? Explain your answer.
2. Explain why the first ionisation energy of helium is higher than that of hydrogen.
3. Explain why the first ionisation energy of lithium is much lower than that of helium, even though the lithium atom has a greater nuclear charge.
4. Would you expect the first ionisation energy of gallium (Ga) to be higher or lower than that of calcium (Ca)? Explain your answer.
5. Why does neon have the highest first ionisation energy of all the elements in Period 2?

SUBJECT VOCABULARY

periodic properties (periodicity) regularly repeating patterns of atomic, physical and chemical properties, which can be predicted using the Periodic Table and explained using the electron configurations of the elements

ELEMENTAL FINGERPRINTS

SKILLS CREATIVITY

Is it possible to know what stars in distant galaxies are made from?

ELEMENTAL 'FINGERPRINTS'

In 1825, the French philosopher Auguste Comte said that there are some things we will never know, among these the chemical composition of the stars. This was an unfortunate example. We do not need actual material from the stars in order to analyse them. What we do need, of course, is information, and this they send us plentifully, in the form of light. To reveal this information, we must separate the light into its different wavelengths or colours.

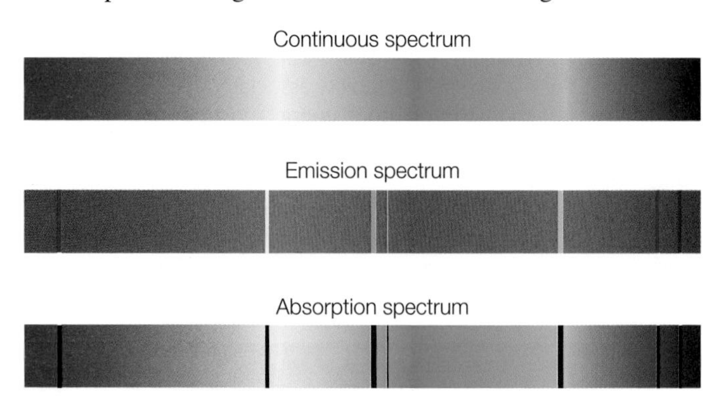

fig A

From *From Stars to Stalagmites: How Everything Connects* by Paul Braterman

Many years earlier Isaac Newton had passed sunlight through a glass prism, and found that this gave him all the colours of the rainbow. In 1802 William Wollaston in England … repeated the experiment, using an up-to-date high quality glass prism, and discovered that the continuous spectrum was interrupted by narrow dark lines. (Rather unfairly these are now known as Frauenhofer lines, after the German physicist Joseph von Frauenhofer, who confirmed and extended Wollaston's observations). These lines later proved to match exactly the light given out by different elements when heated or in electric discharges. A now familiar example is provided by the element sodium, whose yellow emission is used in street lamps. The lines can be used as a kind of fingerprint for each element. We can take this fingerprint using electrical discharges here on Earth and compare it with the lines found in the spectrum of the Sun. In this way we can get a good chemical analysis of the surface layer of the Sun or of any other star whose light we can collect and astronomers now extend the same process to the most distant galaxies.

THINKING BIGGER TIPS

An emission spectrum shows the frequencies of electromagnetic radiation that are emitted by a substance. An absorption spectrum shows the frequencies of electromagnetic radiation that are absorbed by the substance. This means that the absorption spectrum is essentially the 'negative' of the emission spectrum.

DID YOU KNOW?

Many images of galaxies taken by the Hubble telescope use the line emission spectra of elements such as hydrogen, sulfur and oxygen to build up a colour image of the galaxy. The different elements are assigned red, green and blue frequencies to build up a full colour image.

SCIENCE COMMUNICATION

The extract is taken from the book *From Stars to Stalagmites: How Everything Connects* by Paul Braterman (2012).

1. (a) Do you think this article is aimed at scientists, the general public, or people who are not scientists but have an interest in science? Look back through the extract and find examples to support your answer.

(b) Explain how the book's subtitle *How Everything Connects* is supported by the extract.

THINKING BIGGER TIP

As well as considering the complexity of the scientific ideas, think about the level of detail in which they are discussed. Notice how many different branches of natural science are mentioned in the extract. Are all of the scientists mentioned chemists or did they work in different scientific disciplines?

CHEMISTRY IN DETAIL

2. (a) Describe the structure of the sodium atom in terms of protons, neutrons and electrons.

(b) Give the electronic structure of a sodium atom using s, p, d notation.

(c) What is giving rise to (i) the absorption and (ii) the emission spectrum of sodium?

(d) Can you suggest why sodium vapour is used in preference to potassium vapour in street lamps?

3. Would you expect heavier isotopes of sodium to give a different 'fingerprint'? Explain your answer.

4. The element helium was first discovered in the outer layer of the Sun. Suggest why it was discovered there before it was discovered here on Earth. Think back to work you have done on identifying alkali metals using flame tests.

WRITING SCIENTIFICALLY

You need to use reasoning and maybe examples to give an explanation of your answer and support your point. Consider how the particles involved (e.g. atoms, ions, molecules, electrons, etc.) affect your answer.

ACTIVITY

fig B A picture taken by the Hubble Telescope showing over 10 000 galaxies!

Today, the elemental fingerprints in the light collected from distant galaxies have provided evidence for an expanding universe. The elemental fingerprints are subject to a phenomenon called 'red shift'.

Explain in 200–300 words how red shift has provided evidence for an expanding universe. Try to structure your explanation so that the concepts can be understood by an audience of 16-year-old students.

THINKING BIGGER TIP

The Activity provides opportunity for creative explanation of advanced concept to younger students.

2 EXAM PRACTICE

1 The relative atomic mass of boron is 10.8.

A sample of boron contains the isotopes $^{10}_{5}B$ and $^{11}_{5}B$. What is the percentage of $^{11}_{5}B$ atoms in the isotopic mixture of this sample?

A 0.8% **B** 8.0% **C** 20% **D** 80% [1]

(Total for Question 1 = 1 mark)

2 Which of the following elements has no paired p electrons in a single uncombined atom of the element?

A carbon **B** oxygen **C** fluorine **D** neon [1]

(Total for Question 2 = 1 mark)

3. Which of the following electronic configurations is that of an atom of an element which forms a simple ion with a charge of −3?

A $1s^2\ 2s^2\ 2p^6\ 3s^2\ 3p^1$ **B** $1s^2\ 2s^2\ 2p^6\ 3s^2\ 3p^3$

C $1s^2\ 2s^2\ 2p^6\ 3s^2\ 3p^6\ 3d^1\ 4s^2$ **D** $1s^2\ 2s^2\ 2p^6\ 3s^2\ 3p^6\ 3d^3\ 4s^2$ [1]

(Total for Question 3 = 1 mark)

4 A sample of chlorine contains isotopes of mass numbers 35 and 37.

The sample is analysed in a mass spectrometer. How many peaks corresponding to Cl_2^+ are recorded?

A 1 **B** 2 **C** 3 **D** 4 [1]

(Total for Question 4 = 1 mark)

5 What is the atomic number of an element that contains atoms which have four unpaired electrons in their ground state?

A 6 **B** 16 **C** 22 **D** 26 [1]

(Total for Question 5 = 1 mark)

6 Which of the following ions has more electrons than protons, and also has more protons than neutrons?

($H = {}^{1}_{1}H$ $D = {}^{2}_{1}H$ $He = {}^{4}_{2}He$ $O = {}^{16}_{8}O$)

A OD^- **B** D_3O^+ **C** He^+ **D** OH^- [1]

(Total for Question 6 = 1 mark)

7 A sample of helium from a rock was found to contain two isotopes with the following composition by mass: ^{3}He, 0.992%; ^{4}He, 99.008%.

(a) State what is meant by isotopes. [1]

(b) State the difference in the atomic structures of ^{3}He and ^{4}He. [1]

(c) (i) Which isotope is used as the basis for relative atomic mass measurements? [1]

(ii) Calculate the relative atomic mass of helium in the rock sample. [2]

(d) Helium has the largest first ionisation energy of all the elements.

(i) State what is meant by first ionisation energy. [2]

(ii) Write an equation, including state symbols, to represent the first ionisation energy of helium. [2]

(iii) Explain why the first ionisation energy of helium is larger than that of hydrogen. [2]

(Total for Question 7= 12 marks)

8 The five ionisation energies of boron are:

801 2427 3660 25 026 32 828

(a) State and justify the group in the Periodic Table in which boron is placed. [2]

(b) Which of the following represents the **second** ionisation energy of boron?

A $B(g) \rightarrow B^{2+}(g) + 2e^-$ $\Delta H = +2427\ kJ\ mol^{-1}$

B $B^+(g) \rightarrow B^{2+}(g) + e^-$ $\Delta H = +2427\ kJ\ mol^{-1}$

C $B(g) \rightarrow B^{2+}(g) + 2e^-$ $\Delta H = -2427\ kJ\ mol^{-1}$

D $B^+(g) \rightarrow B^{2+}(g) + e^-$ $\Delta H = -2427\ kJ\ mol^{-1}$ [1]

(c) Explain why the second ionisation energy of boron is larger than the first. [2]

(d) Give the electronic configuration of a boron atom. [1]

(e) Is boron classified as an s-, a p- or a d-block element? Justify your answer. [2]

(f) Explain why the first ionisation energy of boron is less than that of beryllium, even though a boron atom has a greater nuclear charge. [2]

(Total for Question 8 = 10 marks)

9 (a) State what is meant by periodic property. [1]

(b) Explain the **general** trend in first ionisation energy of the elements Na to Ar in Period 3 of the Periodic Table. [3]

(c) Explain the trend in atomic radii of the elements sodium to chlorine in Period 3 of the Periodic Table. [4]

(d) The table shows the melting temperatures of the elements sodium to chlorine in Period 3 of the Periodic Table.

Symbol of element	**Na**	**Mg**	**Al**	**Si**	**P**	**S**	**Cl**
melting temperature / °C	98	649	660	1410	44	113	−101
bonding							
structure							

Complete the table using the following guidelines.

(i) Complete the 'bonding' row using **only** the words 'metallic' or 'covalent'.

(ii) Complete the 'structure' row using **only** the words 'simple molecular' or 'giant lattice'.

(iii) Explain why the melting temperature of phosphorus is different from that of silicon. [5]

(Total for Question 9 = 13 marks)

10 (a) Complete the table to show the properties of the three major sub-atomic particles. [3]

Particle	Relative charge	Relative mass
proton		
neutron		
electron		

(b) The particles in each pair below differ **only** in the number of protons **or** the number of neutrons **or** the number of electrons that they contain. State the difference in each pair.

(i) ^{16}O and ^{17}O

(ii) ^{24}Mg and $^{24}Mg^{2+}$

(iii) $^{39}K^{+}$ and $^{40}Ca^{2+}$ [3]

(Total for Question 10 = 6 marks)

11 (a) State what is meant by an orbital. [2]

(b) Draw the shapes of:

(i) an s-orbital

(ii) a p-orbital. [2]

(c) How many electrons can occupy

(i) an s-orbital

(ii) a p-subshell? [2]

(d) Using s and p notation, give the electronic configurations of each of the following atoms and ions:

(i) Na

(ii) O^{2-}

(iii) Mg^{2+}

(iv) Cl [4]

(e) Which of these ions contains the fewest number of electrons?

A NH_4^+ **B** P^{3-} **C** S^{2-} **D** Cl^- [1]

(Total for Question 11 = 11 marks)

12 Chlorine has two isotopes, ^{35}Cl and ^{37}Cl.
The diagram shows part of the mass spectrum of a sample of chlorine gas.

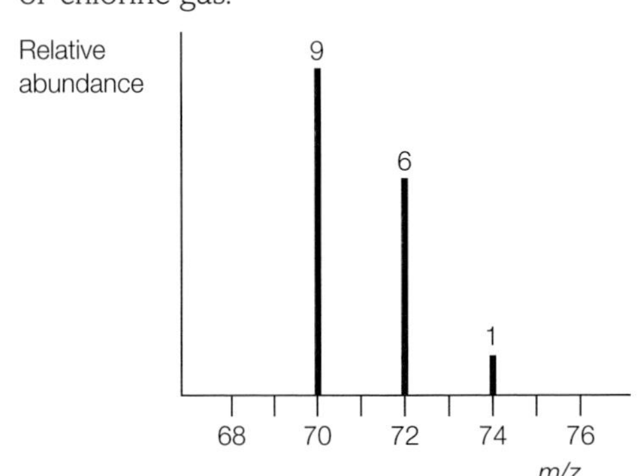

(a) Give the formula of the species responsible for each peak at m/z of 70, 72 and 74. [3]

(b) Give the m/z ratio for the two other peaks in the mass spectrum of chlorine. Give the formula of the species responsible for each peak and state the relative abundance of these two species. [5]

(Total for Question 12 = 8 marks)

13 The table gives the first four ionisation energies of the elements sodium, magnesium and aluminium.

Ionisation energy/kJ mol^{-1}				
Element	**1st**	**2nd**	**3rd**	**4th**
sodium	496	4563	6913	9544
magnesium	738	1451	7733	10 541
aluminium	578	1817	2745	11 578

Explain why:

(a) the first ionisation energy of sodium is lower than that of the first ionisation energy of magnesium. [2]

(b) the first ionisation energy of magnesium is higher than the first ionisation energy of aluminium. [2]

(c) the second ionisation energy of magnesium is lower than the second ionisation energy of aluminium. [2]

(d) the fourth ionisation energy of aluminium is higher than its third ionisation energy. [2]

(Total for Question 13 = 8 marks)

14 Lithium is an element in the s-block of the Periodic Table. Naturally occurring lithium contains a mixture of two isotopes, ^{6}Li and ^{7}Li.

(a) Complete the table to show the atomic structure of each of the two isotopes.

Isotope	Number of portions	Number of neutrons	Number of electrons
^{6}Li			
^{7}Li			

[2]

(b) A sample of lithium was found to contain 7.59% of ^{6}Li and 92.41% of ^{7}Li.

(i) Give the name of the instrument that can be used to determine this information. [1]

(ii) The ^{7}Li isotope has a relative isotopic mass of 7.016 and the ^{6}Li isotope has a relative isotopic mass of 6.015. State what is meant by relative isotopic mass. [2]

(iii) Use the relative isotopic masses to calculate the relative atomic mass of lithium. Give your answer to three significant figures. [2]

(c) (i) Give the electronic configuration of an atom of lithium. [1]

(ii) State why lithium is described as an s-block element. [1]

(iii) Would you expect the two isotopes of lithium to have the same chemical properties? Justify your answer. [1]

(Total for Question 14 = 10 marks)

TOPIC 3 BONDING AND STRUCTURE

A IONIC BONDING | B COVALENT BONDING | C SHAPES OF MOLECULES | D METALLIC BONDING | E SOLID LATTICES

Now that we know something of the structure of atoms, in particular their electronic configurations, we can look at how atoms combine. Knowing about the way substances are bonded, and the effect that the bonding and structure has on chemical and physical properties, is essential to the development of new materials. These new materials are then used to refine and improve products. Computers and mobile phones are smaller, lighter and faster than ever before. New and better materials for clothes and shoes, particularly for use outdoors in bad weather conditions, are constantly being developed. Polymers and composites have almost completely taken the place of more traditional materials such as wood and metal for many uses.

Knowledge of the shapes of molecules is of fundamental importance in understanding how medicines work and for the future design of new medicines. Knowing the shapes of enzymes is important in the development of biochemical catalysts for chemical reactions. The shapes of macromolecules such as DNA and proteins are extremely complicated. However, the rules that govern their shapes are the same as the rules used to predict the shapes of simple molecules such as methane, water and ammonia.

MATHS SKILLS FOR THIS TOPIC

- Use angles and shapes in regular 2D and 3D structures
- Visualise and represent 2D and 3D forms including two-dimensional representations of 3D objects
- Understand the symmetry of 2D and 3D shapes

What prior knowledge do I need?

- Metallic, ionic and covalent bonding
- Using dot-and-cross diagrams to represent ions and molecules
- The physical properties of metals, ionic compounds and covalent compounds, both simple molecular and giant structures
- The electronic configurations of the first 36 elements in the Periodic Table

What will I study in this topic?

- The nature of metallic, ionic, covalent, polar covalent and dative covalent bonding
- The nature of intermolecular interactions, including hydrogen bonding
- The shapes of discrete (simple) molecules
- Electronegativity and polarity of molecules
- An explanation of the physical properties of substances based on their bonding and structure

What will I study later?

Topic 4

- The existence of isomerism in organic compounds

Topics 4 and 5

- The mechanisms of some reactions of alkanes, alkenes, halogenoalkanes and alcohols

Topic 8

- Trends in the properties of Group 2 and Group 7 elements

Topics 15 and 19 (Book 2: IAL)

- The mechanisms of some reactions of carbonyls, carboxylic acids, arenes and organic nitrogen compounds

Topic 17 (Book 2: IAL)

- The nature of the bonding in, and the shapes of, transition metal complexes
- The characteristic properties of transition metals, e.g. the ability to have more than one oxidation state and the ability to act as catalysts

3A 1 THE NATURE OF IONIC BONDING

SPECIFICATION REFERENCE 3.1 PART 3.2 3.3 3.5

LEARNING OBJECTIVES

- Describe the formation of ions in terms of loss or gain of electrons.
- Draw dot-and-cross diagrams to show electrons in cations and anions.
- Know that ionic bonding is the result of strong net electrostatic attraction between oppositely charged ions.
- Know and be able to interpret evidence for the existence of ions using electron density maps and from the migrations of ions.

THE FORMATION OF CATIONS AND ANIONS

Some ionic compounds can be formed by the direct combination of two elements.

FORMATION OF SODIUM AND CHLORIDE IONS

For example, sodium chloride can be formed by burning sodium in chlorine:

$$2Na(s) + Cl_2(g) \rightarrow 2NaCl(s)$$

We can represent the reaction that occurs by two ionic half-equations:

$$2Na \rightarrow 2Na^+ + 2e^- \text{ and } Cl_2 + 2e^- \rightarrow 2Cl^-$$

Each sodium atom has lost one electron to become a positive sodium ion. The chlorine molecule has gained two electrons to become two chloride ions.

We can represent the electronic changes involved by dot-and-cross diagrams.

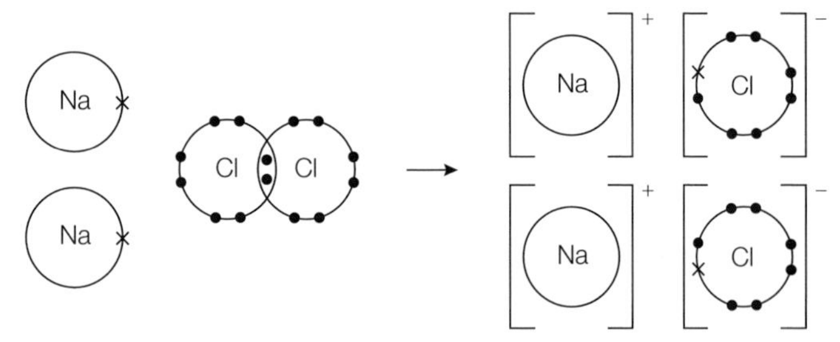

fig A Dot-and-cross diagram showing the formation of sodium and chloride ions.

LEARNING TIP

It is important when drawing a dot-and-cross diagram to represent the electrons of one atom using a cross and the other using a dot. You only need to show the outer electrons.

FORMATION OF MAGNESIUM AND OXIDE IONS

Here is the equation for the formation of magnesium oxide:

$$2Mg(s) + O_2(g) \rightarrow 2MgO(s)$$

A dot-and-cross diagram for the reaction between magnesium and oxygen is shown in **fig B**.

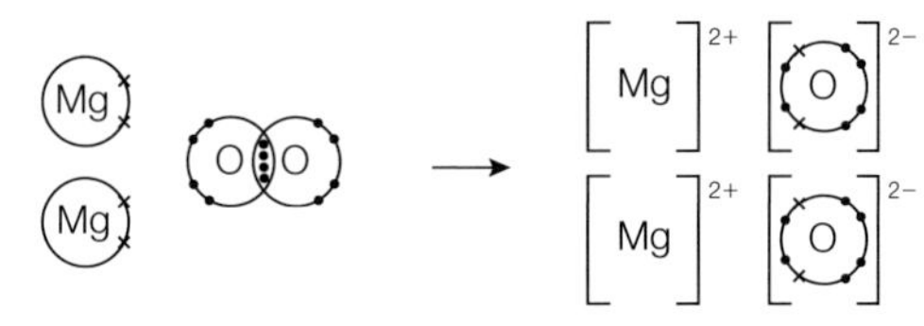

fig B Dot-and-cross diagram showing the formation of magnesium and oxide ions.

THE NATURE OF IONIC BONDING

Ionic bonding occurs in solid materials consisting of a regular array of oppositely charged ions extending throughout a giant lattice network.

The most familiar ionic compound is sodium chloride, NaCl. It consists of a regular array of sodium ions, Na^+, and chloride ions, Cl^-, as shown in **fig C**.

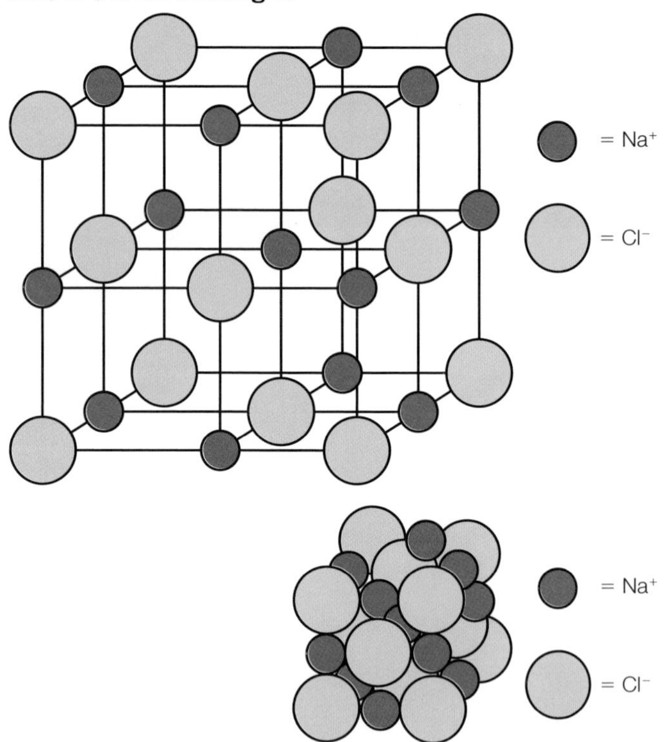

fig C Structure of sodium chloride.

The diagram on the left is an 'exploded' version of the structure, which is often drawn for the sake of clarity. In practice, the ions are touching one another, as shown in the diagram on the right.

EXAM HINT

It is worth practising drawing a section of an ionic crystal as you may be asked to do so in an exam. Remember to include a key to show which ion is which.

In an ionic solid, there are strong electrostatic interactions between the ions. The ions are arranged in such a way that the electrostatic attractions between the oppositely charged ions are greater than the electrostatic repulsions between ions with the same charge. The electrostatic interaction between ions is not directional: all that matters is the distance between two ions, not their orientation with respect to one another. (Compare this with covalent bonding in **Topic 3B**.)

When ions are present, the electrostatic interaction between them tends to be dominant. However, it is possible for there to be significant covalent interactions between ions, so you should think of pure ionic bonding as an idealised bonding situation. We will develop this concept further in **Topic 12B (Book 2: IAL)**.

THE STRENGTH OF IONIC BONDING

You can determine the strength of ionic bonding by calculating the amount of energy required in one mole of solid to separate the ions to infinity (i.e. in the gas phase). When they are at an infinite distance from one another, the ions can no longer interact.

Table A below shows the energy required to separate to infinity the ions in one mole of various alkali metal halides.

	AMOUNT OF ENERGY REQUIRED TO SEPARATE THE IONS TO INFINITY / kJ mol^{-1}			
	F^-	Cl^-	Br^-	I^-
Li^+	1031	848	803	759
Na^+	918	780	742	705
K^+	817	711	679	651
Rb^+	783	685	656	628

table A Energy required to break up a lattice of an ionic compound.

For ions of the same charge, the smaller the ions the more energy is required to overcome the electrostatic interactions between the ions and to separate them.

The size of the ions is one factor that affects the strength of ionic bonding, which in turn determines how closely packed the ions are in the lattice.

The lattice energy for lithium fluoride, Li^+F^-, is 1031 kJ mol^{-1}. The equivalent energy for magnesium fluoride, $Mg^{2+}(F^-)_2$, is 2957 kJ mol^{-1}. The radius of the Mg^{2+} ion (0.072 nm) is very similar to the radius of the Li^+ ion (0.074 nm). The increased charge of the Mg^{2+} ion compared to the Li^+ ion results in a significant increase in the strength of the ionic bonding.

When both cation and anion are doubly charged, the energy required to separate the ions is even larger. For magnesium oxide, $Mg^{2+}O^{2-}$, the value is 3791 kJ mol^{-1}.

There is no simple mathematical relationship to describe the effects that ionic radius and ionic charge have on the strength of ionic bonding. The situation is complicated by the way in which the ions pack together to form the lattice, and by the extent to which there are covalent interactions between the ions. In general, however, the smaller the ions and the larger the charge on the ions, the stronger the ionic bonding.

EVIDENCE FOR THE EXISTENCE OF IONS

Ionic compounds can conduct electricity and undergo electrolysis when either molten or in aqueous solution. This is the most convincing evidence for the existence of ions.

For example, when you pass a direct electric current through molten sodium chloride (**fig D**), sodium is formed at the negative electrode and chlorine is formed at the positive electrode.

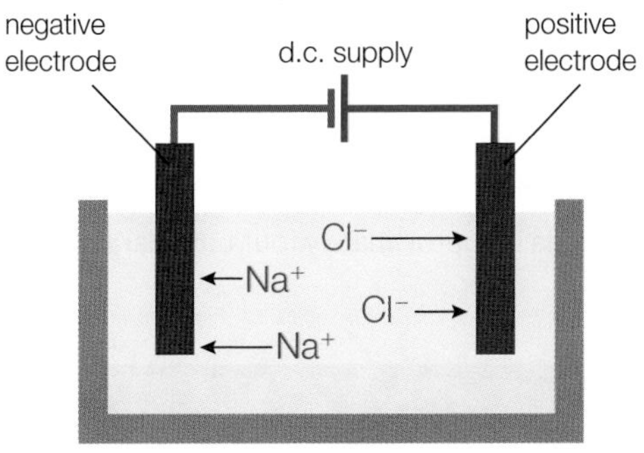

▲ **fig D** Electrolysis of molten sodium chloride.

LEARNING TIP

Avoid saying that there is an 'ionic bond' between two ions.

This is because in an ionic solid, each ion interacts with many other ions, of both the same and opposite charge to itself. The energy that binds the structure does not come from single interactions between ions (so-called ionic bonds), but from the interactions between all of the ions in the lattice. This energy is called the 'lattice energy' and you will meet this in **Topic 12B (Book 2: IAL)**.

The explanation for this phenomenon is that:

- the positive sodium ions migrate towards the negative electrode where they gain electrons and become sodium atoms
- the negative chloride ions migrate towards the positive electrode where they lose electrons and become chlorine molecules.

At the negative electrode: $2Na^+ + 2e^- \rightarrow 2Na$

At the positive electrode: $2Cl^- \rightarrow Cl_2 + 2e^-$

Overall equation: $2NaCl \rightarrow 2Na + Cl_2$

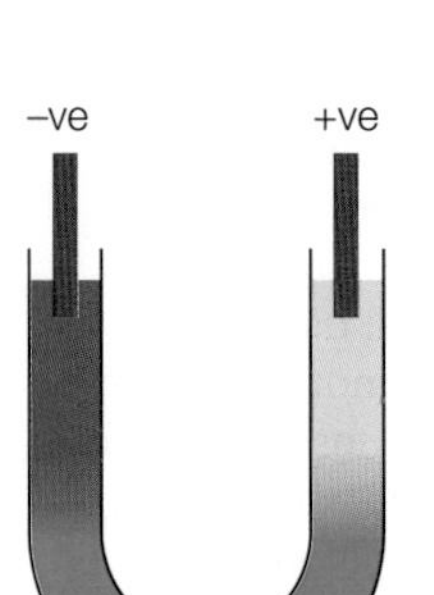

▲ **fig E** The effect of passing an electric current through aqueous copper(II) chromate.

We can demonstrate the movement of ions by passing a direct current through copper(II) chromate(VI) solution (**fig E**). Aqueous copper(II) ions, Cu^{2+}(aq), are blue and aqueous chromate(VI) ions, CrO_4^{2-}(aq), are yellow.

The Cu^{2+}(aq) ions migrate towards the negative electrode and the solution around this electrode turns blue. The CrO_4^{2-}(aq) ions migrate towards the positive terminal and the solution around this electrode turns yellow.

Further evidence for the existence of ions is supplied by electron density maps.

Fig F is an electron density map for sodium chloride.

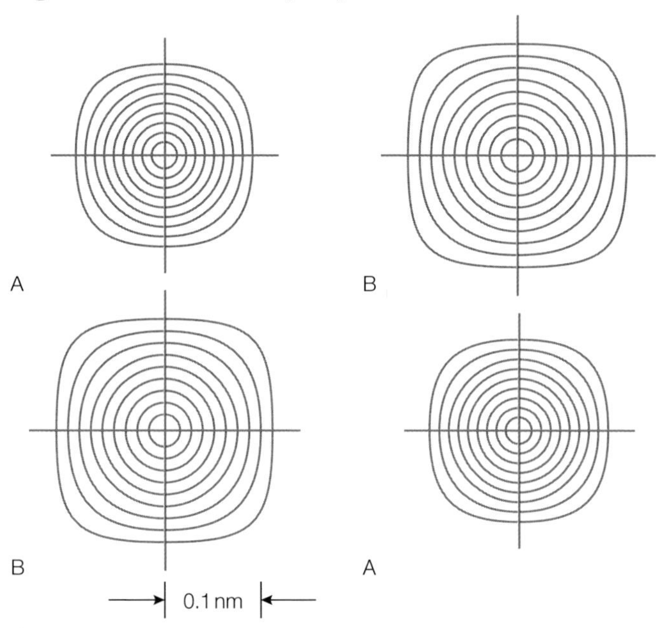

▲ **fig F** Electron density map of sodium chloride produced from X-ray diffraction patterns.

The electron density map clearly shows separate ions. A represents a sodium ion and B represents a chloride ion.

CHECKPOINT

1. Explain what is meant by the term 'ionic bonding'.
2. Calcium reacts with fluorine to form the ionic compound calcium fluoride:

   $Ca(s) + F_2(g) \rightarrow CaF_2(s)$

   Use a dot-and-cross diagram to show the electronic changes that occur in this reaction.

SKILLS REASONING

3. (a) Suggest why the strength of ionic bonding is greater in sodium fluoride than in potassium fluoride.

   (b) Suggest why the strength of ionic bonding in calcium oxide is approximately four times larger than that in potassium fluoride.

SUBJECT VOCABULARY

ionic bonding the electrostatic attraction between oppositely charged ions

3A 2 IONIC RADII AND POLARISATION OF IONS

SPECIFICATION REFERENCE 3.6 3.7 3.8 3.9

LEARNING OBJECTIVES

- Understand the effects of ionic radius and ionic charge on the strength of ionic bonding.
- Understand reasons for the trends in ionic radii down a group in the Periodic Table, and for a set of isoelectronic ions, e.g. N^{3-} to Al^{3+}.
- Understand the meaning of the term polarisation as applied to ions.

TRENDS IN IONIC RADII

Ionic radii are difficult to measure accurately, and vary according to the environment of the ion. For example, it is important how many oppositely charged ions are touching it (i.e. the co-ordination number). The nature of the ions is also important.

There are several different ways of measuring ionic radii and they all produce slightly different values. If you are going to make reliable comparisons using ionic radii, all the values must come from the same source.

Remember that there are quite large uncertainties when using ionic radii. Trying to explain things in detail is made difficult because of those uncertainties.

GROUP 1		
ION	ELECTRONIC CONFIGURATION	IONIC RADIUS / nm
Li^+	2	0.076
Na^+	2.8	0.102
K^+	2.8.8	0.138
Rb^+	2.8.18.8	0.152

GROUP 7		
ION	ELECTRONIC CONFIGURATION	IONIC RADIUS / nm
F^-	2.8	0.133
Cl^-	2.8.8	0.181
Br^-	2.8.18.8	0.196
I^-	2.8.18.18.8	0.220

table A Trends in ionic radii in Groups 1 and 7.

As you go down each group, the ions have more electron shells; therefore, the ions get larger.

PERIOD 2	N^{3-}	O^{2-}	F^-
Number of protons	7	8	9
Electronic configuration	2.8	2.8	2.8
Ionic radius/nm	0.146	0.140	0.133

PERIOD 3	Na^+	Mg^{2+}	Al^{3+}
Number of protons	11	12	13
Electronic configuration	2.8	2.8	2.8
Ionic radius/nm	0.102	0.072	0.054

table B Trends in ionic radii across a period.

All six of the ions listed in **table B** are isoelectronic. In other words, they have the same number of electrons and therefore the same electronic configuration.

The ionic radius decreases as the number of protons increases.

As the positive charge of the nucleus increases, the electrons are attracted more strongly and are therefore pulled closer to the nucleus.

POLARISATION AND POLARISING POWER OF IONS

In an ionic lattice, the positive ion will attract the electrons of the anion. If the electrons are pulled towards the cation, the anion is polarised since the even distribution of its electron density has been distorted.

The extent to which an anion is polarised by a cation depends on several factors. The two main factors are known as Fajan's rules and are summarised here.

DID YOU KNOW?

When you compare chemical values, you should make sure all the data you are comparing come from the same source. For example, the values in **tables B** and **C** have all been taken from the Database of Ionic Radii from Imperial College London. However, Chemistry Data Book (JG Stark and HG Wallace) gives a value of 0.171 for the ionic radius of the nitride ion, N^{3-}. In either case, the value is larger than that for the oxide ion, O^{2-}, as expected.

Polarisation will be increased by:
- high charge and small size of the cation (i.e. high charge density of the cation)
- high charge and large size of the anion.

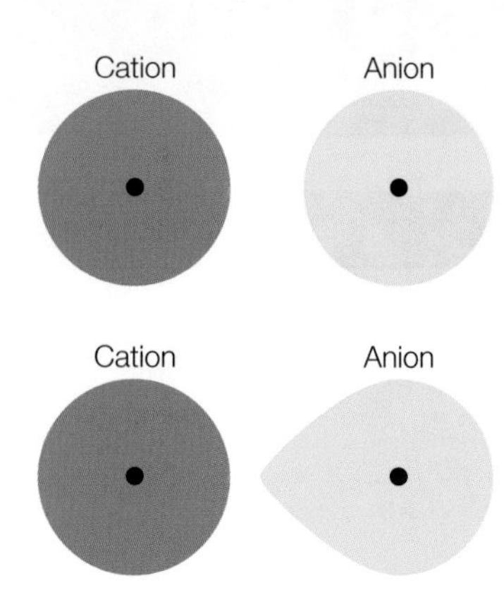

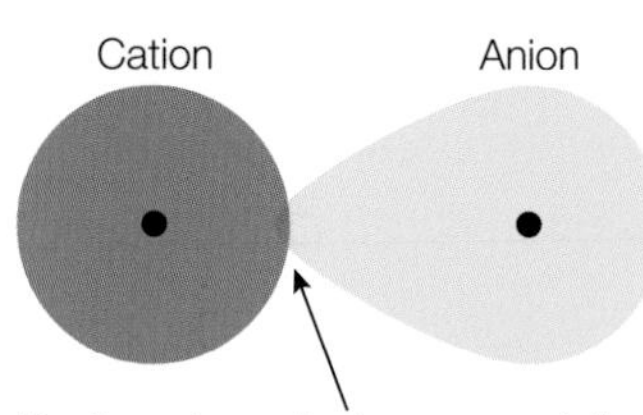

▲ **fig A** A representation of a cation attracting the electrons of an anion in an ionic lattice.

HIGH CHARGE AND SMALL SIZE OF CATIONS

The ability of a cation to attract electrons from the anion towards itself is called its **polarising power**. A cation with a high charge and a small radius has a large polarising power. An approximate value for the polarising power of a cation can be obtained by calculating its charge density. The charge density of a cation is the charge divided by the surface area of the ion. If the ion is assumed to be a sphere, its surface area is equal to $4\pi r^2$, where r is the ionic radius.

An approximation to the charge density can be determined by dividing the charge by the square of its ionic radius.

$$\text{charge density} \sim \frac{\text{charge}}{r^2}$$

HIGH CHARGE AND LARGE SIZE OF ANIONS

The ease with which an anion is polarised depends on its charge and its size. Anions with a high charge and a large size are polarised the most easily.

In an ionic lattice, the polarisation of the anions creates some degree of sharing of electrons between the two nuclei. That is, some degree of covalent bonding exists. You will learn more about this concept in **Topic 12B (Book 2: IAL)**.

CHECKPOINT

1. Explain the trend in the following ionic radii:

(a) $Ca^{2+} > Mg^{2+} > Be^{2+}$

(b) $P^{3-} > S^{2-} > Cl^-$

2. The table gives the ionic radii of some ions.

FORMULA OF ION	IONIC RADIUS / nm
Li^+	0.076
Na^+	0.102
Mg^{2+}	0.072
Al^{3+}	0.054

Arrange the ions in order of their polarising power. Show how you arrived at your answer.

SUBJECT VOCABULARY

polarising power the ability of a positive ion (cation) to distort the electron density of a neighbouring negative ion (anion)
polarisation the distortion of the electron density of a negative ion (anion)

3A 3 PHYSICAL PROPERTIES OF IONIC COMPOUNDS

SPECIFICATION REFERENCE 3.1 PART 3.4

LEARNING OBJECTIVES

- Be able to explain the physical properties of ionic compounds in terms of their bonding and structure.

Ionic compounds typically have the following physical properties:
- high melting temperatures
- brittleness
- poor electrical conductivity when solid but good when molten
- often soluble in water.

HIGH MELTING TEMPERATURES

Ionic solids consist of a giant lattice network of oppositely charged ions (see **Topic 2B.2**). There are many ions in the lattice and the combined electrostatic forces of attraction among all of the ions is large.

A large amount of energy is required to overcome the forces of attraction sufficiently for the ions to break free from the lattice and slide past one another.

BRITTLENESS

If a stress is applied to a crystal of an ionic solid, then the layers of ions may slide over one another.

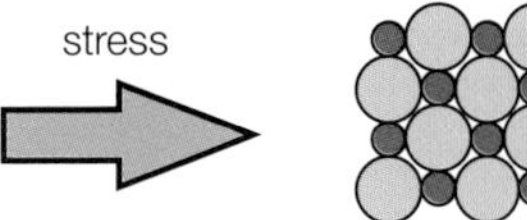

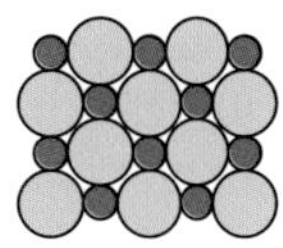
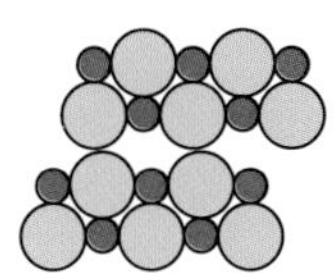
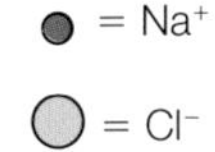

fig A Effect of stress on an ionic crystal.

Ions of the same charge are now side by side and repel one another. The crystals break apart.

ELECTRICAL CONDUCTIVITY

Solid ionic compounds do not, in general, conduct electricity. This is because there are no delocalised electrons and the ions are also not free to move under the influence of an applied potential difference.

However, molten ionic compounds *will* conduct since the ions are now mobile and will migrate to the electrodes of opposite sign when a potential difference is applied. If direct current is used, the compound will undergo electrolysis as the ions are discharged at the electrodes.

EXAM HINT

Remember, oxidation always takes place at the anode.

Aqueous solutions of ionic compounds also conduct electricity and undergo electrolysis, since the lattice breaks down into separate ions when the compound dissolves.

DID YOU KNOW?

As nearly always in chemistry, there are exceptions. Some ionic compounds do conduct electricity when solid. For example, solid lithium nitride (Li_3N) will conduct electricity and is used in batteries for this reason.

SOLUBILITY

Many ionic compounds are soluble in water. We will explain this solubility more fully in **Topic 12A (Book 2: IAL)**, including the part played by entropy changes.

At the moment, you just need to understand that the energy required to break apart the lattice structure and separate the ions can, in some instances, be supplied by the **hydration** of the separated ions produced. Both positive and negative ions are attracted to water molecules because of the polarity that water molecules possess (see **Topic 3B** for an explanation of polarity).

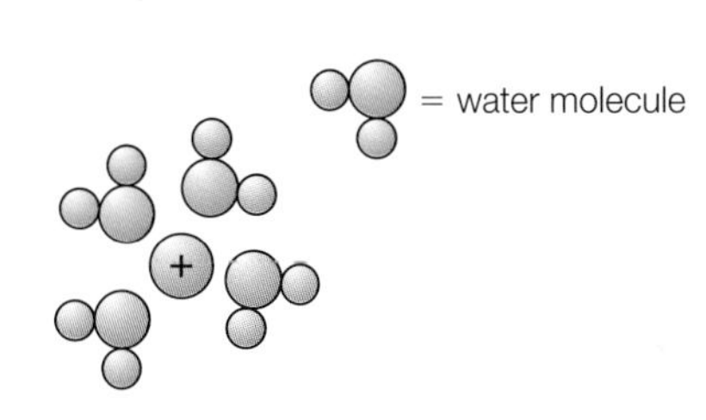

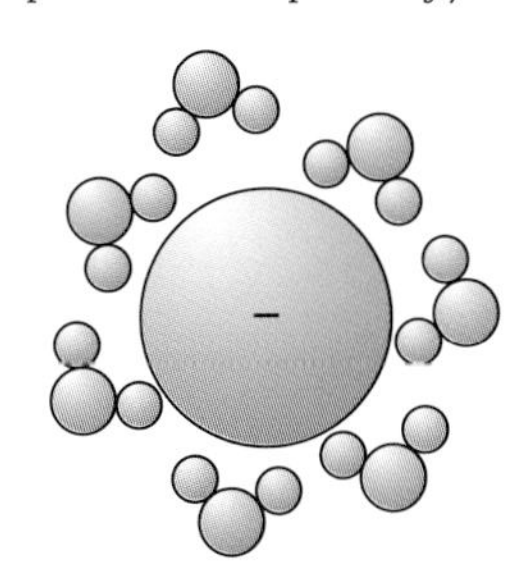

The oxygen ends of the water molecules are attracted to positive ions.

The hydrogen ends of the water molecules are attracted to negative ions.

fig B Hydration of ions.

CHECKPOINT

1. Explain why sodium chloride:
   (a) does not conduct electricity when solid, but does when molten
   (b) has a high melting temperature
   (c) is soluble in water.

SUBJECT VOCABULARY

hydration the process of water molecules being attracted to ions in solution and surrounding the ions; the oxygen ends of the water molecules are attracted to the positive ions (cations); the hydrogen ends of the water molecules are attracted to the negative ions (anions); hydration of ions is an exothermic process (i.e. heat energy is released)

3B 1 COVALENT BONDING

SPECIFICATION REFERENCE 3.10

LEARNING OBJECTIVES

- Know that a covalent bond is formed by the overlap of two atomic orbitals each containing a single electron.
- Know that a covalent bond is the strong electrostatic attraction between the nuclei of two atoms and the bonding (shared) pair of electrons.
- Understand the relationship between bond length and bond strength for covalent bonds.

FORMATION OF COVALENT BONDS

A covalent bond forms between two atoms when an atomic orbital containing a single electron from one atom overlaps with an atomic orbital, which also contains a single electron, of another atom. The two electrons in the area of overlap are the bonding electrons. They are sometimes referred to as a 'shared pair of electrons'. The covalent bond is the electrostatic attraction between the two nuclei of the bonded atoms and the pair of electrons shared between them.

The atomic orbitals involved can be any of those found in the atoms, but we shall limit our discussion to those involving only s- and p-orbitals.

Fig A shows three ways in which these orbitals may overlap.

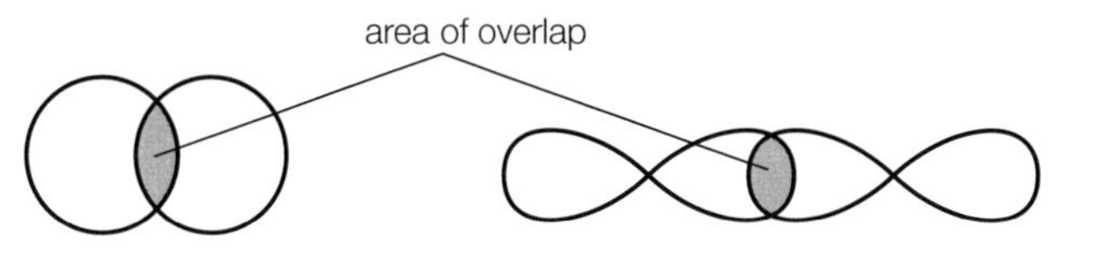

▲ **fig A** Formation of sigma bonds by end-on overlap of atomic orbitals and a pi bond by sideways overlap of p-orbitals.

An end-on overlap leads to the formation of sigma (σ) bonds. This leads to the formation of a single covalent bond between the two atoms.

A sideways overlap of two p-orbitals leads to the formation of a pi (π) bond. A feature of a π bond is that it cannot form until a σ bond has been formed. For this reason π bonds only exist between atoms that are joined by double or triple bonds.

The different types of orbital overlap are shown in the following examples.

EXAMPLE 1. HYDROGEN

A hydrogen atom has an electronic configuration of $1s^1$.

When two hydrogen atoms bond together to form a hydrogen molecule, the two s-orbitals overlap to form a new molecular orbital. The two electrons then exist in this new orbital. The highest electron density is between the two nuclei.

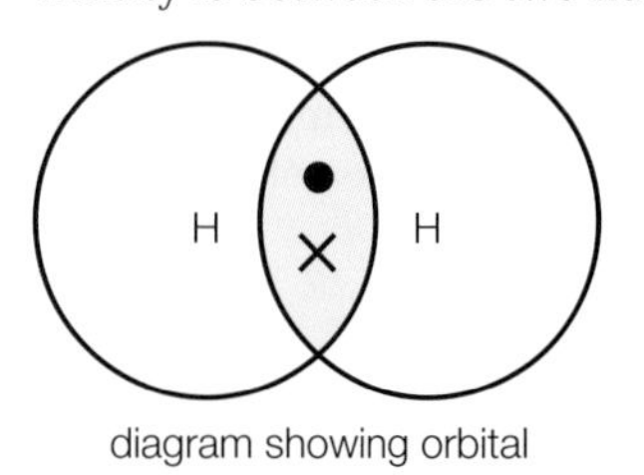

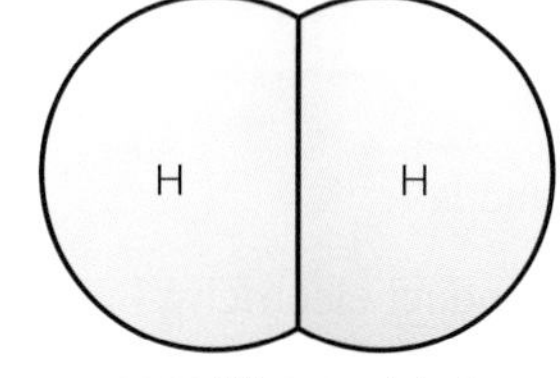

▲ **fig C** Formation of the σ bond in hydrogen.

DID YOU KNOW?

There is a fourth way that overlap can occur. An s- and p-orbital can overlap end on.

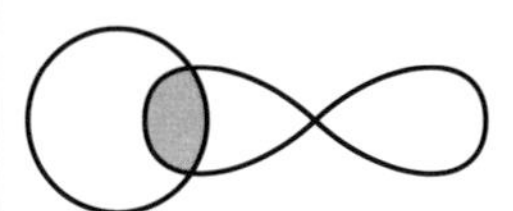

▲ **fig B** Overlap of an s- and a p-orbital to form a sigma bond.

This overlap, however, can only result from atoms of two different elements. This almost always leads to the formation of a variant of a covalent bond, known as a 'polar' covalent bond (see Topic 3B.2).

EXAMPLE 2. CHLORINE

A chlorine atom has an electronic configuration of $1s^2\ 2s^2\ 2p^6\ 3s^2\ 3p_x^{\ 2}\ 3p_y^{\ 2}\ 3p_z^{\ 1}$.

When two chlorine atoms bond together, the two p orbitals (each containing a single electron) overlap.

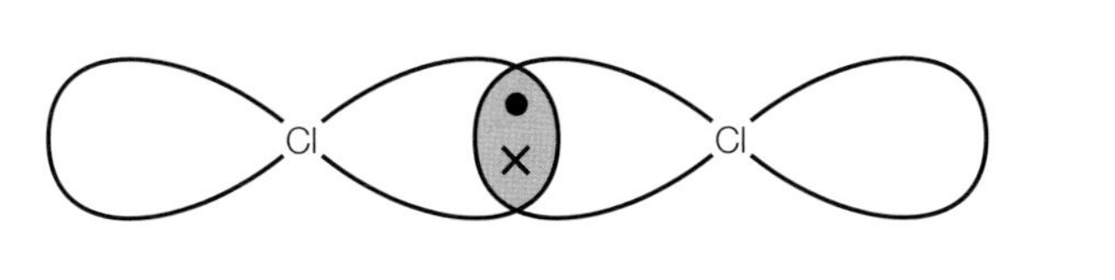

diagram showing the orbital overlap

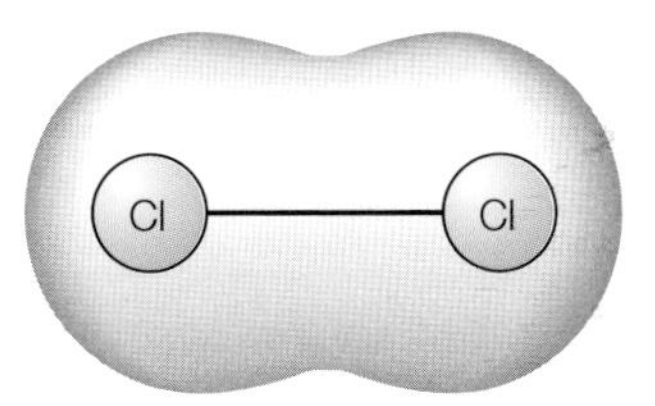

space filling model of a chlorine molecule

▲ **fig D** Formation of the σ bond in chlorine.

> **DID YOU KNOW?**
> This view of the bonding in chlorine is very simple.
> An alternative theory describes the bonding as the overlap between two sp^3 hybrid orbitals. This theory is not included in this book, but we would encourage you to carry out your own research if you are interested. Look up 'orbital hybridisation'.

EXAMPLE 3. π BOND FORMATION

Once a σ bond has been formed, it is possible, in certain circumstances, for a π bond to form.

The π bond results in a high electron density both above and below the molecule, as shown in **fig E**.

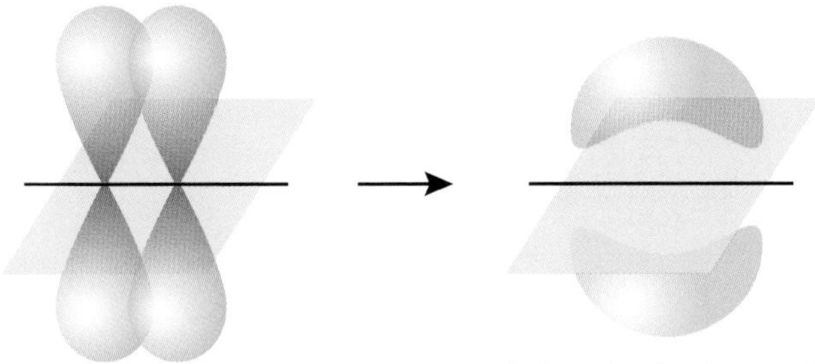

sideways overlap of p-orbitals

electron density above and below the molecule

▲ **fig E** Formation of a π bond.

This is what happens in the ethene molecule. One of the bonds between the carbon atoms is a σ bond; the other is a π bond.

The π bond in ethene is weaker than the σ bond. This is the reason for the increased reactivity of alkenes compared with alkanes, and why alkenes can easily undergo addition reactions. (See **Topic 5** for more information.)

The triple bond in the nitrogen molecule (N≡N) is made up of one σ bond and two π bonds.

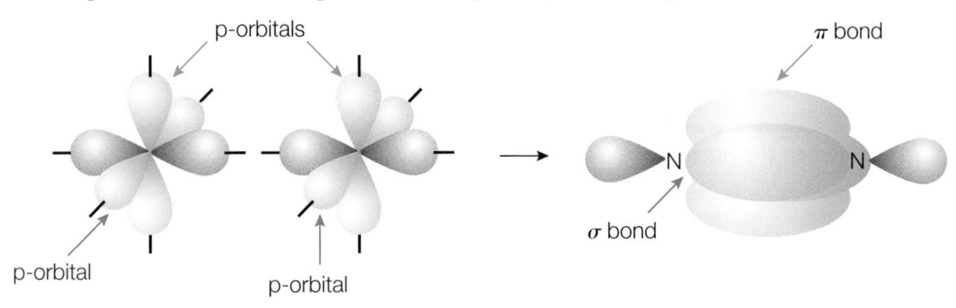

The p-orbitals marked are those that are involved in the formation of the pi bonds.

▲ **fig F** Formation of the two π bonds in nitrogen.

BOND LENGTH AND BOND STRENGTH

The **bond length** is the distance between the nuclei of two atoms that are covalently bonded together.

The strength of a covalent bond is measured in terms of the amount of energy required to break one mole of the bond in the gaseous state (see **Topic 6**).

Table A shows the relationship between the bond length and bond strength of a selection of covalent bonds.

LEARNING TIP

When making a comparison between bond length and bond strength, it is important to compare 'like with like', in other words, compare things that are the same or very similar.

For example, the strength of the C–C bond (347 kJ mol^{-1}) is greater than that of the N–N bond (158 kJ mol^{-1}) despite being longer (0.154 nm compared with 0.145 nm).
In a molecule such as hydrazine ($H_2N–NH_2$), each nitrogen atom has a non-bonding (lone) pair of electrons and these repel one another, weakening the bond.
In a molecule such as ethane ($H_3C–CH_3$), the carbon atoms do not have any lone pairs.

BOND	BOND LENGTH / nm	BOND STRENGTH / kJ mol^{-1}
Cl-Cl	0.199	242
Br–Br	0.228	193
I–I	0.267	151
C-C	0.154	347
C=C	0.134	612
C≡C	0.120	838
N-N	0.145	158
N=N	0.120	410
N≡N	0.110	945
O-O	0.148	144
O=O	0.121	498

table A Relationship between bond length and bond strength for a range of covalent bonds.

The general relationship between bond length and bond strength, for bonds that are of a similar nature, is the shorter the bond, the greater the bond strength. This is a result of an increase in electrostatic attraction between the two nuclei and the electrons in the overlapping atomic orbitals.

CHECKPOINT

1. Suggest a reason for the following trend in bond strengths:

   C–C > Si–Si > Ge–Ge
2. The F–F bond in fluorine is much shorter (0.142 nm) than the Cl–Cl (0.199 nm) bond in chlorine, and yet it is much weaker (158 kJ mol^{-1} compared with 242 kJ mol^{-1}). Suggest a reason for this.
3. Suggest a reason why the sigma (σ) bond between the two carbon atoms in the ethene molecule is stronger that the pi (π) bond.

SUBJECT VOCABULARY

bond length the distance between the nuclei of two atoms that are covalently bonded together

2 ELECTRONEGATIVITY AND BOND POLARITY

SPECIFICATION REFERENCE
3.10(ii) 3.13 3.14

LEARNING OBJECTIVES

- Know that electronegativity is the ability of an atom to attract a bonding pair of electrons.
- Know that ionic and covalent bonding are the extremes of a continuum of bonding type and that electronegativity differences lead to bond polarity.
- Understand what a polar covalent bond is.
- Understand that electron density maps for discrete (simple) molecules show that there is a high electron density between the nuclei of two covalently bonded atoms.

WHAT IS ELECTRONEGATIVITY?

Electronegativity is the ability of an atom to attract a bonding pair of electrons.

The electronegativity of elements, in general:

- decreases down a group of the Periodic Table, that is, from top to bottom
- increases from left to right across a period.

This is demonstrated in the following section of the Periodic Table (**fig A**).

									H 2.1								He
Li 1.0	Be 1.5											B 2.0	C 2.5	N 3.0	O 3.5	F 4.0	Ne
Na 0.9	Mg 1.2											Al 1.5	Si 1.8	P 2.1	S 2.5	Cl 3.0	Ar
K 0.8	Ca 1.0	Sc 1.3	Ti 1.5	V 1.6	Cr 1.6	Mn 1.5	Fe 1.8	Co 1.8	Ni 1.8	Cu 1.9	Zn 1.6	Ga 1.6	Ge 1.8	As 2.0	Se 2.4	Br 2.8	Kr
Rb 0.8	Sr 1.0	Y 1.2	Zr 1.4	Nb 1.6	Mo 1.8	Tc 1.9	Ru 2.2	Rh 2.2	Pd 2.2	Ag 1.9	Cd 1.7	In 1.7	Sn 1.8	Sb 1.9	Te 2.1	I 2.5	Xe
Cs 0.7	Ba 0.9	La 1.1	Hf 1.3	Ta 1.5	W 1.7	Re 1.9	Os 2.2	Ir 2.2	Pt 2.2	Au 2.4	Hg 1.9	Tl 1.8	Pb 1.8	Bi 1.9	Po 2.0	At 2.2	Rn

▲ **fig A** able of electronegativities.

DISTRIBUTION OF ELECTRON DENSITY

If two atoms of the same element are bonded together by the overlap of atomic orbitals, the distribution of electron density between the two nuclei will be symmetrical. This is because the ability of each atom to attract the bonding pair of electrons is identical.

The diagram in **fig B** is an electron density map for chlorine (Cl_2):

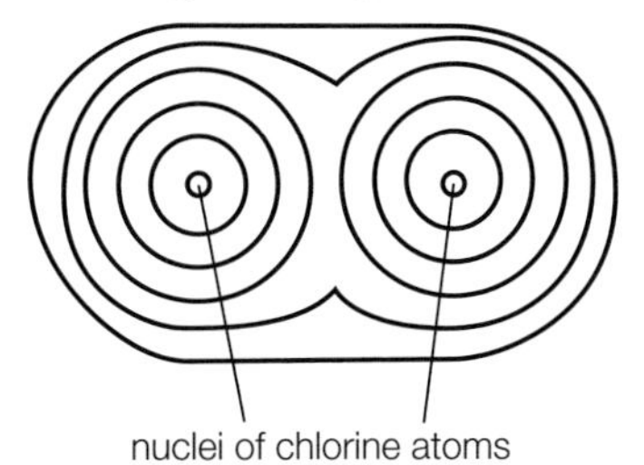

▲ **fig B** Electron density map of a chlorine molecule.

The diagram looks like a contour map. The contour lines correspond to electron density. You can think of them as showing how likely it is that a bonding electron will fall within that contour at a given instant in time. For a normal covalent bond, the contours are symmetrical around the nuclei.

POLAR COVALENT BONDS

However, if the two atoms bonded together are from elements that have different electronegativities, then the distribution of electron density will not be symmetrical about the two nuclei. This is shown in **fig C** by the electron density map for the hydrogen chloride (HCl) molecule.

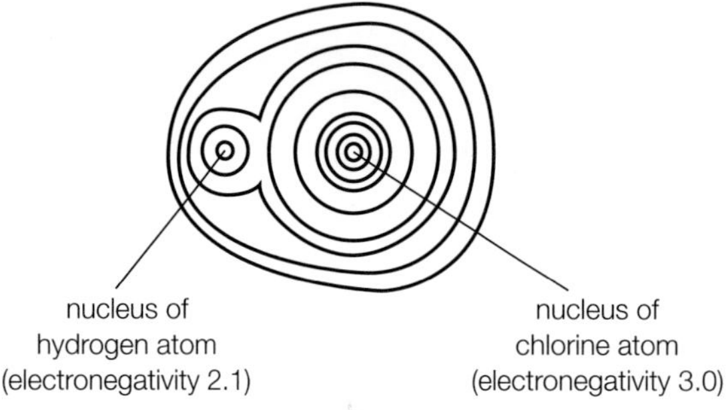

▲ **fig C** Electron density map of a hydrogen chloride molecule.

The contour lines are more closely spaced near to the chlorine atom, which is the atom with the higher electronegativity.

LEARNING TIP

Covalent bonds that are non-polar are sometimes called 'normal' covalent bonds or 'pure' covalent bonds. This is to distinguish them from polar covalent bonds.

Both types of bond are formed by the overlap of atomic orbitals.

In both cases the overlapping area contains two electrons per bond formed.

Since the electron density is higher around the chlorine atom, that end of the molecule has acquired a slightly negative charge. This is represented by the symbol δ–. The other end of the molecule carries a slightly positive charge, represented by the symbol δ+.

$$H^{\delta+} — Cl^{\delta-}$$

A bond like this is called a **polar covalent bond** or sometimes just a 'polar bond'.

Another way of representing a polar covalent bond is to use an arrow to show the direction of electron drift.

$$H \rightarrow\!\!- Cl$$

Other examples of polar covalent bonds are:

$C^{\delta+} \rightarrow\!\!- Cl^{\delta-}$ $H^{\delta+} \rightarrow\!\!- O^{\delta-}$ $H^{\delta+} \rightarrow\!\!- N^{\delta-}$ $H^{\delta+} \rightarrow\!\!- C^{\delta-}$

CONTINUUM OF BONDING TYPE

Polar covalent bonds can be thought of as being between two ideals of bonding types. These ideals are:
- pure (100%) covalent
- pure (100%) ionic.

Consider a polar covalent bond as a covalent bond that has some degree of ionic character.

If the electronegativity difference is large enough, then the main type of bonding is ionic.

A very approximate measure of the degree of ionic bonding in a compound is given in **table A**.

ELECTRONEGATIVITY DIFFERENCE	0	0.1	0.2	0.3	0.4	0.5	0.6	0.7	0.8	0.9	1.0	1.1
APPROXIMATE % IONIC CHARACTER	0	0.5	1	2	4	6	9	12	15	19	22	26

ELECTRONEGATIVITY DIFFERENCE	1.2	1.3	1.4	1.5	1.6	1.7	1.8	1.9	2.0	2.1	2.2	2.3
APPROXIMATE % IONIC CHARACTER	30	34	39	43	47	51	55	59	63	67	70	74

ELECTRONEGATIVITY DIFFERENCE	2.4	2.5	2.6	2.7	2.8	2.9	3.0	3.1	3.2	3.3
APPROXIMATE % IONIC CHARACTER	76	79	82	84	86	88	89	90	91	92

table A Relationship between percentage ionic character and difference in electronegativity.

CHECKPOINT

1. Suggest why the electronegativity of fluorine is greater than that of chlorine, despite the fact that the nucleus of a chlorine atom contains more protons.

2. Ionic bonding and covalent bonding are two extremes of chemical bonding. Many compounds have bonding that is intermediate in character.

(a) Giving an example in each case, explain what is meant by the terms:

(i) ionic bonding, and (ii) covalent bonding.

(b) Select a compound that has bonding of an intermediate character and explain why it has this type of bonding.

3. Place the following bonds in order of decreasing polarity (i.e. place the most polar first).

C–Br C–Cl C–F C–I

Explain how you arrived at your answer.

SUBJECT VOCABULARY

electronegativity the ability of an atom to attract a bonding pair of electrons in a covalent bond

polar covalent bond a type of covalent bond between two atoms where the bonding electrons are unequally distributed; because of this, one atom carries a slight negative charge and the other a slight positive charge

3 BONDING IN DISCRETE (SIMPLE) MOLECULES

LEARNING OBJECTIVES

- Understand what is meant by the term discrete (simple) molecule.
- Draw dot-and-cross diagrams to show electrons in discrete molecules with single, double and triple bonds.
- Draw displayed formulae to represent the bonding in discrete molecules.

DID YOU KNOW?

Molecules are common in organic substances (and therefore biochemistry). They also make up most of the oceans and the atmosphere.

However, ionic crystals (salts) and giant covalent crystals (network solids) are often made up of repeating unit cells that extend either in a plane (such as in graphene) or three-dimensionally (such as in diamond or sodium chloride). These substances are not composed of molecules. Solid metals are also not composed of molecules.

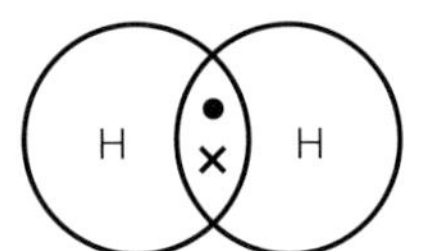

fig A Dot-and-cross diagram for hydrogen with overlapping circles.

DISCRETE MOLECULES

A **discrete (simple) molecule** is an electrically neutral group of two or more atoms held together by covalent bonds.

DOT-AND-CROSS DIAGRAMS

Covalent and polar covalent bonding in discrete molecules can be shown by dot-and-cross diagrams.

Fig A shows the example of hydrogen, H_2.

Further examples of dot-and-cross diagrams are shown in **table A**.

SUBSTANCE	DOT-AND-CROSS DIAGRAM	
Water, H_2O	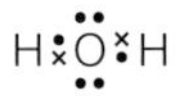 H O H	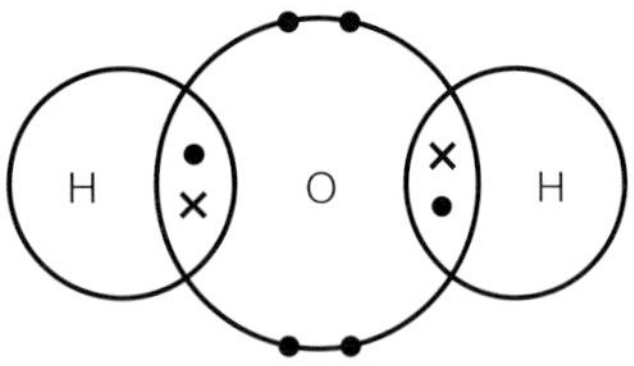 H O H
Ammonia, NH_3	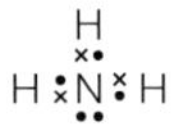 H H N H nucleus of hydrogen atom (electronegativity 2.1)	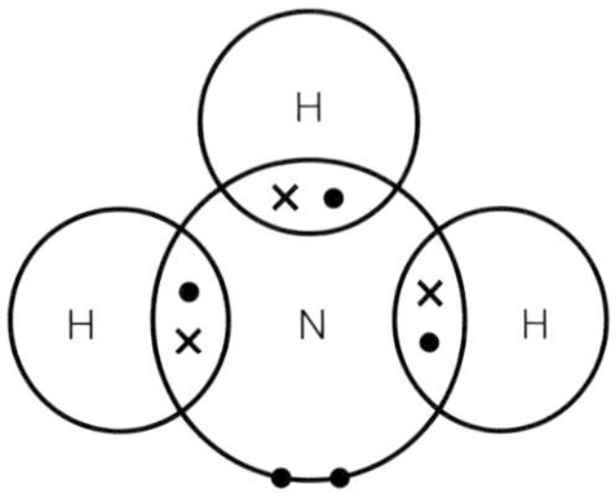 H H N H
Methane, CH_4	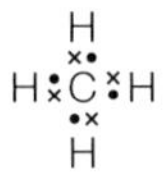 H H C H H	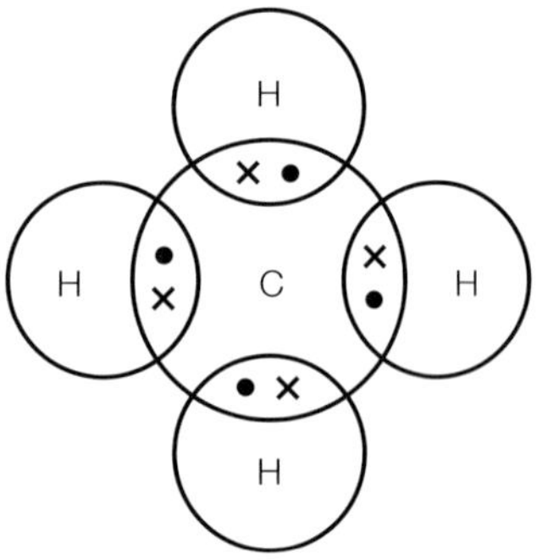 H H C H H

table A Dot-and-cross diagrams for water, ammonia and methane.

EXAM HINT

When drawing a molecule such as chloromethane, do not forget to show all of the non-bonding electrons on the chlorine atom.

THE OCTET RULE

You might read that in order to form a stable compound, the outer shell of each atom must have the same number of electrons as the outer shell of a noble gas. In most cases this will be eight electrons. This has led to a rule that is often referred to as the 'octet rule'.

This is not always true, as you can see from the examples in **table B**. In each case, the outer shell of the central atom of the molecule does not contain eight electrons.

SUBSTANCE	DOT-AND-CROSS DIAGRAM	NUMBER OF ELECTRONS AROUND CENTRAL ATOM
Beryllium chloride, $BeCl_2$	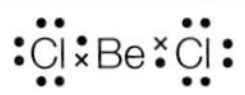 Cl Be Cl	4
Boron trichloride, BCl_3	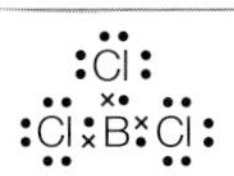 Cl / Cl B Cl	6
Phosphorus(V) chloride, PCl_5	 Cl Cl / Cl P Cl / Cl	10
Sulfur hexafluoride, SF_6	 F / F F / S / F F / F	12

table B Examples breaking the octet rule.

DOT-AND-CROSS DIAGRAMS OF MOLECULES CONTAINING MULTIPLE BONDS

Table C shows the dot-and-cross diagrams for three molecules (O_2, N_2, CO_2) that contain a double or triple bond.

DISPLAYED FORMULAE (FULL STRUCTURAL FORMULAE)

A **displayed (full structural) formula** shows each bonding pair as a line drawn between the two atoms involved.

Table C gives some examples of dot-and-cross diagrams together with the displayed formulae.

SUBSTANCE	DOT-AND-CROSS DIAGRAM	DISPLAYED FORMULA
Water, H_2O	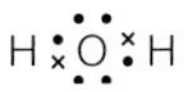 H O H	H–O–H
Ammonia, NH_3	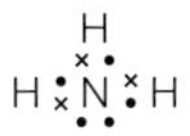 H / H N H	H / | / H–N–H
Oxygen, O_2	 O O	O=O
Nitrogen, N_2	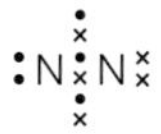 N N	N≡N
Carbon dioxide, CO_2	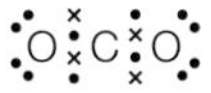 O C O	O=C=O

table C Examples of displayed formulae with the corresponding dot-and-cross diagram.

CHECKPOINT

1. Draw a dot-and-cross diagram for each of the following molecules:
   (a) H_2S (b) PH_3 (c) PF_3 (d) SCl_2 (e) AsF_5 (f) HCN (g) SO_2
2. Draw the displayed formula for each of the molecules in Question 1.

LEARNING TIP

Although it is essential to show all of the non-bonding (lone) pairs of electrons in a dot-and-cross diagram, it is not necessary to show them in a displayed formula.

SUBJECT VOCABULARY

discrete (simple) molecule an electrically neutral group of two or more atoms held together by covalent bonds

displayed (full structural) formula a formula that shows each bonding pair as a line drawn between the two atoms involved

3B 4 DATIVE COVALENT BONDS

SPECIFICATION REFERENCE 3.11(ii)

LEARNING OBJECTIVES

- Know that in a dative covalent bond both electrons in the bond are supplied by only one of the atoms involved in forming the bond.
- Be able to draw dot-and-cross diagrams for some molecules and ions that contain dative covalent bonds, including Al_2Cl_6 and the ammonium ion.

DATIVE COVALENT BOND FORMATION

A **dative covalent bond** is formed when an empty orbital of one atom overlaps with an orbital containing a non-bonding pair (lone pair) of electrons of another atom.

The bond is often represented by an arrow from the atom providing the pair of electrons, to the atom with the empty orbital. Below are three examples of dative covalent bonds.

THE HYDROXONIUM ION, H_3O^+

The dot-and-cross diagram and the displayed formula of a hydroxonium ion are shown in **fig A**.

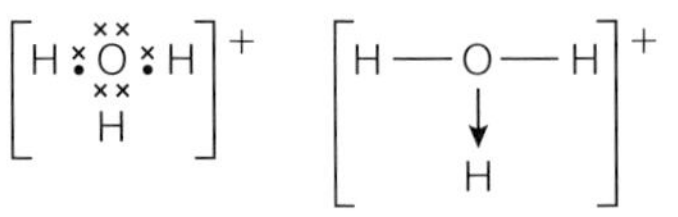

▲ **fig A** Dot-and-cross diagram and displayed formula for the hydroxonium ion.

The empty 1s orbital of the H^+ ion overlaps with the orbital of the oxygen atom that contains the lone pair of electrons.

THE AMMONIUM ION, NH_4^+

The dot-and-cross diagram and the displayed formula of an ammonium ion are shown in **fig B**.

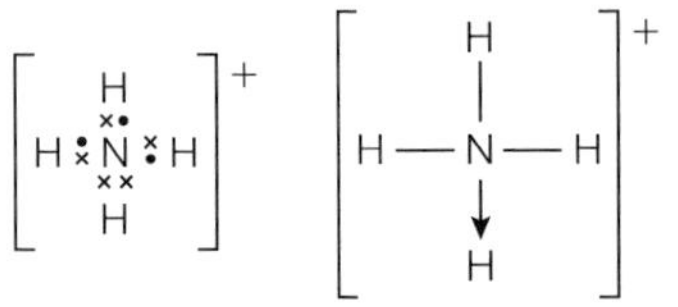

▲ **fig B** Dot-and-cross diagram and displayed formula for the ammonium ion.

The empty 1s orbital of the H^+ ion overlaps with the orbital of the nitrogen atom that contains the lone pair of electrons.

ALUMINIUM CHLORIDE, Al_2Cl_6

The aluminium atom in the $AlCl_3$ molecule has only six electrons in its outer shell and so has an empty orbital (**fig C**).

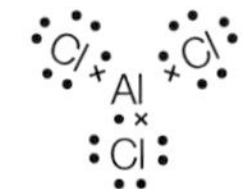

▲ **fig C** Dot-and-cross diagram for aluminium chloride.

In the gas phase, just above its sublimation temperature, aluminium chloride exists as Al_2Cl_6 molecules (**fig D**).

Two $AlCl_3$ molecules bond together. One of the atomic orbitals of a chlorine atom of one $AlCl_3$ molecule that contains a lone pair overlaps with the empty orbital of the aluminium atom of a second $AlCl_3$ molecule. The same happens between the chlorine atom of the second molecule and the aluminium atom of the first molecule.

One chlorine atom from each molecule acts as a bridge connecting the two molecules with dative covalent bonds.

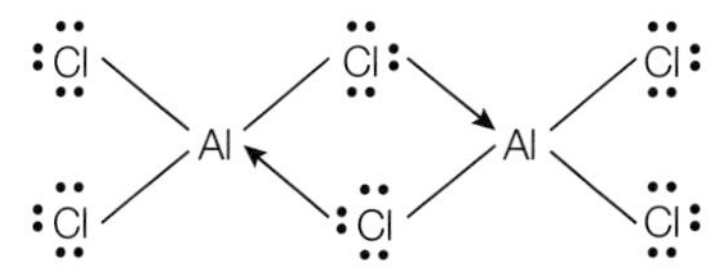

fig D Displayed formula for the aluminium dimer.

CHECKPOINT

1. (a) Draw a dot-and-cross diagram for a molecule of NH_3 and a molecule of BF_3.
   (b) Draw a dot-and-cross diagram for a molecule of $NH_3{\cdot}BF_3$.
2. Draw a dot-and-cross diagram and displayed formula for the $AlCl_4^-$ ion and identify the dative covalent bond.
3. One way of describing the bonding in a molecule of carbon monoxide (CO) is to state that it contains two covalent bonds and one dative bond. Using this description, draw a dot-and-cross diagram and displayed formula for a molecule of carbon monoxide.

SUBJECT VOCABULARY

dative covalent bond the bond formed when an empty orbital of one atom overlaps with an orbital containing a lone pair of electrons of another atom

3C 1 SHAPES OF MOLECULES AND IONS

SPECIFICATION REFERENCE 3.16 3.17 PART 3.18 3.19

LEARNING OBJECTIVES

- Understand the principles of the electron-pair repulsion theory, used to interpret and predict the shapes of simple molecules and ions.
- Understand the term bond angle.
- Know and be able to explain the shapes of, and bond angles in, $BeCl_2$, BCl_3, CH_4, NH_3, NH_4^+, H_2O, CO_2, gaseous PCl_5, SF_6 and C_2H_4.
- Be able to apply the electron-pair repulsion theory to predict the shapes of, and bond angles in, molecules and ions analogous to those mentioned above.

ELECTRON PAIR REPULSION THEORY

The **electron pair repulsion (EPR) theory** states that:

- the shape of a molecule or ion is caused by repulsion between the pairs of electrons, both bond pairs and lone (non-bonding) pairs, that surround the central atom
- the electron pairs arrange themselves around the central atom so that the repulsion between them is at a minimum
- lone pair–lone pair repulsion > lone pair–bond pair repulsion > bond pair–bond pair repulsion.

LEARNING TIP

This theory is sometimes also called the valence shell electron pair repulsion theory, abbreviated to VSEPR.

The first two rules are used to obtain the basic shape of the molecule or ion. The third rule is used to estimate values for the bond angles.

THE SHAPES OF MOLECULES AND IONS

To obtain the shape of a molecule or ion it is first necessary to obtain the number of bond pairs and lone pairs of electrons around the central atom.

The easiest way to do this is by drawing a dot-and-cross diagram. You can then apply the guidelines listed in **table A**.

MOLECULES WITH MULTIPLE BONDS

To determine the shape of a molecule containing one or more multiple bonds, treat each multiple bond as if it contained only one pair of electrons.

EXAMPLE 1. CARBON DIOXIDE, CO_2

The displayed formula for carbon dioxide is O=C=O. There are no lone pairs on the carbon atom.

If each double bond is treated as an electron pair, then the molecule is linear, like $BeCl_2$.

NUMBER OF BOND PAIRS	NUMBER OF LONE PAIRS	SHAPE	EXAMPLE
2	0	linear	Cl—Be—Cl
3	0	trigonal planar	Cl, B, Cl, Cl
4	0	tetrahedral	H, C, H, H, H
5	0	trigonal bipyramidal	Cl, Cl, Cl—P, Cl, Cl
6	0	octahedral	F, F, F, S, F, F, F
3	1	trigonal pyramidal	N, H, H, H
2	2	V-shaped	O, H, H

table A Shapes of molecules.

EXAMPLE 2. ETHENE, C_2H_4

The displayed formula of ethene is:

```
H   H
|   |
C = C
|   |
H   H
```

There are no lone pairs on either carbon atom.

Treating each double bond as an electron pair produces a planar molecule with 120° bond angles.

```
H       H
 \     /
  C = C
 /     \
H       H
```

THE BOND ANGLES IN MOLECULES AND IONS

Table B shows the bond angles of a range of molecules and ions.

Linear, e.g. $BeCl_2$ The bond angle is 180°.	Trigonal planar, e.g. BCl_3 The bond angle is 120°.
Tetrahedral, e.g. CH_4 The bond angle is 109.5°.	Trigonal pyramidal, e.g. NH_3 The bond angle is 107°. Lone pair–bond pair repulsion is greater than bond pair–bond pair repulsion, so the angle is slightly less than 109.5°.
V-shaped, e.g. H_2O The bond angle is 104.5°. Lone pair–lone pair repulsion is greater than lone pair–bond pair repulsion, so the bond angle is even further depressed from 109.5°, and is slightly less than the 107° in NH_3.	Trigonal bipyramidal, e.g. PCl_5 There are two bond angles: 90° and 120°.
Octahedral, e.g. SF_6 There are two bond angles: 90° and 180°. The angle between the bonds of two fluorine atoms opposite one another is 180°.	Tetrahedral, e.g. NH_4^+ As with CH_4, the bond angles are 109.5°. Note the change from 107° in ammonia to 109.5° in the ammonium ion.

table B The bond angles of a range of molecules and ions.

CHECKPOINT

1. (a) Draw a diagram to show the shape of each of the following molecules:
   (i) H_2S (ii) PH_3 (iii) PF_3 (iv) SCl_2 (v) AsF_5 (vi) HCN (vii) SO_2
   (b) Give the name of each shape.
2. Solid phosphorus pentachloride has the formula $[PCl_4]^+(PCl_6]^-$.
   (a) Draw a diagram to show the shape of each ion.
   (b) State the bond angles present in each ion.
3. Two possible ways of arranging the bonding pairs and lone pairs of electrons in a molecule of XeF_4 are:

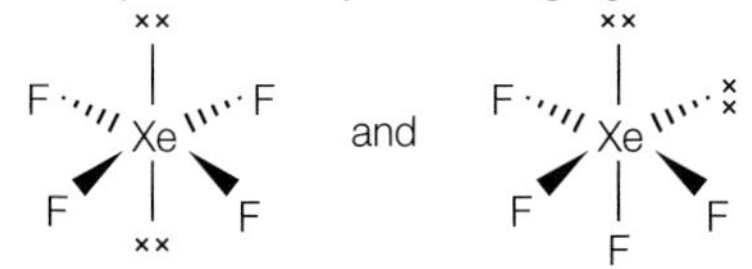

and

Suggest which of these two arrangements is the more likely and justify your answer.

SUBJECT VOCABULARY

electron pair repulsion (EPR) theory the electron pairs on the central atom of a molecule or ion arrange themselves in order to create the minimum repulsion between them; lone pair-lone pair repulsion is greater than lone pair-bond repulsion, which in turn is greater than bond pair-bond pair repulsion

3C 2 NON-POLAR AND POLAR MOLECULES

SPECIFICATION REFERENCE 3.15 PART

LEARNING OBJECTIVES

- Understand the difference between non-polar and polar molecules and be able to predict whether or not a given molecule is likely to be polar.

SHAPE AND POLARITY

The drift of bonded electrons towards the more electronegative element (see **Topic 3B.2**) results in a separation of charge. This separation of charge is called a **dipole**.

Each of the bonds in a molecule has its own dipole associated with it. The overall dipole of a molecule depends on its shape. Depending on the relative angles between the bonds, the individual dipoles can either reinforce one another or cancel out each other.

- If the cancellation is complete, the resulting molecule will have no overall dipole and is said to be 'non-polar'.
- If the dipoles reinforce one another, the molecule will possess an overall dipole and is said to be 'polar'.

DIATOMIC MOLECULES

Hydrogen and chlorine are examples of diatomic molecules that are non-polar. The two atoms in each molecule are the same and so have the same electronegativity. The distribution of electron density of the bonding electrons in either molecule is totally symmetrical (see **Topic 3B.2**). The bond in each is therefore non-polar, making the molecules non-polar.

However, the bond in the hydrogen chloride molecule is polar because the electronegativity of chlorine (3.0) is greater than that of hydrogen (2.1).

$H^{\delta+}$—$Cl^{\delta-}$

Since this is the only polar bond in the molecule, the molecule itself is polar.

The following symbol is used to represent a dipole: ⟶ (crossed arrow)

The dipole in the hydrogen chloride molecule is shown as:

$H^{\delta+}$ — $Cl^{\delta-}$
⟶

POLYATOMIC MOLECULES

1. LINEAR MOLECULES

Example: carbon dioxide, CO_2

Both bonds in the carbon dioxide molecule are polar, but the dipoles cancel out one another.

$O^{\delta-}{=}C^{\delta+}{=}O^{\delta-}$
⟵ ⟶

▲ **fig A** Dipoles in carbon dioxide.

The carbon dioxide molecule is therefore non-polar.

2. TRIGONAL PLANAR MOLECULES

Example: boron chloride, BCl_3

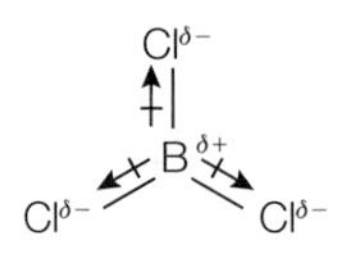

▲ **fig B** Dipoles in boron trichloride.

All three B—Cl bonds are polar, but because the molecule is symmetrical the dipoles cancel out one another. The molecule is non-polar.

3. TETRAHEDRAL MOLECULES

Example 1: tetrachloromethane, CCl_4

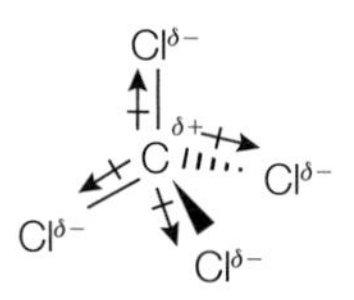

▲ **fig C** Dipoles in tetrachloromethane.

All four C—Cl bonds are polar, but because the molecule is symmetrical the dipoles cancel out one another. The molecule is non-polar.

Example 2: trichloromethane, $CHCl_3$

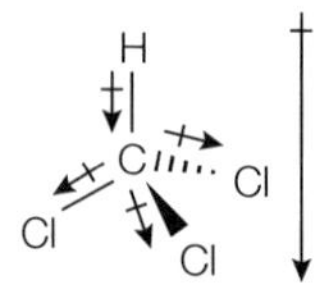

▲ **fig D** Dipoles in trichloromethane.

All four bonds are polar but, although the molecule is symmetrical, the dipoles reinforce one another and so the molecule is polar.

4. TRIGONAL PYRAMIDAL MOLECULES

Example: ammonia, NH_3

All three N—H bonds are polar and the dipoles reinforce one another. The molecule is polar.

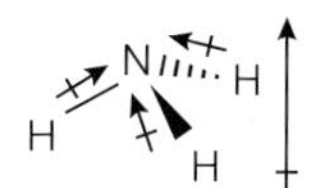

▲ **fig E** Dipoles in ammonia.

5. V-SHAPED MOLECULES

Example: water, H_2O

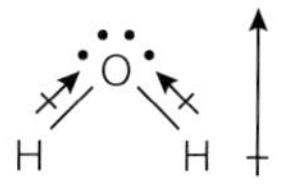

▲ **fig F** Dipoles in water.

Both O—H bonds are polar and the dipoles reinforce one another. The molecule is polar.

ADDITIONAL READING

Dipole moments

The polarity of the molecule is measured by its **dipole moment**.

For a diatomic molecule such as hydrogen chloride, the dipole moment is defined as the difference in charge (i.e. the difference in magnitude between δ+ and δ−) multiplied by the distance of separation between the charges.

For a polyatomic molecule, it is more complicated because the polarities of each bond have to be taken into account, as well as any lone pairs on the central atom.

Table A gives the dipole moments of a number of molecules.

MOLECULE	DIPOLE MOMENT / D
H_2	0
Cl_2	0
HCl	1.05
CO_2	0
BCl_3	0
CCl_4	0
$CHCl_3$	1.02
NH_3	1.48
H_2O	1.84

table A Dipole moments of some molecules.

The unit of dipole moment is the Debye, symbol D. You do not need to understand this unit; just focus on the magnitude of the numbers. The larger the number, the more polar the molecule.

CHECKPOINT

1. A bond between two atoms in a molecule may possess a dipole.

(a) Explain how this dipole arises.

(b) Some bonds that you are likely to meet in organic chemistry are listed. Which of these bonds are likely to possess a dipole? In each case indicate which atom is δ+ and which is δ−.

C–Cl O–H C–C C–O C=C C–N N–H

2. State whether each of the following molecules are non-polar or polar. In each case, explain your reasoning.

(a) H_2S

(b) CH_4

(c) SO_2

(d) SO_3

(e) $AlBr_3$

(f) PBr_3

3. There are two stereoisomers of dichloroethene. The structure and shape of each molecule is:

H H H Cl
C=C C=C
Cl Cl Cl H

cis-dichloroethene *trans*-dichloroethene

Suggest why the *cis* isomer is polar, while the *trans* isomer is non-polar.

SUBJECT VOCABULARY

dipole exists when two charges of equal magnitude but opposite signs are separated by a small distance

dipole moment the difference in magnitude between δ+ and δ− multiplied by the distance of separation between the charges

3D METALLIC BONDING

SPECIFICATION REFERENCE 3.20 3.21 3.22

LEARNING OBJECTIVES

- Understand that metals consist of giant lattices of metal ions (cations) in a sea of delocalised electrons.
- Know that metallic bonding is the strong electrostatic attraction between cations and the delocalised electrons.
- Be able to use the above two models to interpret simple properties of metals, e.g. electrical conductivity and high melting temperature.

THE NATURE OF METALLIC BONDING

Metals typically have the following physical properties:

- high melting temperatures
- good electrical conductivity
- good thermal conductivity
- malleability
- ductility.

Any theory of the way that the atoms in a metal are bonded together must explain the above properties.

Metals typically have one, two or three electrons in the outer shell of their atoms and have low ionisation energies. The electrical conductivity of a metal generally increases as the number of outer-shell electrons increases.

Since electrical conductivity depends on the presence of mobile carriers of electric charge, we can build a picture of a metal as consisting of an array of atoms with at least some of their outer-shell electrons removed and free to move throughout the structure. These **delocalised electrons** are largely responsible for the characteristic properties of metals.

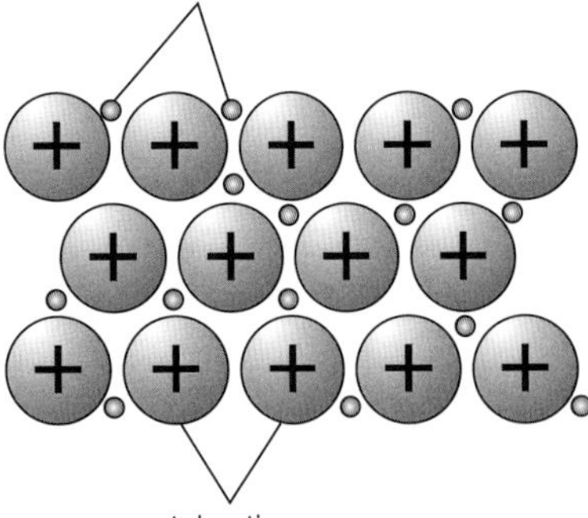

▲ **fig A** Diagram showing the particles present in a metal.

EXAM HINT

If asked to draw a diagram of a metallic bond in an exam, make sure you include a key showing ions and electrons.

The electrons are said to be delocalised since they are free to move throughout the structure and are not confined, i.e. localised, between any pair of cations.

There are electrostatic forces of attraction between the cations and the delocalised electrons. This is known as **metallic bonding**.

EXPLAINING THE PHYSICAL PROPERTIES OF METALS

MELTING TEMPERATURE

In order to melt a metal, it is necessary to partially overcome the forces of attraction between the cations and the delocalised electrons to such an extent that the cations are free to move around the structure. Metals have a giant lattice structure where many of these forces must be overcome. The energy required to do this is usually very large, so the melting temperatures are typically high.

The number of delocalised electrons per cation plays a part in determining the melting temperature of a metal.

- Group 1 metals have low melting temperatures.
- Group 2 metals have higher melting temperatures.
- Metals in the d-block typically have high melting temperatures because they have more delocalised electrons per cation.

Another factor that affects the melting temperature is the charge-to-radius ratio of the cation. The greater the charge-to-radius ratio, the stronger the attraction for the delocalised electrons. Therefore, for two cations of the same charge, the smaller cation will attract the delocalised electrons more strongly. This is why, for example, the melting temperature of lithium is greater than that of sodium.

> **DID YOU KNOW?**
> The argument relating charge-to-radius ratio of the cation to the melting temperature is over-simplified. The way the ions are arranged in the lattice also affects the melting temperature. This is why there is no regular trend in the melting temperatures of the Group 2 metals.
> The way in which the ions are arranged in the lattice changes down the group from beryllium to radium. This is not included in this book and is not required for IAS or IAL studies.

ELECTRICAL CONDUCTIVITY

When a potential difference is applied across the ends of a metal, the delocalised electrons will be attracted to, and move towards, the positive terminal of the cell. This flow of electrons constitutes an electric current.

THERMAL CONDUCTIVITY

Two factors contribute to the ability of metals to transfer heat energy.

1 The free-moving delocalised electrons pass kinetic energy along the metal.

2 The cations are closely packed and pass kinetic energy from one cation to another.

The conduction by the delocalised electrons is by far the more significant of the two factors.

MALLEABILITY AND DUCTILITY

Metals can be hammered or pressed into different shapes (malleability). They can also be drawn into a wire (ductility). Both of these properties depend on the ability of the delocalised electrons and the cations to move throughout the structure of the metal.

When a stress is applied to a metal, the layers of cations may slide over one another (**fig B**).

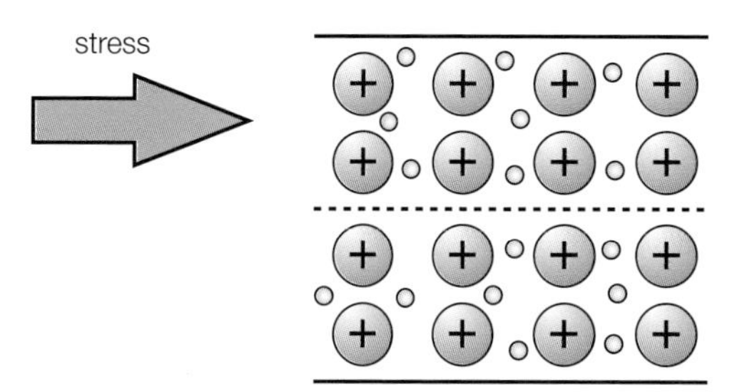

layers of cations slide and electrons move with them

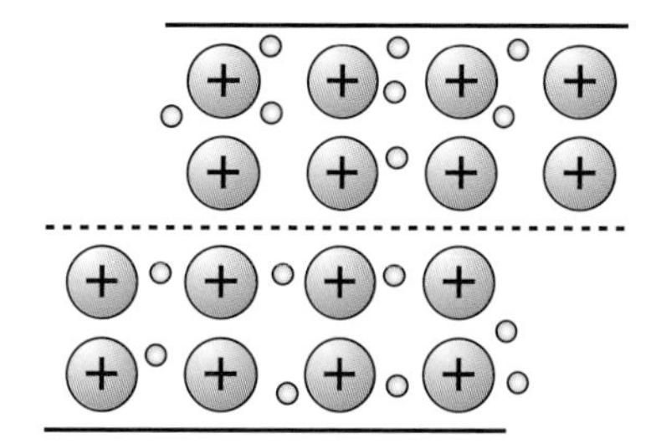

▲ **fig B** The effect of stress on a metal.

However, because the delocalised electrons are free moving, they move with the cations and prevent strong forces of repulsion forming between the cations in one layer and the cations in another layer.

CHECKPOINT

1. Explain what is meant by the term 'metallic bonding'.

2. Suggest why the melting temperatures and electrical conductivities of sodium, magnesium and aluminium are in the order Na $<$ Mg $<$ Al.

SUBJECT VOCABULARY

delocalised electrons electrons that are not associated with any single atom or any single covalent bond

metallic bonding the electrostatic force of attraction between the metal cations and delocalised electrons

3E 1 INTRODUCTION TO SOLID LATTICES

SPECIFICATION REFERENCE 3.12

LEARNING OBJECTIVES

- Know that giant lattices are present in: solid metals (giant metallic lattices); ionic solids (giant ionic lattices); and covalently bonded solids, such as diamond, graphite and silicon(IV) oxide (giant covalent lattices).
- Know the different structures formed by carbon atoms, including graphite, diamond and graphene.
- Know that the structure of solid iodine is discrete (simple) molecular.

The four types of solid structures we shall deal with in this topic are:
- giant metallic lattices
- giant ionic lattices
- giant covalent lattices
- discrete (simple) molecular lattices.

DID YOU KNOW?

There is one other important type of solid structure described as 'polymeric'. These solids are composed of macromolecules such as natural polymers and synthetic (made by humans) polymers. We will look at these in Topic 5B.

METALLIC LATTICES

Metallic lattices are composed of a regular arrangement of positive metal ions (cations) surrounded by delocalised electrons.

Substances that have a giant metallic lattice typically have the following properties:
- high melting and boiling temperatures
- good electrical conductivity
- good thermal conductivity
- malleability
- ductility.

We explained these properties in **Topic 3D**.

GIANT IONIC LATTICES

Giant ionic lattices are composed of a regular arrangement of positive and negative ions.

Substances that consist of giant ionic lattices typically have the following properties:
- fairly high melting temperatures
- brittleness
- poor electrical conductivity when solid but good when molten
- often soluble in water.

We explained these properties in **Topic 3D**.

GIANT COVALENT LATTICES

Giant covalent lattices are sometimes called network covalent lattices. They consist of a giant network of atoms linked to each other by covalent bonds.

Four of the most common giant covalent substances are:
- diamond
- graphite
- graphene
- silicon(IV) oxide (not discussed below).

DIAMOND

In diamond, each carbon atom forms four sigma (δ) bonds to four other carbon atoms, in a giant three-dimensional tetrahedral arrangement (**fig A**). All bond angles are 109.5°.

Diamond is extremely hard because of the very strong C—C bonding throughout the structure. It also has a very high melting temperature because a great number of strong C—C bonds have to be broken in order to melt it. This requires a large amount of heat energy.

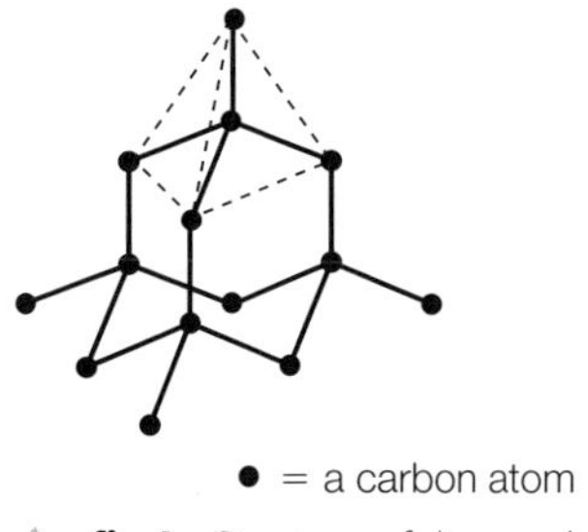

▲ **fig A** Structure of diamond.

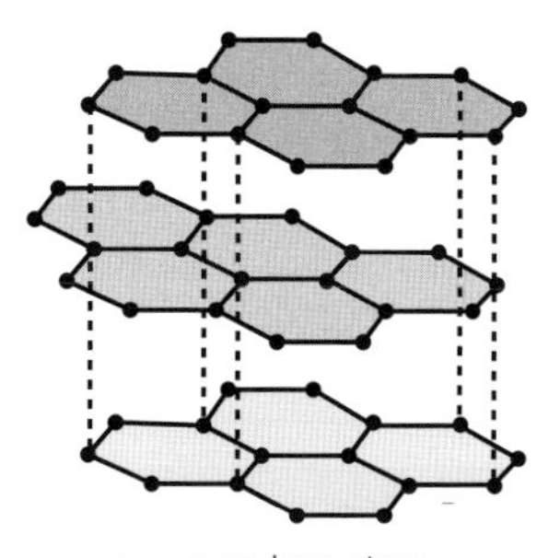

▲ **fig B** Structure of graphite.

GRAPHITE

Graphite has a layered structure as shown in **fig B**.

Each carbon atom is bonded to three others by sigma bonds, forming interlocking hexagonal rings. The fourth electron on each carbon atom is in a p-orbital. The carbon atoms are close enough for the p-orbitals to overlap with one another to produce a cloud of delocalised electrons, both above and below the plane of the rings. We will compare this with the structure of benzene in **Topic 18 (Book 2: IAL)**.

ADDITIONAL READING

Graphite can be used as a solid lubricant since the layers slide easily over one another. Although there are only relatively weak London forces of attraction between the layers (see **Topic 7**), this does not account for its lubricating properties. Graphite is five times less able to act as a lubricant at high altitude, and eight times less in a vacuum. Its lubricating properties are as a result of adsorbed gases on the surface of the carbon atoms. Graphite's inability to act as an effective lubricant in a vacuum is the reason why it is not used as a lubricant in spacecraft. Either molybdenum disulfide (MoS_2) or hexagonal boron nitride (BN) can be used instead.

Graphite is a fairly good conductor of electricity. The delocalised electrons between the layers are free to flow under the influence of an applied potential difference. An interesting feature of graphite is that it can only conduct electricity parallel to its layers. The delocalised electrons are not free to move from one layer to the next. Compare this with a metal, which is able to conduct electricity in all directions throughout the structure.

Graphite has a high melting temperature for the same reason as diamond.

GRAPHENE

Graphene is pure carbon in the form of a very thin sheet, one atom thick (**fig C**). The carbon atoms are bonded in exactly the same way as in graphite and it can, therefore, be described as a one-atom thick layer of graphite.

▲ **fig C** Structure of graphene.

DID YOU KNOW?

You may have heard diamond referred to as 'ice'. Diamond first acquired this nickname because of its properties rather than its appearance. Owing to the wave nature of electrons, the electron wave is able to travel more smoothly through a perfect crystal without bouncing, in much the same way as light travels through a clear crystal. In addition, thermal conduction is improved by the network of strong covalent bonds. The use of a heat probe is one of the major tests in determining if a diamond is authentic. This is because real diamonds are able to conduct heat and are actually one of the best conductors of heat. Even as a diamond heats up, it is cool to the touch, just like ice. This is why the points of diamond cutting tools do not become overheated.

DID YOU KNOW?

Graphene is the thinnest material on Earth

- It is one million times thinner than a human hair.
- It is 200 times stronger than steel.
- It is an excellent thermal conductor, better even than diamond.
- It absorbs light.
- It can be considered the basic unit of most other forms of carbon.
- Graphite consists of layers of graphene on top of one another joined by London forces.
- A sheet of graphene can be rolled into a ball to produce fullerene molecules.
- A sheet of graphene can be rolled into a cylinder to produce a carbon nanotube.
- Graphene can self-repair holes in its sheets when exposed to molecules containing carbon, such as hydrocarbons. When bombarded with pure carbon atoms, the atoms perfectly align into hexagons, completely filling the holes.

ADDITIONAL READING

Molecular lattices

A common solid molecular lattice is iodine.

Iodine exists as diatomic molecules, I_2. In solid iodine these molecules are arranged in a regular pattern (**fig D**), which explains its crystalline nature.

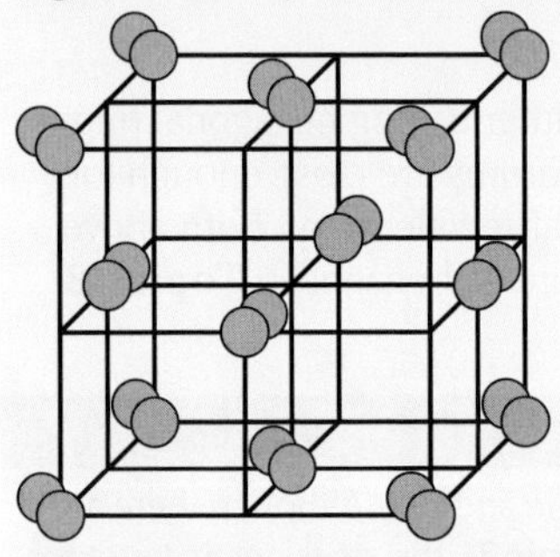

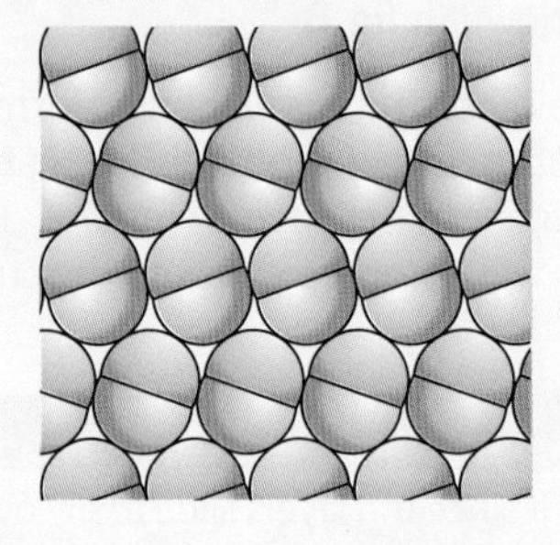

▲ **fig D** Structure of iodine.

The diagram on the left of **fig D** shows the arrangement of the iodine molecules. The structure is described as 'face-centred cubic'. In practice, the iodine molecules will be touching one another (right-hand diagram); they have been drawn apart for the sake of clarity. The molecules of iodine are held together by London forces (see **Topic 7**).

▲ **fig E** Large crystals of iodine.

▲ **fig F** Sucrose.

Other molecular solids include sulfur (S_8), white phosphorus (P_4), Buckminsterfullerene (C_{60}, **fig G**), dry ice (solid carbon dioxide, CO_2, **fig H**), sucrose ($C_{12}H_{22}O_{11}$, **fig F**) and solid alkanes (e.g. paraffin wax).

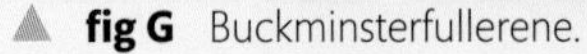

▲ **fig G** Buckminsterfullerene.

▲ **fig H** Dry ice.

PHYSICAL PROPERTIES OF MOLECULAR SOLIDS

Molecular solids will, in general, have low melting and boiling temperatures. In order to melt a molecular solid, it is not necessary to break the covalent bonds within the molecule (the intramolecular bonds); it is only necessary to overcome the intermolecular forces of attraction.

Since intermolecular forces of attraction tend to be much weaker than covalent bonds, little energy is required to either break down the lattice structure of the solid and cause it to melt, or to separate the molecules and cause the liquid to boil and vaporise.

Intermolecular forces tend to increase with both an increase in the number of electrons per molecule and also with increasing length of molecule. This means that a macromolecular solid such as poly(ethene) will have a much higher melting temperature than its monomer, ethene.

$H_2C{=}CH_2$ — ethene, melting point = −169°C

$-\!\!\left[CH_2-CH_2\right]\!\!-_n$ — poly(ethene), melting point typically 120 to 180°C

fig I Ethene and (poly)ethene.

CHECKPOINT

1. Magnesium oxide ($Mg^{2+}O^{2-}$) has the same structure as sodium chloride ($Na^{+}Cl^{-}$).

(a) Draw a diagram to show the arrangement of the ions in a crystal of magnesium oxide.

(b) Explain why the melting temperature of magnesium oxide is higher than that of sodium chloride.

2. Explain the following observations:

(a) Magnesium and magnesium fluoride both have giant lattice structures containing ions. Solid magnesium conducts electricity, but solid magnesium fluoride does not.

(b) Silicon and phosphorus are both described as covalent substances, but silicon has a much higher melting temperature than phosphorus.

3. Hexagonal boron nitride has a structure similar to that of graphite.

Suggest why solid boron nitride can act as a lubricant in a vacuum, whereas graphite cannot.

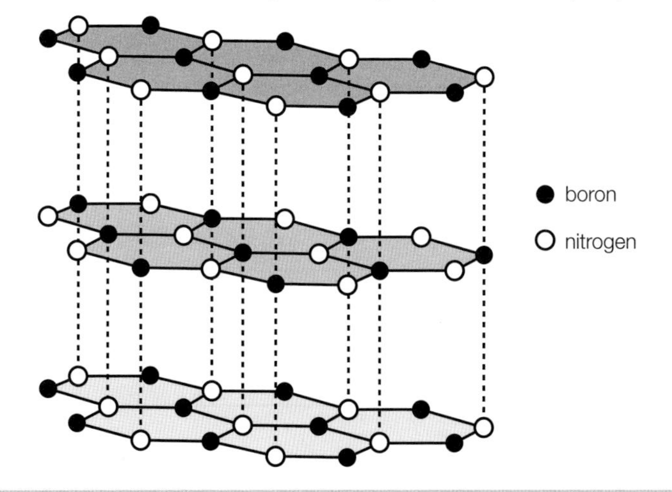

4. Explain, in terms of their bonding and structure, why

(a) diamond is hard and graphite is soft, and

(b) diamond does not conduct electricity but graphite does.

3E 2 STRUCTURE AND PROPERTIES

SPECIFICATION REFERENCE
3.12

LEARNING OBJECTIVES

- Predict the type of structure and bonding present in a substance from numerical data and/or other information.
- Predict the physical properties of a substance, including melting and boiling temperature, electrical conductivity and solubility in water, in terms of:
  (i) the types of particle present (atoms, molecules, ions, electrons)
  (ii) the structure of the substance
  (iii) the type of bonding and the presence of intermolecular forces, where relevant.

TYPES OF BONDING AND STRUCTURE

Table A shows the types of bonding and structure that exist in elements and compounds.

BONDING	STRUCTURE	EXAMPLES
Metallic	giant lattice	Mg, Al, Cu, Zn
Ionic	giant lattice	NaCl, MgO, CsF
Covalent (including polar covalent)	giant lattice molecular macromolecular	C (diamond), C (graphite), Si, SiO_2, BN H_2O (ice), I_2, P_4, S_8, C_{60}, $C_{12}H_{22}O_{11}$ (sucrose) polymers (e.g. poly(ethene)), proteins, DNA

table A Types of bonding and structure.

EXAM HINT

If a question asks you to explain the properties of a substance, it is not enough simply to describe the bonding. You must refer to the particles involved and how they are arranged or behave. For example, metals conduct electricity because the delocalised electrons are free to move through the metallic crystal when a potential difference is applied.

PREDICTING PHYSICAL PROPERTIES

The physical properties of a substance are determined by the type of bonding and structure it has.

Fig A allows you to determine the type of bonding and structure in a substance by considering some of its properties.

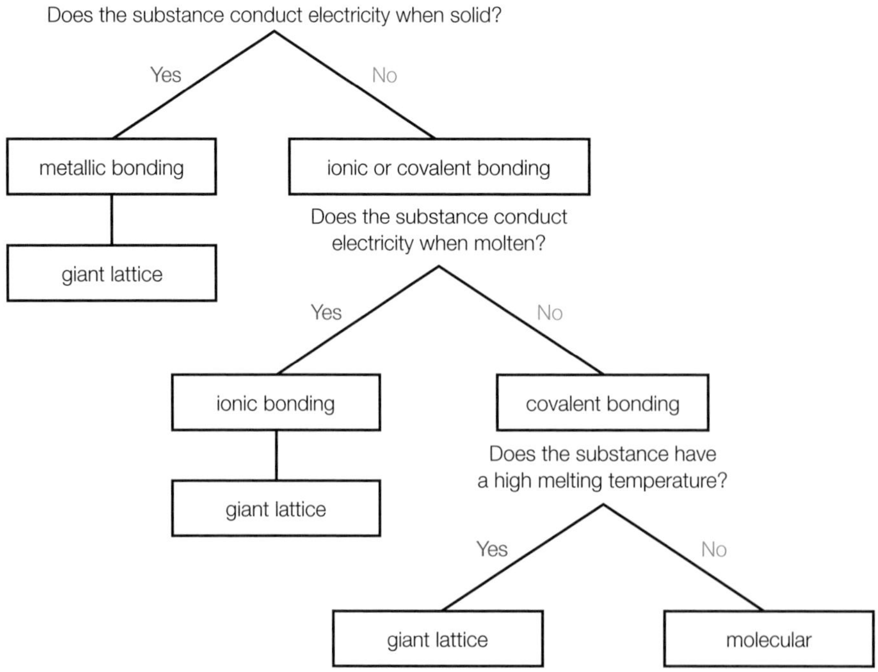

fig A Flow chart for determining bonding and structure.

There will, of course, always be exceptions. For example, graphite has a giant covalent structure and yet it is a relatively good conductor of electricity when solid.

Table B gives a summary of the major properties of each type of structure.

	GIANT METALLIC	GIANT IONIC	GIANT COVALENT	MOLECULAR
Particles present	positive ions and delocalised electrons	positive and negative ions	atoms	molecules
Type of bonding	metallic	ionic	covalent	covalent
Are there any intermolecular forces of attraction?	no	no	no	yes
Melting and boiling temperatures	fairly high to high	fairly high to high	high to very high	generally low
Electrical conductivity	good when solid and when molten	non-conductor when solid; good when molten	non-conductor	non-conductor
Solubility in water	insoluble. However, some metals, e.g. Na, react with water to form a metal hydroxide that then dissolves	generally soluble, but with notable exceptions, e.g. AgCl, AgBr, AgI and $BaSO_4$	insoluble	generally insoluble, but may dissolve if hydrogen bonding is possible (e.g. sucrose), or if the substance reacts with water (e.g. Cl_2) (See **Topic 7** for an explanation of hydrogen bonding.)

table B Structure and properties of a substance.

CHECKPOINT

1. The table below gives some properties of four substances: A, B, C and D.

Analyse the data and then decide what type of bonding and structure is likely to be present in each substance.

	MELTING TEMPERATURE / °C	ABILITY TO CONDUCT ELECTRICITY WHEN SOLID	ABILITY TO CONDUCT ELECTRICITY WHEN MOLTEN	SOLUBILITY IN WATER
A	1083	good	good	insoluble
B	119	poor	poor	insoluble
C	2230	poor	poor	insoluble
D	801	poor	good	soluble

2. Each of the substances listed below has at least one property that is unusual for its type of bonding and structure. In each case, state the most likely bonding and structure, and identify the unusual property or properties. Suggest a possible identity for each substance.

(a) Substance **P**.

Melting temperature 813 °C; good conductor of electricity when solid and when molten; reacts with water to form ammonia gas and an alkaline solution.

(b) Substance **Q**.

Melting temperature 1414 °C; semi-conductor of electricity when solid; insoluble in water.

(c) Substance **R**.

Melting temperature 98 °C; good conductor of electricity when solid; reacts with water to form hydrogen gas and an alkaline solution.

A CHEMICAL INTERVENTION

In the following story, an attempt to ensure ideal skiing conditions led to wider environmental concerns.

AMMONIUM NITRATE GOES ON THE SLOPES

Swiss winter sports event organisers have caused an environmental stir by using chemical fertilisers to maintain their precious slopes. It emerged that up to 1.5 tonnes of ammonium nitrate was used to prepare and protect the piste for the Lauberhorn downhill ski race in Switzerland. Environmental researchers are now investigating the extent and effects of this practice.

As with any salt, dissolving ammonium nitrate involves breaking it into its constituent ammonium and nitrate ions, which takes in energy from its surroundings. The formation of new bonds between these ions and surrounding water molecules then releases energy. But since ammonium and nitrate ions are relatively large, the water molecules have relatively weak interactions with their diffuse charges. So with little thermodynamic payback during this bond formation, the immediate effect of adding ammonium nitrate to slushy snow is to cool it down.

Christian Rixen from the Swiss Federal Research Institute reported that ammonium nitrate is used for races where wet snow can slow skiers down. His research highlighted some of the environmental risks of this practice.

'This is a very strong fertilizer. At high elevation, you have alpine meadows with vegetation that lives within nutrient-poor soil. If this is highly fertilized it can have negative effects,' said Rixen. 'We have already seen that a single application of ammonium nitrate can cause a reduction in biodiversity,' he said.

The practice has also raised questions about the contamination of waterways. 'It could even be a problem for organic farmers in the area, who are not allowed to use any chemicals at all on their land,' added Rixen.

He is now working closely with the Swiss Federal Office for the Environment and says that more information is needed about how widespread this practice is. Agricultural use of chemical fertilisers is already subject to detailed guidelines, including regulations barring its use at high altitudes and on snow-covered fields.

However, Daniel Hartmann from the Swiss Federal Office for the Environment reported that no specific rules exist controlling the use of fertilizers on ski slopes.

'There are many rules to control its agricultural use because farming is such a chemically-intensive practice,' he explained. 'We need to find out more about how widely these chemicals are used on ski slopes and, if necessary, review the policy and define rules to be put in place to protect the environment.'

SCIENCE COMMUNICATION

SKILLS PROBLEM SOLVING

1. Imagine you are writing an article for an organic farming magazine. How could you re-write this story to emphasise the environmental issues without changing the factual content of the article?

CHEMISTRY IN DETAIL

2. (a) Write down the chemical formula of ammonium nitrate and calculate the percentage by mass of nitrogen in the compound.
   (b) Ammonium nitrate is an ionic compound. Describe a test that you could carry out in the lab to show this and state the expected results.
   (c) Describe the bonding in the NH_4^+ ion and explain its shape.
3. Ammonium nitrate decomposes on heating to produce nitrogen, oxygen and water.
   (a) Write a balanced chemical equation for this decomposition.
   (b) Using your answer to part (a), calculate the volume of oxygen (measured at r.t.p) that would be produced if one tonne of ammonium nitrate decomposed. (Assume 1 mole of any gas occupies 24 dm3 at r.t.p.)
   (c) Using information from the article and your answers to the questions above suggest which **two** hazard symbols you would expect to find on a container of ammonium nitrate. Explain your choices.

ACTIVITY

INTELLECTUAL INTEREST AND CURIOSITY, ADAPTABILITY

The article raises the issue of '... a possible problem for organic farmers in the area'. Carry out an internet research investigation into the following questions:

i) What is meant by 'organic farming'?
ii) What are the possible advantages and disadvantages of this approach to farming?
iii) After you have carried out your research, prepare a presentation either for or against organic farming. Your teacher may allocate you the side of the argument that you may not feel you agree with so you will need to prepare carefully.

INTERPRETATION NOTE

Be constructively critical about the nature of the source material on the web. What constitutes a reliable resource?

3 EXAM PRACTICE

1 Which of the following statements about the properties associated with ionic and covalent compounds is correct?

A An ionic compound cannot undergo electrolysis.

B The only covalently bonded substances with high melting temperatures are those in which hydrogen bonds are present.

C A compound cannot contain both ionic bonding and covalent bonding.

D Ionic compounds differ from metals because they do not conduct electricity in the solid state. [1]

(Total for Question 1 = 1 mark)

2 Which is the best explanation for the ability of graphite to act as a solid lubricant?

A It has delocalised electrons that are able to flow parallel to the layers.

B The covalent bonds within the layers are strong.

C Gas molecules are trapped between the layers.

D The forces of attraction between the layers are very weak. (1)

(Total for Question 2 = 1 mark)

3 Which of the following molecules contains six bonding electrons?

A C_2H_4 **B** CO_2 **C** H_2S **D** NCl_3 [1]

(Total for Question 3 = 1 mark)

4 A solid melts just above 100 °C. It does not conduct electricity even when molten.

Which of the following structures is the compound most likely to have?

A giant ionic

B giant covalent

C simple molecular

D giant metallic [1]

(Total for Question 4 = 1 mark)

5 The species Ar, K^+ and Ca^{2+} are isoelectronic.

In what order do their radii increase?

	smallest	----→	largest
A	Ar	Ca^{2+}	K^+
B	Ar	K^+	Ca^{2+}
C	Ca^{2+}	Ar	K^+
D	Ca^{2+}	K^+	Ar

[1]

(Total for Question 5 = 1 mark)

6 Which of the following molecules does not have a permanent dipole?

A CF_4 **B** CHF_3 **C** CH_2F_2 **D** CH_3F [1]

(Total for Question 6 = 1 mark)

7 Lithium, magnesium and sodium all exhibit metallic bonding.

(a) State what is meant by metallic bonding. [2]

(b) State how a metal can conduct electricity. [1]

(c) Explain why magnesium is a better conductor of electricity than sodium. [2]

(d) Predict whether or not sodium is a better electrical conductor than lithium. Justify your answer. [2]

(Total for Question 7 = 7 marks)

8 The table shows some of the properties of four substances, **A**, **B**, **C** and **D**.

State the type of bonding and structure that is likely to be present in each of the substances.

In each case justify your answer. [8]

Substance	Solubility in water	Electrical conductivity		Melting temperature / °C
		of solid	in aqueous solution	
A	insoluble	poor	—	1610
B	soluble	poor	good	801
C	insoluble	good	—	1083
D	soluble	poor	fair	–78

(Total for Question 8 = 8 marks)

9 The electron pair repulsion (EPR) theory can be used to predict the shape of simple molecules and ions.

(a) State the main assumptions of the EPR theory. [3]

(b) Draw a dot-and-cross diagram for each of the following molecules/ions.

(i) BF_3 (ii) NH_3 (iii) NH_4^+ (iv) SF_6 [4]

(c) Predict the shape and bond angles in each of the molecule/ions in part (b). [5]

(Total for Question 9 = 12 marks)

10 Nitrogen and carbon monoxide are both gases consisting of diatomic molecules. Both gases are colourless, odourless and tasteless, but unlike nitrogen, carbon monoxide is extremely toxic.

(a) The dot-and-cross diagram for carbon monoxide,, showing only the outer electrons, is:

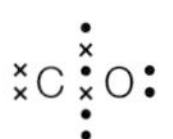

Copy the diagram and add labels to identify:

(i) a lone pair of electrons

(ii) a covalent bonding pair of electrons and

(iii) a dative covalent bonding pair of electrons. [3]

(b) The nitrogen and carbon monoxide molecules are isoelectronic.

(i) State what is meant by isoelectronic. [1]

(ii) The HCN molecule is also isoelectronic with N_2 and CO.

Draw a dot-and-cross diagram for HCN, showing only the outer shell electrons. [2]

(iii) Suggest why nitrogen is much less reactive than either carbon monoxide or hydrogen cyanide. [1]

(Total for Question 10 = 7 marks)

11 Sodium chloride is an ionic compound containing sodium ions (Na^+) and chloride ions (Cl^-).

It has a fairly high melting temperature, is soluble in water and is a poor conductor of electricity when solid, but good when molten.

(a) Draw dot-and-cross diagrams to show the changes in arrangement of electrons that take place when sodium (Na) reacts with chlorine (Cl_2) to form sodium chloride. Show only the outer electrons in each case. [4]

(b) Complete the diagram to show the structure of sodium chloride using the key provided. [2]

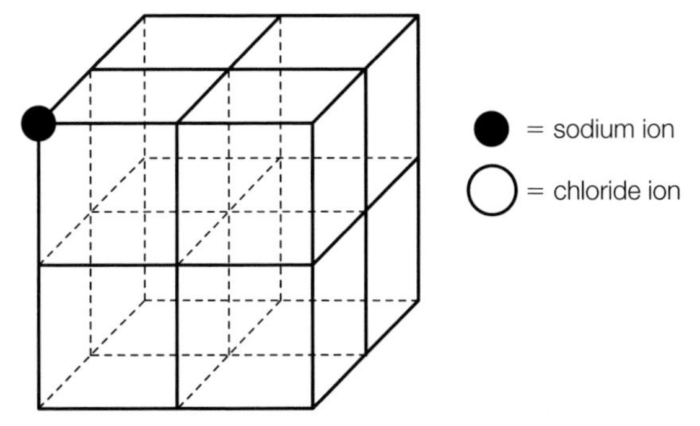

(c) Explain why sodium chloride

(i) has a fairly high melting temperature [3]

(ii) is soluble in water [3]

(iii) is a poor conductor of electricity when solid, but a good conductor when molten. [2]

(Total for Question 11 = 14 marks)

12 This question is about four simple molecular compounds: water, ammonia, methane and boron trichloride.

(a) Complete the table to show the number of bonding pairs and non-bonding (lone) pairs of electrons in one molecule of each compound. [4]

Formula of molecule	H_2O	NH_3	CH_4	BCl_3
Number of bonding pairs				
Number of non-bonding pairs				

(b) The ammonia molecule and the boron trichloride molecule both contain polar bonds.

(i) State what is meant by a polar bond, and explain how the polarity arises. [3]

(ii) Explain why the ammonia molecule is polar, but the boron trichloride molecule is not. [3]

(Total for Question 12 = 10 marks)

TOPIC 4 INTRODUCTORY ORGANIC CHEMISTRY AND ALKANES

A INTRODUCTION TO ORGANIC CHEMISTRY | B ALKANES

Organic chemistry is one of the traditional branches of chemistry, like physical and inorganic chemistry. Students of biology will understand its importance because most types of compounds in this topic are found in, or are formed in, plants and animals, including the human body. Many aspects of our lives have been revolutionised by the production of new organic compounds, for example:

- new polymers with special properties
- more effective drugs to treat diseases
- the ongoing search for new antibiotics
- sustainable fuels to replace fossil fuels.

Fertilisers and pesticides have increased crop yields to feed the world's growing population, but this is an example of where the application of the knowledge of chemistry has caused unexpected problems. Some people prefer food grown naturally, without the use of human-made chemicals – ironically these foods are often described as 'organic'.

In this topic, you will learn about the basics of organic chemistry, for example

- homologous series (compounds that are very similar to each other)
- nomenclature (a systematic way of naming organic compounds)
- isomerism (two or more compounds that have the same molecular formula but are not the same)

You will then look at simple hydrocarbons called alkanes, including their uses as fuels.

MATHS SKILLS FOR THIS TOPIC

- Use ratios to construct and balance equations
- Represent chemical structures using angles and shapes in 2D and 3D structures

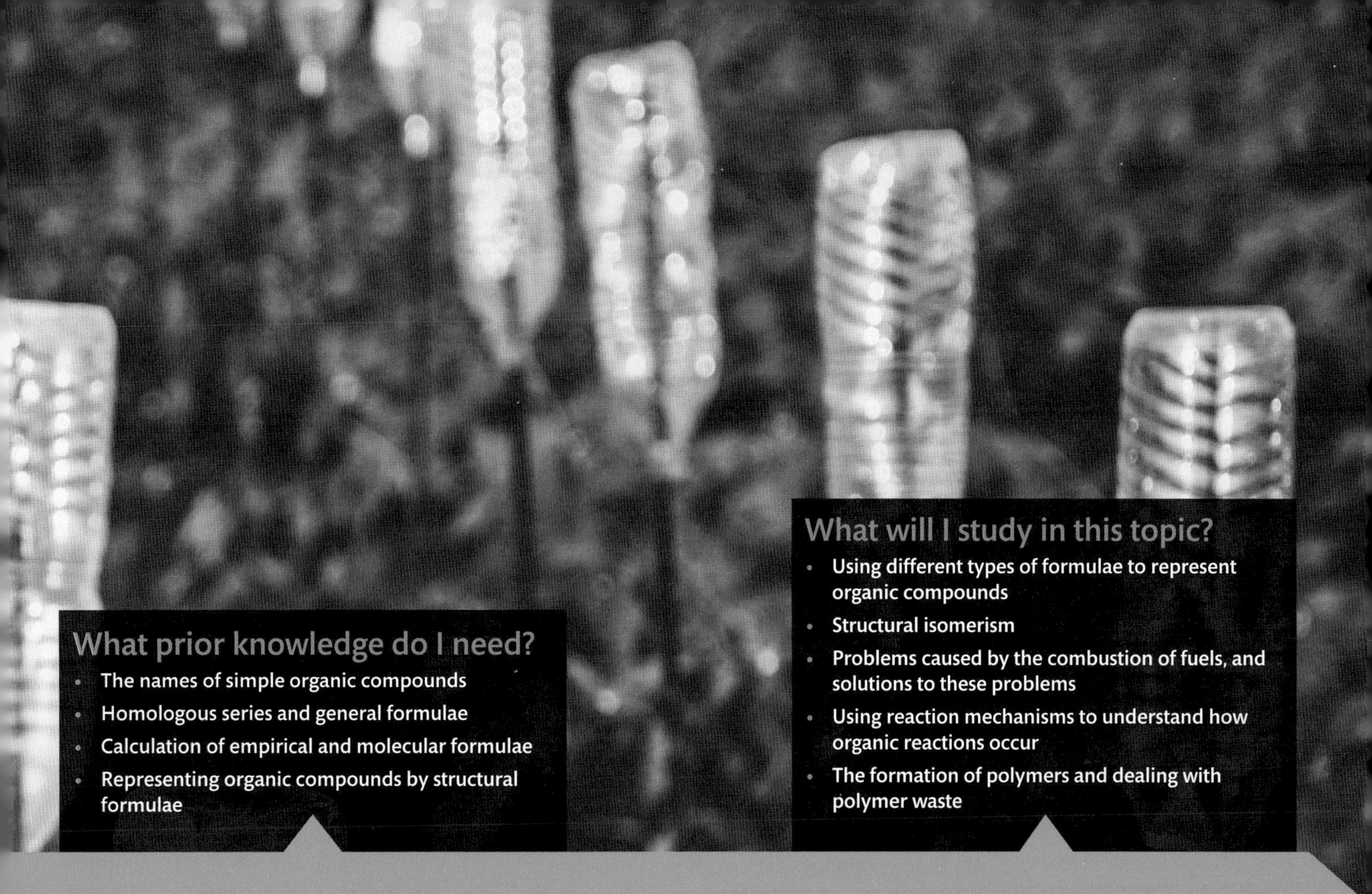

What prior knowledge do I need?

- The names of simple organic compounds
- Homologous series and general formulae
- Calculation of empirical and molecular formulae
- Representing organic compounds by structural formulae

What will I study in this topic?

- Using different types of formulae to represent organic compounds
- Structural isomerism
- Problems caused by the combustion of fuels, and solutions to these problems
- Using reaction mechanisms to understand how organic reactions occur
- The formation of polymers and dealing with polymer waste

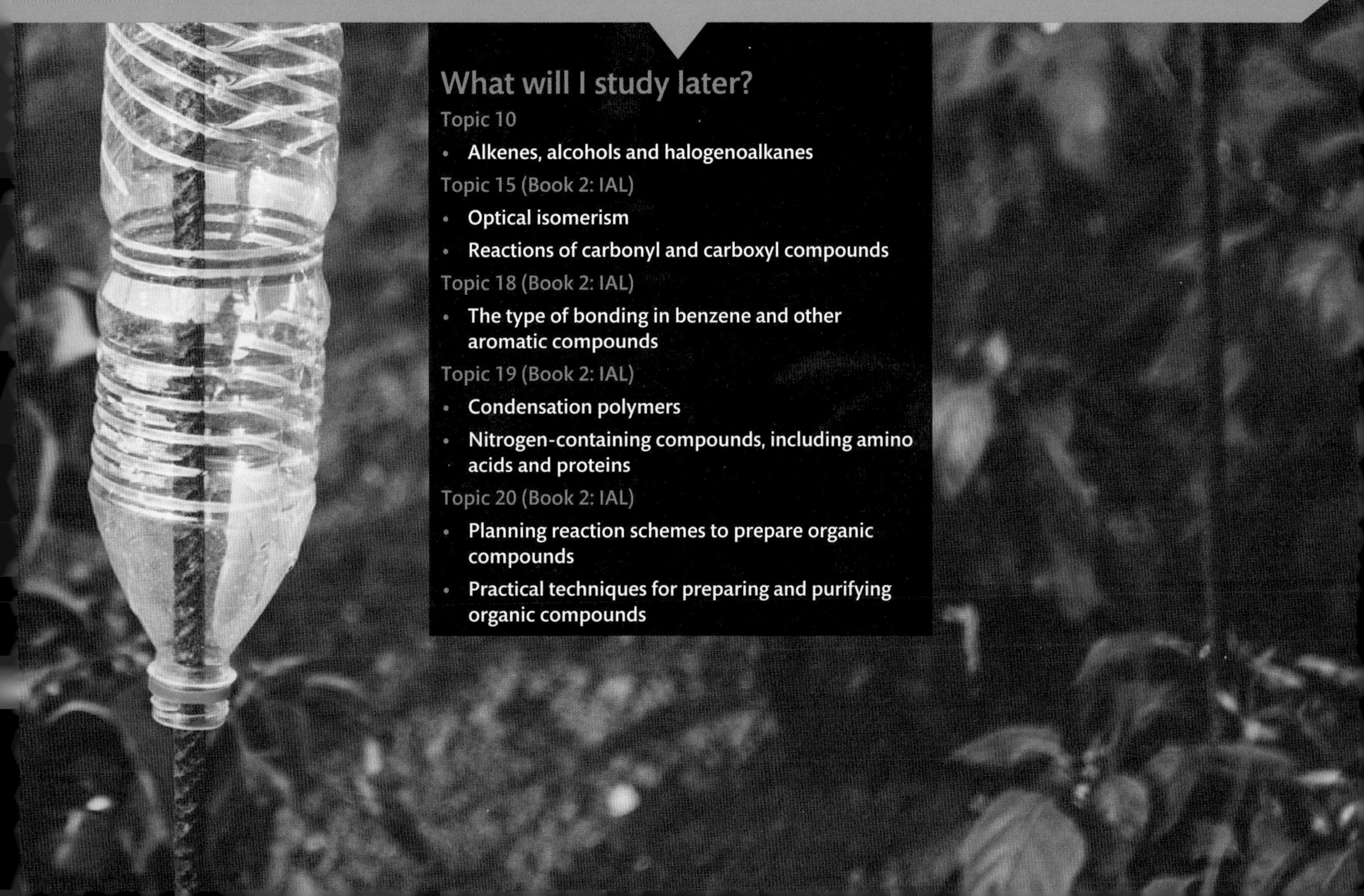

What will I study later?

Topic 10

- Alkenes, alcohols and halogenoalkanes

Topic 15 (Book 2: IAL)

- Optical isomerism
- Reactions of carbonyl and carboxyl compounds

Topic 18 (Book 2: IAL)

- The type of bonding in benzene and other aromatic compounds

Topic 19 (Book 2: IAL)

- Condensation polymers
- Nitrogen-containing compounds, including amino acids and proteins

Topic 20 (Book 2: IAL)

- Planning reaction schemes to prepare organic compounds
- Practical techniques for preparing and purifying organic compounds

4A 1 WHAT IS ORGANIC CHEMISTRY?

SPECIFICATION REFERENCE 4.9

LEARNING OBJECTIVES

- Understand that alkanes and cycloalkanes are hydrocarbons which are saturated.

▲ **fig A** Friedrich Wöhler was a German chemistry professor and an early pioneer in the field of organic chemistry. He is best remembered for making urea (an organic compound) starting only from inorganic compounds.

EARLY DAYS

Since the 1800s, our knowledge and understanding of chemistry has grown rapidly. To make sense of all this knowledge, chemists divided chemistry into three main categories: inorganic, organic and physical chemistry. Each category is equally important. There are millions of different compounds in existence, and the vast majority are organic compounds. Now, what is organic chemistry?

Today, the word 'organic' has a very different meaning in everyday life, and it is often applied to farming and food. Going back to the 1800s, people believed that there was something special about some substances. They were only made in plants or animals. One example is the compound called urea, which is present in human urine. People used to believe that it could only be produced in the human body. In 1828, the German chemist Friedrich Wöhler discovered that urea could be made by heating a compound (ammonium cyanate) that was not organic. This meant that the idea of organic compounds only coming from living things was no longer correct.

HYDROCARBONS

The main feature of an organic compound is that it contains carbon. Almost all of these compounds also contain hydrogen. Some of the most important compounds contain elements such as nitrogen and oxygen as well. In this topic, we look at some of the large numbers of compounds that contain only carbon and hydrogen. These compounds are called **hydrocarbons**.

If an organic compound contains other elements as well as carbon and hydrogen, then it is not a hydrocarbon. For example, many foods contain a sugar called sucrose. Sucrose contains carbon and hydrogen, but also oxygen, so it is not a hydrocarbon. It is an example of a carbohydrate – the *-ate* ending shows that it contains oxygen.

DID YOU KNOW?

In the 1960s, the organic chemist Stephanie Kwolek invented an extremely strong polymer called Kevlar®. Kevlar fibre is now widely used in many applications, including heat resistant gloves, smartphones, sports equipment and body armour.

▲ Award-winning US chemist Stephanie Louise Kwolek was inducted into the National Inventors Hall of Fame in 1995.

SATURATED OR UNSATURATED?

Although there are many thousands of different hydrocarbons, most of them are classed as **saturated** or **unsaturated**. Like many chemical terms, these words have a very different meaning in everyday life. Someone who has been caught in a heavy rain shower may say that their clothes are saturated, which means that they have absorbed as much water as they possibly can.

In organic chemistry, these terms have nothing to do with water, although there is a connection. A hydrocarbon that is saturated contains as much hydrogen as possible, which depends on the number of carbon atoms in the molecule. If a hydrocarbon has fewer hydrogen atoms than the maximum, then it is not saturated, we say it is unsaturated.

The formula of the simplest hydrocarbon, containing only one carbon atom, is:

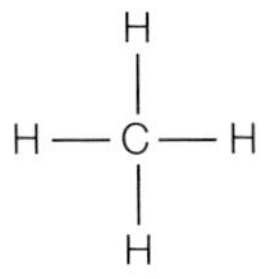

When a hydrocarbon contains two carbon atoms, there is a maximum of six hydrogen atoms.

There is a hydrocarbon that contains two carbon atoms and six hydrogen atoms, but also one that contains two carbon atoms and only four hydrogen atoms. The formulae of these hydrocarbons are:

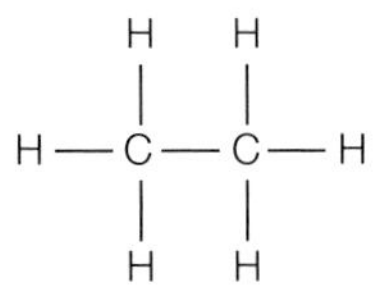

saturated 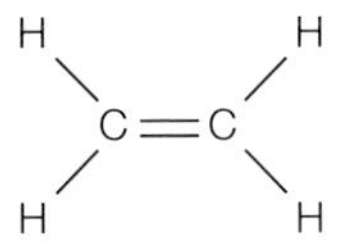 unsaturated

You can see that in all three examples, each carbon atom has four bonds to other atoms. This is a general rule for organic compounds. In most cases, every carbon atom has four bonds.

The difference between a saturated hydrocarbon and an unsaturated hydrocarbon is to do with whether or not there is room, for more hydrogen atoms.

- If there is no room, then the hydrocarbon is saturated.
- If there is room, then the hydrocarbon is unsaturated.

One easy way to decide whether a hydrocarbon is saturated or unsaturated is to look at structures like the ones above.

- If there are two bonds (a double bond) drawn between one or more carbon atoms, then the hydrocarbon is unsaturated.
- If there are only single bonds, then the hydrocarbon is saturated.

ALKANES AND CYCLOALKANES

Look at the formulae in **fig B**. These are alkanes.

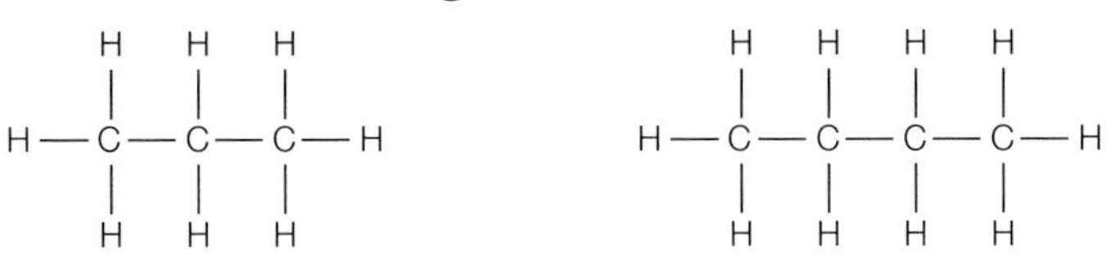

fig B Propane and butane.

Both structures have only single bonds and so are saturated.

Now look at the formulae in **fig C**. These are cycloalkanes.

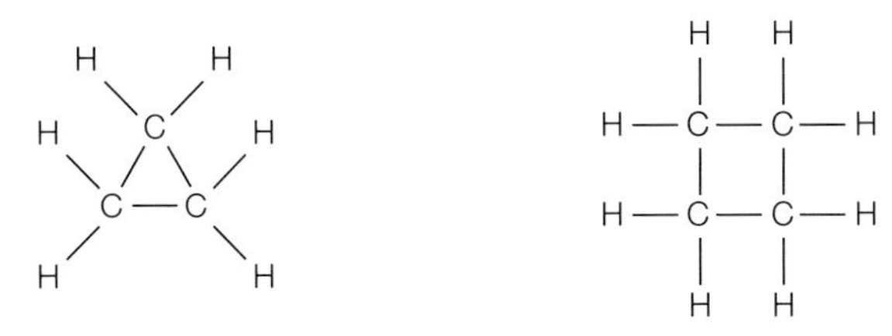

fig C Cyclopropane and cyclobutane.

Both structures have only single bonds and so are saturated. They both have carbon atoms joined in a ring structure. The first has a triangular arrangement and the second has a square arrangement. These ring structures mean that they are cyclic compounds. We will look at alkanes and cycloalkanes in more detail in **Topic 4B**.

EXAM HINT

It is important to remember that these structures are actually three dimensional. The bond angle in cyclobutane looks like it is 90 degrees as it is drawn in 2D but in fact is closer to 109.5 degrees as it would be in methane.

LEARNING TIP

The formulae of the hydrocarbons in this topic show every bond and every atom. These formulae can also be written as molecular formulae. The first one can be written as CH_4. Try writing molecular formulae for the other hydrocarbons shown in this topic.

CHECKPOINT

1. A saturated hydrocarbon contains five carbon atoms in its molecule. What are the two possible molecular formulae for this hydrocarbon?

2. Consider the compound with the formula $C_2H_4O_2$. Is it an organic compound? Explain your answer.

SUBJECT VOCABULARY

hydrocarbon a compound that contains only carbon and hydrogen atoms
saturated a compound containing only single bonds
unsaturated a compound containing one or more double bonds

4A 2 DIFFERENT TYPES OF FORMULAE

SPECIFICATION REFERENCE 4.5(ii)

LEARNING OBJECTIVES

- Draw compounds and represent organic molecules using structural, displayed and skeletal formulae.

EXAM HINT

If an exam question asks for a particular type of formula in an answer, it is important that you use it. If it asks for a displayed formula, then a skeletal formula will not gain full marks.

USING DIAGRAMS TO REFER TO ORGANIC COMPOUNDS

There are many millions of organic compounds, so making clear which ones we are referring to can be challenging.

There are two main ways to refer to organic compounds. We can use:

- formulae
- names.

In this topic we will consider how to refer to organic compounds using formulae.

DISPLAYED FORMULAE

The formulae you saw in **Topic 4A.1** are all **displayed formulae**. They show (display) every atom and every bond separately. In many situations, these are the best type of formulae to use, but sometimes it is better to simplify them.

Consider the hydrocarbon with this displayed formula, its name is butane:

```
    H   H   H   H
    |   |   |   |
H — C — C — C — C — H
    |   |   |   |
    H   H   H   H
```

STRUCTURAL FORMULAE

One way to simplify this displayed formula is to group all the atoms joined to a particular carbon atom together. We can choose to show the bonds between the carbons, or we can leave them out. These are both **structural formulae** of butane:

$CH_3—CH_2—CH_2—CH_3$ and $CH_3CH_2CH_2CH_3$

SKELETAL FORMULAE

Another way to represent a compound is by a **skeletal formula**. The word skeletal is connected with the word skeleton, which, as you know, shows only the bones in a human or animal body.

A skeletal formula is a zig-zag line that shows only the bonds between the carbon atoms. Every change in direction and every ending means that there is a carbon atom (with as many hydrogen atoms as needed). Atoms other than carbon and hydrogen need to be shown.

This is the skeletal formula of butane:

The start and end both represent CH_3, and the two junctions between lines each represent CH_2.

MOLECULAR FORMULAE

The displayed, structural and skeletal formulae above show the structures of the molecules unambiguously. In other words, each formula represents only one compound. With a displayed formula, this is very clear. With a structural formula, you have to imagine how the atoms are joined together in groups such as CH_2 and CH_3, but that is very straightforward. With a skeletal formula, once you know the rules, you can be sure how every atom is arranged in the molecule.

Now consider the formula C_3H_7Cl. This is clearly not a displayed formula or a skeletal formula, but it is also not a structural formula. This is because, with three carbon atoms, the chlorine atom could be attached to the middle carbon atom or to either of the end carbon atoms. The formula C_3H_7Cl actually represents two different compounds.

Formulae like these are called **molecular formulae.** They only show the numbers of each type of atom in the molecule, and not its structure. Of course, in very simple molecules such as CH_3Cl, the molecular formula can be used to work out the displayed, structural and skeletal formulae because there is only one way in which these five atoms can be joined together.

EMPIRICAL FORMULAE

Another type of formula is an **empirical formula**. This shows the compound like a molecular formula, but the numbers of each atom are in their simplest possible whole-number ratio. This means that butane (molecular formula C_4H_{10}) has an empirical formula of C_2H_5.

In chemistry, the word empirical usually means 'as found from practical evidence'. You would normally work out this type of formula mathematically from the results of an experiment. You can see how we do this in **Topic 1D.1**.

DIFFERENT TYPES OF FORMULA FOR CHLOROETHANE

Until now, we have only considered the different types of formula using a hydrocarbon as the example.

Consider an example containing a third element (chlorine) – chloroethane. **Table A** shows the different types of formula for chloroethane.

TYPE OF FORMULA	FORMULA
displayed formula	H H | | H—C—C—Cl | | H H
structural formula	$CH_3–CH_2–Cl$ or CH_3CH_2Cl
skeletal formula	Cl
molecular formula	C_2H_5Cl
empirical formula	C_2H_5Cl

table A Different types of formula for chloroethane.

LEARNING TIP

For some compounds, the numbers in the molecular formula cannot be simplified. This means that the molecular formula and the empirical formula are identical.

CHECKPOINT

1. The displayed formula of a compound is:

```
    H   Cl  H
    |   |   |
H — C — C — C — H
    |   |   |
    H   Cl  H
```

Use a table (like the one for chloroethane) to show the structural, skeletal, molecular and empirical formulae of the compound above.

2. The skeletal formula of a compound is:

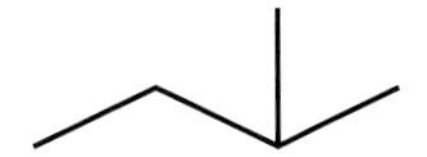

Use a table (like the one for chloroethane) to show the displayed, structural, molecular and empirical formulae of the compound above.

SUBJECT VOCABULARY

displayed formula shows every atom and every bond

structural formula shows (unambiguously) how the atoms are joined together

skeletal formula shows all the bonds between carbon atoms

molecular formula shows the actual numbers of each atom in the molecule

empirical formula shows the numbers of each atom in the simplest whole-number ratio

4A 3 FUNCTIONAL GROUPS AND HOMOLOGOUS SERIES

SPECIFICATION REFERENCE 4.4

LEARNING OBJECTIVES

- Understand the concepts of homologous series and functional groups.

FUNCTIONAL GROUP

A **functional group** in a molecule is an atom or group of atoms that gives the compound some distinctive and predictable properties. For example, the functional group of atoms shown as COOH gives substances such as vinegar a sour, acidic taste.

There are many organic compounds containing this group. Here are some examples:

$HCOOH$ CH_3COOH CH_3CH_2COOH $CH_3CH_2CH_2COOH$

HOMOLOGOUS SERIES

If you look at the formulae above, you can see that each formula has one more carbon atom and two more hydrogen atoms than the previous one, they differ by CH_2. These compounds are the first four members of what is called a **homologous series**. A homologous series is a set of compounds with the same functional group, similar chemical properties, and physical properties that show a gradation (a gradual change from one to the next).

ALKANES

The organic compounds that are mainly used as fuels are the alkanes (you will learn more about alkanes in **Topic 4B**). Alkanes are not considered to contain a functional group, but otherwise they form a homologous series. The displayed formulae of some alkanes are:

```
    H            H   H             H   H   H              H   H   H   H
    |            |   |             |   |   |              |   |   |   |
H — C — H    H — C — C — H     H — C — C — C — H      H — C — C — C — C — H
    |            |   |             |   |   |              |   |   |   |
    H            H   H             H   H   H              H   H   H   H
```

GENERAL FORMULAE

In **Topic 4A.2,** we looked at five different types of formulae. Now we are going to look at another type of formula. For the compounds in a homologous series, we can use a general formula to represent all of them. This is done by using the letter *n* for the number of carbon atoms, excluding any in the functional group. For the compounds with formulae ending in COOH, the general formula is $C_nH_{2n+1}COOH$.

Table A shows the formulae for some of the homologous series in this book.

NAME	GENERAL FORMULA	EXAMPLE
alkane	C_nH_{2n+2}	CH_4
alkene	C_nH_{2n}	C_2H_4
halogenoalkane	$C_nH_{2n+1}X$	CH_3CH_2Br
alcohol	$C_nH_{2n+1}OH$	CH_3CH_2OH

table A Examples of homologous series used in this book.

PROPERTIES OF A HOMOLOGOUS SERIES

ALKANES

We can use the alkanes to illustrate the similarity in chemical properties of a homologous series. For example, when alkanes are burned completely in air, they all form the same two products: carbon dioxide and water.

The commonest alkane is methane. The equation for the complete combustion of methane is:

$$CH_4 + 2O_2 \rightarrow CO_2 + 2H_2O$$

ALCOHOLS

We can use the alcohols to illustrate the gradation in physical properties of a homologous series. For example, the boiling temperatures of the first four alcohols are shown in **table B**.

FORMULA	BOILING TEMPERATURE / °C
CH_3OH	65
CH_3CH_2OH	79
$CH_3CH_2CH_2OH$	97
$CH_3CH_2CH_2CH_2OH$	117

table B You can see that as the number of carbon and hydrogen atoms increases, so does the boiling temperature.

EXAM HINT

An exam question may ask you to plot the boiling point of organic compounds (such as alcohols) with carbon number on the *x*-axis. Be sure to use a suitable scale and at least half of the axes provided in both directions.

LEARNING TIP

When looking at molecular models, remember that different elements are represented by different colours. The most common colours are black for carbon, white for hydrogen and red for oxygen.

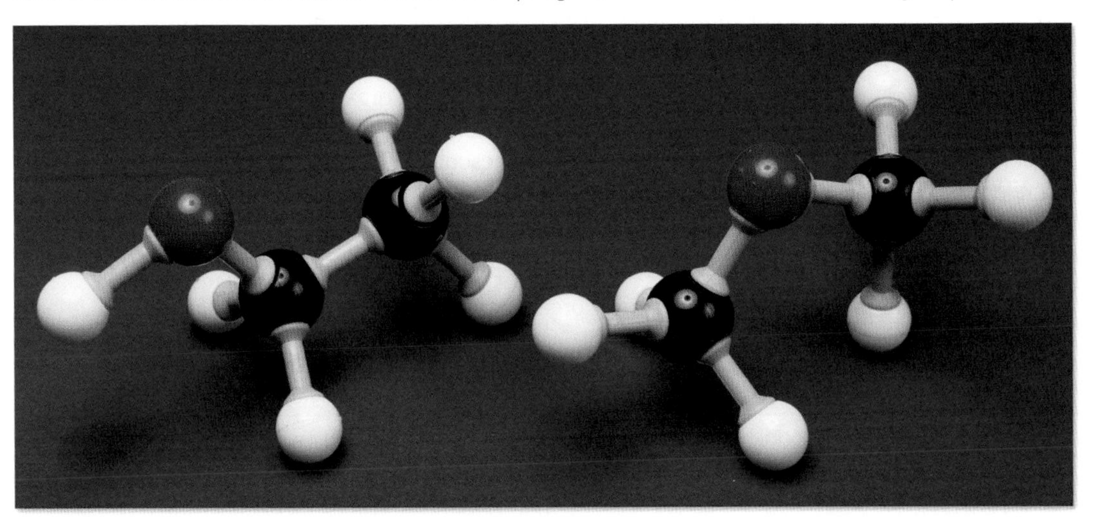

fig A Molecular models are very useful in organic chemistry. Both of these structures contain an oxygen atom (shown in red), but you can see that they belong to different homologous series.

CHECKPOINT

1. The equation for the complete combustion of propane is:

   $$C_3H_8 + 5O_2 \rightarrow 3CO_2 + 4H_2O$$

   What is the equation for the complete combustion of the alkane with five carbon atoms in its molecule?

2. The structural formula of a compound is CH_3CH_2CHO. What are the formulae of the two simpler compounds in the same homologous series?

SUBJECT VOCABULARY

functional group an atom or group of atoms in a molecule that is responsible for its chemical reactions

homologous series a family of compounds with the same functional group, which differ in formula by CH_2 from the next member

4A 4 NOMENCLATURE

SPECIFICATION REFERENCE
4.5(i)

LEARNING OBJECTIVES

- Apply the rules of IUPAC nomenclature to name compounds relevant to this course.

WHY DO WE NEED RULES FOR NAMING ORGANIC COMPOUNDS?

As the number of known organic compounds has increased, it has become harder to continue to find new names for them. In **Topic 4A.3**, we referred to the simplest organic compound (CH_4) as methane, but it was originally known as marsh gas (because it was found in marshes, where it was formed by the decay of plants). Many other organic compounds were named in similar ways.

An organisation called the International Union of Pure and Applied Chemistry (usually abbreviated to IUPAC ('eye-you-pack')) made some rules about how to name organic compounds. These rules are known as 'nomenclature'. The detailed rules needed for naming very complicated compounds are complex, but the simpler rules for the compounds described in your IAS course are much easier to understand and apply.

THE SIMPLE RULES OF NOMENCLATURE

Table A summarises the principles of naming organic compounds, including rules for **prefixes**, **suffixes** and **locants**.

THE PART OF THE NAME	HOW TO WRITE IT	EXAMPLE
Number of carbon atoms	This is shown by using a letter code (usually three or four letters).	meth = one carbon atom
Prefixes Suffixes	The presence of atoms other than carbon and hydrogen is shown by adding other letters before or after the code for the number of carbon atoms.	bromo = an atom of bromine ol = a hydroxyl group (OH)
Multiplying prefixes	The presence of two or more identical groups is shown by using the prefixes di-, tri-, etc.	di = two
Locants	Where atoms and groups can have different positions in a molecule, numbers and hyphens are used to show their positions. The numbers represent the carbon atoms in the longest chain that the atoms and groups are attached to.	2- = the atom or group is attached to the second carbon atom in the chain

table A The principles of naming organic compounds.

The letter codes for the number of carbon atoms (up to ten) are shown in **table B**.

NUMBER	CODE	PREFIX
1	meth	methyl
2	eth	ethyl
3	prop	propyl
4	but	butyl
5	pent	pentyl
6	hex	hexyl
7	hept	heptyl
8	oct	octyl
9	non	nonyl
10	dec	decyl

table B

APPLYING THE RULES TO WRITE NAMES

ALKANES

We can see how these rules work for some of the alkanes (**table C**). The names of all the alkanes end in *-ane*.

STRUCTURAL FORMULA	NAME
$CH_3—CH_2—CH_3$	propane
$CH_3—CH—CH_3$ with CH_3 on C2	methylpropane The locant 2- is not needed because if the methyl group below the horizontal chain were attached to one of the carbon atoms at either end of the chain, then there would be a sequence of four carbon atoms, and the compound would be named butane.
$CH_3—CH_2—CH—CH_2—CH_3$ with CH_3 on C3	3-methylpentane The longest carbon chain contains five carbon atoms, and there is a methyl group attached to the third one.
$CH_3—CH_2—CH_2—CH—CH_3$ with CH_3 on C4	2-methylpentane This is not 4-methylpentane because another rule is that the lowest locant numbers should be used.
$CH_3—CH—CH—CH_3$ with CH_3 on C2 and CH_3 on C3	2,3-dimethylbutane This example shows the use of a comma between the locants when the attached groups are the same.
$CH_3—CH_2—CH—CH—CH_3$ with $CH_2—CH_3$ on C3 and CH_3 on C4	3-ethyl-2-methylpentane This example illustrates the rule about prefixes being in alphabetical order. Ethyl comes before methyl because e comes before m in the alphabet. Notice also that it is not called 3-ethyl-4-methylpentane because these numbers (3 + 4) total more than the numbers 3 + 2 in the correct name.

table C Naming alkanes from structural formulae using the rules of IUPAC nomenclature.

ALCOHOLS

Next, look at the alcohols in **table D**. The rules for these are a bit different because the presence of the alcohol functional group is indicated by a suffix, not a prefix. The names for all the alcohols end in *-ol*.

STRUCTURAL FORMULA	NAME
$CH_3—CH_2—OH$	ethanol
$CH_3—CH_2—CH_2—OH$	propan-1-ol This time, the locant appears near the end of the name, but, as before, it appears directly before the letters representing the group.
$CH_3—CH—CH—CH_3$ with OH on C2 and CH_3 on C3	3-methylbutan-2-ol This example illustrates the use of both a prefix and a suffix. This is not called 2-methylbutan-3-ol because the lowest number locant should be used for the suffix functional group (-2-ol not -3-ol).
$CH_3—C—CH_2—CH_2—OH$ with two CH_3 on C	3,3-dimethylbutan-1-ol This example shows the use of prefixes, a suffix, locants and a comma. As with alkanes, the name uses locants that add up to the smallest possible number.

table D Naming alcohols from structural formulae using the rules of IUPAC nomenclature.

APPLYING THE RULES TO WRITE FORMULAE

Table E gives some examples of applying the rules the other way round, i.e. writing a structural formula for a compound from its IUPAC name.

LEARNING TIP

When you practise writing names from structural formulae, always check that the code you have used for the longest carbon chain is for the longest chain. This may not be the one shown horizontally. When you practise writing structural formulae from names, always check that each carbon has only four bonds. Showing three or five bonds is a common error.

NAME	STRUCTURAL FORMULA
dimethylpropane	*prop* indicates a chain of three carbon atoms *dimethyl* indicates two methyl groups attached to the chain No locants are used, so the two methyl groups must be attached to the carbon chain in a way that does not make the longest carbon chain any longer than three carbon atoms. So the structural formula is: CH_3 | $CH_3 — C — CH_3$ | CH_3
3-methylbutan-1-ol	*but* indicates a chain of four carbon atoms *methyl* indicates a CH_3 group *1-* and *3-* indicate attachments to the first and third carbon atoms in the chain So the structural formula is: $CH_2 — CH_2 — CH — CH_3$ | | OH CH_3

table E Writing structural formulae from IUPAC names.

CHECKPOINT

SKILLS REASONING

1. Write IUPAC names for the compounds with these structural formulae.

(a) $CH_2 — CH — CH_3$
| |
Br Br

(b) $CH_2 — CH — CH_3$
| |
Br OH

2. Write structural formulae for the compounds with these IUPAC names.

(a) 2,2-dimethylpentane (b) 2,3-dimethylbutan-2-ol

SUBJECT VOCABULARY

prefix a set of letters written at the beginning of a name
suffix a set of letters written at the end of a name
locant a number used to indicate which carbon atom in the chain an atom or group is attached to

4A 5 STRUCTURAL ISOMERISM

SPECIFICATION REFERENCE 4.10 4.11

LEARNING OBJECTIVES

- Understand the term 'structural isomerism' and draw the structural isomers of organic molecules, given their molecular formula.
- Draw the structural isomers of alkanes and cycloalkanes with up to six carbon atoms.

STRUCTURAL ISOMERS

Consider these two structures:

```
CH3— CH2— CH2— CH3          CH3— CH — CH3
                                 |
                                 CH3

      butane                 methylpropane
```

You can see that they are different compounds because their names and structures are different. However, their molecular formulae are the same. They can both be represented by C_4H_{10}. These two compounds are simple examples of **structural isomers**. In other words, they have the same molecular formula but different structural formulae.

▲ **fig A** By counting the atoms you can see that both structures have the molecular formula C_4H_{10}.

The molecular formula C_4H_{10} represents only two possible structures, but more complicated molecular formulae can be represented by several possible structures. As the number of carbon atoms increases, the number of possible structures can be hundreds or thousands.

TYPES OF STRUCTURAL ISOMERISM

There are two important types of structural isomerism.

CHAIN ISOMERISM

Chain isomerism refers to molecules with different carbon chains. Butane and methylpropane (shown in **fig A** above) are examples of chain isomers because their carbon chains are different.

POSITION ISOMERISM

Position isomerism refers to molecules with the same functional group attached in different positions on the same carbon chain. Propan-1-ol and propan-2-ol are simple examples of position isomerism:

```
CH2— CH2— CH3          CH3— CH — CH3
|                           |
OH                          OH

  propan-1-ol            propan-2-ol
```

They are examples of position isomers because the carbon chains are the same, but the OH groups are attached to different carbon atoms in the chain.

You might see examples where both of these types of isomerism are present:

```
CH2— CH2— CH — CH3          CH3— CH2— CH — CH2— CH3
|         |                           |
OH        CH3                         OH
```

DRAWING STRUCTURAL ISOMERS FROM MOLECULAR FORMULAE

You will remember that a molecular formula shows the actual numbers of atoms of each element, but does not show how the atoms are arranged. In some cases (such as C_2H_6) there is only one possible structure that can be drawn. In many cases there are lots of structures that can be drawn.

When you draw a structure, you could use a displayed formula, but with many bonds and atoms the formula would look complicated. At this stage, it is best to draw structural formulae. These formulae show every carbon (and other atoms such as oxygen) separately, with bonds between these atoms. The hydrogen atoms are grouped together to keep the structure simpler.

A good way to begin is by showing all the carbon atoms separately, in a straight line and with bonds between them. Then show the other atoms in as many different ways as possible, starting with atoms other than hydrogen.

Next see if there is another way to show the carbon atoms, not as a straight chain but as a shorter chain with one or more branches. Then add the other atoms in as many different ways as possible.

Time for some examples.

It is useful to refer to each carbon in a chain as C1, C2, etc.

WORKED EXAMPLE

Draw the structural isomers for the molecular formula $C_4H_8Cl_2$.

Step 1 The four carbon atoms can all be in a straight line, or in a straight line of three with a branch from the middle carbon to the fourth carbon. Let's start with the straight line of four:
C– C– C– C

Step 2 There are two Cl atoms. They could both be attached to C1 or to C2 in this chain. You might think that they could be attached to C3 and C4 as well. They could be, but counting from the right, C1 and C4 are the same, and C2 and C3 are the same.

The two Cl atoms could also be attached to different carbon atoms, such as C1 and C2, or C2 and C3. It will take a lot of practice before you decide on the correct number of different structures. You might miss some possible structures, or write two structures that look different but are actually the same.

Step 3 Finally, add the necessary hydrogen atoms, making sure that every carbon atom has four bonds.

Table A shows the approach.

CARBON CHAIN	ATTACHING THE TWO Cl ATOMS	RESULT
C– C– C– C	Both on C1	Cl | CH — CH_2 — CH_2 — CH_3 | Cl
	Both on C2	Cl | CH_3 — C — CH_2 — CH_3 | Cl
	On C1 and C2	Cl Cl | | CH_2 — CH — CH_2 — CH_3
	On C1 and C3	Cl Cl | | CH_2 — CH_2 — CH — CH_3
	On C1 and C4	Cl Cl | | CH_2 — CH_2 — CH_2 — CH_2
C — C — C | C	Both on C1	Cl | CH — CH — CH_3 | | Cl CH_3
	On C1 and C2	Cl Cl | | CH_2 — C — CH_3 | CH_3
	On C1 and C3	Cl Cl | | CH_2 — CH — CH_2 | CH_3

table A

You may think there are other possible structures. These others may look different on paper, but careful checking will show that they have already been used. Here is a useful way to be sure whether two structures are different, or whether they look different but are actually the same. Apply the IUPAC rules to name the structures. If this gives you two completely different names, then the structures are different. If it gives you the same names, then the structures are the same.

ALKANES AND CYCLOALKANES

You need to be able to draw the structures of all the alkanes and cycloalkanes with up to six carbon atoms. You have already seen structures of some examples of these homologous series, but now let us adapt the method we used for the structures of $C_4H_8Cl_2$. We can do this for the compounds with five carbon atoms as examples.

First, the alkanes. With 5 carbon atoms, there are only three different ways to arrange the carbon atoms. These are: (1) a line of 5; (2) a line of 4 with the 5th carbon branching off one of the central carbons; and (3) a line of 3 with the 4th and 5th carbons both branching off the middle one. Once you have realised this, you can just write these three possibilities, making sure that each carbon has the correct number of hydrogen atoms. Remember that each carbon atom must have 4 bonds, and the total number of hydrogen atoms must be 12.

$CH_3—CH_2—CH_2—CH_2—CH_3$

$CH_3—CH(CH_3)—CH_2—CH_3$

$CH_3—C(CH_3)_2—CH_3$

Now for the cycloalkanes. There can be a ring of 5 carbon atoms; or a ring of 4 carbon atoms with the 5th carbon atom attached to any one of the 4; or a ring of 3 carbon atoms with the 4th and 5th attached to the same or different carbons in the ring. Remember that each carbon atom must have 4 bonds, and the total number of hydrogen atoms must be 10. These are the possibilities.

CH_2 ring: CH_2, CH_2, CH_2, $CH_2—CH_2$ (five-membered ring)

CH_3 on CH in ring: $CH—CH_2$, $CH_2—CH_2$ (four-membered ring)

CH_3, CH_3 on C in ring: C, $CH_2—CH_2$ (three-membered ring)

CH_3 on CH in ring: CH, $CH_2—CH—CH_3$ (three-membered ring)

LEARNING TIP

Try drawing structures for all the possible isomers of the alkanes with six carbon atoms in the molecule. You will probably draw more than there really are. You can then decide which ones you don't need by naming them. If they look different but have the same name, then they are the same.

CHECKPOINT

1. Draw a structure for each of the alkane isomers with the molecular formula $C_3H_6Br_2$.

2. Draw a structure for each of the cycloalkane isomers with the molecular formula $C_3H_4Br_2$.

SUBJECT VOCABULARY

structural isomers compounds with the same molecular formula but different structural formulae

4A 6 TYPES OF REACTION

SPECIFICATION REFERENCE 4.6 4.7 4.8

LEARNING OBJECTIVES

- Classify reactions as addition, substitution, oxidation, reduction or polymerisation.
- Understand that bond breaking can be homolytic (to produce free radicals) or heterolytic (to produce ions).
- Know the definitions of the terms 'free radical' and 'electrophile'.

REACTIONS IN ORGANIC CHEMISTRY

In later topics of this book, you will learn about many different reactions and see many equations to represent these reactions. It will help if you can recognise these reactions as belonging to one of five main types. There is much more detail about these reactions in later topics (for example, **Topic 10**).

▲ **fig A** What types of reaction are occurring in these flasks?

ADDITION REACTIONS

In an **addition reaction**, two reactant species combine together to form a single product species. Usually all the species are molecules. A general equation for this type of reaction is:

$$A + B \rightarrow C$$

One example is the reaction between ethene and bromine:

$$C_2H_4 + Br_2 \rightarrow C_2H_4Br_2$$

SUBSTITUTION REACTIONS

In a **substitution reaction**, two reactant species combine together to form two product species. Usually all the species are molecules or ions. A general equation for this type of reaction is:

$$A + B \rightarrow C + D$$

One example is the reaction between bromoethane and potassium hydroxide. Potassium hydroxide is an ionic compound, and as the potassium ion is a spectator ion, the reaction only involves the hydroxide ion. An equation for this reaction is:

$$C_2H_5Br + OH^- \rightarrow C_2H_5OH + Br^-$$

In this reaction, the OH group has taken the place of, or *substituted*, the Br atom.

OXIDATION REACTIONS

In an **oxidation reaction**, one organic compound is oxidised, usually by an inorganic reagent. This means that the organic compound can either lose hydrogen or gain oxygen. There is not a suitable general equation that can be used for this type of reaction, but here is one example you will find in **Topic 10** – the oxidation of ethanol by a mixture of potassium dichromate(VI) and sulfuric acid. The equation is not written to include the inorganic reagent as it would be very complicated. Usually the oxygen atoms produced by the oxidising agent are shown using the symbol [O], so the equation then becomes:

$$C_2H_5OH + [O] \rightarrow CH_3CHO + H_2O$$

This reaction is classified as oxidation because the ethanol molecule loses two hydrogen atoms.

EXAM HINT

It is important that you use the symbol [O] for an oxidising agent. O_2 would be incorrect in this case as it suggests that the oxidising agent is molecular oxygen from the air.

REDUCTION REACTIONS

In a **reduction reaction**, one organic compound is reduced, sometimes by hydrogen gas and a catalyst and sometimes by an inorganic reagent. This means that the organic compound can either gain hydrogen or lose oxygen. There is not a suitable general equation that can be used for this type of reaction, but here is one example you will find in **Topic 5A.3**. This is the reduction of an alkene to an alkane by hydrogen gas and a nickel catalyst. The equation for the reaction is:

$$C_2H_4 + H_2 \rightarrow C_2H_6$$

You can now see why this reaction is classified as reduction. The ethene molecule gains two hydrogen atoms. Note that this is also an example of an addition reaction.

POLYMERISATION REACTIONS

In this book, all the **polymerisation reactions** you will meet are examples of addition polymerisation. In addition polymerisation, very large numbers of a reactant molecule (sometimes of two different reactant molecules) react together to form one very large product molecule. A general equation for this type of reaction is:

$$n\ \mathrm{WY}C{=}C\mathrm{XZ} \longrightarrow \left[-\overset{W}{\underset{Y}{C}}-\overset{X}{\underset{Z}{C}}- \right]_n$$

A familiar example of this type of reaction is the polymerisation of ethene to poly(ethene).

BOND BREAKING IN ORGANIC REACTIONS

Organic compounds contain covalent bonds, for example between

- two carbon atoms
- a carbon atom and a hydrogen atom
- a carbon atom and a halogen atom.

There are two different ways for the covalent bond to break. These are **homolytic fission** and **heterolytic fission**.

HOMOLYTIC FISSION

Like many other scientific terms, 'homolytic' comes from Greek: 'homo' indicates 'same' and 'lytic' indicates 'splitting'.

In homolytic fission, the shared pair of electrons in the covalent bond divide equally between the two atoms. This can be shown like this:

$$C\overset{\times}{\bullet}C \longrightarrow C^{\times} + {\bullet}C$$

Each product species keeps one of the electrons from the covalent bond. These species are called **free radicals**. Each free radical has an unpaired electron and is uncharged.

Homolytic fission usually occurs when the two atoms bonded together are identical or when they have similar electronegativities.

HETEROLYTIC FISSION

'Heterolytic' is another term from Greek: 'hetero' indicates 'different' and 'lytic' indicates 'splitting'.

In heterolytic fission, both electrons of the shared pair in the covalent bond are kept by one of the atoms. This can be shown like this:

$$C\overset{\times}{\bullet}C \longrightarrow C\overset{\times}{\bullet} + \ C$$

The left-hand product species keeps both of the electrons from the covalent bond. This species is a negative ion. The right-hand product species does not keep either of the electrons from the covalent bond. This species is a positive ion.

Heterolytic fission could also occur like this:

$$C\overset{\times}{\bullet}C \longrightarrow C \ + \overset{\times}{\bullet}C$$

Here there are still two ions formed, but this time the right-hand product is the negative ion.

Heterolytic fission usually occurs when the two atoms bonded together have different electronegativities. The atom with the higher electronegativity is the one that keeps both electrons from the bond.

ELECTROPHILES

The origin of the term **electrophile** is 'electron', which indicates negative charge, and 'phile', which means liking. You may have come across the word 'bibliophile', which means a person who likes books. An electrophile refers to a chemical species that 'likes negative charge'. So how does this term fit in with this topic? You remember that opposite charges (positive and negative) attract each other. The positive ion produced by heterolytic fission will be attracted to a region of high electron density in another molecule. This region is often labelled with the symbol $\delta-$. You will learn much more about electrophiles and their importance in reaction mechanisms in **Topic 5A.4**.

LEARNING TIP

Look at some equations in **Topic 4** and try to classify at least two as addition, substitution, oxidation, reduction or polymerisation.

CHECKPOINT

1. What type of reaction is shown by each of the following equations?

(a) $CH_3CHO + 2[H] \rightarrow CH_3CH_2OH$

(b) $CH_3CHO + [O] \rightarrow CH_3COOH$

(c) $CH_3CHO + HCN \rightarrow CH_3CH(OH)CN$

(d) $CH_3Br + KOH \rightarrow CH_3OH + KBr$

2. Write an equation to represent each of the following.

(a) Homolytic fission of the carbon-to-carbon bond in C_2H_6.

(b) Heterolytic fission of the carbon-to-bromine bond in C_2H_5Br.

SUBJECT VOCABULARY

addition reaction reaction in which two molecules combine to form one molecule

substitution reaction reaction in which one atom or group is replaced by another atom or group

oxidation reaction reaction in which a substance gains oxygen or loses hydrogen

reduction reaction reaction in which a substance loses oxygen or gains hydrogen

polymerisation reaction reaction in which a large number of small molecules react together to form one very large molecule

homolytic fission the breaking of a covalent bond where each of the bonding electrons leaves with one species, forming a free radical

heterolytic fission the breaking of a covalent bond so that both bonding electrons are taken by one atom

free radical a species that contains an unpaired electron

electrophile a species that is attracted to a region of high electron density

7 HAZARDS, RISKS AND RISK ASSESSMENTS

LEARNING OBJECTIVES

- Understand the difference between hazard and risk.
- Understand the hazards associated with organic compounds and why it is necessary to carry out risk assessments when dealing with potentially hazardous materials.
- Suggest ways in which risks can be reduced and reactions carried out safely.

SAFETY IN CHEMISTRY LABORATORIES

Incidents that cause harm to people are rare in school and college laboratories. One of the reasons for this is that laboratories need to consider the hazards of doing chemistry experiments and use safe methods of working.

This applies to all chemistry experiments, but especially to those involving organic compounds. It is particularly important in experiments that you plan yourself, but also for those where you are following a method you have been given.

When you plan an organic synthesis, you need to consider the hazards associated with the reactants, the substance you are synthesising, and also any intermediates formed.

HAZARD AND RISK

The **hazard** of a chemical substance relates to the inherent properties of the substance. The **risk** is more to do with how you plan to use it and the chance of it causing harm. Most people would consider that water is completely safe and has no hazards. In most situations, this is the case. However, consider a beaker of water being boiled on a tripod and gauze. The steam coming from the water, and the boiling water itself, could both cause harm if they came into contact with your skin.

Now consider a substance that most people would consider hazardous. You have probably used hydrochloric acid in experiments involving marble chips. You will have been told to use eye protection when using the acid because of the harm it could do, especially if it got into your eyes.

The hazard exists because hydrochloric acid is corrosive and so could damage your eyes..

The risk depends on how likely it is that the acid can get into your eyes.

Using eye protection does not affect the hazard, but greatly decreases the risk.

HAZARD WARNING SYMBOLS

A long time ago, bottles containing certain chemicals were labelled with the word POISON. This early attempt to prevent harm to laboratory workers was well-intentioned, but some people might think that bottles without such a label contained harmless substances.

More recently, symbols (sometimes called pictograms) have been used to label substances but also to identify the actual hazard of the substance inside. The actual symbols have changed over the years, and you may see older and newer types together in the same laboratory.

Older ones are often square in shape with an orange background such as these:

The symbols in current use are red diamond shapes. One department of the United Nations Organisation has developed these GHS labels for international use. 'GHS' is an abbreviation for Globally Harmonised System of Classification and Labelling of Chemicals, and the use of these labels is spreading throughout the world. **Table A** shows some of the more common ones, and includes a short description of their meanings.

SYMBOL	MEANING	
	Health hazard	includes warning on skin rashes, eye damage and ingestion
	Corrosive	can cause skin burns and permanent eye damage
	Flammable	can catch fire if heated or comes into contact with a flame
	Acute toxicity	can cause life-threatening effects, even in small quantities

table A Some pictograms used for chemical hazards.

In some cases, the substance may have more than one symbol, especially when it is an aqueous solution. For example, you are likely to use hydrochloric acid in three different concentrations:

- in a titration, it may have a concentration of about 0.1 mol dm^{-3}
- as a general laboratory reagent, it may have a concentration of 1 or 2 mol dm^{-3}
- for some purposes, you may use concentrated hydrochloric acid with a concentration of more than 10 mol dm^{-3}.

These very different concentrations have different hazards.

RISK ASSESSMENTS AND CONTROL MEASURES

The person in charge of a laboratory (or any other place of work) is responsible for making **risk assessments**. First, they look at the hazards of all the chemical substances, being guided by the hazard symbols. Then they consider the ways in which these substances will be used (this is assessing the risk). Finally, they write some guidelines for those who use the laboratory (these are the control measures).

EXAM HINT

If an exam question asks you to suggest a precaution when tackling a practical assignment, it is NOT enough to say 'wear a lab coat and goggles'. This should be standard practice for any practical assignment. You should comment on a particular procedure in the practical, such as, 'HCl gas is produced and so this step should be carried out in a fume cupboard'.

Such guidelines will consider many different factors, including:

- the amount used
- the age and experience of the person using it
- whether it will be heated
- whether ventilation or a fume cupboard should be used.

The control measures may refer to:

- the type of eye protection that should be worn
- the need to wear gloves
- keeping the cap on the bottle after removing some of the substance
- keeping the substance away from a source of heat
- what to do if some of the substance is spilled on the floor or gets on the skin.

Remember that there are also hazards and risks in your home. Many cleaning materials contain hazardous materials. **Fig A** shows a warning label for a household oven cleaner.

OVEN CLEANER
DANGER

HAZARD STATEMENTS:
Causes severe skin burns and eye damage

PRECAUTIONARY STATEMENTS:
- Wear protective gloves, protective clothing, eye protection, face protection.
- Wash hands thoroughly after handling.

- IF SWALLOWED: Rinse mouth Do NOT induce vomiting.
- IF IN EYES: Rinse cautiously with water for several minutes. Remove contact lenses if present and easy to do; continue rinsing.
- IF ON SKIN: Remove/Take off immediately all contaminated clothing. Rinse skin with water/shower.
- Dispose of contents/container in accordance with local regulations.

SEE SDS FOR MORE INFORMATION

▲ **fig A** You may find examples of hazards, risks and control measures in your home.

APPARATUS

Don't forget about the apparatus you might use in chemistry laboratories. Some of this apparatus may also have hazards.

For example, mercury thermometers obviously contain mercury, which is a hazardous substance. This is why you are more likely to use spirit thermometers and digital thermometers, their hazards are much lower.

Another example is methods of heating. Traditionally, heating involved the use of a Bunsen burner, tripod and gauze. Perhaps an electrical heating mantle would be safer, as well as easier to use.

In organic chemistry, some experiments involve glass apparatus for techniques such as distillation. A long time ago, the traditional way to connect the different pieces of glass apparatus was by using connecting glass tubing with corks or rubber bungs. More recently, apparatus made only of glass, connected mainly by ground glass joints, has been used. This is less hazardous because it does not involve the risks involved in assembling the apparatus.

LEARNING TIP

Consider the hazards of the substances you will use in an experiment of your choice. Decide what the risks of doing the experiment are, and suggest what control measures you think should be in place.

CHECKPOINT

1. Explain the difference between hazard and risk, using dilute sulfuric acid as an example.
2. A substance to be heated in an experiment is in a bottle labelled 'flammable'.

   Suggest two control measures that could be used to reduce the risk to a laboratory worker using this substance.

SUBJECT VOCABULARY

hazard something that could cause harm to a user
risk the chance of a hazard causing harm
risk assessment the identification of the hazards involved in carrying out a procedure and the control measures needed to reduce the risks from those hazards

4B 1 ALKANES FROM CRUDE OIL

LEARNING OBJECTIVES

- Know that alkanes are used as fuels and obtained from the fractional distillation, cracking and reforming of crude oil, and be able to write equations for these reactions.

THE NEED FOR FUELS

The worldwide demand for energy is huge and steadily rising. At present, most of this energy comes from burning fossil fuels, in the form of coal, crude oil and natural gas. Most compounds in crude oil and natural gas are alkanes.

▲ **fig A** Petrol is just one product of crude oil, but is perhaps the best known.

In this topic we will look at the three main processes used to convert crude oil into fuels. They are:

- **fractional distillation**
- **cracking**
- **reforming.**

These processes are used in oil refineries located all over the world. You will already know something about fractional distillation, but you may be less familiar with cracking and reforming.

▲ **fig B** Several different processes take place in oil refineries. This distillation plant is just one small part of a large refinery.

FRACTIONAL DISTILLATION

Crude oil is a complex mixture of compounds, mostly hydrocarbons. The composition of the mixture varies quite a lot depending on which part of the world the crude oil comes from. The process is sometimes called 'fractionation' because it involves converting the crude oil into a small number of fractions. The number of fractions varies between different refineries but is typically six. Fractionation is done in a distillation column.

EXAM HINT

It is important to remember that a specific fraction from crude oil is still a mixture of compounds. It is just a smaller number of compounds within a defined range of boiling points.

THE PROCESS

The crude oil is first heated in a furnace, which turns most of it into vapour, which is then passed into the column near the bottom. There is a **temperature gradient** in the column: it is hotter near the bottom and cooler near the top.

As the vapour passes up the column through a series of bubble caps, different fractions condense at different heights in the column, depending on the boiling temperature range of the molecules in the fraction.

- Near the bottom of the column, the fractions contain larger molecules with longer chains and higher boiling temperatures.
- Near the top of the column, the fractions contain smaller molecules with shorter chains and lower boiling temperatures.
- Some of the hydrocarbons in crude oil are dissolved gases, and they rise to the top of the column without condensing.

Some fractions still contain many different compounds, so they may undergo further fractional distillation separately.

CRACKING

The world has fewer uses for longer-chain hydrocarbons so there is a surplus of these. The demand for shorter-chain hydrocarbons is much higher because they are better fuels and can be used to make other substances such as polymers. Unfortunately, there are not enough of these to satisfy the demand. The solution is to convert the longer chains into shorter chains, which is what happens in cracking.

THE PROCESS

Cracking is done by passing the hydrocarbons in the heavier (longer chain) fractions through a heated catalyst, usually of zeolite, which is a compound of aluminium, silicon and oxygen. This causes larger molecules to break up into smaller ones. From one large molecule, at least two smaller molecules are formed.

A good example is the cracking of decane into octane and ethene:

decane

$C_{10}H_{22}$

octane

C_8H_{18}

+

ethene

C_2H_4

This is a good example because the two smaller molecules that are formed have familiar uses. Octane is one of the hydrocarbons in petrol and ethene is used to make polymers.

REFORMING

So far, we haven't mentioned one important point about the alkanes used as fuels. During the very rapid combustion that occurs in vehicle engines, not all hydrocarbons of the right size burn in the same way. Those with straight chains burn less efficiently than those with branched chains and those with rings (cyclic compounds). The process of reforming is used to convert straight-chain alkanes into branched-chain alkanes and cyclic hydrocarbons by heating them with a catalyst, usually platinum. This helps them to burn more smoothly in the engine.

EXAMPLES

Here are some examples of reforming reactions, using skeletal formulae.

In the first one, pentane (C_5H_{12}) is converted into a cyclic alkane:

pentane → cyclopentane + H_2

In the second one, heptane (C_7H_{16}) is converted into methylbenzene, which is a cyclic hydrocarbon but not an alkane. You will learn the meaning of the circle inside the hexagon later.

heptane → methylbenzene + $4H_2$

In each example, hydrogen is formed. It is a useful by-product.

LEARNING TIP

Be careful not to confuse fractional distillation and cracking. Fractional distillation involves separating existing compounds, not making new ones in a chemical reaction.
Cracking involves a chemical reaction in which new compounds are formed.

CHECKPOINT

SKILLS PROBLEM-SOLVING, ADAPTIVE LEARNING

1. One molecule of the alkane $C_{12}H_{26}$ is cracked to form two molecules of ethene and one molecule of a different alkane. What is the molecular formula of the alkane formed?
2. The products of a cracking reaction are two molecules of ethene and one molecule of pentane. What is the molecular formula of the alkane that is cracked?

SUBJECT VOCABULARY

fractional distillation the process used to separate a liquid mixture into fractions by boiling and condensing
cracking the breakdown of molecules into shorter ones by heating with a catalyst
reforming the conversion of straight-chain hydrocarbons into branched-chain and cyclic hydrocarbons
temperature gradient the way in which the temperature changes up and down the column

4B 2 ALKANES AS FUELS

SPECIFICATION REFERENCE 4.13 4.14 4.17(i)

LEARNING OBJECTIVES

- Know that pollutants, including carbon monoxide, oxides of nitrogen and sulfur, carbon particulates, and unburned hydrocarbons are emitted during the combustion of alkane fuels.
- Understand the problems arising from pollutants from the combustion of alkane fuels, limited to the toxicity of carbon monoxide and why it is toxic, and the acidity of oxides of nitrogen and sulfur.
- Understand the reactions of alkanes with oxygen in the air (combustion).

THE COMPLETE COMBUSTION OF ALKANES

As we mentioned earlier, alkanes can burn. They are burned in vast quantities to provide the world's energy. For example, propane is sold in containers at high pressure for use as a fuel, both in homes and when camping. The equation for its **complete combustion** is:

$$C_3H_8 + 5O_2 \rightarrow 3CO_2 + 4H_2O$$

EXAM HINT

If you are asked to write an equation for the complete combustion of a hydrocarbon, balance the carbon and hydrogen atoms first and finish with the required amount of O_2. Do not include state symbols unless you are specifically asked to.

THE PRODUCTS OF COMBUSTION

One of the products of combustion is water, which is not a problem as it simply adds to the total quantity of global H_2O.

The other product is carbon dioxide. As you know, this is a greenhouse gas and most scientists consider its increasing production to be responsible for global warming, climate change and other problems.

EXAM HINT

Remember, carbon dioxide works by trapping infrared radiation that is emitted from the Earth's surface and so prevents it from escaping back into space.

Unfortunately, other problems are caused by using alkanes as fuels. The water and carbon dioxide formed are not considered to be pollutants by most people, but some other compounds formed during the combustion of alkanes are definitely pollutants.

INCOMPLETE COMBUSTION

Sometimes the combustion of an alkane is incomplete because there is not enough oxygen present, or because the combustion is very rapid. All of the hydrogen atoms in an alkane molecule are converted into water, but some of the carbon atoms can form gaseous carbon monoxide or solid carbon. These products can cause problems.

CARBON

You can often see when **incomplete combustion** forms solid carbon. This can be seen as smoke in the air or soot on the burner. One example of an equation for a reaction in which carbon is formed is:

$$C_3H_8 + 4O_2 \rightarrow C + 2CO_2 + 4H_2O$$

Notice that in this reaction two of the carbon atoms in propane undergo complete combustion and one does not. Tiny particles of carbon in the atmosphere can be harmful, but there is another product of combustion that can be fatal.

CARBON MONOXIDE

Carbon monoxide is a toxic gas that causes the death of many people each year. It acts by preventing the transport of oxygen around the body. It is colourless and odourless, so people breathe it into their lungs without knowing, which is why it is sometimes described as 'the silent killer'. Here is an example of an equation for a reaction in which carbon monoxide is formed:

$$C_3H_8 + 4O_2 \rightarrow 2CO + CO_2 + 4H_2O$$

UNBURNED HYDROCARBONS

The ultimate example of incomplete combustion is when the hydrocarbon does not burn at all. A small proportion of the hydrocarbons in a fuel are released into the atmosphere unchanged. They are known as unburned hydrocarbons (sometimes abbreviated to UHC).

OXIDES OF SULFUR

Some of the molecules in crude oil contain atoms of sulfur, and these may not be removed by the fractional distillation, cracking or reforming processes. During the combustion of alkanes, these atoms of sulfur form sulfur dioxide gas, and then can react in the atmosphere to form sulfur trioxide gas. The equations for these reactions are:

$$S + O_2 \rightarrow SO_2 \quad \text{and} \quad 2SO_2 + O_2 \rightarrow 2SO_3$$

Both sulfur gases are acidic oxides. When they dissolve in water in the atmosphere, they form sulfurous acid and sulfuric acid:

$$SO_2 + H_2O \rightarrow H_2SO_3 \quad \text{and} \quad SO_3 + H_2O \rightarrow H_2SO_4$$

Both acids contribute to the formation of acid rain. Acid rain is responsible for a lot of environmental damage, including damage to aquatic life in lakes and rivers, and damage to crops and forests.

OXIDES OF NITROGEN

Although very few molecules used as alkane fuels contain atoms of nitrogen, their combustion occurs at very high temperatures. Under these conditions, especially around the spark plugs in cars, these very high temperatures cause nitrogen molecules in the air to react with oxygen molecules. These reactions lead to the formation of what are collectively known as oxides of nitrogen. They are represented by the formula NO_x. There are several of these oxides, but the main ones are nitrogen monoxide (NO) and nitrogen dioxide (NO_2).

At very high temperatures, the main reaction is:

$$N_2 + O_2 \rightarrow 2NO$$

However, nitrogen monoxide can then react with more oxygen in the atmosphere as follows:

$$2NO + O_2 \rightarrow 2NO_2$$

Nitrogen dioxide is acidic and can dissolve in water in the atmosphere, forming nitrous acid and nitric acid:

$$2NO_2 + H_2O \rightarrow HNO_2 + HNO_3$$

Both acids contribute to environmental damage in the same way as sulfurous acid and sulfuric acid.

fig A Air pollution is a growing problem in many cities.

CATALYTIC CONVERTERS TO THE RESCUE

Cars and other road vehicles are responsible for a lot of air pollution. The widespread use of catalytic converters fitted to exhaust systems has made pollution less of a problem.

There are different types of catalytic converter, but they all use small quantities of precious metals such as platinum, rhodium and palladium. These metals are spread thinly over a honeycomb mesh to increase the surface area for reaction (and also to save money). One common type is known as a three-way catalyst because it can remove three different pollutants: carbon monoxide, unburned hydrocarbons and oxides of nitrogen.

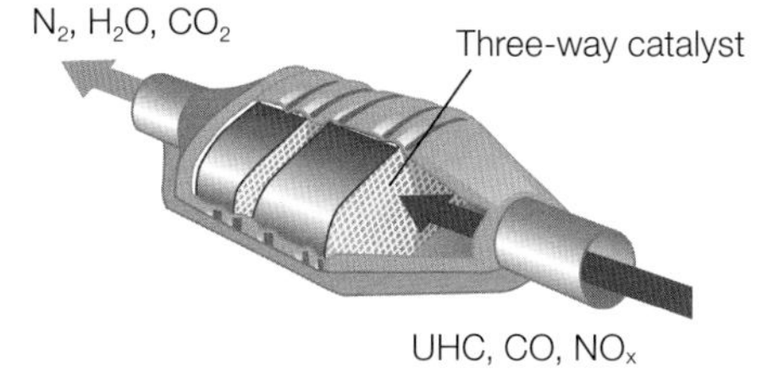

fig B A three-way catalytic converter.

EXAM HINT

Also note that like many (but not all) catalysts, platinum, rhodium and palladium are found in the d-block of the Periodic Table.

As the exhaust gases from the engine pass through the catalytic converter, several reactions can occur.

Examples are the oxidation of the carbon monoxide and the oxidation of unburned hydrocarbons:

$$2CO + O_2 \rightarrow 2CO_2 \quad \text{and} \quad C_8H_{18} + 12\tfrac{1}{2}O_2 \rightarrow 8CO_2 + 9H_2O$$

Here is another useful reaction that removes two pollutants at the same time:

$$2NO + 2CO \rightarrow N_2 + 2CO_2$$

The catalysts currently used are not very good at removing sulfur compounds. The best way to prevent sulfur-based pollution is to remove the sulfur compounds from the fuel before the fuel is burned. This is done in some countries, where the resulting fuel is described as low sulfur or ultra-low sulfur fuel.

LEARNING TIP

Remember that even in the incomplete combustion of an alkane, all of the hydrogen atoms are completely oxidised to water.

CHECKPOINT

1. Summarise information about the products of combustion of alkanes, including names and whether complete or incomplete combustion was involved.
2. In a table, summarise the substances that react in a catalytic converter, and the products formed from them.

SUBJECT VOCABULARY

combustion a chemical reaction in which a compound reacts with oxygen
complete combustion reaction in which all of the atoms in the fuel are fully oxidised, producing only carbon dioxide and water
incomplete combustion reaction in which some of the atoms in the fuel are not fully oxidised, producing carbon dioxide, carbon monoxide and soot (unburnt carbon)

4B 3 ALTERNATIVE FUELS

SPECIFICATION REFERENCE 4.15 4.16

LEARNING OBJECTIVES

- Discuss the reasons for developing alternative fuels in terms of sustainability and reducing emissions, including the emission of CO_2 and its relationship to climate change.
- Apply the concept of carbon neutrality to different fuels, such as petrol, bioethanol and hydrogen.

THE NEED FOR ALTERNATIVE FUELS

There are serious concerns about relying on the combustion of fossil fuels to produce energy. We have already considered the pollution caused by the combustion of alkanes.

The other concerns are:

- the depletion of natural resources
- global warming and climate change.

Recently, there have been attempts to produce new fuels as alternatives to fossil fuels. Most of these fuels can be described as **biofuels**, which means that they are obtained from living matter that has died recently, rather than having died many millions of years ago. A wide definition of biofuels would include wood, which has been used as a fuel for many centuries and is still important in some countries today.

The terms 'renewable' and 'non-renewable' are often used when discussing energy sources. Non-renewable usually refers to coal, oil and natural gas. Renewable sources include biofuels, but also sunlight, wind, waves and tides, and geothermal energy.

CARBON NEUTRALITY

Fuels can be considered in terms of their carbon neutrality. Ideally, a fuel should be completely carbon neutral, although few are. The closer a fuel is to being carbon neutral, the better.

WHAT DOES CARBON NEUTRAL MEAN?

'**Carbon neutral**' is a term used to represent the idea of carbon dioxide neutrality. For example, when a tree grows, it absorbs carbon dioxide from the atmosphere, and the carbon atoms become part of the structure of the tree. If the tree is cut down and the wood is burned, then carbon dioxide is formed during its combustion. If the amount of carbon dioxide formed in the combustion is the same as the amount absorbed during the tree's growth, then the wood used is described as carbon neutral. This is because, over the time period between the tree starting to grow and the use of its wood as a fuel, the amount of carbon dioxide in the atmosphere has not been altered by its combustion.

You might imagine that fossil fuels, such as those formed from trees, could be described as carbon neutral because they too absorbed carbon dioxide during their growth and form the same amount of carbon dioxide when they are burned. The reason that they are not considered to be carbon neutral is that the carbon dioxide was absorbed from the atmosphere millions of years ago, when the amount of carbon dioxide in the atmosphere was much higher. When fossil fuels are burned, this increases the amount of carbon dioxide in today's atmosphere.

BIOALCOHOLS

You might think that **bioalcohols** are carbon neutral fuels because they are made from recently-grown plants that have absorbed carbon dioxide from the atmosphere. However, this does not recognise the fact that the plants have to be harvested, transported to a factory and processed in the factory, and the products transported to a point of sale. All these stages involve the use of energy, much of which involves the formation of carbon dioxide. Overall, the use of bioalcohols involves forming more carbon dioxide than is absorbed. Even so, bioalcohols are closer to being carbon neutral than fossil fuels.

▲ **fig A** Bioethanol (on the left) is sold next to unleaded petrol and diesel.

BIOETHANOL

Currently, the commonest bioalcohol is bioethanol. Remember that bioethanol and ethanol are not different compounds. Bioethanol is identical to ethanol. The 'bio' part of the name only refers to the method of production. For centuries, ethanol has been produced by the fermentation of sugars. This involves the use of yeasts that contain enzymes, but there is an upper limit to the concentration of the ethanol in the solution. The ethanol has to be separated from the much larger amount of water before it can be used as a fuel, and this separation requires energy.

It is now possible to use a wide range of plants, and also plant waste to produce ethanol. . The upper limit to the amount of ethanol that can be obtained from a given amount of starting material is increasing, and is much higher than in traditional fermentation. In the United States, corn is the main source of ethanol used in cars.

▲ **fig B** Not everyone agrees that it is a good idea for corn to be used as a fuel when it could be used to feed people.

EXAM HINT

Exam questions asking you to compare fuels usually carry 4 or more marks and are worth preparing on rough paper first before transferring to your exam paper. Avoid poorly presented and poorly structured answers.

COMPARING FUELS

BIOFUELS

The choice of alternative fuels is continually changing, as new sources of starting material and new processing methods are investigated. There are many factors to consider in any comparison, but for biofuels such as bioethanol these include the following.

- Land use – how much land is used to grow the crop? Should the land be used for other purposes, especially to grow food to feed people?
- Yield – how much of a crop can be grown on a given piece of land, and how quickly does it grow? What percentage of the carbon and hydrogen atoms in the crop ends up in the fuel?
- Manufacture and transport – how much energy is used in growing (including any fertilisers), processing and transporting the crop?
- Carbon neutrality – how close is the fuel to being carbon neutral?

HYDROGEN

For a long time, hydrogen has been considered an ideal alternative fuel, although it is not a biofuel. There are two main ways in which hydrogen can be used as a fuel in cars:

- it can be burned instead of a fossil fuel such as petrol or natural gas
- it can be used in a fuel cell to generate electricity that powers an electric motor.

Using a hydrogen fuel cell or burning hydrogen instead of a hydrocarbon in a car seems promising. No carbon dioxide is produced, which suggests no increase in the greenhouse effect. Hydrogen is very common, and is much more abundant than carbon in the Earth's crust. However, nearly all hydrogen is present in the water molecules in the oceans. Obtaining hydrogen gas from water is not difficult, but it requires energy, and where does the energy come from to do this? It could come from electricity, but how is the electricity generated? If the electricity comes from power stations that burn fossil fuels, then much of the advantage of using hydrogen instead of hydrocarbons is lost. Another major problem is hydrogen storage. The gas has a very low density, so storing enough inside a car needs a container under very high pressure. This means a very strong container and therefore one that is very heavy. It could be stored as a liquid, but as its boiling temperature is –253 °C, the container would need to be very cold and very well insulated.

Battery technology has improved greatly in recent years, and the range of battery-powered cars is continually increasing. Many scientists believe that this is a more practical alternative than using hydrogen as a fuel.

FOSSIL FUELS AND BIOFUELS

You should also be able to compare fossil fuels with biofuels. Here is one example of a comparison between such fuels.

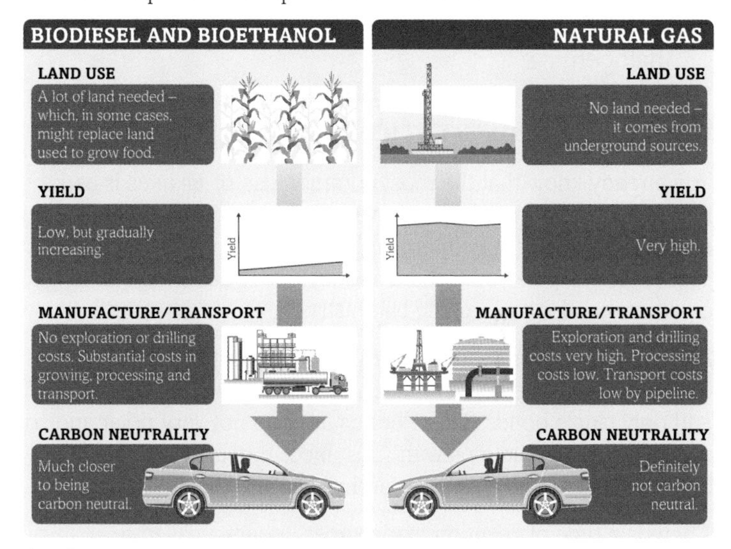

▲ **fig C** Comparing bioethanol with natural gas as fuels for cars.

DID YOU KNOW?

One problem with using bioethanol as a fuel in vehicles is that internal combustion engines were designed to use petrol and not bioethanol. Currently very few cars use 100% bioethanol as their fuel. Normally the bioethanol sold is a blend of bioethanol and petrol, often identified by its E-number. Two common blends are E10 (10% bioethanol and 90% petrol) and E85 (85% bioethanol and 15% petrol).

LEARNING TIP

Focus on the bigger picture of comparing fuels, not on the chemical reactions that occur in their manufacture.

CHECKPOINT

SKILLS REASONING, ARGUMENTATION

1. No carbon dioxide is formed when hydrogen is used as a fuel. Suggest why hydrogen is not a carbon neutral fuel.
2. Summarise reasons why a biofuel may not be carbon neutral.

SUBJECT VOCABULARY

biofuel fuel obtained from living matter that has died recently
bioalcohol fuel made from plant matter, often using enzymes or bacteria
carbon neutral a considered net zero effect on the amount of carbon dioxide in the atmosphere

4B 4 SUBSTITUTION REACTIONS OF ALKANES

SPECIFICATION REFERENCE 4.17(ii) 4.18

LEARNING OBJECTIVES

- Understand the reactions of alkanes with halogens.
- Understand the mechanism of the free radical substitution reaction between an alkane and a halogen.

WHAT IS A SUBSTITUTION REACTION?

You already know that the most common use of alkanes is as fuels. Combustion reactions are very important in producing energy, but are not very interesting from a chemist's point of view. This topic will give you a chance to increase your understanding of other reactions in organic chemistry.

Alkanes, apart from readily undergoing combustion, are fairly unreactive because they contain only carbon and hydrogen atoms and only single bonds. These bonds are also not very polar and so do not undergo reactions with substances that are considered to be very reactive, such as acids and alkalis and reactive metals.

There is a type of reaction that alkanes undergo, called a **substitution reaction,** which we will now look at in detail. Here is the equation for a reaction of the simplest alkane, methane:

$$CH_4 + Cl_2 \rightarrow CH_3Cl + HCl$$

You can see from the equation that one of the hydrogen atoms in methane has been replaced (substituted) by an atom of chlorine. The reaction can be described as chlorination or, in general (if another halogen were used), halogenation.

MECHANISMS

As you study organic chemistry more thoroughly, you will come across reactions that have been carefully studied, and for which there are explanations of exactly how they occur. The equation shown above for the reaction of methane with chlorine only shows the formulae of the reactants and products. It does not show how or why the reaction occurs.

A **mechanism** tries to explain the actual changes that occur during a reaction, especially in the bonding between the atoms. A mechanism is a sequence of two or more steps, each one represented by an equation, that shows how a reaction takes place.

THE CHLORINATION OF METHANE

When methane is only mixed with chlorine, no reaction occurs. If the temperature is increased, a reaction eventually occurs. However, the reaction will occur at room temperature if the mixture is exposed to ultraviolet radiation (or sunlight). We know that alkanes are not affected by ultraviolet radiation, but that ultraviolet radiation can affect chlorine. What happens in this reaction?

STEP 1

Ultraviolet radiation breaks the chlorine molecule into chlorine atoms. As the bond in the chlorine molecule consists of a shared pair of electrons, which are equally shared between the two atoms, then each chlorine atom takes one electron from the shared pair. This kind of bond breaking is called **homolytic fission**, a type of reaction you met in **Topic 4A.6.**

Step 1 can be represented in an equation:

$$Cl_2 \rightarrow Cl\bullet + Cl\bullet$$

However, it is better to use a different type of equation that shows clearly what happens to the electrons in the Cl–Cl bond.

$$Cl\!:\!Cl \rightarrow Cl^{\bullet} + {}_{\bullet}Cl$$

The dots on the products each represent an unpaired electron. The formula Cl• does not represent an ion or a molecule; the term **free radical** (sometimes just radical) is used for it. A free radical is a species with an unpaired electron.

This equation above shows the original shared pair of electrons in the Cl–Cl bond. Each curly half-arrow shows what happens to the electrons. The upper arrow shows that one electron stays with the left-hand chlorine. The lower arrow shows that the other electron stays with the right-hand chlorine. Notice that the equation involves one molecule forming two free radicals. This type of reaction is called **initiation**, which means it starts the sequence of steps that forms the overall reaction.

STEP 2

Chlorine free radicals are very reactive species and when they collide with methane molecules they react by removing a hydrogen atom. An equation for this process is:

$$Cl\bullet + CH_4 \rightarrow HCl + CH_3\bullet$$

Notice that $CH_3\bullet$ is formed. This is a methyl free radical and, like Cl•, it is also very reactive. It can then react with chlorine molecules as follows:

$$CH_3\bullet + Cl_2 \rightarrow CH_3Cl + Cl\bullet$$

In this reaction, the methyl free radical removes a chlorine atom from a chlorine molecule, forming chloromethane and a chlorine free radical.

Notice that these two equations both involve one free radical reacting with one molecule, and that the products are also one free radical and one molecule. This type of reaction is called **propagation**, which means that the two steps considered together result in the conversion of CH_4 into the product CH_3Cl.

STEP 3

With all these free radicals being formed, it is likely that two of them will collide with each other. When this happens, they react to form a molecule, as the two unpaired electrons are shared to form a covalent bond. As there are two different free radicals available, this means that there are three possibilities:

$$Cl\bullet + Cl\bullet \rightarrow Cl_2$$

$$Cl\bullet + CH_3\bullet \rightarrow CH_3Cl$$

$$CH_3\bullet + CH_3\bullet \rightarrow C_2H_6$$

These three equations all involve two free radicals reacting with each other to form one molecule. This type of reaction is called **termination**. This means that the sequence of reactions comes to an end because two reactive species are converted into unreactive species.

FURTHER SUBSTITUTION REACTIONS

You now know that a hydrogen atom in methane can be replaced by a chlorine atom in a substitution reaction. You can also see that the product chloromethane (CH_3Cl) still contains hydrogen atoms. These three hydrogen atoms can also be replaced, one by one, by chlorine atoms in similar substitution reactions.

fig A Free radicals are important for understanding many chemical reactions, but free radicals in the human body can be harmful. Fortunately, antioxidants in fruit help protect the body from these harmful effects.

It is not easy to prevent these further substitution reactions from occurring. As well as the formation of chloromethane, these other reactions occur and other products are formed:

the formation of dichloromethane

$$CH_3Cl + Cl_2 \rightarrow CH_2Cl_2 + HCl$$

the formation of trichloromethane

$$CH_2Cl_2 + Cl_2 \rightarrow CHCl_3 + HCl$$

the formation of tetrachloromethane

$$CHCl_3 + Cl_2 \rightarrow CCl_4 + HCl$$

Each of these overall reactions can be represented by the same sequence of initiation, propagation and termination steps as for the formation of chloromethane.

The likelihood of these further reactions occurring means that this is not a good method for the preparation of chloromethane or other halogenoalkanes. This is because the yield will be low because of these further reactions, and also because several products have to be separated.

LEARNING TIP

Focusing on the three types of reaction will help you to understand radical mechanisms.

The three types are:

- initiation - one molecule becomes two free radicals
- propagation - a molecule and a free radical become a different free radical and molecule, and there are two reactions in this step
- termination - two free radicals become one molecule.

CHECKPOINT

1. Write the six equations for the mechanism of the reaction between methane and bromine.

2. Classify each of these reactions as initiation, propagation or termination. Explain your choice in each case.

   (a) $C_2H_5\bullet + CH_3\bullet \rightarrow C_3H_8$

   (b) $2O\bullet \rightarrow O_2$

   (c) $F\bullet + CH_4 \rightarrow HF + CH_3\bullet$

SUBJECT VOCABULARY

substitution reaction reaction in which an atom or group is replaced by another atom or group

mechanism the sequence of steps in an overall reaction; each step shows what happens to the electrons involved in bond breaking or bond formation

homolytic fission the breaking of a covalent bond where each of the bonding electrons leaves with one species, forming a free radical

free radical a species that contains an unpaired electron

initiation first step that starts the reaction, involving the formation of free radicals, usually as a result of bond breaking caused by ultraviolet radiation

propagation the two steps that, when repeated many times, convert the starting materials into the products of a reaction

termination final step that involves the formation of a molecule from two free radicals, halting the reaction

ALKANES, ALKANES EVERYWHERE

The article below considers alkanes as they occur in nature.

ALKANES: NATURAL PRODUCTS

Alkanes are widespread in nature, originating mostly from biological processes. For example, odd-numbered, unbranched alkanes can be found in the spores of fungi while even-numbered alkanes are contained in sedimentary rocks. A theory stating that large amounts of methane found on Earth and Jupiter are of non-biogenic origin has not been proven as yet.

In nature, methane is mostly produced by bacteria, for example in the intestines of cows. Additionally, the simplest alkane is also produced by bacteria in wetlands, so for a long time it was known as marsh gas.

fig A Methane is naturally produced in animals' digestive tracts.

Methane also exists in coal mines (mine gas) and because of the very explosive nature of an air and methane mixture it is responsible for mine gas explosions. The output of methane has doubled in the last hundred years making it one of the most important greenhouse gases which are responsible for global warming.

Scientific theories state that methane played an essential role in the origin of life on Earth. Methane and ammonia were the main components of the primordial atmosphere. Under the influence of UV irradiation, they form hydrogen cyanide (HCN) which subsequently could polymerise to adenine, an important building block of ribonucleic acid. Reaction of methane and ammonia in the presence of water leads to amino acids.

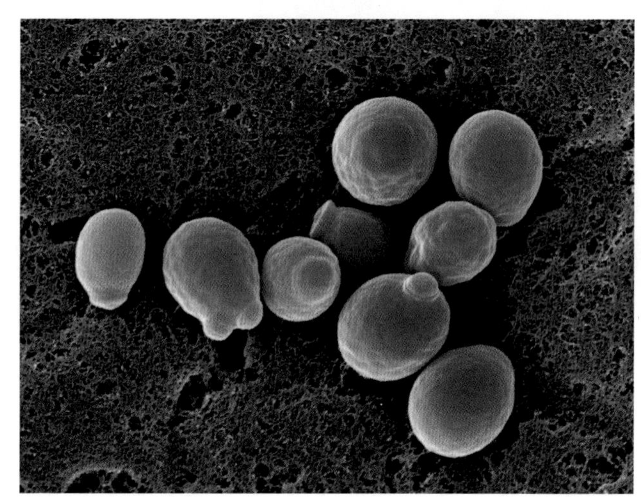

fig B *Candida albicans* is a single-celled fungus that is commonly found on the skin and mucous membranes of the mouth, and in digestive and respiratory tracts.

Some microorganisms, e.g. fungi, can metabolise alkanes. These microorganisms become increasingly important as a resource for degrading polluted soil caused by crude oil spills, i.e. the spoiled soil does not have to be removed and deposited elsewhere. The microorganisms detoxify the soil by breaking down the contaminants into harmless or less harmful substances.

SCIENCE COMMUNICATION

1. Humans rely on alkanes as a source of energy. Many people think this use of alkanes is damaging our planet. In what ways does the author of this article suggest that this is not a serious problem?

CHEMISTRY IN DETAIL

2. (a) Give the displayed and molecular formulae for the straight chain alkane nonane, which has 9 carbon atoms.

(b) Give the skeletal structures of 3 isomers of nonane. Name each structure.

3. Give a balanced equation for the complete combustion of methane. Explain why it can be said that methane is oxidised in this reaction.

4. Explain what is meant by the term polymerisation. Give an example to illustrate your answer.

ACTIVITY

SKILLS REASONING, DECISION MAKING

What is the impact of methane on climate change? Working in pairs, prepare a Powerpoint® presentation to inform your class about current understanding of this issue. You should look to include data from reliable sources and show consideration of any possible bias in sources consulted.

DID YOU KNOW?

Scientists now believe that there are lakes of hydrocarbons on the surface of Titan, one of Saturn's moons.

4 EXAM PRACTICE

1 An organic compound is shown by this formula.

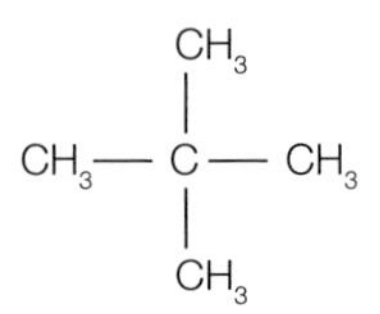

What type of formula is this?

A displayed formula

B general formula

C molecular formula

D structural formula [1]

(Total for Question 1 = 1 mark)

2 Which statement is correct for the members of a homologous series?

A They have similar boiling points

B Their molecular formulae differ by CH_3

C They contain the same functional group

D Their chemical properties are different [1]

(Total for Question 2 = 1 mark)

3 What is the IUPAC name for this compound?

Cl Cl

CH_2 — CH — CH_2 — CH_3

A 3,4-dichlorobutene

B 1,2-dichlorobutane

C 3,4-dichlorobutane

D 1,2-dichlorobutene [1]

(Total for Question 3 = 1 mark)

4 Which fuel can be carbon neutral?

A ethanol

B natural gas

C petrol

D wood [1]

(Total for Question 4 = 1 mark)

5 Which equation represents a substitution reaction?

A $C_2H_4 + H_2O \rightarrow C_2H_5OH$

B $C_2H_4 + Br_2 \rightarrow C_2H_4Br_2$

C $C_2H_6 + F_2 \rightarrow C_2H_5F + HF$

D $C_2H_6 \rightarrow C_2H_4 + H_2$ [1]

(Total for Question 5 = 1 mark)

6 Which equation represents a termination step in the reaction between chlorine and chloromethane?

A $\bullet CH_2Cl + Cl\bullet \rightarrow CH_2Cl_2$

B $H\bullet + Cl\bullet \rightarrow HCl$

C $\bullet CH_3 + Cl_2 \rightarrow CH_3Cl + Cl\bullet$

D $CH_3Cl + Cl_2 \rightarrow CH_2Cl_2 + HCl$ [1]

(Total for Question 6 = 1 mark)

7 The table lists the boiling temperatures of some alkanes.

Alkane	Molecular formula	Boiling temperature / K
butane	C_4H_{10}	273
pentane	C_5H_{12}	309
hexane	C_6H_{14}	342
heptane	C_7H_{16}	372
octane		399
nonane	C_9H_{20}	
decane	$C_{10}H_{22}$	447

(a) Give the molecular formula of octane. [1]

(b) (i) Explain the trend in boiling temperature of the alkanes. [2]

(ii) Predict a value for the boiling temperature of nonane. [1]

(c) Long chain alkanes, such as decane, can be cracked into shorter chain alkanes and alkenes.

(i) Write an equation for the cracking of decane into octane and ethene. [1]

(ii) The ethene produced can be converted into ethanol by direct hydration with steam.

Write an equation for this reaction and state the conditions that are used in industry. [4]

(d) Reforming is a process used in the production of petrol. Unbranched-chain alkanes can be reformed to produce either branched-chain alkanes or cycloalkanes.

The equation shows the reforming of decane into 2-methylnonane.

→

(i) Using skeletal formulae, write an equation for the reforming of decane into 2,3-dimethyloctane. [1]

(ii) Using skeletal formulae, write an equation for the reforming of heptane into methylcyclohexane. [2]

(iii) State why reforming is used in the production of petrol. [1]

(Total for Question 7 = 13 marks)

8 Compound **Y** is a hydrocarbon containing 85.7% of carbon by mass.

(a) (i) Calculate the empirical formula of **Y**. [2]

(ii) The molar mass of **Y** is 56 g mol^{-1}. Show that the molecular formula of **Y** is C_4H_8. [1]

(b) There are six isomers for compound **Y**; four unsaturated molecules and two saturated molecules.

Draw a displayed formula for each of the six isomers and name each compound. [6]

(Total for Question 8 = 9 marks)

9 The structural formulae of four hydrocarbons, **A**, **B**, **C** and **D**, are shown.

A $CH_3 — CH_2$
$\quad\quad\quad\quad |$
$\quad\quad\quad CH_2 — CH_3$

B $CH_3 — CH_2$
$\quad\quad\quad\quad |$
$\quad\quad\quad CH_2 — CH_2 — CH_3$

C $CH_3 — CH_2 — CH_2$
$\quad\quad\quad\quad\quad\quad\quad |$
$CH_3 — CH_2 — CH_2$

D $CH_3 — CH_2$
$\quad\quad\quad\quad |$
$\quad\quad\quad CH_2 — CH_2 — CH_2 — CH_3$

(a) Identify the homologous series to which all these hydrocarbons belong.

Give a reason for your answer. [2]

(b) Explain which structural formulae represent only one compound. [2]

(c) Give the structural formulae for the two isomers of **B**.

Give the IUPAC name for each isomer. [4]

(Total for Question 9 = 8 marks)

TOPIC 5 ALKENES

A ALKENES | B ADDITION POLYMERS

In **Topic 4**, you learned about the basics of organic chemistry and a homologous series called the alkanes, which are mostly used as fuels. In this topic, you will learn about a second homologous series called the alkenes. Alkenes can be burned to produce heat energy, but are far too valuable to waste in this way, as they have more important uses.

In particular, alkenes are used to make polymers (often called plastics). These substances have transformed the way we use materials. Many objects that used to be made of wood or metal are now made of polymers. Some polymers have replaced clothing that used to be made of wool, cotton or silk.

One disadvantage of polymers is that in some situations they have been viewed as disposable, so many plastic shopping bags and drink bottles are intended to be used once only. When they are then thrown away, they cause litter or can harm animals and sea life.

Chemists are working on ways to solve these problems by:

- developing biodegradable polymers
- finding ways to convert polymer waste into useful materials such as clothing.

MATHS SKILLS FOR THIS TOPIC

- Use ratios to construct and balance equations
- Represent chemical structures using angles and shapes in 2D and 3D structures

What prior knowledge do I need?

Topic 1D

- Calculation of empirical and molecular formulae

Topic 4

- The names of simple organic compounds
- Homologous series and general formula
- Representing organic compounds by structural formulae

What will I study in this topic?

- Geometric isomerism
- Reaction mechanisms using full curly arrows to understand how organic reactions occur
- The formation of polymers
- Dealing with polymer waste

What will I study later?

Topic 10

- Alcohols and halogenoalkanes

Topic 15 (Book 2: IAL)

- Optical isomerism
- Reactions of carbonyl and carboxyl compounds

Topic 18 (Book 2: IAL)

- The type of bonding in benzene and other aromatic compounds

Topic 19 (Book 2: IAL)

- Condensation polymers
- Nitrogen-containing compounds, including amino acids and proteins

Topic 20 (Book 2: IAL)

- Planning reaction schemes to prepare organic compounds
- Practical techniques for preparing and purifying organic compounds

5A 1 ALKENES AND THEIR BONDING

LEARNING OBJECTIVES

- Know the general formula for alkenes and understand that alkenes and cycloalkenes are hydrocarbons which are unsaturated (have a carbon-carbon double bond which consists of a sigma bond and a pi bond).

WHAT ARE ALKENES?

You have already discovered a lot of information about alkanes, but so far you have only come across alkenes as examples used to illustrate nomenclature and isomerism.

The main difference between alkanes and alkenes is that alkanes contain only single bonds, but alkenes contain at least one C=C double bond, so they are unsaturated. Alkenes are much less common than alkanes, but they can be made from alkanes in cracking reactions.

GENERAL FORMULA FOR ALKENES

The structures and names of some common alkenes are shown in **table A**.

STRUCTURE	NAME
$CH_2{=}CH_2$	ethene
$CH_2{=}CH{-}CH_3$	propene
$CH_2{=}CH{-}CH_2{-}CH_3$	but-1-ene
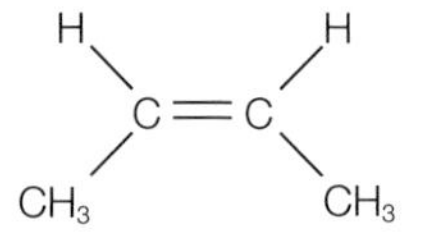 H, H / C=C / CH_3, CH_3	cis-but-2-ene
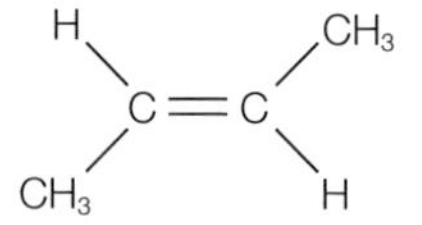 H, CH_3 / C=C / CH_3, H	trans-but-2-ene
$CH_2{=}C(CH_3){-}CH_3$	methylpropene

table A The structures and names of some common alkenes.

If you count the number of carbon and hydrogen atoms in each structure, you can see that there are twice as many hydrogen atoms as carbon atoms. This means that the general formula for the alkene homologous series is C_nH_{2n}.

EXAM HINT

When drawing an alkene, draw all of the carbon atoms first. Then fill in the double bond. Then complete the diagram with H atoms making sure each carbon has **no more** than 4 bonds.

When drawing the structures of alkenes, you normally show the bonds at angles of 120°. For example, ethene would be shown as:

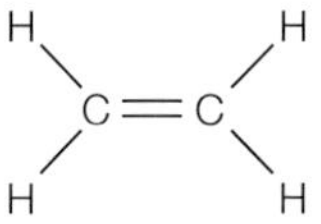

H, H / C=C / H, H

Just as some alkanes are cyclic, so are some alkenes. A common cyclic alkene is cyclohexene. The structure of cyclohexene is:

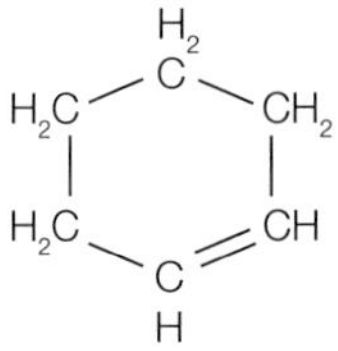

H_2C, CH_2 (ring with H_2C, C–H, CH, H_2C; C=CH double bond)

Note that cyclohexene (and other cyclic alkenes) does not have the same general formula as non-cyclic alkenes. The molecular formula of cyclohexene is C_6H_{10}. Compared to hexene, it has two fewer hydrogen atoms. The general formula of cycloalkenes is C_nH_{2n-2}.

What makes alkenes more interesting than alkanes is the C=C double bond. This makes them more reactive than alkanes and so they can be used in many more useful reactions. Before we look at these reactions, we need to consider exactly what a C=C double bond is.

WHAT IS A C=C DOUBLE BOND?

In some ways, the use of the symbol C=C to represent a double bond is very useful, for example, it makes writing the structures of alkenes straightforward. However, it can be somewhat misleading, because it implies that the two bonds between the carbon atoms are the same. They are not.

In an alkene molecule, both carbon atoms in the C=C double bond are joined to only three other atoms (in alkanes, it is four other atoms). You may remember from **Topics 2A.3** and **3C.1** that:

- electrons can exist in s orbitals and p orbitals
- pairs of electrons around an atom can be represented by a balloon shape.

SIGMA BONDS

All of the single covalent bonds you have met so far involve the merging or overlapping of the orbitals of two different atoms.

They may involve the overlapping of two s orbitals, or one s orbital and one p orbital, or two p orbitals. All of these bonds are represented by a straight line. The covalent bond between the two hydrogen atoms in a hydrogen molecule can be shown as H—H. All of these types of bond can be referred to as **sigma bonds** (σ-bonds). When these sigma bonds are formed between carbon atoms in an alkene, their formation is sometimes described as formation by axial overlap, or end-on overlap.

PI BONDS

Now, consider what happens when the three sigma bonds around each carbon atom are formed in ethene. After they are formed, each carbon atom has one electron in a p orbital that has not been used in bond formation. These p orbitals are parallel to each other and are not able to overlap in the same way as in the formation of sigma bonds. When they do overlap, this results in the formation of two regions of negative charge above and below the C—C sigma bond. This type of bond is formed by the sideways overlap of orbitals, and bonds of this type are referred to as **pi bonds** (π-bonds). Each of these regions of negative charge contains one electron, so together they make up a second shared pair of electrons between the two carbon atoms.

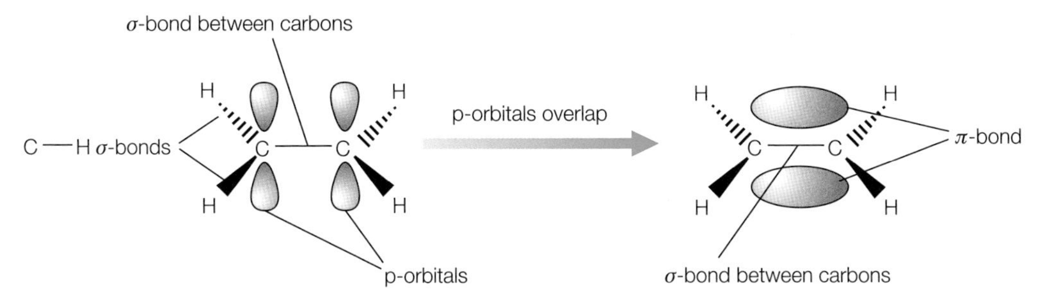

The diagram shows that these electrons seem to be further away from the carbon atoms than the electrons in the sigma bonds are. This means that they can be thought of as less under the control of the carbon atoms and so more available for reactions. We will look at these reactions in **Topic 5A.3**.

LEARNING TIP

You can use the C=C symbol to represent double bonds in molecules when showing their structures. If you are explaining their reactions, it is better to refer to the separate sigma and pi bonds in C=C.

CHECKPOINT

SKILLS REASONING, ARGUMENTATION

1. The general formula for alkenes is C_nH_{2n}.
   Why does cyclohexene (a cyclic hydrocarbon with five C–C single bonds and one C=C double bond) not have this general formula?

2. Alkenes are not used as fuels because they are more valuable for other purposes, but they do burn very well. Write an equation for the complete combustion of propene.

SUBJECT VOCABULARY

sigma bonds covalent bonds formed when electron orbitals overlap axially (end-on)
pi bonds covalent bonds formed when electron orbitals overlap sideways

5A 2 GEOMETRIC ISOMERISM

SPECIFICATION REFERENCE 5.2 5.3

LEARNING OBJECTIVES

- Explain geometric isomerism in terms of restricted rotation around a C=C double bond and the nature of the substituents on the carbon atoms.
- Understand the *E-Z* naming system for geometric isomers and why it is necessary to use this when the cis- and trans- naming system breaks down.

STEREOISOMERISM

You learned about structural isomerism in **Topic 4A.5**. Structural isomers are compounds with the same molecular formula but with different structural formulae. Now we are going to consider a different type of isomerism called stereoisomerism. This is the overall term for two types of isomerism that you need to know about. You will recognise the 'stereo' part of the word from its use as a description of headphones. Stereo headphones allow you to hear some sounds through your left ear and other sounds through your right ear, and stereo is a reference to three dimensions.

One type of stereoisomerism is optical isomerism, which you will meet in **Topic 15**. The other type is geometric isomerism, which we will look at in this topic.

All the types of isomerism you will meet in this course are shown in **fig A** in the form of a family tree.

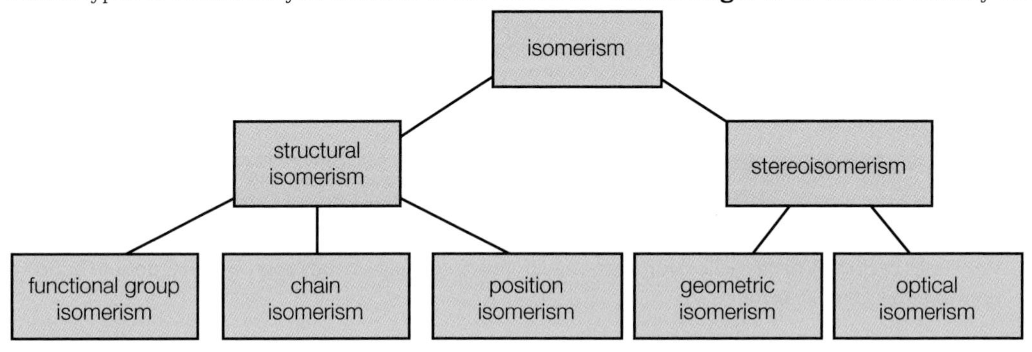

▲ **fig A** The isomerism family tree.

So what are **stereoisomers**? Stereoisomers have the same structural formula but differ from each other because their atoms or groups are arranged differently in three dimensions.

GEOMETRIC ISOMERISM

Geometric isomers differ from each other because their atoms or groups are attached at different positions on opposite sides of a C=C double bond.

To understand this, start by considering the alkene but-2-ene. Its structural formula can be shown as:

$$CH_3—CH=CH—CH_3$$

The name comes from four carbon atoms in a chain (but-), a carbon-carbon double bond (-ene), with 2- showing that the double bond comes after the second carbon atom in the chain.

Unfortunately, showing the structure like this (with the atoms in a straight line) does not help us understand what geometric isomerism is. We need to show the bonds at angles of 120° to each other. Now you can see that there can be two different arrangements:

CH_3 H CH_3 CH_3
C=C C=C
H CH_3 H H

The left-hand structure shows the two CH_3 groups further apart from each other. They are across the molecule. We add the abbreviation *trans-* (Latin for 'across') to the beginning of the name to indicate this. Think of the word transatlantic, which means at opposite ends of the Atlantic Ocean. The complete name is *trans*-but-2-ene.

In the right-hand structure, the two CH_3 groups are still on opposite sides of the C=C bond, but they are both shown above the double bond and not at opposite ends of the molecule. We add the term *cis-* (Latin for 'on this side') to the name of this compound. The complete name is *cis*-but-2-ene.

These two compounds are known as **geometric isomers**. This type of stereoisomerism is described as *cis-trans* isomerism or geometric isomerism.

This type of isomerism can exist in alkenes but not in alkanes. This is because there needs to be a C=C double bond for *cis-* and *trans*-isomers to occur. The presence of a C=C double bond leads to restricted rotation, so that there cannot be any rotation around the double bond. The groups attached to each C in C=C can only be in one of two positions.
In alkanes, which do not have any double bonds, carbon atoms and their attached hydrogen atoms can rotate freely, without restriction. This is much easier to understand if you have access to molecular models.

E-Z NAMING SYSTEM

There is a problem with the *cis-trans* naming system. It only works with some compounds. Consider these two examples:

H, F on left C; Cl (upper), Br (lower) on right C: (H)(F)C=C(Cl)(Br)

H, F on left C; Br (upper), Cl (lower) on right C: (H)(F)C=C(Br)(Cl)

Because there are four different atoms attached to the C=C, the idea of two identical groups (or atoms in this example) being in a *cis-* or *trans-* arrangement cannot work. We need a different system for deciding the names that will clearly distinguish between each of these compounds. This is where the *E-Z* system is useful. Using the *E-Z* system to work out names is more complicated than using *cis-trans* notation, so let us break it down into steps (**table A**).

STEP	WHAT TO DO
1	Work out the part of the name that can be used for both isomers using the normal nomenclature rules. In this example, the name is 1-bromo-1-chloro-2-fluoroethene.
2	Use the priority rules to decide which of the two atoms on the left of the double bond has the higher priority. Priority is decided by which atom has the higher atomic number. You can check this if you are not sure by using the Periodic Table. In this example, H = 1 and F = 9, so fluorine has the higher priority.
3	Do the same as in Step 2 for the two atoms on the right of the double bond. In this case, Cl = 17 and Br = 35, so bromine has the higher priority.
4	Now decide where the two atoms with the higher priorities from steps 2 and 3 are in relation to each other. If both are above (or both are below) the double bond, then this is the Z-isomer. If one is above and the other is below the double bond, then this is the E-isomer.

table A Step-by-step guide to applying the *E-Z* naming system.

In the examples above, in the molecule on the left, F and Br are both below the double bond, so it is the *Z*-isomer.

The molecule on the right has F below the double bond and Br above the double bond, so it is the *E*-isomer.

WHERE DO *E* AND *Z* COME FROM?

There are several ways to remember the difference between *E* and *Z*, and you might be able to work out a memorable way yourself. For now, try thinking about enemies, a word that begins with E. Enemies are far apart. You don't need a separate way to remember Z, as the *Z*-isomer is the one that isn't the *E*-isomer.

If you understand German, you might not need to remember this, because the letters *E* and *Z* come from German words that have opposite meanings: *E* = entgegen, or opposite, and *Z* = zusammen, or together.

EXAM HINT

Practise drawing skeletal structures for *E/Z* isomers.

LEARNING TIP

Try working things out the other way around, writing structures for compounds named using the *E-Z* system.

The rules are a bit more complicated when there are groups instead of single atoms attached to the C=C bond, so here is an example of one of those.

WORKED EXAMPLE

Consider the structure of this compound:

CH_3CH_2 (upper left), Cl (lower left); CH_2OH (upper right), CH_3 (lower right): C=C

Table B shows the step-by-step guide to naming this compound.

STEP	APPLYING THE RULES
1	The longest carbon chain containing the OH functional group has 5 carbons, so the name contains 'pent'. The C of CH_2OH will be the first carbon in the chain, therefore the double bond comes after the second carbon, so '2-en' should be in the name. A chlorine atom is attached to C3 and a CH_3 group to C2, so the name contains '3-chloro' and '2-methyl'. Putting all these together gives 3-chloro-2-methylpent-2-en-1-ol.
2	Now apply the priority rules to the left-hand C of C=C. The upper left atom is C, which has an atomic number of 6. The lower left atom is Cl, which has an atomic number of 17, so Cl has priority.
3	The upper and lower right atoms are both C, which therefore have the same priority, so what do we do now? Look at the next atom joined to the CH_2 groups. The upper one is O (atomic number 8), which has a higher priority than H (atomic number 1).
4	You can see that the rules become more complicated as the compounds become more complicated. However, you should now be able to see that the groups with the highest priority are Cl (bottom left) and CH_2OH (top right). As one of these is below and the other is above the double bond then this is the E-isomer.

table B

CHECKPOINT

1. Explain why the halogenoalkene with the molecular formula C_2HBrCl_2 does not need to be named using the *E-Z* system.
2. There are three ways to attach two CH_3 groups and two Cl atoms to C=C. Draw these structures and use your knowledge of nomenclature and *cis-trans* isomerism to name them.

SUBJECT VOCABULARY

stereoisomers compounds with the same structural formula (and the same molecular formula), but with the atoms or groups arranged differently in three dimensions

geometric isomers compounds containing a C=C bond with atoms or groups attached at different positions

5A 3 ADDITION REACTIONS OF ALKENES

SPECIFICATION REFERENCE 5.4 5.5

LEARNING OBJECTIVES

- Know the qualitative test for a C=C double bond using bromine or bromine water.
- Describe the reactions of alkenes with hydrogen, halogens, steam, hydrogen halides and potassium manganate(VII).

WHY DO ADDITION REACTIONS OCCUR?

In **Topic 5A.1** we learned that the C=C double bond in an alkene is made up of two different single bonds, a sigma bond and a pi bond. Because the sigma bond electrons are more tightly held between the two carbon atoms, a sigma bond is stronger than a pi bond. This means that a double bond is stronger than a single bond, but it is not twice as strong.

Most reactions of alkenes involve the double bond becoming a single bond. In these reactions, the sigma bond remains unchanged, but the pi-bond electrons are used to form new bonds with an attacking molecule. This reaction forms a product that is saturated. As the product contains sigma bonds, and not pi bonds, the bonds in the product are stronger, and so the product is more stable.

EXAM HINT

The electrons in a pi bond are more exposed because of the way the p orbitals overlap and so are more likely to undergo electrophilic attack.

The equation for a typical addition reaction of an alkene, between ethene and bromine, is:

$$C_2H_4 + Br_2 \rightarrow C_2H_4Br_2$$

This reaction is used as a chemical test for the presence of C=C in a compound, because the products of these reactions are colourless. When this reaction occurs, the colour of bromine disappears. We often say that the bromine is decolorised. The bromine is normally used as an aqueous solution (bromine water), but the result is still decolorisation.

You can see why this is called an **addition reaction**. Two molecules become one molecule. However, this equation does not show the mechanism of the reaction, i.e. how it occurs, in terms of the movement of electrons.

EXAM HINT

It is important that this test for the presence of a C=C bond is carried out away from sunlight. This is because alkanes will react with bromine water by a free radical type reaction in the presence of sunlight. (See **Topic 4**.)

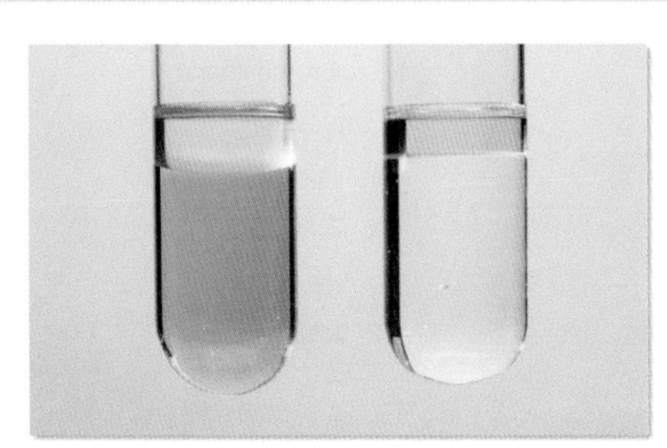

▲ **fig A** The test for a C=C double bond. The tube on the left shows coloured bromine water with a layer of an organic compound on top. The tube on the right shows the mixture after shaking and leaving to settle. The bromine has been decolorised by the C=C bond.

HYDROGENATION

Hydrogenation is an addition reaction in which hydrogen is added to an alkene. The simplest example is the hydrogenation of ethene:

```
H         H                        H   H
 \       /                         |   |
  C == C        +   H2   --Ni-->  H—C—C—H
 /       \                         |   |
H         H                        H   H
```

This reaction forms ethane, which is an alkane, and is done using heat and a nickel catalyst. You might be wondering why we would want to convert a useful alkene into an alkane, which has few uses except as a fuel. In fact, this particular reaction would never be done.

MANUFACTURE OF MARGARINE

Hydrogenation is extensively used in industry to manufacture margarine. Naturally occurring vegetable oils are unsaturated and so contain C=C double bonds. When these react with hydrogen, some of the C=C double bonds become C—C single bonds. This process changes the properties of the vegetable oil and converts it into a solid: margarine.

There is much concern about fats in the human diet, and many people consider that monounsaturated fats (one C=C double bond per molecule) and polyunsaturated fats (two or more C=C double bonds per molecule) are better than saturated fats (no C=C double bonds):

```
  H H H H H H H H H H H H H H H H H H H H H H H
  | | | | | | | | | | | | | | | | | | | | | | |
H—C—C—C—C=C—C—C—C—C—C—C=C—C—C—C—C—C—C—C—C—C—C—C—COOH
  | | |     | | | | |     | | | | | | | | | | |
  H H H     H H H H H     H H H H H H H H H H H
```

margarine

HALOGENATION

Reactions between alkenes and bromine are examples of **halogenation**. The products of these reactions are dihalogenoalkanes. Reactions with chlorine are examples of chlorination (forming dichloroalkanes), and so on. Here are some examples of halogenation reactions:

$$H_2C{=}CH_2 + Br_2 \longrightarrow H{-}CH(Br){-}CH(Br){-}H$$

1, 2-dibromoethane

$$H_2C{=}CH_2 + Cl_2 \longrightarrow H{-}CH(Cl){-}CH(Cl){-}H$$

1, 2-dichloroethane

HYDRATION

Hydration should not be confused with hydrogenation. Hydration means adding water, but you should consider it as adding H and OH to the two atoms in a C=C double bond. This reaction is usually done by heating the alkene with steam and passing the mixture over a catalyst of phosphoric acid. The reaction with ethene can be represented as:

$$H_2C{=}CH_2 + H_2O \xrightarrow{H_3PO_4} H{-}CH_2{-}CH_2(OH)$$

ethene steam ethanol

Unlike the hydrogenation of ethene, the hydration of ethene forms ethanol, which is a useful product. This reaction, and other similar ones to make propanol, are extensively used in industry.

ADDITION OF HYDROGEN HALIDES

Another example of an addition reaction is the addition of a hydrogen halide (often hydrogen bromide or hydrogen chloride) to form a halogenoalkane (more specifically, a bromoalkane or a chloroalkane). Here is one example:

$$CH_2{=}CH_2 + H{—}Br \rightarrow CH_3{—}CH_2Br$$

The product is bromoethane. This reaction looks a lot like the one between ethene and bromine, but it cannot be used as a test for C=C because the reactants and the product are all colourless, so there is no colour change to observe.

$$H_2C{=}CH_2 + HBr \longrightarrow H{-}CH_2{-}CH_2(Br)$$

ethene bromoethane

OXIDATION TO DIOLS

This heading suggests a different type of reaction to the previous ones. In fact, the reaction involves both addition and oxidation. A **diol** is a compound containing two OH (alcohol) groups.

The oxidising agent is potassium manganate(VII) in acid conditions (usually dilute sulfuric acid). Although you do not need to know the full details of how this reaction occurs, you can think of the reaction as oxidation followed by addition. The potassium manganate(VII) provides an oxygen atom (oxidation) and the water in the solution provides another oxygen atom and two hydrogen atoms, so there is the addition of two OH groups across the double bond.

The equation for the reaction of ethene can be represented like this:

$$CH_2{=}CH_2 + [O] + H_2O \rightarrow CH_2OH{—}CH_2OH$$

The symbol [O] represents the oxygen supplied by the oxidising agent. You do not need to show the potassium manganate(VII) in the equation or know how it supplies the oxygen for the oxidation. The product is ethane-1,2-diol.

During the reaction, the colour of the potassium manganate(VII) solution changes from purple to colourless. This colour change means that this reaction can be used like bromine to distinguish alkenes from alkanes (alkanes do not have double bonds and so are not oxidised in this way).

EXAM HINT

Make sure you do not confuse hydrogenation with hydration. The term containing 'hydrogen' refers to adding hydrogen. Terms that do not contain the word 'hydrogen' but start with 'hydr' refer to water. (Think of words such as hydrated, hydraulic and hydro-electric).

CHECKPOINT

1. Write equations for the reactions between but-2-ene and:
   (a) hydrogen
   (b) bromine
   (c) hydrogen chloride.
2. Use IUPAC rules to write the names of the products of each of the reactions in question 1.

SUBJECT VOCABULARY

addition reaction a reaction in which two molecules combine to form one molecule
hydrogenation a reaction involving the addition of hydrogen
halogenation a reaction involving the addition of a halogen
hydration a reaction involving the addition of water (or steam)
diol a compound containing two OH (alcohol) groups

5A 4 THE MECHANISMS OF ADDITION REACTIONS

SPECIFICATION REFERENCE 5.6

LEARNING OBJECTIVES

- Describe the mechanism (including diagrams), giving evidence where possible, of the electrophilic addition of bromine and hydrogen bromide to ethene, and the addition of hydrogen bromide to propene.

BACKGROUND

We have already looked at reaction mechanisms with alkanes (initiation, propagation and termination). We saw how a curly half-arrow was used to represent the movement of a single electron. Now we can look in some detail at how addition reactions occur, which will involve the use of (full) **curly arrows** to represent the movement of a pair of electrons.

We will start with the reaction between ethene and hydrogen bromide, which we looked at in **Topic 5A.3**.

You already know that an alkene such as ethene has a pi bond, which is a region of high electron density (we could say that the molecule is electron rich around the C=C double bond). This makes an alkene molecule attractive to other species that are electron deficient, including molecules with polar bonds. Hydrogen bromide is a polar molecule because bromine is more electronegative than hydrogen, and can be shown with partial charges as:

$$\overset{\delta+}{\text{H}}\text{—}\overset{\delta-}{\text{Br}}$$

WHY DO ELECTROPHILES ATTACK ALKENES?

When a hydrogen bromide molecule approaches an ethene molecule, the slightly positive end of the HBr molecule is attracted to the electrons in the pi bond in C=C. The HBr molecule is described as an **electrophile** when it does this. Remember that electrophiles attack centres of negative charge.

The curly arrows used in reactions of this type must either:

- start from a bond and move to an atom, or
- start from a lone pair of electrons and move to an atom.

ELECTROPHILIC ADDITION OF HYDROGEN HALIDES

The complete name of this reaction is **electrophilic addition**. It involves addition and it involves attack by an electrophile. This is the mechanism of the reaction between ethene and hydrogen bromide. Notice that the hydrogen bromide molecule breaks so that both electrons in the H—Br bond go to one atom (in this case, bromine, because it is more electronegative than hydrogen).

This kind of bond breaking is called **heterolytic fission** (compare this with homolytic fission in the substitution reactions of alkanes).

Step 1

$H_2C{=}CH_2 + H^{\delta+}\text{—}Br^{\delta-} \longrightarrow H_3C\text{—}CH_2^{+} + :Br^{-}$

a carbocation is formed

In this step, two ions are formed. The positive ion has its charge on a carbon atom, so it is known as a **carbocation**.

In these diagrams, the charges are sometimes shown in circles to avoid possible confusion between + signs used to separate reactants and products, and − signs that could be confused with covalent bonds. Using these circles is a good idea, although it is not essential.

The bromide ion is shown with four lone pairs, which is quite correct, but often only the lone pair that moves (where the arrow starts from) is shown.

Step 2

The two oppositely charged ions attract each other and react to form a new covalent bond as one of the lone pairs of electrons forms a covalent bond with the carbon atom in the carbocation.

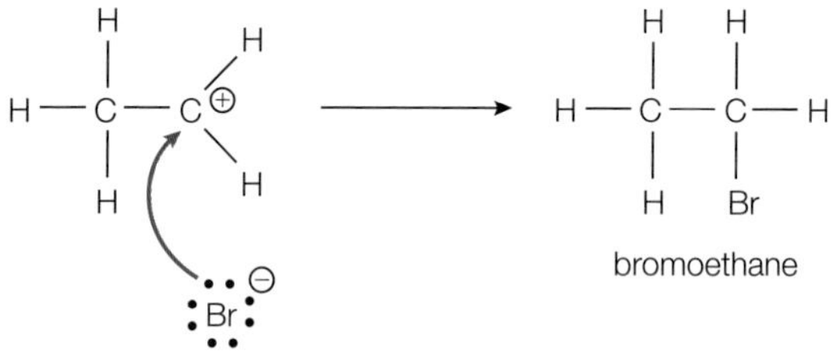

bromoethane

ELECTROPHILIC ADDITION OF HALOGENS

This reaction is very similar to the reaction of hydrogen halides. The only difference is that the attacking bromine molecule does not have a polar bond. However, as it approaches the C=C bond, the electrons in the pi bond repel the electrons in the Br—Br bond and induce (cause) the molecule to become polar. After that happens, the mechanism is just the same as for hydrogen bromide.

Step 1

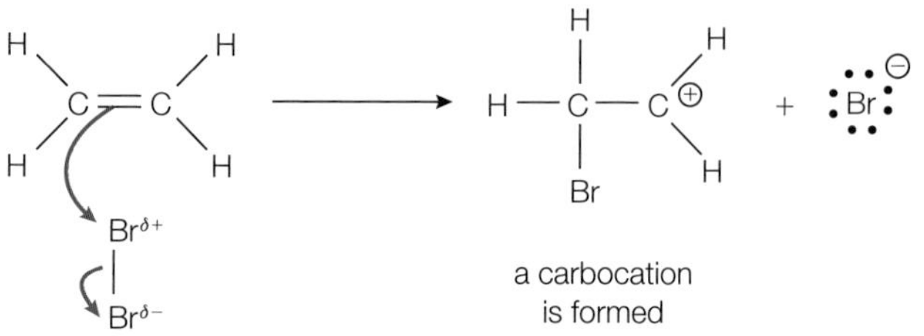

$Br^{\delta+}$—$Br^{\delta-}$

a carbocation is formed

Step 2

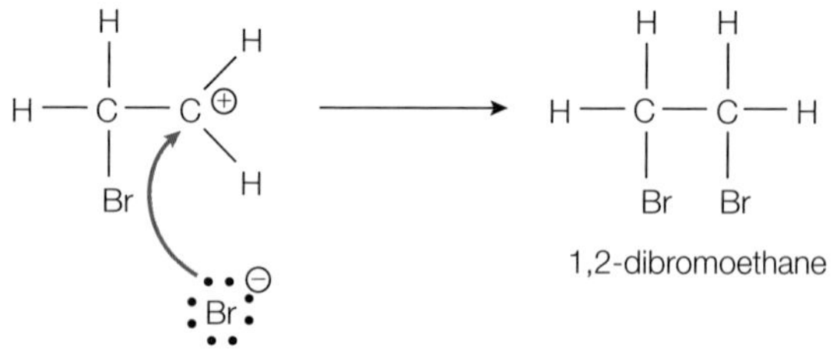

1,2-dibromoethane

ASYMMETRICAL MOLECULES

There is one more point to consider before leaving this topic. When a molecule such as H—Br or Br—Br reacts with ethene in an addition reaction, there can only be one product. If both the alkene and the attacking molecule are asymmetrical, then there are two possible products. This is because the atoms in the attacking molecule can be added in two different places.

An asymmetrical alkene is one in which the atoms on either side of the C=C bond are not the same. An asymmetrical attacking molecule is one in which the atoms are different.

A good example is the reaction between propene and hydrogen bromide.

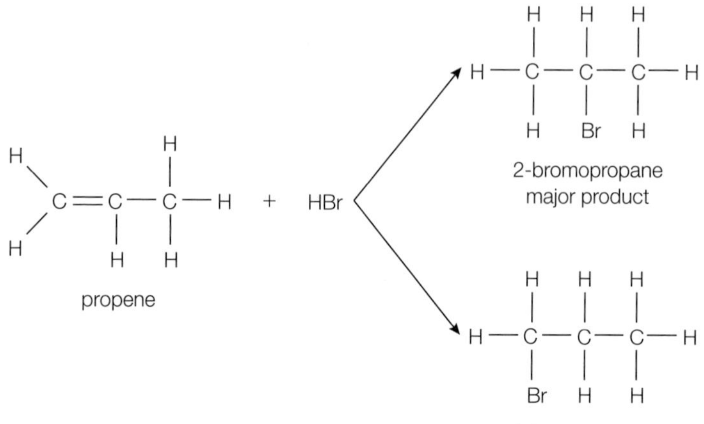

propene + HBr

2-bromopropane
major product

1-bromopropane
minor product

In reactions of this type, one product is formed in greater amounts than the other. There is a major product and a minor product. Now we need to explain why. We can do this by considering the two possible carbocations formed in Step 1 of the reaction.

$CH_3—\overset{\oplus}{C}H—CH_3$ structure A

$\overset{\oplus}{C}H_2—CH_2—CH_3$ structure B

Structure A shows the carbon atom with the positive charge joined to two alkyl groups. This is called a secondary carbocation.

In structure B, there is only one alkyl group joined to the carbon atom with the positive charge, so B is a primary carbocation.

A general principle to consider is that a carbocation in which the charge can be spread over more atoms is more stable than one in which there are fewer atoms available to spread the charge. Alkyl groups are **electron-releasing groups**, so when there are two of them, the charge on the carbocation is spread more than when there is only one.

In some reactions, there might be a tertiary carbocation (with three alkyl groups joined to the carbon atom with the positive charge), and this would be more stable than a secondary carbocation.

EXAM HINT

Practise drawing skeletal structures of both the major and minor products of this type of reaction and name them.

LEARNING TIP

When drawing organic structures, you already know that each carbon atom has four bonds to other atoms. This does not apply to carbocations. They only have three bonds around the carbon atom with the positive charge.

SUMMARY

Reactions involving asymmetrical molecules are complex, so here is a quick summary.

- Electrophilic addition reactions proceed via carbocations.
- Carbocations can be primary, secondary or tertiary.
- The stability of a carbocation is greatest for tertiary carbocations and least for primary carbocations.
- Carbocations are more stable when there are more electron-releasing alkyl groups attached to the carbon with the positive charge.
- The major product is formed from the more stable carbocation.

CHECKPOINT

1. Name the types of reactions that occur when:

(a) alkanes react with halogens

(b) alkenes react with halogens.

2. What is the name of the major product formed when but-1-ene reacts with hydrogen chloride?

SUBJECT VOCABULARY

curly arrows (full ones, not half-arrows) represent the movement of electron pairs

electrophile a species that is attracted to a region of high electron density

electrophilic addition a reaction in which two molecules form one molecule and the attacking molecule is an electrophile

heterolytic fission the breaking of a covalent bond so that both bonding electrons are taken by one atom

carbocation a positive ion in which the charge is shown on a carbon atom

electron-releasing group a group that pushes electrons towards the atom it is joined to

5B 1 POLYMERISATION REACTIONS

SPECIFICATION REFERENCE 5.7

LEARNING OBJECTIVES

- Describe the addition polymerisation of alkenes and draw the repeat unit given the monomer, and vice versa.

ALKENES USED IN ADDITION POLYMERISATION

Many compounds containing the C=C double bond can be polymerised. There is no need to know the mechanisms of these reactions, but you should describe them as addition reactions because the alkene molecules add together in vast numbers to form the polymer. You also do not need to know the exact conditions used in polymerisation reactions, but normally they use a combination of high pressure and high temperature, which varies depending on the polymer.

You know that alkenes are hydrocarbons with the general formula C_nH_{2n}. When considering polymers, the term 'alkenes' is often widened to include other compounds containing C=C attached to other hydrocarbon groups and to halogens.

NAMING POLYMERS

When alkene molecules are used in polymerisation, they are often referred to as **monomers**. The standard way to name a polymer is by writing 'poly', followed by the name of the monomer in brackets. The obvious example is the use of ethene to form poly(ethene). Most people abbreviate this name to polythene. Even though the polymer formed is saturated, the 'ene' ending is still used.

Table A shows information about some common polymers.

MONOMER	POLYMER	COMMON NAME
ethene	poly(ethene)	polythene
propene	poly(propene)	polypropene or polypropylene
chloroethene	poly(chloroethene)	polyvinyl chloride or PVC
tetrafluoroethene	poly(tetrafluoroethene)	PTFE or Teflon®
phenylethene	poly(phenylethene)	polystyrene

table A Information about polymers, including their common name.

EQUATIONS FOR POLYMERISATION REACTIONS

Because the polymers formed do not have a fixed molecular formula (their molecular masses can be anything from many tens of thousands to millions), we need to find a different way to show what happens in the reaction. The usual way to do this is to use the letter 'n' to represent the number of monomer molecules reacting, then to show the **repeat unit** of the polymer inside a bracket (curved or square). The letter n is shown as a subscript after the bracket, and there are covalent bonds shown passing through the brackets to indicate that there is another repeat unit joined on to each side.

A general equation that you can modify for use with all addition polymerisation reactions is:

n WYC=CXZ → [–CWY–CXZ–]$_n$

repeat unit

monomer – any alkene

polymer

EXAMPLES OF EQUATIONS

Here is the equation showing the formation of poly(ethene):

n $H_2C=CH_2$ → [–CH_2–CH_2–]$_n$

ethene

poly(ethene) – polythene

Here is the equation showing the formation of poly(propene):

n $H_2C=CH(CH_3)$ → [–CH_2–$CH(CH_3)$–]$_n$

propene

poly(propene) – polypropylene

This equation shows the formation of poly(chloroethene), better known as PVC:

n $H_2C=CHCl$ → [–CH_2–CHCl–]$_n$

chloroethene

poly(chloroethene) – PVC

Finally, this equation shows the formation of poly(phenylethene), better known as polystyrene:

n $H_2C=CH(C_6H_5)$ → [–CH_2–$CH(C_6H_5)$–]$_n$

phenylethene

poly(phenylethene) – polystyrene

IDENTIFYING THE MONOMER

If you are given the repeat unit of a polymer, or a section of the polymer that contains several repeat units, you can work out the structure of the corresponding monomer. You need to identify the part of the structure that is repeated. This will be two carbon atoms in the chain and the four atoms or groups joined to them. The monomer structure is all of these atoms, but with a double bond between the two carbon atoms.

This is part of the structure of poly(methyl methacrylate), better known as Perspex:

```
        H   CH3          H   CH3          H   CH3
        |   |            |   |            |   |
 -------C---C------------C---C------------C---C-------
        |   |            |   |            |   |
        H   COOCH3       H   COOCH3       H   COOCH3
```

There are no brackets and no subscript n, because this shows part of the structure and not just the repeat unit. You can see that on alternate carbon atoms in the chain, there are two hydrogen atoms, one methyl group and one $COOCH_3$ group. It doesn't matter if you don't recognise the $COOCH_3$ group, you can still work out the monomer structure. The monomer structure is:

```
H         CH3
 \       /
  C == C
 /       \
H         COOCH3
```

methyl methacrylate

LEARNING TIP

In previous topics, you have seen alkene molecules drawn with angles of 120° between the bonds. When drawing the structures of polymers, use angles of 90°.

CHECKPOINT

1. The formula of a monomer used to make a polymer called PVA is:

```
H   H
|   |
C = C
|   |
H   OH
```

Draw the structure of the repeat unit of PVA.

2. Part of the structure of a polymer is:

```
   H   H   H   H   H   H
   |   |   |   |   |   |
 --C---C---C---C---C---C--
   |   |   |   |   |   |
   H   CN  H   CN  H   CN
```

Draw the structure of the monomer used to make this polymer.

SUBJECT VOCABULARY

monomers the small molecules that combine together to form a polymer

repeat unit the set of atoms that are joined together in large numbers to produce the polymer structure

5B 2 DEALING WITH POLYMER WASTE

SPECIFICATION REFERENCE
5.8

LEARNING OBJECTIVES

- Understand how chemists limit the problems caused by polymer disposal by developing biodegradable polymers and by removing toxic waste gases produced by the incineration of polymers.

BACKGROUND

A hundred years ago, traditional materials used to make everyday objects were substances such as wood, metal, glass, wool and paper. Although these materials are still used today, we have become increasingly reliant on polymers (plastics) for many everyday objects. Reasons for the increasing use of polymers include the following.

- They can be manufactured on a large scale in a variety of complex shapes and with a wide range of physical properties. Think of plastic bottles that can be rigid when used to hold bleach or flexible to hold washing-up liquid.
- They are often lighter in weight than traditional alternatives. Think of milk in a glass bottle compared to a plastic bottle.
- They are unreactive and so they can be used to contain many substances safely for long periods. Think of how metals corrode and wood rots.

Polymers are also relatively cheap to make when they are mass produced and many people see them as disposable. Many years ago, 'disposable' meant that after use, the only thing to do was to throw the object away, perhaps in a bin, to be forgotten and taken away by the refuse collection service. Until a few years ago, most polymer waste ended up in landfill, in other words, buried in the ground, but this method of disposal is now used much less. This is partly because of the limited space available in landfill sites, but also because in many countries there are now rules (supported by financial penalties) to decrease the use of landfill. In many countries there is growing awareness of the value of waste such as plastic bottles, and these are separated for **recycling**.

▲ **fig A** There must be a way to reduce how many single-use plastic bottles of water are thrown away.

SOLUTIONS TO POLYMER WASTE

There are several ways to limit the problems caused by the disposal of polymer waste.

- One way is not to use polymers unnecessarily. In some countries, the use of single-use plastic bags by supermarkets has been banned.

- Another way is recycling, which means converting the polymer waste into new materials that are useful. For example, poly(ethylene terephthalate), better known as PET or PETE, is widely used in plastic bottles, and this polymer is now recycled on a large scale to make carpets.
- **Incineration** (burning the polymers) is widely used. Although this method gets rid of the polymer waste, unfortunately it leads to the formation of carbon dioxide (a greenhouse gas) and also some toxic gases.
- Biodegradable polymers are considered by many scientists to be worth developing.

INCINERATION

The elements present in polymer waste are mostly hydrogen and carbon, so they can be used as fuels, in a similar way to other hydrocarbons. An incinerator takes in polymer waste and converts it into heat energy that can be used to heat homes and factories, or used to generate electricity. There is very little solid waste left after incineration, but that is because most of the atoms in the polymers end up in gaseous products which pass into the atmosphere via a chimney.

There is often local opposition when there is a proposal to build an incinerator because of concerns about air pollution. This is because, as well as hydrogen and carbon, there are other elements in the polymer waste: PVC which contains chlorine; and small amounts of toxic heavy metals from the pigments used to colour plastics. These pollutants are difficult to remove from the waste gases released into the atmosphere. In recent years, chemists have made progress in finding better ways to remove toxic waste gases produced by incineration.

▲ **fig B** This incineration plant in the Maldives gets rid of polymer waste, but what is coming out of the chimney?

BIODEGRADABLE POLYMERS

Traditional plastics put into landfill do not break down. The idea of using **biodegradable** polymers (sometimes described as biopolymers) is to allow them to be broken down by microbes in the environment. This sounds like a good idea, and some of them are used on a small scale in medicine (for sutures (stitches) and in drug delivery). However, there are some disadvantages.

- They are often made from plant material, so there is the same issue to consider as with biofuels, that is, land is needed to grow the plants.
- They are designed to break down in the environment, so when they do, the hydrogen and carbon atoms they contain cannot be directly used.

▲ **fig C** This looks like an ordinary plastic cup, but it is actually biodegradable and made from plant material.

LEARNING TIP

Try to use the terms 'disposal' and 'incineration' correctly.

Disposal just refers to removing or throwing away the polymer without indicating how, while incineration is a specific method of polymer disposal.

CHECKPOINT

1. Summarise the advantages of polymers over traditional materials.
2. Summarise the advantages and main disadvantage of incineration as a method of disposing of polymer waste.

SUBJECT VOCABULARY

recycling converting polymer waste into other materials that are useful
incineration converting polymer waste into energy by burning
biodegradable a biodegradable substance is one that can be broken down by microbes

TOWARDS A GREENER ENVIRONMENT

SKILLS INITIATIVE, SELF DIRECTION

The extract below considers examples of the roles of catalysts in the modern petroleum industry.

CATALYSTS FOR A GREEN INDUSTRY

Important catalytic reactions

Today, the industrial world relies upon an enormous number of chemical reactions and an even greater number of catalysts. A selection of important reactions reveals the scope of modern catalysis and demonstrates how crucial it will be for chemists to achieve their environmental objectives.

A sacrifice: worst case catalyst

A sacrificial, or stoichiometric, catalyst is used once and discarded. The amount of waste produced is not insignificant since these catalysts are used in stoichiometric amounts. For example, the catalyst may typically be in a 1 : 1 mole ratio with the main reactant.

In the manufacture of anthraquinone for the dyestuffs industry, for example, aluminium chloride is the sacrificial catalyst in the initial step, the acylation of benzene, see **fig A**. This is a type of Friedel–Crafts reaction in which the spent catalyst is discarded along with waste from the process. Fresh catalyst is required for the next batch of reactants. The problem is that the aluminium chloride complexes strongly with the products, i.e. Cl^-, forming $[AlCl_4]^-$ and cannot be economically recycled, resulting in large quantities of corrosive waste.

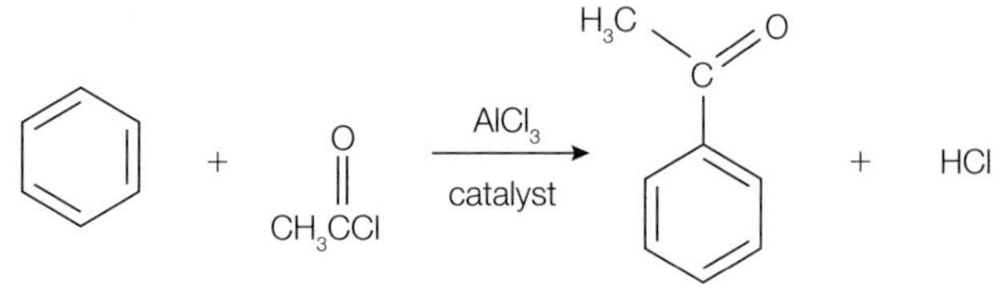

+ CH_3CCl (O) — $AlCl_3$ catalyst → (H_3C, C=O) + HCl

fig A Acylation of benzene

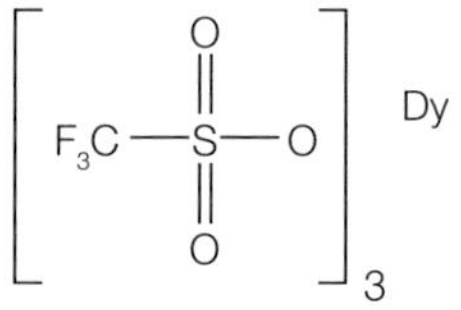

$[F_3C{-}SO_2{-}O]_3\,Dy$

fig B Dysprosium trifluoromethane sulfonate.

New catalysts, with better environmental credentials, are now being tried out. Compounds, such as the highly acidic dysprosium(III) triflate (trifluoromethane sulfonate, see **fig B**) offer the possibility of breaking away from the sacrificial catalyst by enabling the catalyst to be recycled.

Low sulfur fuels: desulfurisation catalysis

Petroleum-derived fuels contain a small amount of sulfur. Unless removed, this sulfur persists throughout the refining processes and ends up in the petrol or diesel. Pressure to decrease atmospheric sulfur has driven the development of catalytic desulfurisation. One of the problems was that much of the sulfur present was in compounds such as the thiophenes, which are stable and resistant to breakdown.

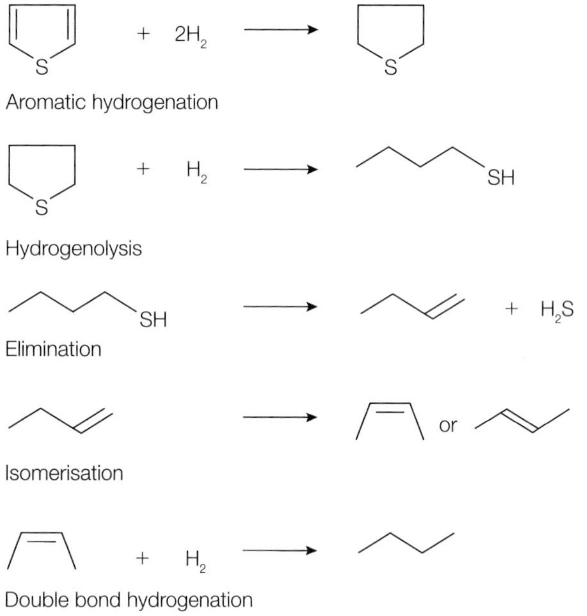

S + $2H_2$ → S

Aromatic hydrogenation

S + H_2 → SH

Hydrogenolysis

SH → + H_2S

Elimination

→ or

Isomerisation

+ H_2 →

Double bond hydrogenation

fig C Desulfurisation of thiophenic compounds from petroleum.

The catalyst molybdenum disulfide coated on an alumina support provided one solution. Cobalt is added as a promoter, suggesting that the active site is a molybdenum-cobalt sulfide arrangement. In the catalytic reaction (see **fig C**), which is essentially a hydrogenation sequence, the adsorbed thiophene molecule is hydrogenated and its aromatic stability destroyed. This enables the C–S bond to break and release the sulfur as hydrogen sulfide. This is an interesting example of a catalyst performing different types of reactions: hydrogenation, elimination and isomerisation.

From an article in *Education in Chemistry* magazine, published by the Royal Society of Chemistry

SCIENCE COMMUNICATION

1. (a) What do you understand by the term 'scientific literacy'? Do you think scientific literacy is required to read this extract? Explain your answer.
   (b) Imagine that you are required to convince the general public that removing the sulfur from fuels is worth the extra cost. Consider what information from the extract will be useful, and how you will present it. Also think about whether you need to do extra research to prepare your arguments. Design a pamphlet to present the case as strongly as possible.

INTERPRETATION NOTE

Think about how illustrations can elicit very emotive responses. How could you use illustrations in your pamphlet?

CHEMISTRY IN DETAIL

2. (a) Work out the molecular formula of thiophene (shown below).

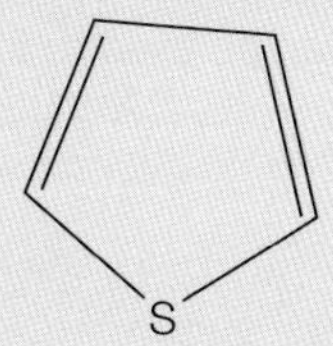

   (b) Calculate the percentage by mass of sulfur in the molecule.
   (c) Write a balanced equation for the complete combustion of thiophene. (You can assume the oxidised product of sulfur is SO_2 only.)
3. (a) During the elimination (**fig C**) part of the reaction sequence, butan-1-thiol is converted into two products. Name them.
   (b) The isomerisation process gives rise to two stereoisomers. Explain what is meant by a geometric isomer and name both stereoisomers in this case. Note that this type of isomerism occurs due to the lack of free rotation about a C=C bond.
   (c) Why can the double bond hydrogenation reaction be considered to have 100% atom economy?

THINKING BIGGER TIP

In Chemistry, you will often need to make assumptions that allow simplification of calculations. In general, these assumptions make little difference to the answers in the real world.

ACTIVITY

SKILLS RESPONSIBILITY, ETHICS

You may wonder why sulfur appears in fossil fuels at all! The chemistry of sulfur gives it some special properties and, apart from carbon, hydrogen, oxygen and nitrogen, it is the only other element present in the building blocks of all proteins: amino acids. Prepare a 5 minute presentation to the class on the importance of sulfur in proteins. Your presentation should include:

- which amino acids contain sulfur
- what properties of sulfur make it so important in protein structure
- what the consequences of a diet low in sulfur can be.

DID YOU KNOW?

Hydrogen sulfide (H_2S) is highly toxic but, luckily, most humans can detect it at concentrations of less than 0.5 parts per billion (or ppb; that's 1 molecule in 2×10^9 air molecules)! It smells like rotten eggs so we get plenty of warning before the level of 800 000 ppb, which can be fatal, is reached.

5 EXAM PRACTICE

1 A saturated hydrocarbon has the molecular formula C_4H_8.

Which is a possible name for this hydrocarbon?

A butane

B butene

C cyclobutane

D methylpropane [1]

(Total for Question 1 = 1 mark)

2 Which statement about the bonding in a propene molecule is correct?

A the C–C bond is a pi bond only

B the C=C bond is a pi bond only

C the C=C bond is a sigma bond only

D the C–H bond is a sigma bond only [1]

(Total for Question 2 = 1 mark)

3 What is the IUPAC name for $(CH_3)_2C{=}CHCH_2CH_3$?

A 2-methyl-3-ethylprop-2-ene

B 2-methylpent-2-ene

C 2,2-dimethylbut-2-ene

D 4-methylpent-3-ene [1]

(Total for Question 3 = 1 mark)

4 How many different compounds can be represented by the formula C_3H_6?

A 1

B 2

C 3

D 4 [1]

(Total for Question 4 = 1 mark)

5 An equation for the reaction of an alkene is

$CH_3CH_2CH{=}CH_2 + H_2O \rightarrow CH_3CH_2CH(OH)CH_3$

Which is a correct name for this reaction?

A addition

B hydrogenation

C redox

D substitution [1]

(Total for Question 5 = 1 mark)

6 A polymer can be represented by this repeat unit:

$$\left[-\overset{CH_3}{\underset{CH_3}{C}} - \overset{H}{\underset{Cl}{C}} - \right]_n$$

What is the name of the monomer that can be used to form this polymer?

A 1-chloro-1,2-dimethylethene

B 1-chloro-2-methylpropene

C 2-chlorobut-2-ene

D 2,2-dimethylchloroethene [1]

(Total for Question 6 = 1 mark)

7 But-1-ene can be produced by the cracking of alkanes.

(a) Draw a dot and cross diagram to show the bonds in a molecule of but-1-ene. [2]

(b) Write an equation for the reaction in which a molecule of dodecane ($C_{12}H_{26}$) is cracked to form two molecules of but-1-ene and one molecule of a saturated hydrocarbon. [1]

(c) A polymer can be made from the monomers ethene and but-1-ene.

Draw the two possible repeat units of the polymer formed from one molecule of each monomer. [2]

(d) But-1-ene can be converted into butane-1,2-diol, $CH_2(OH)CH(OH)CH_2CH_3$

State the reagents needed for this conversion, and the colour change that occurs. [3]

(e) But-1-ene can be distinguished from butane using bromine.

Draw the displayed formula of the organic product formed in the reaction between but-1-ene and bromine.

State the name of this product. [2]

(f) Write the mechanism for the formation of the major product of the reaction between but-1-ene and hydrogen bromide. Include relevant dipoles. [4]

(Total for Question 7 = 14 marks)

8 The structural formulae of two unsaturated compounds, A and B, are shown.

Compound A: CH_3 and H on one carbon, CH_3 and Cl on the other (C=C)

Compound B: CH_3 and H on one carbon, Br and Cl on the other (C=C)

(a) Draw the formula of a structural isomer of compound A and deduce its name. [2]

(b) Draw the formula of a geometric isomer of compound A and state its name using the *cis-trans* notation. [2]

(c) Deduce the name of compound B, using the *E-Z* notation. [1]

(d) Compound B reacts with chlorine in an addition reaction.

Draw the displayed formula of the product and state its name. [2]

(Total for Question 8 = 7 marks)

9 Ethane-1,2-diol is a product of these reactions.

Reaction 1 $C_2H_4 + [O] + H_2O \rightarrow C_2H_6O_2$

Reaction 2 $C_2H_4Cl_2 + 2NaOH \rightarrow C_2H_6O_2 + 2NaCl$

(a) Using Reaction 1, 34.5 kg of ethene were converted into 57.6 kg of ethane-1,2-diol.

Calculate the percentage yield in this conversion. [3]

(b) Calculate the atom economy of Reaction 2. [3]

(c) Give two reasons why Reaction 1 is more likely to be used to manufacture ethane-1,2-diol. [2]

(Total for Question 9 = 8 marks)

10 The structures of two alkenes are shown.

A: $CH_3—CH=CH—CH_3$

B: $CH_3—C(=CH_2)—CH_3$

(a) A student wrote this mechanism for the reaction between alkene **A** and hydrogen bromide.

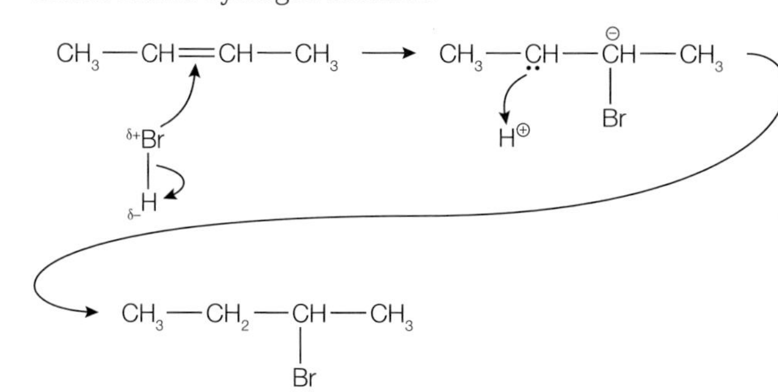

Identify **six** mistakes in the mechanism. [6]

(b) Draw the two possible structures for the carbocation formed in the reaction between alkene **B** and hydrogen chloride.

Explain which carbocation is the more stable of the two.

Give the name of the major product of the reaction. [5]

(Total for Question 10 = 11 marks)

TOPIC 6 ENERGETICS

A INTRODUCING ENTHALPY AND ENTHALPY CHANGE | B ENTHALPY LEVEL DIAGRAMS | C STANDARD ENTHALPY CHANGE OF COMBUSTION | D STANDARD ENTHALPY CHANGE OF NEUTRALISATION | E STANDARD ENTHALPY CHANGE OF FORMATION | F BOND ENTHALPY AND HESS'S LAW | G USING MEAN BOND ENTHALPIES

You will be familiar with many exothermic reactions in everyday life even if you do not always realise it. Exothermic reactions can be easily identified because the reaction mixture gets hot. The heat energy generated can then be used for heating. Perhaps the most common example of this is burning natural gas: the heat energy generated can then be used to cook food.

If your hands have got very cold while outdoors, you may have used a chemical hand warmer. One type of chemical hand warmer uses anhydrous calcium oxide (CaO) and water. These are kept in separate compartments and then mixed by breaking the seal. When the two chemicals mix, an exothermic reaction takes place and the mixture gets hot.

In contrast, endothermic reactions can often be recognised by the reaction mixture getting cold. They are less common, but you may be familiar with sherbet, which contains citric acid and sodium hydrogencarbonate. Sherbet is the effervescent powder that children eat as a sweet and which can also be made into a drink. When you add water to this mixture, an endothermic reaction takes place and the temperature of the mixture drops, which is why the inside of your mouth feels cold when you eat sherbet.

Interestingly, the chemical reactions that take place when an egg is cooked are also endothermic. Another way of recognising an endothermic reaction is that it needs a constant supply of energy for the reaction to continue. If you take the frying pan off the hob, then the egg will stop cooking.

MATHS SKILLS FOR THIS TOPIC

- Recognise and make use of appropriate units in calculations
- Recognise and use expressions in decimal and ordinary form
- Use the appropriate number of significant figures
- Change the subject of an equation
- Substitute numerical values into algebraic equations using appropriate units for physical quantities
- Solve algebraic equations

What prior knowledge do I need?

- Exothermic and endothermic reactions
- Energy level diagrams
- Simple experiments to determine temperature changes in chemical reactions such as dissolving and neutralisation

What will I study in this topic?

- Enthalpy change as the heat energy change measured at constant pressure
- The importance of standard conditions when comparing enthalpy changes
- Enthalpy changes of formation, combustion and neutralisation
- Experiments to obtain data required to calculate the enthalpy changes of reactions including combustion and neutralisation
- Calculations to determine the enthalpy changes of reactions from experimental data
- How Hess's Law can be used to determine enthalpy changes of reactions that cannot be determined directly
- Bond enthalpies and their use to calculate enthalpy changes of reactions, and mean bond enthalpies from enthalpy changes of reactions

What will I study later?

Topic 10

- The significance of bond enthalpy in determining the relative rates of reaction of different halogenoalkanes

Topic 12B (Book 2: IAL)

- Lattice energy, electron affinity, enthalpy change of hydration and enthalpy change of solution
- Born–Haber cycles

6A INTRODUCING ENTHALPY AND ENTHALPY CHANGE

SPECIFICATION REFERENCE 6.1 6.4 PART

LEARNING OBJECTIVES

- Know that the enthalpy change, ΔH, is the heat energy change measured at constant pressure and that standard conditions are 100 kPa and a specified temperature, usually 298 K.
- Be able to define what we mean by standard enthalpy of reaction.

CHEMICAL AND HEAT ENERGY

The first law of thermodynamics states that, during a chemical reaction, energy cannot be created or destroyed. However, one form of energy can be transferred into another form.

Various forms of energy are interesting to a chemist. Two of the most important ones are:

- chemical energy
- heat energy.

CHEMICAL ENERGY

Chemical energy is made up of two components:

- Kinetic energy, which is a measure of the motion of the particles (atoms, molecules or ions) in a substance.
- Potential energy, which is a measure of how strongly these particles interact with one another (i.e. both attract and repel one another).

HEAT ENERGY

Heat energy is the portion of the potential energy and the kinetic energy of a substance that is responsible for the temperature of the substance.

The heat energy of a substance is directly proportional to its absolute temperature (i.e. the temperature measured in Kelvin).

ADDITIONAL READING

Enthalpy is the sum of the internal energy of a system (the energy required to create the system) and the amount of energy required to make room for the system by displacing its environment and establishing its volume and pressure.

Enthalpy has the symbol H and internal energy has the symbol U. This equation shows their relationship:

$$H = U + pV$$

where p is the pressure of the system and V is its volume.

ENTHALPY AND ENTHALPY CHANGES

Enthalpy is a measure of the total energy of a system. When considering a chemical reaction, the 'system' refers to the reaction mixture. Everything outside of the system is called the 'surroundings', which in practice is the air in the room in which the reaction is taking place.

You cannot directly determine the enthalpy of a system, but you can measure the enthalpy change (ΔH) that takes place during a physical or a chemical change.

The enthalpy change of a process is the heat energy that is transferred between the system and the surroundings at *constant pressure.*

LEARNING TIP

Do not confuse heat energy with heating. Heating is the result of a transfer of heat energy from one system to another, which in turn produces a change of temperature.

For example, when water at 60 °C is put in contact with air at 20 °C, heat energy will be transferred from the water to the air. This will result in an increase in temperature of the air, with a subsequent decrease in temperature of the water. This is why a cup of hot coffee gets cold when you forget to drink it.

EXOTHERMIC AND ENDOTHERMIC PROCESSES AND REACTIONS

Two types of process can take place. These are:

- **exothermic** – where heat energy is transferred from the system to the surroundings
- **endothermic** – where heat energy is transferred from the surroundings to the system.

Examples of exothermic and endothermic processes are given in **table A**.

EXOTHERMIC	ENDOTHERMIC
Freezing water	Melting ice
Condensing water vapour	Evaporating water
Dissolving sodium hydroxide in water	Dissolving ammonium nitrate in water
Reaction between dilute hydrochloric acid and aqueous sodium hydroxide	Reaction between dilute ethanoic acid and solid sodium hydrogencarbonate
Combustion of petrol	Photosynthesis

table A Examples of exothermic and endothermic processes.

EXAM HINT

When water condenses, hydrogen bonds are formed between water molecules. Bond formation is an exothermic process. See **Topic 7** for an explanation of hydrogen bonds.

An example of an exothermic reaction is:

$$HCl(aq) + NaOH(aq) \rightarrow NaCl(aq) + H_2O(l)$$
$$\Delta H = -57.1\,kJ\,mol^{-1}$$

An example of an endothermic reaction is:

$$C_6H_8O_7(aq) + 3NaHCO_3(s) \rightarrow C_6H_5O_7^{3-}(aq) + 3Na^+(aq) + 3CO_2(g) + 3H_2O(l)$$
(citric acid)

$$\Delta H = +70\,kJ\,mol^{-1}$$

LEARNING TIP

The negative (–) sign indicates that heat energy is transferred from the system to the surroundings.

The positive (+) sign indicates that heat energy is transferred from the surroundings to the system.

Exothermic reactions can usually be recognised because they result in an immediate increase in temperature. For example, when hydrochloric acid is added to aqueous sodium hydroxide, the temperature of the reaction mixture increases. Similarly, when natural gas burns in oxygen, the flame produced is hot.

Conversely, endothermic reactions often produce a decrease in temperature of the reaction mixture, for example when solid sodium hydrogencarbonate is added to aqueous citric acid.

Any reaction that has to be continually heated in order for it to take place is endothermic. For instance, the thermal decomposition of calcium carbonate into calcium oxide and carbon dioxide is an endothermic reaction:

$$CaCO_3(s) \rightarrow CaO(s) + CO_2(g) \qquad \Delta H = +178\,kJ\,mol^{-1}$$

STANDARD CONDITIONS

In 1982, the International Union of Pure and Applied Chemistry (IUPAC) recommended that all enthalpy changes should be quoted using standard conditions of 100 kPa pressure and a stated temperature. The temperature most commonly used is 298 K.

Under these conditions, the enthalpy change measured is called the 'standard enthalpy change', and is given the symbol $\Delta H^\ominus_{298K}$ or simply $\Delta H^\ominus$.

LEARNING TIP

Some data books and textbooks quote standard enthalpy changes at a standard pressure of one atmosphere, i.e. 1 atm (1 atm = 101.325 kPa). This is not in agreement with IUPAC recommendations. For this reason, we will use 100 kPa as the standard pressure for thermochemical measurements.

STANDARD ENTHALPY CHANGE OF REACTION, $\Delta_r H^\ominus$

When looking at **standard enthalpy change of reaction** ($\Delta_r H^\ominus$), it is important to recognise that the enthalpy change is for the reaction *as written*.

For the reaction

$$N_2(g) + 3H_2(g) \rightarrow 2NH_3(g) \qquad \Delta_r H^\ominus = -92\ kJ\ mol^{-1}$$

But for the reaction when written as:

$$\tfrac{1}{2}N_2(g) + 1\tfrac{1}{2}H_2(g) \rightarrow NH_3(g) \qquad \Delta_r H^\ominus = -46\ kJ\ mol^{-1}$$

In each case the 'per mole' refers to *one mole of equation*, and not to one mole of any reactant or product.

EXAM HINT

Be sure to include the correct state symbols when writing an equation for standard enthalpy change.

CHECKPOINT

1. Classify each of the following as exothermic or endothermic processes:
   (a) cooking an egg
   (b) formation of snow in clouds
   (c) burning candle wax
   (d) forming a cation from an atom in the gas phase
   (e) baking bread.
2. Sherbet is a solid mixture of sodium hydrogencarbonate and citric acid. If you put sherbet in your mouth and mix it with saliva, your mouth will feel cold. Explain why.

SUBJECT VOCABULARY

exothermic a reaction where heat energy is transferred from the system to the surroundings

endothermic a reaction where heat energy is transferred from the surroundings to the system

standard enthalpy change of reaction the enthalpy change which occurs when equation quantities of materials react under standard conditions

6B ENTHALPY LEVEL DIAGRAMS

SPECIFICATION REFERENCE 6.2 6.3

LEARNING OBJECTIVES

- Know that, by convention, exothermic reactions have a negative enthalpy change and that endothermic reactions have a positive enthalpy change.
- Be able to construct and interpret enthalpy level diagrams showing an enthalpy change, including appropriate signs for exothermic and endothermic reactions.

HOW TO DRAW AND INTERPRET ENTHALPY LEVEL DIAGRAMS

In an exothermic reaction, the final enthalpy of the system is less than its initial enthalpy. The reverse is true for an endothermic process. This is shown in the two **enthalpy level diagrams** in **fig A**.

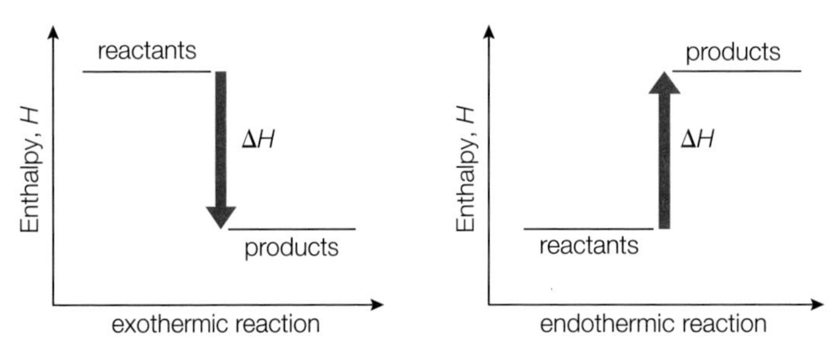

fig A Enthalpy level diagrams for exothermic and endothermic reactions.

The change in enthalpy, ΔH, is given by:

$$\Delta H = H_{\text{products}} - H_{\text{reactants}}$$

For an exothermic reaction, $H_{\text{reactants}} > H_{\text{products}}$, so ΔH is negative.

For an endothermic reaction, $H_{\text{reactants}} < H_{\text{products}}$, so ΔH is positive.

LEARNING TIP

You must get two things right in labels for an enthalpy level diagram:

1. The + / − sign of ΔH must be correct
2. The direction of the arrow for ΔH must be correct.

The arrow should always point towards the products, as the change is from the reactants to the products.

WORKED EXAMPLE 1

Draw an enthalpy level diagram for the following reaction:

$$C(s) + O_2(g) \rightarrow CO_2(g) \qquad \Delta H = -394\,\text{kJ mol}^{-1}$$

Answer

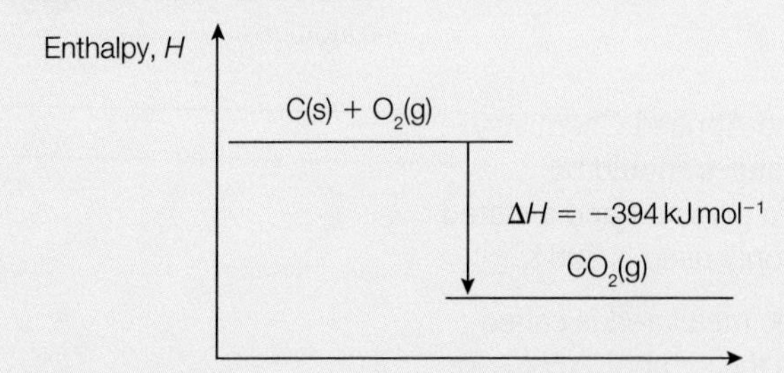

fig B Enthalpy level diagram for the reaction of carbon with oxygen.

WORKED EXAMPLE 2

Draw an enthalpy level diagram for the following reaction:

$C(s) + CO_2(g) \rightarrow 2CO(g) \quad \Delta H = +172\,kJ\,mol^{-1}$

Answer

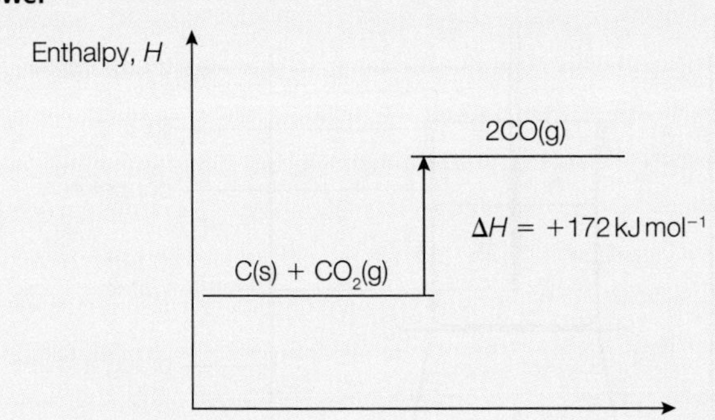

▲ **fig C** Enthalpy level diagram for the reaction of carbon with carbon dioxide.

Here are some points to remember when constructing enthalpy level diagrams.

- You only need to label the vertical axis. It is not necessary to label the horizontal axis in an enthalpy level diagram (if you did want to label it, you could use either 'Extent of reaction' or 'Progress of reaction'). It is, however, essential to label the horizontal axis in an enthalpy profile diagram (see **Topic 9** on reaction kinetics).
- The formulae for both reactants and products should be given, including their state symbols.
- The values for ΔH should be given, including the correct sign.
- It is not essential to show the activation energy in an enthalpy level diagram, but it should be shown in an enthalpy profile diagram (see **Topic 9** on reaction kinetics).

CHECKPOINT

1. Draw an enthalpy level diagram for the following reaction:

   $CH_4(g) + 2O_2(g) \rightarrow CO_2(g) + 2H_2O(l) \quad \Delta H = -890\,kJ\,mol^{-1}$

2. (a) What information can be obtained from the following enthalpy level diagram?

SKILLS INTERPRETATION

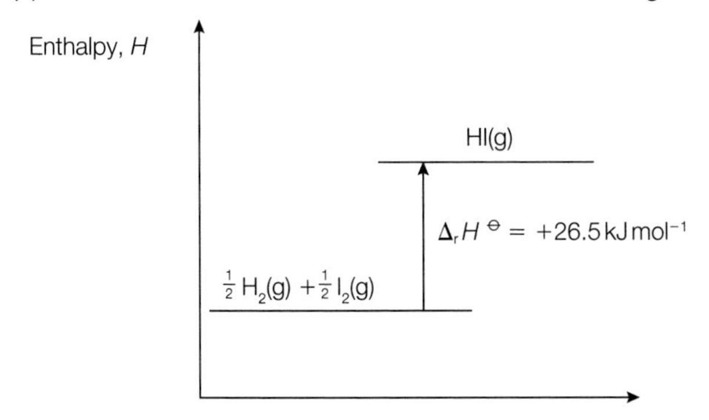

   (b) What would be the value of $\Delta_r H$ for this reaction?

   $H_2(g) + I_2(g) \rightarrow 2HI(g)$

SUBJECT VOCABULARY

enthalpy level diagram a diagram that shows the relationship between the enthalpy of the reactants and the enthalpy of the products in a chemical reaction

6C STANDARD ENTHALPY CHANGE OF COMBUSTION

SPECIFICATION REFERENCE 6.4 PART 6.5 6.8 PART

LEARNING OBJECTIVES

- Be able to define standard enthalpy change of combustion.
- Understand simple experiments to measure enthalpy changes in terms of evaluating sources of error and assumptions made in the experiments.
- Be able to process results from experiments to calculate a value for the enthalpy change of combustion of a substance.

WHAT IS MEANT BY STANDARD ENTHALPY CHANGE OF COMBUSTION

The **standard enthalpy change of combustion** ($\Delta_c H^\ominus$) is the enthalpy change measured at 100 kPa and a specified temperature, usually 298 K, when one mole of a substance is completely burned in oxygen.

When you are writing an equation to represent the standard enthalpy change of combustion, it is important that you specify it is one mole of the substance that is being burned. Two common ways of writing the combustion of hydrogen are:

$H_2(g) + \frac{1}{2}O_2(g) \rightarrow H_2O(l)$ and

$2H_2(g) + O_2(g) \rightarrow 2H_2O(l)$

The first equation, in which one mole of hydrogen undergoes combustion, represents $\Delta_c H^\ominus$. The enthalpy change for the second equation is $2 \times \Delta_c H^\ominus$.

EXPERIMENTAL DETERMINATION OF ENTHALPY CHANGE OF COMBUSTION OF A LIQUID

To find the enthalpy change of combustion of a liquid, a known mass of the liquid is burned and the heat energy released is used to heat a known volume of water.

The following procedure is used.

- A spirit burner containing the liquid being tested is weighed.
- A known volume of water is added to a copper can.
- The temperature of the water is measured.
- The burner is lit.
- The mixture is constantly stirred with the thermometer.
- When the temperature of the water has reached approximately 20 °C above its initial temperature, the flame is extinguished and the burner is immediately reweighed.
- The final temperature is measured.

The appropriate laboratory apparatus is shown in **fig A**.

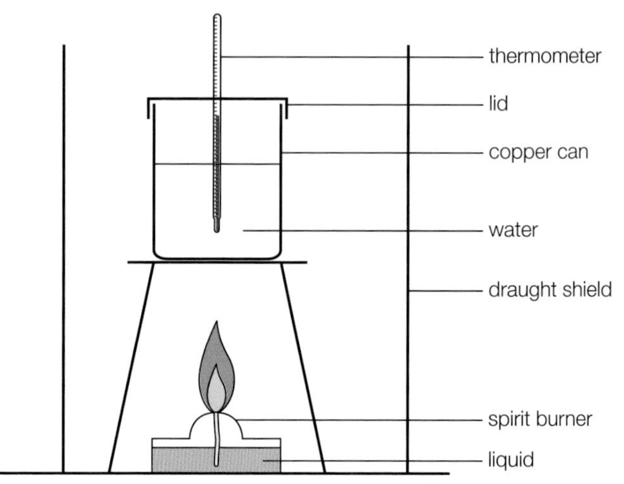

fig A Laboratory apparatus to find the enthalpy change of combustion of a liquid.

A typical set of results for ethanol (C_2H_5OH, molar mass 46.0 g mol^{-1}) are shown in **table A**.

volume of water heated	100.0 cm^3
mass of ethanol burned	0.420 g
temperature change, ΔT	+24.5 °C

table A

CALCULATING ENTHALPY CHANGE OF COMBUSTION

The enthalpy change of combustion is now calculated in three stages.

Stage 1: Calculate the heat energy, Q, transferred to the water using the equation

$Q = mc\Delta T$, where m is the mass of water and c is the specific heat capacity of water.

If m is in grams, then c is quoted as 4.18 J g^{-1} K^{-1}.

Assuming the density of water is 1.00 g cm^{-3}, then m = 100.0 g.

The temperature *change*, ΔT, has the same value in K as it does in °C.

So, $Q = 100.0\,\text{g} \times 4.18\,\text{J g}^{-1}\text{K}^{-1} \times +24.5\,\text{K}$

$= +10\,241\,\text{J} = +10.24\,\text{kJ}$

Stage 2: Calculate the amount, n, of ethanol burned.

$$n(C_2H_5OH) = \frac{0.420\,\text{g}}{46.0\,\text{g mol}^{-1}} = 9.13 \times 10^{-3}\,\text{mol}$$

Stage 3: Calculate $\Delta_c H^\ominus$, using the equation

$$\Delta H = -\frac{Q}{n}$$

$$\Delta_c H^\ominus = -\frac{+10.24\,\text{kJ}}{9.13 \times 10^{-3}\,\text{mol}} = -1120\,\text{kJ mol}^{-1}$$

(to 3 significant figures)

It is very important that you include a sign with any value of ΔH that you quote.

LEARNING TIP

When the two equations from stages 1 and 3 are used together, they result in the correct sign for ΔH. In this example, ΔH is negative, which is consistent with the exothermic reaction that is taking place.

If the temperature had decreased during the reaction, then ΔT would be negative, which in turn would make Q negative. This would lead to a positive value for ΔH, consistent with an endothermic reaction.

If in doubt, always use your common sense. If the temperature increases, the reaction must be exothermic, and vice versa.

EVALUATING SOURCES OF ERROR AND ASSUMPTIONS MADE IN THE EXPERIMENTS

The value obtained from the above experiment is in reasonable agreement with the standard enthalpy change of combustion of ethanol, as obtained from a data book, of $-1367\,kJ\,mol^{-1}$. This means that the errors in procedure were minimal.

Here are some possible sources of error.

- Some of the heat energy produced in burning is transferred to the air and not the water.
- Some of the ethanol may not burn completely to form carbon dioxide and water. (Incomplete combustion would produce less heat energy and also cause soot to form on the bottom of the copper can.)
- Some of the heat energy produced in burning is transferred to the copper can and not to the water.
- The conditions are not standard. For example, water vapour, not liquid water, is produced.
- The experiment takes a long time. This means that not all of the heat energy transferred from the water to the surroundings is compensated for.

CHECKPOINT

1. Ethanol and methoxymethane have the same molecular formula, C_2H_6O.

The standard enthalpy change of combustion at 298 K of ethanol gas and methoxymethane (CH_3OCH_3) gas are $-1367\,kJ\,mol^{-1}$ and $-1460\,kJ\,mol^{-1}$ respectively.

(a) Write an equation to represent the standard enthalpy change of combustion of: (i) ethanol, and (ii) methoxymethane.

(b) Suggest why the two compounds have different standard enthalpy changes of combustion despite having the same molecular formula.

2. The table shows the results of separately combusting 1.00 g of each of four alcohols and determining the amount of energy required to produce the same temperature rise in each reaction.

SKILLS ADAPTIVE LEARNING

ALCOHOL	MOLAR MASS / $g\,mol^{-1}$	ENERGY REQUIRED / $kJ\,g^{-1}$	ENERGY REQUIRED / $kJ\,mol^{-1}$
methanol	32.0	22.34	
ethanol	46.0	29.80	
propan-1-ol	60.0	33.50	
butan-1-ol	74.0	36.12	

(a) Complete the table by calculating the energy required per mole for each of the four alcohols. Quote the values for $\Delta_c H$ for each of them.

(b) Draw a graph of $\Delta_c H$ (vertical axis) against the number of carbon atoms in one molecule and use it to estimate a value for the enthalpy change of combustion of pentan-1-ol under the same conditions.

(c) Extrapolate the graph line to 0 on the *x*-axis and comment on the value of $\Delta_c H$ you read off from the graph.

EXAM HINT

To *extrapolate* means to extend the graph on the basis that the trend shown in the points recorded continues. So, if the points recorded suggest a curve then you should attempt to continue the curve.

SUBJECT VOCABULARY

standard enthalpy change of combustion ($\Delta_c H^{\ominus}$) the enthalpy change measured at 100 kPa and a stated temperature, usually 298 K, when one mole of a substance is completely burned in oxygen

6D STANDARD ENTHALPY CHANGE OF NEUTRALISATION

SPECIFICATION REFERENCE

6.4 PART | 6.5 | 6.8 PART

LEARNING OBJECTIVES

- Be able to define standard enthalpy change of neutralisation.
- Understand simple experiments to measure enthalpy changes in terms of evaluating sources of error and assumptions made in the experiments.
- Be able to process results from experiments to calculate a value for the enthalpy change of neutralisation.

WHAT IS MEANT BY STANDARD ENTHALPY CHANGE OF NEUTRALISATION?

The **standard enthalpy change of neutralisation** ($\Delta_{neut}H^{\ominus}$) is the enthalpy change measured at 100 kPa and a specified temperature, usually 298 K, when one mole of water is produced by the neutralisation of an acid with an alkali.

The following equation represents the standard enthalpy change of neutralisation of hydrochloric acid:

$$HCl(aq) + NaOH(aq) \rightarrow NaCl(aq) + H_2O(l)$$

For sulfuric acid, it is the enthalpy change for:

$$\frac{1}{2}H_2SO_4(aq) + NaOH(aq) \rightarrow \frac{1}{2}Na_2SO_4(aq) + H_2O(l)$$

For certain combinations of acid and alkali, the standard enthalpy change of neutralisation is remarkably constant. Some values are shown in **table A**.

ACID	ALKALI	$\Delta_{neut}H^{\ominus}$ / kJ mol^{-1}
HCl(aq)	NaOH(aq)	-57.9
HBr(aq)	KOH(aq)	-57.6
HNO_3(aq)	NaOH(aq)	-57.6

table A The standard enthalpy change of neutralisation is remarkably constant for certain combinations of acid and alkali.

All of the acids and alkalis in **table A** are classified as strong acids/alkalis. If we make the assumption that strong acids and alkalis are fully ionised in aqueous solution, then the reaction between them is simplified to:

$$H^+(aq) + OH^-(aq) \rightarrow H_2O(l)$$

Because the reaction is essentially the same in each case, it is not surprising that the enthalpy changes are so similar.

EXPERIMENTAL DETERMINATION OF ENTHALPY CHANGE OF NEUTRALISATION

Here is how to find the enthalpy change of neutralisation.

- Wear safety glasses and a lab coat.
- Using a pipette fitted with a safety filler, place 25.0 cm^3 of 1.00 mol dm^{-3} acid into an expanded polystyrene cup.
- Measure the temperature of the acid.
- Using a pipette fitted with a safety filler, place 25.0 cm^3 of the alkali (usually dilute sodium hydroxide of a concentration slightly greater than 1.00 mol dm^{-3} (to make sure all the acid is neutralised)) into a beaker.
- Measure the temperature of the alkali.
- Add the alkali to the acid, stir with the thermometer and measure the maximum temperature reached.
- The appropriate laboratory apparatus is shown in **fig A**.

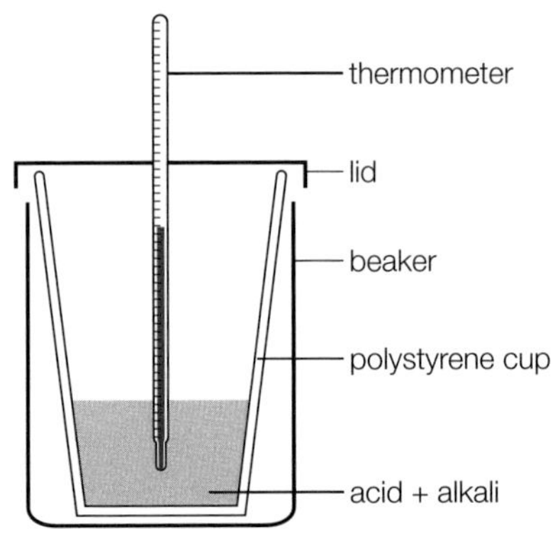

fig A Apparatus used to find the enthalpy change of neutralisation.

WORKED EXAMPLE

Find the enthalpy change of neutralisation.

Volume of 1.00 mol dm^{-3} HCl	25.0 cm^3
Volume of 1.2 mol dm^{-3} NaOH	25.0 cm^3
Initial temperature of acid	18.6 °C
Initial temperature of alkali	18.8 °C
Maximum temperature reached	25.4 °C (298.5 K)

Answer

Mean starting temperature = $\frac{1}{2}$(18.6 + 18.8) = 18.7 °C (291.7 K)

Temperature change, ΔT = (298.5 − 291.7) = +6.7 K

Volume of solution heated = (25.0 + 25.0) = 50.0 cm^3

Mass of solution heated = 50.0 g (assuming that the density of each solution is 1.00 g cm^3)

Assume specific heat capacity of the solution = 4.18 J g^{-1} K^{-1}

Q = 50.0 g × 4.18 J g^{-1} K^{-1} × (+6.7 K) = 1400.3 J = 1.4003 kJ

Amount of acid neutralised = amount of water formed

= 1.00 mol dm^{-3} × 0.025 dm^3

= 0.0250 mol

$\Delta_{neut}H^{\ominus}$ = −1.4003 kJ/0.0250 mol = −56 kJ mol^{-1}

EVALUATING SOURCES OF ERROR AND ASSUMPTIONS MADE IN THE EXPERIMENTS

Answers should be given to only two significant figures. This is because ΔT is given to only two significant figures.

There are the usual uncertainties of measurements involved with the use of the pipette and the thermometer. Additionally, some heat energy will be transferred to the thermometer and the polystyrene cup.

CHECKPOINT

1. A **thermometric titration** was carried out using 1.00 mol dm^{-3} sodium hydroxide solution and dilute hydrochloric acid of unknown concentration.

   50.0 cm^3 of the sodium hydroxide solution was placed in a suitable apparatus and 5.00 cm^3 portions of the hydrochloric acid were added. The mixture was stirred after each addition of acid and the temperature of the mixture was measured. Both solutions were initially at 20 °C.

   The results obtained were as shown in the table.

VOLUME OF ACID / cm^3	5.00	10.00	15.00	20.00	25.00	30.00	35.00	40.00	45.00	50.00
TEMPERATURE / °C	22.80	23.80	24.80	25.80	26.80	27.80	28.10	27.70	27.30	26.80

   (a) Plot a graph of temperature change against volume of acid added. Determine the concentration, in mol dm^{-3}, of the hydrochloric acid.

   SKILLS INTERPRETATION

   (b) Calculate the enthalpy change of neutralisation, per mole of water formed, of the reaction. [Assume the specific heat capacity of the solution formed is 4.18 J g^{-1} K^{-1} and that its density is 1 g cm^{-3}.]

   (c) Suggest suitable apparatus for carrying out the above experiment.

   (d) In further experiments, the following enthalpy changes of neutralisation were determined, again using 1.00 mol dm^{-3} solutions at 20 °C.

   HF(aq) and NaOH(aq) $\Delta_{neut}H = -68.6$ kJ mol^{-1}

   CH_3COOH(aq) and NaOH(aq) $\Delta_{neut}H = -55.2$ kJ mol^{-1}

   Suggest why each of these values is different from the values for HCl(aq) and NaOH(aq).

   Suggest why these values are different from one another.

   [Note: Part (d) of this question is beyond what is expected of you at IAS. However, if you do some research into what is meant by a weak acid, and also consider the energy changes involved when acid molecules ionise in water and also when the ions are subsequently hydrated, you should be able to answer it.]

2. In the experiment described in Question 1, a 50 cm^3 pipette was used to measure the volume of sodium hydroxide solution, and a 50 cm^3 burette was used to add the hydrochloric acid. The measurement uncertainty for the pipette is ±0.10 cm^3. The measurement uncertainty for each reading of the burette is ±0.05 cm^3 and for the thermometer is 0.05 °C.

   (a) Calculate the total percentage measurement uncertainty in the use of:

   (i) the pipette, (ii) the burette and (iii) the thermometer.

   (b) Considering the description of how the experiment was carried out, suggest any possible procedural errors. State how you would carry out the experiment in order to minimise the effect of these errors.

   SKILLS CREATIVITY

SUBJECT VOCABULARY

standard enthalpy change of neutralisation the enthalpy change measured at 100 kPa and a stated temperature, usually 298 K, when one mole of water is produced by the neutralisation of an acid with an alkali

thermometric titration a titration where the endpoint is indicated by a temperature change

6E STANDARD ENTHALPY CHANGE OF FORMATION AND HESS'S LAW

SPECIFICATION REFERENCE 6.4 PART 6.6 CP2

LEARNING OBJECTIVES

- Be able to define standard enthalpy change of formation.
- Be able to construct enthalpy cycles using Hess's Law.
- Be able to calculate enthalpy changes from data using Hess's Law.

WHAT IS MEANT BY STANDARD ENTHALPY CHANGE OF FORMATION?

The **standard enthalpy change of formation** ($\Delta_f H^\ominus$) is the enthalpy change measured at 100 kPa and a specified temperature, usually 298 K, when one mole of a substance is formed from its elements in their standard states.

For the purposes of this definition, the standard state of an element is the form in which it exists at the specified temperature, usually 298 K, and a pressure of 100 kPa.

The standard enthalpy change of formation of gaseous carbon dioxide is the enthalpy change for the reaction:

$$C(s, graphite) + O_2(g) \rightarrow CO_2(g) \quad \Delta_f H^\ominus = -394\,kJ\,mol^{-1}$$

The standard enthalpy change of formation of liquid ethanol is the enthalpy change for the reaction:

$$2C(s, graphite) + 3H_2(g) + \tfrac{1}{2}O_2(g) \rightarrow C_2H_5OH(l) \quad \Delta_f H^\ominus = -278\,kJ\,mol^{-1}$$

LEARNING TIP

Enthalpy changes of formation emphasise the necessity of including state symbols in thermochemical equations. The value for the standard enthalpy change of formation of gaseous ethanol is different from that for liquid ethanol:

$\Delta_f H^\ominus[C_2H_5OH(l)] = -278\,kJ\,mol^{-1}$

$\Delta_f H^\ominus[C_2H_5OH(g)] = -235\,kJ\,mol^{-1}$

Note that the enthalpy change of formation for an element is zero.

HESS'S LAW

Most standard enthalpy changes of formation cannot be determined experimentally. For example, it is impossible to burn carbon in oxygen and form solely carbon monoxide. So, the enthalpy change for the reaction

$$C(s) + \tfrac{1}{2}O_2(g) \rightarrow CO(g)$$

is impossible to determine directly.

Fortunately, we can make use of **Hess's Law**, which is an application of the first law of thermodynamics (the law of conservation of energy).

Hess's Law states that the enthalpy change of a reaction is independent of the path taken in converting reactants into products, provided the initial and final conditions are the same in each case.

EXAM HINT

You should learn the definition of Hess's Law as it is frequently the first part of a question on energetics.

Hess's Law allows us to calculate the standard enthalpy change of formation of carbon monoxide.

The enthalpy changes of combustion of carbon and carbon monoxide can both be determined experimentally. Their values are:

$$C(s, graphite) + O_2(g) \rightarrow CO_2(g) \quad \Delta_c H^\ominus = -394\,kJ\,mol^{-1}$$
$$CO(g) + \tfrac{1}{2}O_2(g) \rightarrow CO_2(g) \quad \Delta_c H^\ominus = -283\,kJ\,mol^{-1}$$

There are two ways to use these data to calculate the standard enthalpy change of formation of carbon monoxide.

METHOD 1: SUBTRACT EQUATIONS

Reverse the second equation above and then add to the first:

$$CO_2(g) \rightarrow CO(g) + \tfrac{1}{2}O_2(g) \quad \Delta_c H^\ominus = +283\,kJ\,mol^{-1}$$
$$C(s, graphite) + O_2(g) \rightarrow CO_2(g) \quad \Delta_c H^\ominus = -394\,kJ\,mol^{-1}$$

Adding the two equations gives:

$$C(s) + \tfrac{1}{2}O_2(g) \rightarrow CO(g) \quad \Delta_c H^\ominus = -111\,kJ\,mol^{-1}$$

So, $\Delta_f H^\ominus[CO(g)] = -111\,kJ\,mol^{-1}$

METHOD 2: CONSTRUCT AN ENTHALPY CYCLE USING HESS'S LAW

An enthalpy cycle using Hess's Law is sometimes also called a Hess's Law cycle.

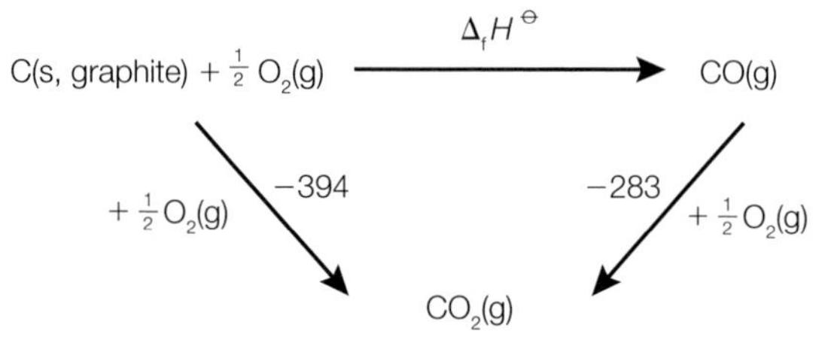

fig A An enthalpy cycle using Hess's Law to calculate the standard enthalpy change of formation of carbon monoxide.

By Hess's law, $\Delta_f H^\ominus + (-283) = -394$

So, $\Delta_f H^\ominus = -394 - (-283) = -111\,kJ\,mol^{-1}$

In general terms, this is the enthalpy cycle used to calculate the enthalpy change of formation of a compound from the relevant enthalpy changes of combustion.

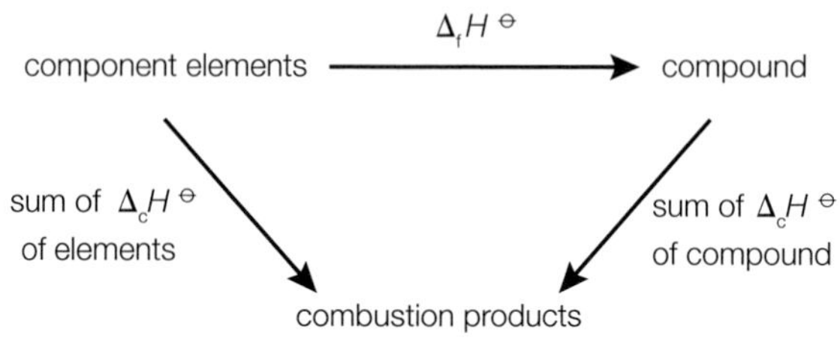

$$\Delta_f H^\ominus = \text{sum of } \Delta_c H^\ominus \text{ of elements} - \Delta_c H^\ominus \text{ of compound}$$

fig B An enthalpy cycle to calculate the enthalpy change of formation from enthalpy changes of combustion.

WORKED EXAMPLE 1

Calculate the enthalpy change of formation of methanol (CH_3OH), given the following enthalpy change of combustion data:

$\Delta_c H^\ominus[CH_3OH(l)] = -726\ kJ\ mol^{-1}$

$\Delta_c H^\ominus[C(s, graphite)] = -394\ kJ\ mol^{-1}$

$\Delta_c H^\ominus[H_2(g)] = -286\ kJ\ mol^{-1}$

$\Delta_f H^\ominus$

$C(s, graphite) + 2H_2(g) + \frac{1}{2}O_2(g) \rightarrow CH_3OH(l)$

$-394 + (-286 \times 2)$ $+1\frac{1}{2}O_2$ $+1\frac{1}{2}O_2$ -726

$CO_2(g) + 2H_2O(l)$

▲ **fig C** An enthalpy cycle using Hess's Law to calculate the enthalpy change of formation of methanol.

$\Delta_f H^\ominus = -394 + (-286 \times 2) - (-726) = -240\ kJ\ mol^{-1}$

USING HESS'S LAW FOR OTHER REACTIONS

Hess's Law can be used to calculate the enthalpy change for many different types of reaction.

WORKED EXAMPLE 2

Calculate $\Delta_r H^\ominus$ for the thermal decomposition of calcium carbonate into calcium oxide and carbon dioxide, given the following data:

$CaCO_3(s) + 2HCl(aq) \rightarrow CaCl_2(aq) + H_2O(l) + CO_2(g)$
$\Delta_r H^\ominus = -17\ kJ\ mol^{-1}$

$CaO(s) + 2HCl(aq) \rightarrow CaCl_2(aq) + H_2O(l)$ $\Delta_r H^\ominus = -195\ kJ\ mol^{-1}$

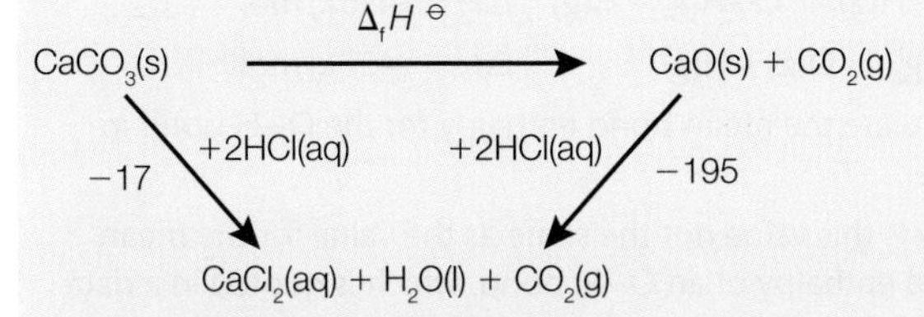

▲ **fig D** An enthalpy cycle using Hess's Law to calculate the enthalpy change of reaction for the thermal decomposition of calcium carbonate.

$\Delta_r H^\ominus = -17 - (-195) = +178\ kJ\ mol^{-1}$

WORKED EXAMPLE 3

Calculate $\Delta_r H^\ominus$ for the hydration of anhydrous copper(II) sulfate, given the following data:

$CuSO_4.5H_2O(s) + aq \rightarrow Cu^{2+}(aq) + SO_4^{2-}(aq) + 5H_2O(l)$
$\Delta_r H^\ominus = +11.3\ kJ\ mol^{-1}$

$CuSO_4(s) + aq \rightarrow Cu^{2+}(aq) + SO_4^{2-}(aq)$ $\Delta_r H^\ominus = -67.0\ kJ\ mol^{-1}$

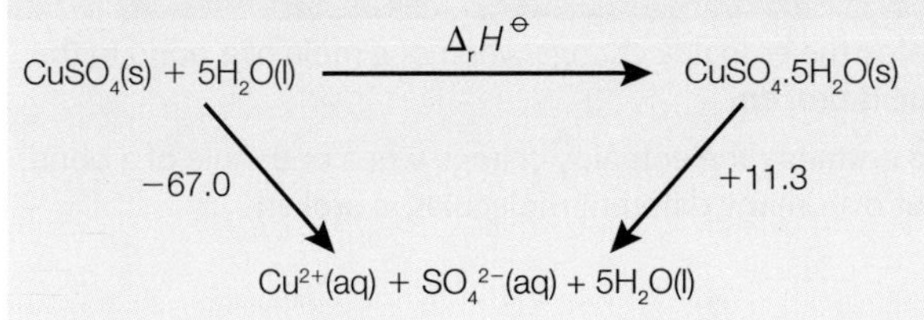

▲ **fig E** An enthalpy cycle using Hess's Law to calculate the enthalpy change of reaction for the hydration of anhydrous copper(II) sulfate.

$\Delta_r H^\ominus = -67.0 - (+11.3) = -78.3\ kJ\ mol^{-1}$

CHECKPOINT

SKILLS PROBLEM SOLVING

1. (a) Write a chemical equation to represent the standard enthalpy change of formation of methane.
   (b) Using the following data, calculate a value for the standard enthalpy change of formation of methane.
   $\Delta_c H^\ominus[C(s, graphite)] = -394\ kJ\ mol^{-1}$
   $\Delta_c H^\ominus[H_2(g)] = -286\ kJ\ mol^{-1}$
   $\Delta_c H^\ominus[CH_4(g)] = -890\ kJ\ mol^{-1}$
2. Using the following data, calculate the standard enthalpy change for the reaction represented by the equation:
   $AlCl_3(s) + 6H_2O(l) \rightarrow AlCl_3.6H_2O(s)$
   $\Delta_f H^\ominus[AlCl_3(s)] = -704\ kJ\ mol^{-1}$
   $\Delta_f H^\ominus[H_2O(l)] = -286\ kJ\ mol^{-1}$
   $\Delta_f H^\ominus[AlCl_3.6H_2O(s)] = -2692\ kJ\ mol^{-1}$
3. The data shows some values of standard enthalpy changes of formation.
   $\Delta_f H^\ominus[LiOH(s)] = -485\ kJ\ mol^{-1}$
   $\Delta_f H^\ominus[H_2O(l)] = -286\ kJ\ mol^{-1}$
   $\Delta_f H^\ominus[Li(s)] = 0.00\ kJ\ mol^{-1}$
   (a) Why is the standard enthalpy change of formation of lithium quoted as zero?
   (b) Write a chemical equation to represent the standard enthalpy change of formation of solid lithium hydroxide.
   (c) Lithium reacts with water according to the following equation:
   $2Li(s) + 2H_2O(l) \rightarrow 2Li^+(aq) + 2OH^-(aq) + H_2(g)$
   Calculate the standard enthalpy change for this reaction. Use the data above and the following information:
   $LiOH(s) + aq \rightarrow Li^+(aq) + OH^-(aq)$ $\Delta H = -21\ kJ\ mol^{-1}$

SUBJECT VOCABULARY

standard enthalpy change of formation the enthalpy change measured at 100 kPa and a specified temperature, usually 298 K, when one mole of a substance is formed from its elements in their standard states

Hess's Law the law states that the enthalpy change of a reaction is independent of the path taken in converting reactants into products, provided the initial and final conditions are the same in each case

6F BOND ENTHALPY AND MEAN BOND ENTHALPY

SPECIFICATION REFERENCE 6.9 PART

LEARNING OBJECTIVES

- Understand the terms bond enthalpy and mean bond enthalpy

WHAT IS BOND ENTHALPY?

Bond enthalpy ($\Delta_B H$) is the enthalpy change when one mole of a bond in the gaseous state is broken.

For a *diatomic* molecule, XY, the bond enthalpy is the enthalpy change for the following reaction:

$$XY(g) \rightarrow X(g) + Y(g)$$

Some examples are:

$Cl_2(g) \rightarrow 2Cl(g)$ $\quad \Delta_B H = +243\,kJ\,mol^{-1}$

$H_2(g) \rightarrow 2H(g)$ $\quad \Delta_B H = +436\,kJ\,mol^{-1}$

$HCl(g) \rightarrow H(g) + Cl(g)$ $\quad \Delta_B H = +432\,kJ\,mol^{-1}$

For *polyatomic* molecules, each bond has to be considered separately. For example, with methane there are four separate bond enthalpies:

$CH_4(g) \rightarrow CH_3(g) + H(g)$ $\quad \Delta_B H = +423\,kJ\,mol^{-1}$

$CH_3(g) \rightarrow CH_2(g) + H(g)$ $\quad \Delta_B H = +480\,kJ\,mol^{-1}$

$CH_2(g) \rightarrow CH(g) + H(g)$ $\quad \Delta_B H = +425\,kJ\,mol^{-1}$

$CH(g) \rightarrow C(g) + H(g)$ $\quad \Delta_B H = +335\,kJ\,mol^{-1}$

WHAT IS MEAN BOND ENTHALPY?

You will notice that the bond enthalpy of the C—H bond varies with its environment. For this reason it is often useful to quote the **mean bond enthalpy**.

The mean bond enthalpy for the C—H bond in methane is approximately $+416\,kJ\,mol^{-1}$.

$$\left[\tfrac{1}{4}(423 + 480 + 425 + 335) = 415.75\right]$$

The mean bond enthalpy for the C—H bond in a large number of organic compounds is $+413\,kJ\,mol^{-1}$. If the bond enthalpy is calculated for a particular compound, it will probably be slightly different from the mean value. For example, the mean bond enthalpy for the C—H bond in ethane is $+420\,kJ\,mol^{-1}$.

Some examples of mean bond enthalpies are given in **table A**.

BOND	MEAN BOND ENTHALPY / $kJ\,mol^{-1}$	BOND	MEAN BOND ENTHALPY / $kJ\,mol^{-1}$
C—C	+347	O—H	+464
C=C	+612	C—F	+467
C≡C	+838	C—Cl	+346
C—O	+358	C—Br	+290
C=O	+743	C—I	+228

table A Examples of mean bond enthalpies.

A quick representation for mean bond enthalpy is to use the letter *E* followed by the bond in brackets. So, the mean bond enthalpy of the C—C bond is written as $E(C—C) = +347\,kJ\,mol^{-1}$.

CHECKPOINT

SKILLS PROBLEM SOLVING

1. (a) State what we mean by the H—I bond enthalpy.
   (b) Write an equation that represents the bond enthalpy of the H—I bond.

2. The two bonds in the water molecule can be separately broken. The enthalpy changes for the two processes are:

   $H–O–H(g) \rightarrow O–H(g) + H(g)$ $\quad \Delta H = +496\,kJ\,mol^{-1}$

   $O–H(g) \rightarrow O(g) + H(g)$ $\quad \Delta H = +432\,kJ\,mol^{-1}$

   (a) Calculate the mean bond enthalpy for the O—H bond in water.
   (b) Why is this value not the same as the value for the mean bond enthalpy of an O—H bond, which is quoted in a data book?

3. Suggest a reason why the mean bond enthalpy of the C—H bond in methane ($+416\,kJ\,mol^{-1}$) is different from the one in ethane ($+420\,kJ\,mol^{-1}$).

4. (a) Write equations to show the successive breaking of the N—H bonds in ammonia, $NH_3(g)$.
   (b) Explain how the enthalpy changes for each reaction are used to calculate the mean bond enthalpy for the N—H bond in ammonia.

SUBJECT VOCABULARY

bond enthalpy the enthalpy change when one mole of a bond in the gaseous state is broken

mean bond enthalpy the enthalpy change when one mole of a bond, averaged out over many different molecules, is broken

6G USING MEAN BOND ENTHALPIES

SPECIFICATION REFERENCE

6.8	6.9
6.10	6.11

LEARNING OBJECTIVES

- Be able to use bond enthalpies to calculate enthalpy changes, understanding the limitations of this method.
- Be able to calculate mean bond enthalpies from enthalpy changes of reaction.
- Understand that bond enthalpy data gives some indication about which bond will break first in a reaction, how easy or difficult it is, and therefore how rapidly a reaction will take place.
- Be able to define standard enthalpy change of atomisation.

CALCULATING AN ENTHALPY CHANGE OF REACTION USING MEAN BOND ENTHALPIES

Here is the process for calculating an enthalpy change of reaction using mean bond enthalpies.

- Step 1: Calculate the sum of the mean bond enthalpies of the bonds broken, Σ(bonds broken).
- Step 2: Calculate the sum of the mean bond enthalpies of the bonds made, Σ(bonds made).
- Step 3: Calculate the enthalpy change of reaction using the equation:

$$\Delta_r H = \Sigma(\text{bonds broken}) - \Sigma(\text{bonds made})$$

WORKED EXAMPLE 1

Calculate the enthalpy change of reaction for:

$H_2(g) + Cl_2(g) \rightarrow 2HCl(g)$

given the following data:

$E(H{-}H) = 436\,kJ\,mol^{-1}$

$E(Cl{-}Cl) = 244\,kJ\,mol^{-1}$

$E(H{-}Cl) = 432\,kJ\,mol^{-1}$

Answer

$\Sigma(\text{bonds broken}) = (436 + 244) = 680\,kJ\,mol^{-1}$

$\Sigma(\text{bonds made}) = (432 \times 2) = 864\,kJ\,mol^{-1}$

$\Delta_r H = (680 - 864) = -184\,kJ\,mol^{-1}$

WORKED EXAMPLE 2

Calculate the enthalpy change of reaction for:

$H_2O_2(l) \rightarrow H_2O(l) + \frac{1}{2}O_2(g)$

given the following data:

$E(O{-}H) = 463\,kJ\,mol^{-1}$

$E(O{-}O) = 146\,kJ\,mol^{-1}$

$E(O{=}O) = 496\,kJ\,mol^{-1}$

If you look at the displayed formula for each substance you will see that the two O–H bonds in hydrogen peroxide, H_2O_2, are also present in water:

H–O–O–H and H–O–H

So we do not have to include these in our calculations.

Answer

$\Sigma(\text{bonds broken}) = 146\,kJ\,mol^{-1}$

$\Sigma(\text{bonds made}) = \frac{1}{2} \times 496\,kJ\,mol^{-1} = 248\,kJ\,mol^{-1}$

$\Delta_r H = (146 - 248) = -102\,kJ\,mol^{-1}$

LEARNING TIP

If you are unsure which bonds break in the reaction, calculate the sum of the bond enthalpies for all of the bonds in both the reactants and products, as in worked example 1.

LIMITATIONS OF THIS METHOD OF CALCULATION

The measured value for the enthalpy change of this reaction is $-98\,kJ\,mol^{-1}$. The reason for the difference is that bond enthalpies are measured in the gaseous state, and both hydrogen peroxide and water are liquids in the reaction. Also, mean bond enthalpies have been used and these may not correspond to the bond enthalpies in the molecules themselves.

CALCULATING MEAN BOND ENTHALPIES FROM ENTHALPY CHANGES OF REACTION

For this type of calculation, you will be supplied with a value for the enthalpy change of a reaction, together with all the relevant mean bond enthalpies except one: the one you are asked to calculate.

To solve the problem, simply substitute the known mean bond enthalpies and the unknown bond enthalpy into the expression:

$$\Delta_r H = \Sigma(\text{bonds broken}) - \Sigma(\text{bonds made})$$

Rearrange the expression to make the unknown bond enthalpy the subject, and solve the problem.

WORKED EXAMPLE 3

Calculate the bond enthalpy of the C=C bond in ethene, given the following data.

$C_2H_4(g) + H_2(g) \rightarrow C_2H_6(g) \qquad \Delta_r H = -147\,kJ\,mol^{-1}$

$E(H{-}H) = 436\,kJ\,mol^{-1}$

$E(C{-}H) = 413\,kJ\,mol^{-1}$

$E(C{-}C) = 347\,kJ\,mol^{-1}$

$\Sigma(\text{bonds broken}) = E(C{=}C) + E(H{-}H) = E(C{=}C) + 436\,kJ\,mol^{-1}$

$\Sigma(\text{bonds made}) = 2E(C{-}H) + E(C{-}C) = (2 \times 413) + 347 = 1173\,kJ\,mol^{-1}$

$\Delta_r H = \Sigma(\text{bonds broken}) - \Sigma(\text{bonds made})$

$-147 = E(C{=}C) + 436 - 1173$

$E(C{=}C) = 1173 - 147 - 436$

$E(C{=}C) = 590\,kJ\,mol^{-1}$

WORKED EXAMPLE 4

In the following example, carbon is in the solid state in the reaction, so you need to think about the enthalpy change for the conversion of the solid into the gas.

Calculate the mean bond enthalpy of the C–H bond in methane using the following data.

$\Delta_f H^{\ominus}[CH_4(g)] = -75\,kJ\,mol^{-1}$

$C(s, \text{graphite}) \rightarrow C(g)$

$\Delta H^{\ominus} = +715\,kJ\,mol^{-1}$

$E(H{-}H) = 436\,kJ\,mol^{-1}$

The equation that represents the standard enthalpy change of formation of methane is:

$C(s, \text{graphite}) + 2H_2(g) \rightarrow CH_4(g) \qquad \Delta_f H^{\ominus} = -75\,kJ\,mol^{-1}$

Answer

So, for the following reaction:

$C(g) + 2H_2(g) \rightarrow CH_4(g)$

$\Delta_r H = -75 + (-715) = -790\,kJ\,mol^{-1}$

$\Sigma(\text{bonds broken}) = 2E(H{-}H) = +872\,kJ\,mol^{-1}$

$\Sigma(\text{bonds made}) = 4E(C{-}H)\,kJ\,mol^{-1}$

$\Delta_r H = \Sigma(\text{bonds broken}) - \Sigma(\text{bonds made})$

$-790 = +872 - 4E(C{-}H)$

$E(C{-}H) = \frac{1}{4}(790 + 872) = 415.5\,kJ\,mol^{-1}$

BOND ENTHALPIES AND EASE OF REACTION

Bond enthalpies can be used to predict which bonds are most likely to break first in a reaction, and how easy it is to break the bond.

Bonds with high bond enthalpies require more energy to break them. Bonds with relatively low bond enthalpies require less energy to break them. This means that they are easier to break, and therefore are more likely to break first in a chemical reaction.

A reaction that involves breaking bonds with low bond enthalpies is more likely to take place at room temperature than reactions involving molecules with high bond enthalpies. Reactions involving the breaking of bonds with high bond enthalpies are more likely to require heating and/or the use of a catalyst.

STANDARD ENTHALPY CHANGE OF ATOMISATION, $\Delta_{at}H^{\ominus}$

The enthalpy change measured at a stated temperature, usually 298 K, and 100 kPa when one mole of gaseous atoms is formed from an element in its standard state is called the **standard enthalpy change of atomisation** of the element. It is given the symbol $\Delta_{at}H^{\ominus}$.

Equations representing some standard enthalpy changes of atomisation at 298 K are given below.

$C(s) \rightarrow C(g) \quad \Delta_{at}H^{\ominus} = +717\ kJ\ mol^{-1}$

$Na(s) \rightarrow Na(g) \quad \Delta_{at}H^{\ominus} = +107\ kJ\ mol^{-1}$

$\frac{1}{2}H_2(g) \rightarrow H(g) \quad \Delta_{at}H^{\ominus} = +218\ kJ\ mol^{-1}$

$\frac{1}{2}Cl_2(g) \rightarrow Cl(g) \quad \Delta_{at}H^{\ominus} = +122\ kJ\ mol^{-1}$

LEARNING TIP

Note that the standard enthalpy change of atomisation is the enthalpy change when one mole of gaseous **atoms** is formed. In the case of elements that exist as polyatomic molecules, it is **not** the enthalpy change when one mole of gaseous **molecules** is atomised.

CHECKPOINT

SKILLS PROBLEM SOLVING

1. Use the following mean bond enthalpies to calculate Δ_rH for each of the following reactions. Assume that all species are in the gaseous state.

BOND	H–H	C–C	C=C	C–H	C=O	O=O
E / kJ mol^{-1}	436	347	612	413	743	498

BOND	H–O	F–F	C–Br	H–Br	H–F	Br–Br
E / kJ mol^{-1}	464	158	290	366	568	193

(a) $H_2 + F_2 \rightarrow 2HF$

(b) $CH_3CH{=}CH_2 + Br_2 \rightarrow CH_3CHBrCH_2Br$

(c) $2CH_3CH_3 + 7O_2 \rightarrow 4CO_2 + 6H_2O$

(d) $CH_2{=}CH_2 + HBr \rightarrow CH_3CH_2Br$

2. (a) Write an equation to represent the standard enthalpy change of formation of ammonia.

(b) Calculate the mean bond enthalpy for the N–H bond in ammonia using this data:

$\Delta_fH^{\ominus}[NH_3(g)] = -46\ kJ\ mol^{-1}$

$E(N{\equiv}N) = 945\ kJ\ mol^{-1}$

$E(H{-}H) = 436\ kJ\ mol^{-1}$

3. The equation for the formation of sulfur hexafluoride is:

$S(s) + 3F_2(g) \rightarrow SF_6(g)$

Use the following data to calculate the mean bond enthalpy for the S–F bond in sulfur hexafluoride.

$\Delta_fH^{\ominus}[SF_6(g)] = -1100\ kJ\ mol^{-1}$

$E(F{-}F) = 158\ kJ\ mol^{-1}$

$\Delta_{at}H^{\ominus}[S(s)] = +223\ kJ\ mol^{-1}$

SUBJECT VOCABULARY

standard enthalpy change of atomisation the enthalpy change measured at a stated temperature, usually 298 K, and 100 kPa when one mole of gaseous atoms is formed from an element in its standard state

6 THINKING BIGGER

WHICH FUEL AND WHY?

SKILLS TEAMWORK

Our dependence on organic fuels has led to the exploitation of renewable biofuels. One of the key aims behind the use of biofuels is to reduce carbon emissions, but with limited land space an increasing area of research is the use of waste biomass.

ORGANIC CHEMISTS CONTRIBUTE TO RENEWABLE ENERGY

Background – Why is this important?

> Biofuels play an essential role in reducing the carbon emissions from transportation. The development of 'drop-in' fuels produced from lignocellulosic raw materials will increase both the availability of biofuels and the sustainability of the biofuel industry.
>
> *Adrian Higson – Energy Consultant*

Biofuels can be either liquid or gaseous. They can be produced from any source that can be replenished rapidly, e.g. plants, agricultural crops and municipal waste. Current biofuels are produced from sugar and starch crops such as wheat and sugar cane, which are also part of the food chain.

One of the key targets for energy researchers is a sustainable route to biofuels from non-edible lignocellulosic (plant) biomass, such as agricultural wastes, forestry residues or purpose-grown energy grasses. These are examples of so-called advanced biofuels.

Current biofuels, such as ethanol, have a lower energy content (volumetric energy density) compared with conventional hydrocarbon fuels, petroleum and natural gas. The aim is to produce fuels that have a high carbon content and therefore have a higher volumetric energy density. This can be achieved by chemical reactions that remove oxygen atoms from biofuel chemical compounds. This process produces a so-called 'drop-in biofuel', i.e. a fuel that can be blended directly with existing hydrocarbon fuels that have similar combustion properties.

What did the organic chemists do?

Efficient synthesis of renewable fuels remains a challenging and important line of research. Levulinic acid and furfural (**fig A**) are examples of potential 'platform molecules', i.e. molecules that can be produced from biomass and converted into biofuels. Levulinic acid can be produced in high yield (>70%) from inedible hexose bio-polymers such as cellulose, which is a polymer of glucose and the most common organic compound on Earth. Furfural has been produced industrially for many years from pentose-rich agricultural wastes and can also act as a platform molecule.

Recent reports have highlighted the use of organic chemistry to convert platform molecules like levulinic acid and furfural into potential advanced biofuels. Specifically, by changing parts of the molecules that are responsible for their structure and function. This process is called 'functional group interconversion' and is part of the basic toolkit of organic chemistry. For example, researchers have described a process for converting levulinic acid into so-called 'valeric biofuels'. One of these biofuels, ethyl valerate, is claimed to be a possible advanced bio-gasoline molecule with several advantages over bio-ethanol.

A second method to create hydrocarbons involves Dumesic's approach via a decarboxylation of gamma-valerolactone, which can be produced in one step from levulinic acid by hydrogenation.

What is the impact?

Biodiesel is likely to be the second most important biofuel after ethanol in the short to medium term. Global production of biodiesel is expected to increase from 11 billion litres to reach 24 billion litres by 2017. For organic chemists, there are significant opportunities associated with further developing energy crops and producing advanced biofuels from new sources such as algae, or industrial or post-consumer waste.

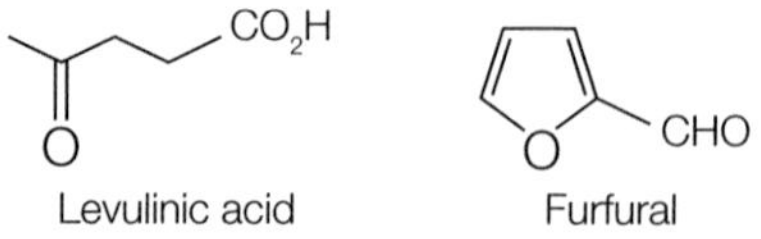

CO_2H O Levulinic acid

O CHO Furfural

fig A Levulinic acid and furfural: two potential platform molecules.

From a case study published by the Organic Division of the Royal Society of Chemistry

SCIENCE COMMUNICATION

1. This article is written for members of an international chemistry association, so it expects a high degree of scientific literacy from the readers. Read the article a few times, then attempt one of the following questions.

(a) Rewrite the article for a less scientifically literate reader. Can you get the main ideas across without using chemical structures and terminology to the same degree?

(b) Rewrite the article in such a way as to present a strongly positive argument in favour of biofuels. Can you present such a positive argument without altering the facts as presented?

INTERPRETATION NOTE

As you read these articles, identify everything that contributes to writing a scientific article. For instance, the vocabulary, the sources, and the way information is presented. Make sure your own answers are written scientifically.

CHEMISTRY IN DETAIL

2. The extract mentions the compounds levulinic acid and furfural.

(a) Calculate the molar mass of each compound.

(b) Calculate the percentage mass of oxygen in each molecule.

(c) Give a chemical test, and its result, that would enable you to distinguish between the two molecules.

3. In this question you will compare four different fuels (shown in **table A**).

1	2	3	4	5	6
Molecule	**Molecular mass in g mol^{-1}**	**Standard molar enthalpy of combustion**	**Number of kJ of energy per mole of CO_2 produced**	**Number of kJ of energy per g of fuel**	**State at 25°C and 1 atm pressure**
CH_4	16	-890			gas
C_8H_{18}	114	-5470			liquid
C_2H_5OH	46	-1367			liquid
$C_{15}H_{32}$	212	-10 047			liquid

table A Thermochemical and physical data for four organic fuels.

(a) Write balanced equations for the complete combustion of C_8H_{18} and C_2H_5OH.

(b) Complete columns 4 and 5 in the table.

4. Why can't the C-15 hydrocarbon synthesised by the process detailed above be described as a completely carbon neutral solution to the energy shortage problem? Refer to the text and any other internet resources to help support your answer.

THINKING BIGGER TIP

Question 4 asks you to think more carefully about the term 'carbon neutral'. Before attempting the question, carry out an internet search for the term 'carbon neutral'. Is there a clear definition for the term?

ACTIVITY

A city borough is about to buy a fleet of 60 buses but must choose one of the four fuels in **table A** on which the buses will run. In groups, select one of the fuels and deliver a presentation on why your chosen fuel is the best option. Think about pros and cons of your chosen fuel. You should attempt to reference a range of source materials in support of your choice. Your presentation should consider the following:

- the pros and cons of your chosen fuel
- energy efficiency
- carbon footprint
- sustainability.

Your presentation should be between 4 and 8 slides and last no more than 10 minutes. However, you should be prepared to justify your presentation in a 5-minute Q&A session!

SKILLS TEAMWORK

THINKING BIGGER TIP

There will be plenty of source material on the internet but be critical about who is presenting the material. For example, companies may want to promote their products and may not present an impartial opinion.

6 EXAM PRACTICE

1 The enthalpy changes of formation of gaseous ethene and gaseous ethane are $52\,kJ\,mol^{-1}$ and $-85\,kJ\,mol^{-1}$ respectively at 298 K.

What is the enthalpy change of reaction at 298 K for the following process?

$C_2H_4(g) + H_2(g) \rightarrow C_2H_6(g)$

A $-137\,kJ\,mol^{-1}$ **B** $-33\,kJ\,mol^{-1}$

C $+33\,kJ\,mol^{-1}$ **D** $+137\,kJ\,mol^{-1}$ [1]

(Total for Question 1 = 1 mark)

2 The enthalpy change for the neutralisation reaction below is $-114\,kJ\,mol^{-1}$.

$2NaOH(aq) + H_2SO_4(aq) \rightarrow Na_2SO_4(aq) + 2H_2O(l)$

Use this information to suggest the most likely value for the enthalpy change for the following neutralisation reaction.

$Ba(OH)_2(aq) + 2HCl(aq) \rightarrow BaCl_2(aq) + 2H_2O(l)$

A $-57\,kJ\,mol^{-1}$ **B** $-76\,kJ\,mol^{-1}$

C $-114\,kJ\,mol^{-1}$ **D** $-228\,kJ\,mol^{-1}$ [1]

(Total for Question 2 = 1 mark)

3 Which of the following processes is endothermic?

A the condensation of steam

B the freezing of water

C the decomposition of calcium carbonate into calcium oxide and carbon dioxide

D the reaction between hydrogen ions and hydroxide ions [1]

(Total for Question 3 = 1 mark)

4 For which of the following reactions does the value of $\Delta H^\ominus$ represent both a standard enthalpy change of formation and a standard enthalpy change of combustion?

A $C(s) + O_2(g) \rightarrow CO_2(g)$ **B** $C(g) + \frac{1}{2}O_2(g) \rightarrow CO(g)$

C $2C(s) + O_2(g) \rightarrow 2CO(g)$ **D** $CO(g) + \frac{1}{2}O_2(g) \rightarrow CO_2(g)$ [1]

(Total for Question 4 = 1 mark)

5 The value of the enthalpy change for the process represented by the equation

$K(s) \rightarrow K^+(g) + e^-$

is equal to

A the electron affinity of potassium

B the enthalpy change of vaporisation of potassium

C the sum of the enthalpy change of atomisation and the first ionisation energy of potassium

D the sum of the enthalpy change of atomisation and the electron affinity of sodium [1]

(Total for Question 5 = 1 mark)

6 Which equation represents the change corresponding to the enthalpy change of atomisation of iodine?

A $I_2(g) \rightarrow 2I(g)$ **B** $I_2(l) \rightarrow 2I(g)$

C $I_2(s) \rightarrow 2I(g)$ **D** $\frac{1}{2}I_2(s) \rightarrow I(g)$ [1]

(Total for Question 6 = 1 mark)

7 Lead forms several solid oxides, the most common of which are PbO, PbO_2 and Pb_3O_4.

This question is about the enthalpy changes that take place during reactions involving these oxides.

(a) (i) State what is meant by enthalpy change of formation. [1]

(ii) State the standard conditions of temperature and pressure that are usually used in calculations involving enthalpy changes. [2]

(iii) Write an equation representing the standard enthalpy change of formation of PbO(s). [2]

(b) 'Red lead' (Pb_3O_4) can be made by heating PbO in oxygen:

$3PbO(s) + \frac{1}{2}O_2(g) \rightarrow Pb_3O_4(s)$

Calculate the standard enthalpy change for this reaction. [3]

$\Delta_f H^\ominus\ PbO(s) = -219\,kJ\,mol^{-1}$

$\Delta_f H^\ominus\ Pb_3O_4(s) = -735\,kJ\,mol^{-1}$

(c) When red lead is heated it decomposes into PbO and PbO_2.

$Pb_3O_4(s) \rightarrow 2PbO(s) + PbO_2(s)\ \Delta_r H^\ominus = +20\,kJ\,mol^{-1}$

Use this information, together with the data supplied in part (b), to calculate the standard enthalpy change of formation of $PbO_2(s)$. [3]

(Total for Question 7 = 11 marks)

8 Propane, C_3H_8, is a gas at room temperature. It is used as a fuel for portable gas cookers.

(a) Give two properties of propane that make it suitable for use as a fuel. [2]

(b) The standard enthalpy change of combustion of propane is represented by the equation:

$C_3H_8(g) + 5O_2(g) \rightarrow 3CO_2(g) + 4H_2O(l)\ \Delta_c H^\ominus = -2220\,kJ\,mol^{-1}$

(i) State what is meant by standard enthalpy change of combustion. [3]

(ii) Complete the enthalpy level diagram for the combustion of propane. Label $\Delta_c H^\ominus$. [2]

Enthalpy

$C_3H_8(g) + 5O_2(g)$

(c) Hess's Law can be used to calculate enthalpy changes of formation from enthalpy changes of combustion.

The equation for the formation of propane is

$3C(s) + 4H_2(g) \rightarrow C_3H_8(g)$

(i) State Hess's law. [2]

(ii) The standard enthalpy changes of combustion of carbon and hydrogen are –394 kJ mol^{-1} and –286 kJ mol^{-1} respectively. Calculate the standard enthalpy change of formation of propane gas. [2]

(Total for Question 8 = 11 marks)

9 Bond enthalpies can be used to calculate enthalpy changes of reaction.

Some bond enthalpies are given in the table.

Bond	**Bond enthalpy / kJ mol^{-1}**
H – H	+436
F – F	+158
H – Cl	+431
H – F	+562

(a) (i) State what is meant by bond enthalpy. [2]

(ii) State why the sign for a bond enthalpy is always positive. [1]

(b) The enthalpy change of formation of hydrogen chloride is represented by the following equation:

$\frac{1}{2}H_2(g) + \frac{1}{2}Cl_2(g) \rightarrow HCl(g) \qquad \Delta_f H^\ominus = -92\ kJ\ mol^{-1}$

Use this information, and the data in the table, to calculate the bond enthalpy of the Cl–Cl bond. [3]

(c) Use the data in the table to calculate the enthalpy change of formation of HF(g). [2]

(Total for Question 9 = 8 marks)

10 The enthalpy change of neutralisation for the reaction between hydrochloric acid, HCl(aq), and sodium hydroxide, NaOH(aq), can be determined using the following method:

- Place 50.0 cm^3 of 2.00 mol dm^{-3} HCl(aq) into a polystyrene cup and measure its temperature.
- Place 50.0 cm^3 of 2.10 mol dm^{-3} NaOH(aq) into another polystyrene cup and measure its temperature.
- Mix the two solutions, stir with the thermometer and record the highest temperature reached.

The results of one experiment were:

- Initial temperature of both HCl(aq) and NaOH(aq) = 20 °C
- Highest temperature reached by the mixture = 33.6 °C

The equation for the reaction is:

$HCl(aq) + NaOH(aq) \rightarrow NaCl(aq) + H_2O(l)$

(a) State what is meant by enthalpy change of neutralisation. [2]

(b) Give the name of a piece of apparatus that is suitable for measuring the volumes of acid and alkali. [1]

(c) State why the NaOH(aq) used had a slightly higher concentration than the HCl(aq). [1]

(d) (i) Calculate the heat energy transferred, using the equation:

$Q = mc\Delta T$ [2]

[Assume the specific heat capacity of the final solution is 4.18 J g^{-1} K^{-1} and that its density is 1 g cm^{-3}.]

(ii) Calculate the enthalpy change of neutralisation for the reaction. [3]

(Total for Question 10 = 9 marks)

11 The table gives the values of some mean bond enthalpies.

Bond	**Bond enthalpy / kJ mol^{-1}**
C – H	+412
O – H	+463
O = O	+496
C = C	+743
C – O	+360

The equation for the combustion of methanol (CH_3OH) in the gaseous state is:

$$H_3C{-}O{-}H + 1\tfrac{1}{2}O{=}O \rightarrow O{=}C{=}O + 2\ H{-}O{-}H$$

(a) (i) Use the data in the table to calculate the enthalpy change of combustion of gaseous methanol. [2]

(ii) Give two reasons why the standard enthalpy change of combustion of methanol is different from the value calculated in part (a)(i). [2]

(b) Which process measures the mean bond enthalpy for the C–H bond in methane? [1]

A $CH_4(g) \rightarrow CH_3(g) + H(g)$
ΔH = A; mean bond enthalpy = A

B $CH_4(g) \rightarrow C(g) + 4H(g)$
ΔH = B; mean bond enthalpy = B/4

C $CH_4(g) \rightarrow C(g) + 2H_2(g)$
ΔH = C; mean bond enthalpy = C/4

D $CH(g) \rightarrow C(g) + H(g)$
ΔH = D; mean bond enthalpy = D

(c) Calculate the mean bond enthalpy of the S–F bond in SF_6 given the following data. [2]

$SF_6(g) \rightarrow S(s) + 3F_2(g) \qquad \Delta H = +1100\ kJ\ mol^{-1}$

$S(s) \rightarrow S(g) \qquad \Delta H = +223\ kJ\ mol^{-1}$

$F_2(g) \rightarrow 2F(g) \qquad \Delta H = +158\ kJ\ mol^{-1}$

(Total for Question 11 = 7 marks)

TOPIC 7 INTERMOLECULAR FORCES

A INTERMOLECULAR INTERACTIONS | B INTERMOLECULAR INTERACTIONS AND PHYSICAL PROPERTIES

The forces of attraction holding molecules together are called intermolecular forces. The energy required to separate molecules in liquids and solids is much smaller than the energy required to break covalent bonds. However, intermolecular forces still play an important role in determining the properties of a substance, such as boiling temperature, miscibility and solubility.

When a capillary, a narrow tube, touches the surface of a liquid, fluid rises into the tube. The extent to which a liquid rises is different for different liquids. When a narrow glass tube is inserted into water, the water rises in the tube. This occurs because the surface of glass is quite polar. As water molecules rise along the inside surface of the capillary, they pull up other water molecules to which they have formed hydrogen bonds. The balance of gravity and the attraction of the water for the glass surface determine the height to which the water rises. Other polar liquids as well as water also rise in capillaries, but some non-polar liquids show the opposite effect, the height of the liquid inside the capillary is less than outside. The molecules of these non-polar liquids are attracted to each other more than they are to the surface of the glass. Capillary action is also responsible for absorption of liquids into paper, such as paper towels.

Life on Earth may exist because of the hydrogen bond. The physical properties of water, which covers about two thirds of Earth's surface and composes a similar proportion of the human body, are largely due to its extensive network of hydrogen bonds. Hydrogen bonding and intermolecular forces are the basis of the genetic code and the unique structures and shapes of the non-aqueous components of life: DNA, RNA, proteins, and other biomolecules making up living systems all owe their form and function to hydrogen bonds. The double helix of DNA is held together by hydrogen bonds. Hydrogen bonds are strong but still at least four times weaker than covalent bonds. Hydrogen bonds are strong enough to hold DNA together under most situations, but are weak enough to form and break readily to enable DNA to untwine for replication.

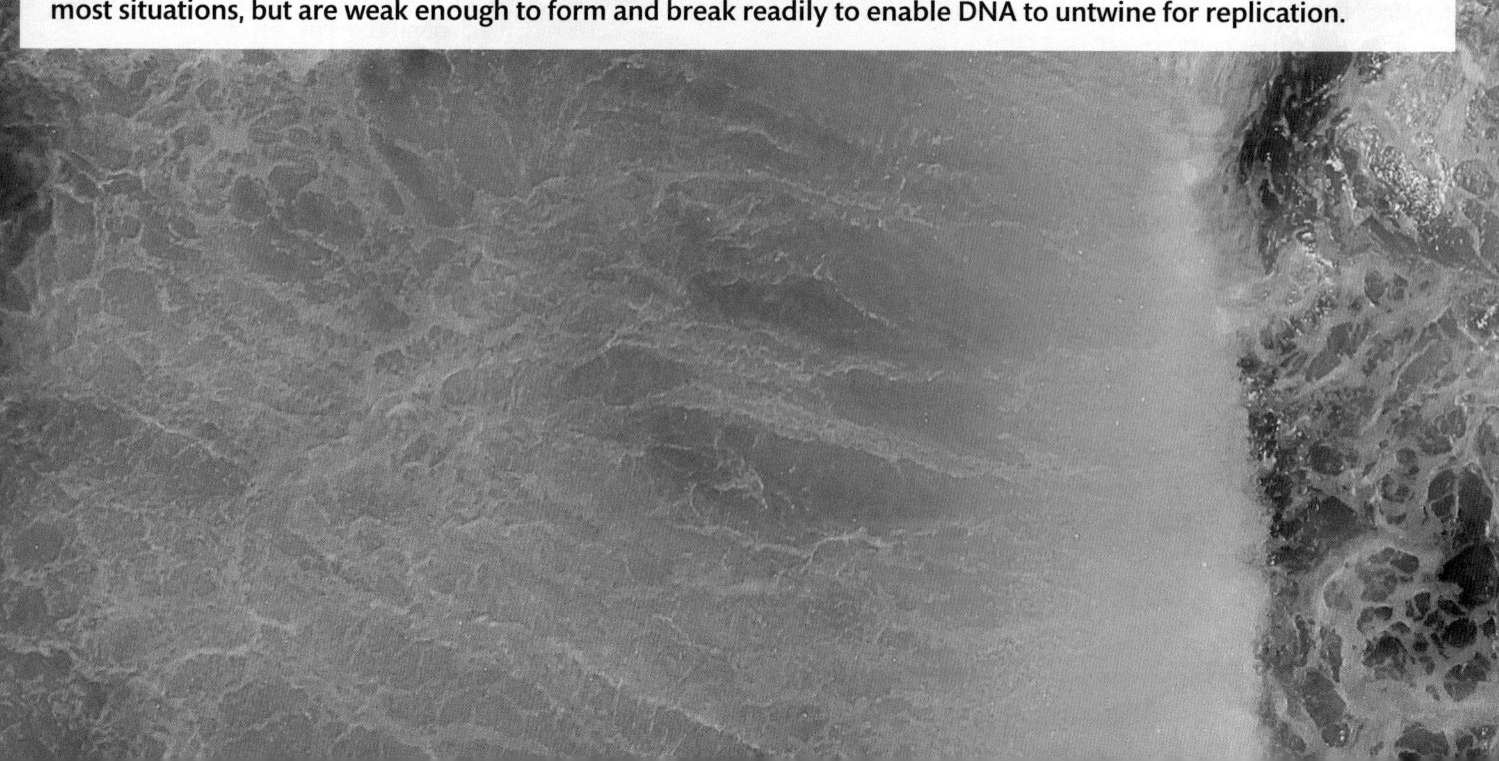

What prior knowledge do I need?
Forces of attraction exist between molecules of simple molecular substances
Simple molecular substances usually have low melting and boiling temperatures because the forces of attraction between the molecules are usually very weak
What will I study in this topic?
The different types of intermolecular forces and how they arise
The factors that determine the strength of intermolecular forces
How intermolecular forces affect the properties of substances
What will I study later?
Topic 10 (Book 1: IAS) and Topics 15 and 19 (Book 2: IAL)
The significance of intermolecular forces in melting and boiling temperatures of organic substances such as alkanes, alcohols, amines and carboxylic acids
The significance of intermolecular forces in explaining the solubility of organic substances such as alkanes, alcohols, amines and carboxylic acids in polar and non-polar liquids

7A INTERMOLECULAR INTERACTIONS

SPECIFICATION REFERENCE 7.1 7.2 7.4

LEARNING OBJECTIVES

- Understand the nature of intermolecular forces resulting from the following interactions:
  (i) London forces
  (ii) permanent dipoles
  (iii) hydrogen bonds.
- Understand the interactions in molecules, such as H_2O, liquid NH_3 and liquid HF, which give rise to hydrogen bonding.
- Predict the presence of hydrogen bonding in molecules analogous to those mentioned above.

BACKGROUND TO NON-BONDED INTERMOLECULAR INTERACTIONS

A number of interactions between molecules are considerably weaker than typical covalent and polar covalent bonds. These interactions are usually described as:

- 'non-bonded interactions' or
- 'intermolecular' because they occur *between* molecules.

The most important non-bonded interactions are 'London forces'. They have this name because their existence was first suggested in 1930 by Fritz London, a German physicist. London forces are also sometimes referred to as 'dispersion forces'. Although they are weaker than covalent and polar covalent bonds, London forces play an important part in determining the physical and chemical properties of many molecules.

Other intermolecular interactions arise from the permanent dipoles that exist in some molecules.

LONDON FORCES

We can describe this interaction by considering two non-polar molecules of nitrogen, labelled A and B in **fig A**.

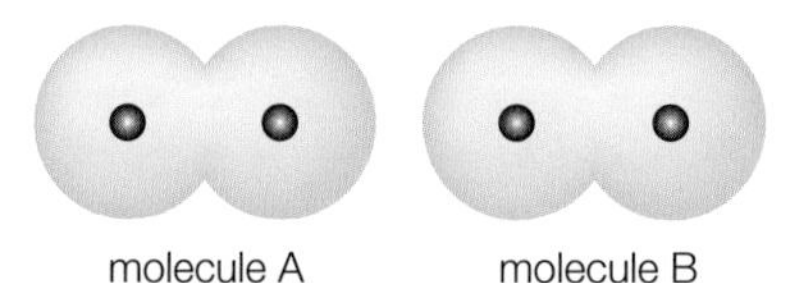

fig A Electron density in nitrogen.

Each molecule is non-polar because, on average, the electron density is symmetrically distributed throughout the molecule (see **Topic 3**).

However, electron density fluctuates over time. If, at any time, the electron density becomes unsymmetrical in molecule A, a dipole will be generated, as shown in **fig B**.

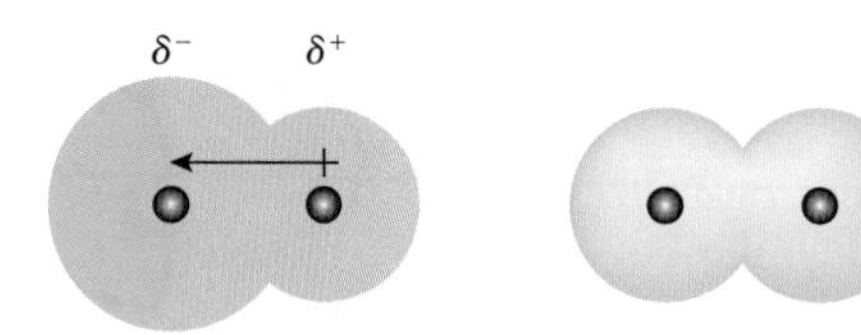

fig B An instantaneous dipole in nitrogen.

The electron density on the left of the molecule has increased, giving that end of the molecule a partial negative charge ($\delta-$). However, the electron density on the right has decreased. This gives that end of the molecule a partial positive charge ($\delta+$). For this reason, an instantaneous dipole is created in molecule A.

The $\delta+$ end of molecule A is closer to molecule B, so the electron density of molecule B is pulled to the left. This generates a partial negative charge on the left-hand end of the molecule, and a partial positive charge on the right-hand end. This creates an induced dipole in molecule B, as shown in **fig C**.

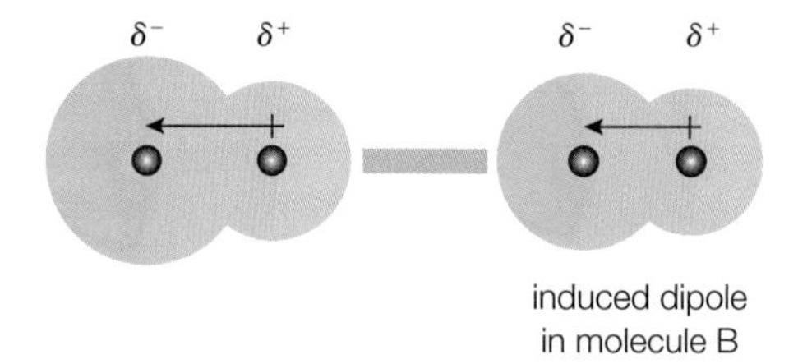

fig C An induced dipole in nitrogen.

Because the dipole of A led to the induction of the dipole in B, the two dipoles are arranged so that they will interact favourably with one another. It is this favourable interaction that is responsible for the London force of attraction between the two molecules.

It is important to realise that the induced dipole will *always* be aligned in such a way that the interaction with the instantaneous dipole is favourable. The fluctuations that lead to the generation of an instantaneous dipole, and the subsequent induction of a dipole in a nearby molecule, are very rapid processes. This is especially the case when they are compared with the rates at which the molecules are moving. This is because of their kinetic energy and rotational energy. As the two molecules move around, they will continue to attract each other regardless of their orientation.

EXAM HINT

If you are asked to explain London forces in an exam question, it is always worth drawing a diagram and then referring to it in your answer. A good diagram is often better than just a written answer.

A feature of London forces is that the attractive force increases with increasing number of electrons in the molecule. We can demonstrate this by observing the boiling temperatures of the noble gases (**table A**). The noble gases all exist as monatomic molecules (i.e. single atom molecules). The London force is the only force of attraction between the molecules. The stronger the force of attraction, the more energy is required to separate the molecules. The boiling temperature increases because of this.

GAS	HELIUM	NEON	ARGON	KRYPTON	XENON	RADON
Boiling temperature / K	4.3	27.1	87.4	121	165	211

table A Boiling temperatures of the noble gases.

The more electrons there are in a molecule, the greater the fluctuation in electron density and the larger the instantaneous and induced dipoles created. There is a similar trend in the boiling temperatures of the halogens from F_2 to I_2 (see **Topic 8**).

A second feature of London forces is that they depend on the shape and size of the molecules. The more points of contact there are between the molecules, the greater the overall London force. We will look at this in more detail in **Topic 7B**.

A third feature is that London forces are always present between molecules. This is the case whether or not they have a permanent dipole and whether or not they form hydrogen bonds with each other (see below).

PERMANENT DIPOLES

If the molecules possess permanent dipoles, they will also interact with one another. If the dipoles are aligned correctly, then there will be a favourable interaction and the two molecules will attract one another (**fig D**).

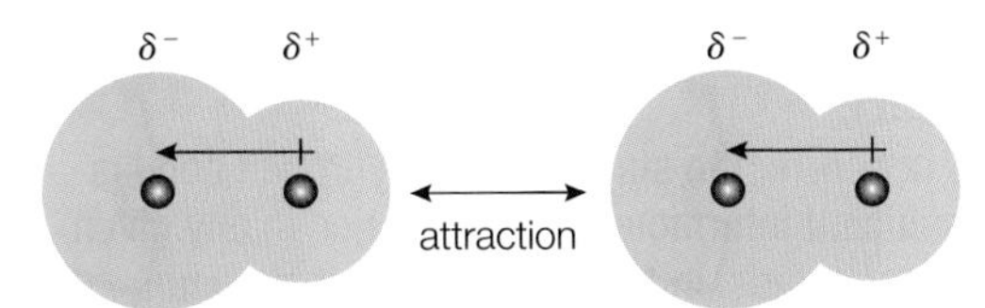

fig D Attraction between permanent dipoles.

The problem here is that, for example, in a liquid, the random movement of the molecules is such that the dipoles are not always aligned to produce a favourable interaction.

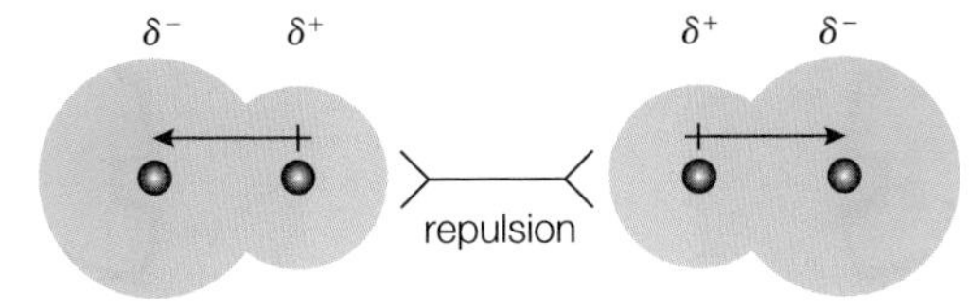

fig E Repulsion between permanent dipoles.

As a result, when averaged out, the interaction between permanent dipoles is usually much less than the interaction between instantaneous and induced dipoles. The London force is usually the more significant interaction between molecules.

LEARNING TIP

Initially, it seems surprising that the interaction between permanent dipoles should be weaker than the interaction between instantaneous and induced dipoles.

The key point is that induced dipoles are always aligned so that their interaction is favourable. This is not true for permanent dipoles.

It is possible for a molecule with a permanent dipole to induce a dipole in a nearby molecule. The two types of interaction, permanent dipole–permanent dipole and permanent dipole–induced dipole, are sometimes put together under the heading permanent dipole–dipole forces.

SUMMARY

Table B is a useful reminder of the origins of non-bonded intermolecular interactions.

NAME OF INTERACTION	ORIGIN
London forces	instantaneous dipole-induced dipole interaction
Permanent dipoles	permanent dipole-permanent dipole interaction

table B Summary of non-bonded intermolecular interactions.

DID YOU KNOW?

Although the strength of the London forces of attraction between the molecules of the noble gases does increase with an increase in the number of electrons, this is **not** the reason for the increase. The strength of the London forces between the molecules depends on the polarisability of the molecules (see **Topic 3A** for a description of polarisation). Molecules that are more easily polarised produce larger instantaneous and induced dipoles. For example, the neon molecule is larger than the helium molecule, so its outer electrons are further away from the nucleus. The outer electrons of the neon molecule are more loosely held than the outer electrons of the helium molecule. Hence, neon molecules are more easily polarised as the electron cloud is more easily distorted. This produces larger instantaneous and induced dipoles and therefore stronger London forces.

It is sometimes stated that molecules with the same number of electrons will have the same London forces of attraction. This is rarely, if ever, true. For example, the London forces between molecules of ammonia (NH_3), water (H_2O) and hydrogen fluoride (HF) may be similar, but they are not the same. This is because the molecules have very different polarisabilities. The water molecule is about three times more easily polarised than the hydrogen fluoride molecule, whereas the ammonia molecules is about four times more easily polarised than the hydrogen fluoride molecule. This is because, of the three molecules, the electrons are most strongly held in HF and least strongly held in NH_3. Hence, the strength of the London forces varies as follows: $NH_3 > H_2O > HF$.

LEARNING TIP

It is important to realise that London forces exist between all types of molecules, whether non-polar or polar.

For example, London forces, as well as permanent dipole-permanent dipole interactions, exist between the polar molecules of hydrogen chloride.

DID YOU KNOW?

The term 'van der Waals force' is often used in conjunction with intermolecular interactions.
The van der Waals force is the sum of all the intermolecular interactions between the molecules. The term includes London forces and permanent dipole-permanent dipole interactions.

THE HYDROGEN BOND

There is one other intermolecular interaction that is, in some cases, very important. It is called a **hydrogen bond**.

The key to understanding the nature of the hydrogen bond is appreciating that the atom bonded to hydrogen has to be more electronegative than hydrogen, and that there must be some evidence of bond formation between the hydrogen and another atom, either within the same molecule ('intramolecular hydrogen bonding') or in a different molecule ('intermolecular hydrogen bonding').

DID YOU KNOW?

We are only looking at *inter*molecular hydrogen bonding, although *intra*molecular hydrogen bonding plays a significant role in many biochemical molecules such as DNA and proteins.
Because there has to be evidence of some bond formation (the nature of the evidence, largely from various types of spectroscopy, is beyond this course), there is an argument that hydrogen bonds could be considered to be bonded intermolecular interactions. However, most academic texts still prefer to use the term non-bonded intermolecular interactions to describe them.

Hydrogen bonding is most significant when hydrogen is bonded to very small, highly electronegative atoms such as oxygen, nitrogen and fluorine, although it is not confined to these atoms.

HYDROGEN BONDING THROUGH OXYGEN

All compounds containing an –O—H group form intermolecular hydrogen bonds.

The most important example is water.

The hydrogen bond forms between the oxygen atom of one water molecule and the hydrogen atom of a second water molecule (**fig F**).

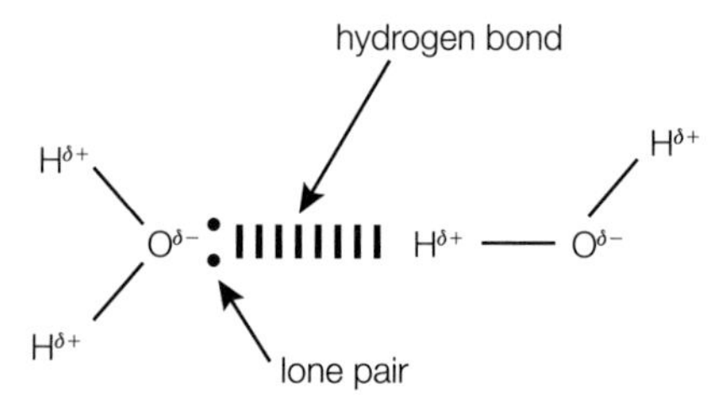

fig F Hydrogen bonding between water molecules.

The interaction is not just that of an extreme dipole–dipole interaction. There is some partial bond formation using a lone pair of electrons on the oxygen atom. Because the oxygen atom has two lone pairs, it can form hydrogen bonds with two other water molecules.

Another feature of a hydrogen bond that indicates partial bond formation is that, like covalent bonds, hydrogen bonds are directional in nature. The bond angle between the three atoms involved is often 180°, or close to it, but this is not always the case.

Alcohols (see **Topic 10**) also form intermolecular hydrogen bonds.

Fig G shows the formation of a hydrogen bond between two ethanol molecules.

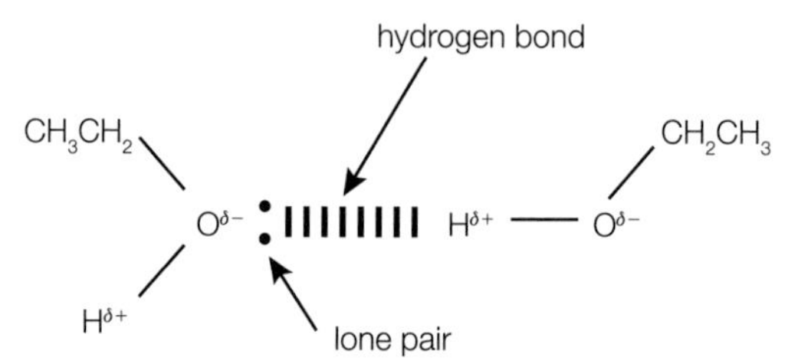

fig G Hydrogen bonding between ethanol molecules.

HYDROGEN BONDING THROUGH NITROGEN

All compounds containing an –N—H group can form intermolecular hydrogen bonds. An example is the organic group of compounds known as primary amines, which have the general formula RNH_2. **Fig H** shows the hydrogen bonding in ammonia, NH_3.

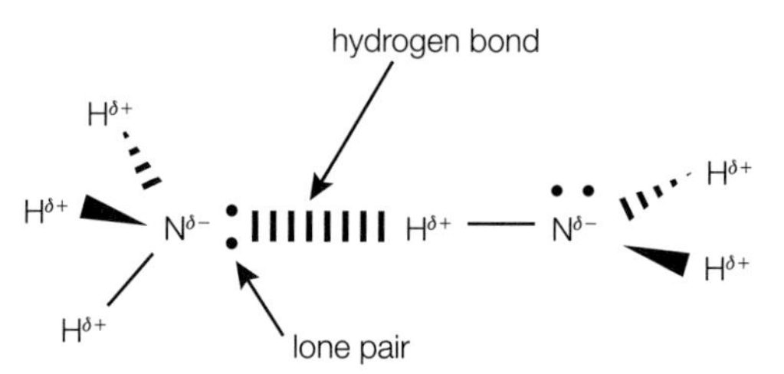

fig H Hydrogen bonding between ammonia molecules.

DID YOU KNOW?

Hydrogen bonds are often quoted as being the strongest of all intermolecular interactions. This is true for some molecules such as water, but it is not true for the majority of molecules containing hydrogen bonds.
For example, although the hydrogen bond is significant in the case of the short chain alcohols, e.g. methanol (CH_3OH) and ethanol (CH_3CH_2OH), it becomes less significant for longer chain alcohols such as pentan-1-ol ($CH_3CH_2CH_2CH_2CH_2OH$) (see Topic 7B). The same trend is exhibited by the amines (see the section on amines in Topic 19 (Book 2: IAL).
For this reason, it is not correct to state that the hydrogen bond is always the most significant intermolecular interaction in any given group of analogous compounds. In fact, we will see that some hydrogen bonds are very weak.

HYDROGEN BONDING THROUGH FLUORINE

The only fluorine compound with intermolecular hydrogen bonding is hydrogen fluoride.

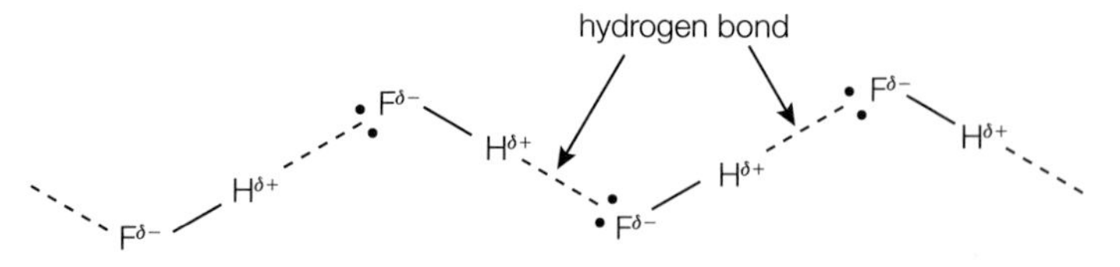

fig I Hydrogen bonding between hydrogen fluoride molecules.

CHECKPOINT

1. Molecules of ethanoic acid dimerise through hydrogen bonding when dissolved in certain organic solvents. The structure and shape of an ethanoic acid molecule is:

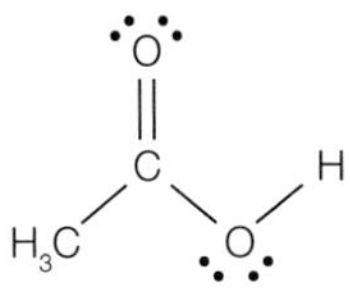

   Draw a diagram to show how hydrogen bonds are formed between two molecules of ethanoic acid.

2. Explain how it is possible for a hydrogen bond to form between a molecule of propanone (CH_3COCH_3) and a molecule of trichloromethane ($CHCl_3$).

   The structures and shapes of the molecules are:

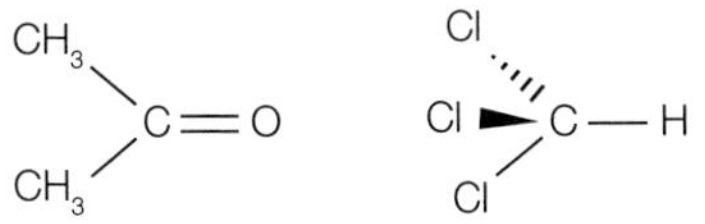

3. The boiling temperature of ethanol (CH_3CH_2OH, 78.5 °C) is considerably higher than the boiling temperature of methoxymethane (CH_3OCH_3, −24.8 °C). The structures and shapes of the molecules are:

   ethanol methoxymethane

   In terms of the intermolecular interactions involved, explain the difference in boiling temperatures.

SUBJECT VOCABULARY

hydrogen bond an intermolecular interaction (in which there is some evidence of bond formation) between a hydrogen atom of a molecule (or molecular fragment) bonded to an atom which is more electronegative than hydrogen and another atom in the same or a different molecule

ADDITIONAL READING

Strong and weak hydrogen bonds

It was once believed that hydrogen bonds were formed only when hydrogen was bonded to oxygen, nitrogen or fluorine. Richard Nelmes of Edinburgh University surprised the world of chemistry when he discovered that solid hydrogen sulfide had an extended system of hydrogen bonding, which was similar to that in ice.

Because sulfur has a low electronegativity, the hydrogen bond between H_2S molecules is considered to be weak. In fact, it has a magnitude of 7 kJ mol^{-1}, compared with 22 kJ mol^{-1} in ice.

Some strengths of hydrogen bonds are shown in **table C** below. For comparison, the strengths of the 'full' bonds in some of these species are also shown.

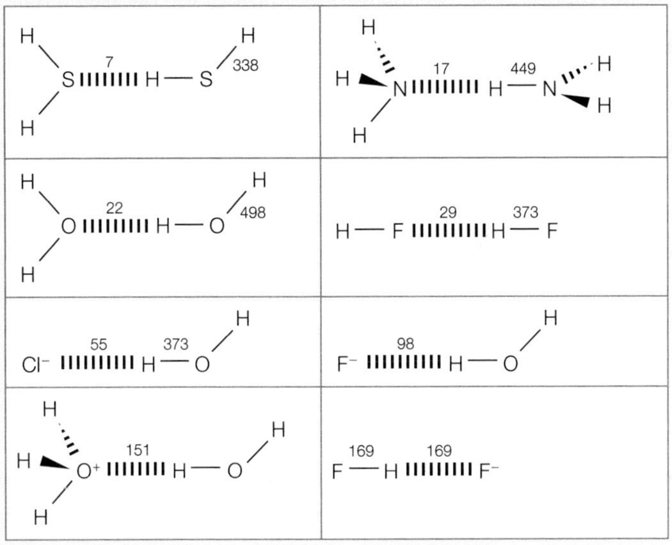

Hydrogen bond	Strengths
H_2S ···· H—SH	7 (hydrogen bond), 338 (H—S)
H_3N ···· H—NH_2	17 (hydrogen bond), 449 (H—N)
H_2O ···· H—OH	22 (hydrogen bond), 498 (H—O)
H—F ···· H—F	29 (hydrogen bond), 373 (H—F)
Cl^- ···· H—OH	55 (hydrogen bond), 373 (H—O)
F^- ···· H—OH	98 (hydrogen bond)
H_3O^+ ···· H—OH	151 (hydrogen bond)
F—H ···· F^-	169 (F—H), 169 (hydrogen bond)

table C Strengths of some hydrogen bonds.

You will notice that, in most cases, the hydrogen bonds are very much weaker than the full bonds. You can also see that the strengths of the hydrogen bonds increase with the electronegativity of the element to which the hydrogen is attached. Hydrogen bonding strength increases as follows:

$$H_2S < NH_3 < H_2O < HF$$

Also note that the hydrogen bond to an ion is much stronger than that to a neutral molecule.

The strongest hydrogen bond shown in the table is that between HF and F^-. In this species the two bond lengths and bond strengths are identical. The resulting ion, $[F–H–F]^-$, is so stable that it is obtainable as the solid sodium salt, sodium hydrogen difluoride, $NaHF_2$.

7B INTERMOLECULAR INTERACTIONS AND PHYSICAL PROPERTIES

SPECIFICATION REFERENCE 7.3 7.5 7.6

LEARNING OBJECTIVES

- Understand in terms of intermolecular interactions, the physical properties shown by materials, including:
  (i) the trends in boiling temperatures of alkanes with increasing chain length
  (ii) the effect of branching in the carbon chain on the boiling temperatures of alkanes
  (iii) the relatively low volatility (higher boiling temperatures) of alcohols compared with alkanes with a similar number of electrons
  (iv) the trends in boiling temperatures of the hydrogen halides HF to HI.
- Understand the following anomalous properties of water resulting from hydrogen bonding:
  (i) its relatively high melting and boiling temperatures
  (ii) the density of ice compared with that of water.
- Understand the reasons for the choice of solvents, including:
  (i) water, to dissolve some ionic compounds, in terms of the hydration of the ions
  (ii) water, to dissolve simple alcohols, in terms of hydrogen bonding
  (iii) water, as a poor solvent for some compounds, in terms of inability to form strong hydrogen bonds
  (iv) non-aqueous solvents, for compounds which have similar intermolecular interactions to those in the solvent.

BOILING TEMPERATURES OF ALKANES AND ALCOHOLS

UNBRANCHED ALKANES

The alkanes are a homologous series of hydrocarbons with the general formula C_nH_{2n+2} (see **Topic 4**).

The graph in **fig A** shows the relationship between the boiling temperature and the relative molecular mass for the first 10 unbranched alkanes (i.e. CH_4 to $C_{10}H_{22}$ inclusive).

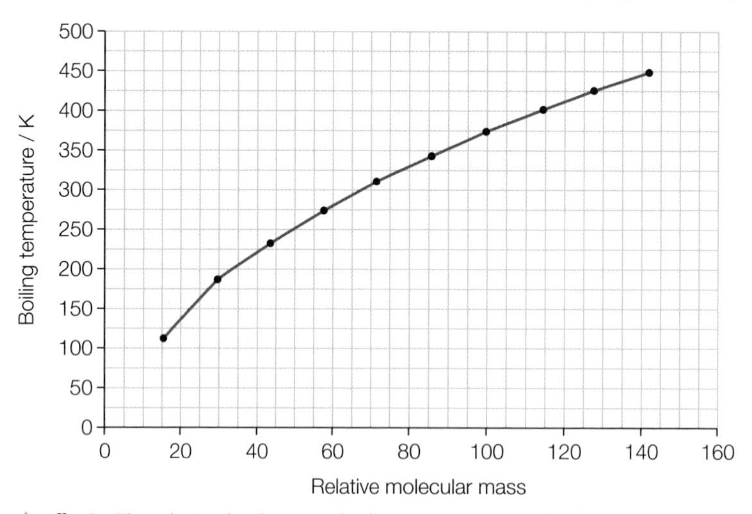

▲ **fig A** The relationship between boiling temperature and relative molecular mass for the first ten alkanes.

EXAM HINT

Think about how important these intermolecular forces are in explaining the properties of polymers.

The only significant intermolecular interaction between alkane molecules is the London force.

There are two reasons for the increase in boiling temperature with increasing molecular mass:

1 As molecular mass increases, the number of electrons per molecule increases and so the instantaneous and induced dipoles also increase (see **Topic 7A.1**).

2 As the length of the carbon chain increases, the number of points of contact between adjacent molecules increases. Instantaneous dipole-induced dipole forces exist at each point of contact between the molecules, so the more points of contact there are, the greater the overall intermolecular (London) force of attraction.

You can see the relationship between chain length and points of contact using the skeletal formulae of the alkanes (**fig B**).

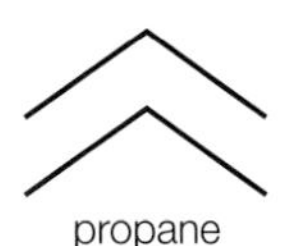

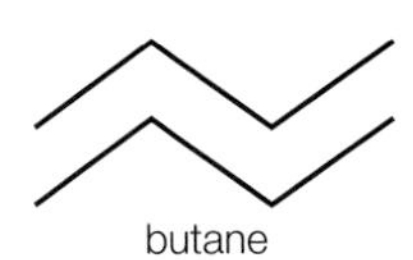

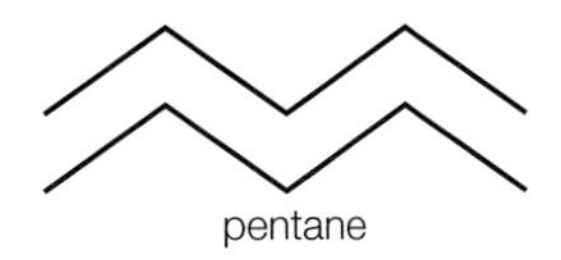

▲ **fig B** Skeletal formulae of propane, butane and pentane

Because of their shapes, the molecules of the alkanes fit together very well and pack very closely. There are points of contact all the way along the chain. The longer the chain, the more points of contact there are.

BRANCHED ALKANES

Branched chain alkanes have lower boiling temperatures than their unbranched isomers (**table A**).

NAME OF ALKANE	STRUCTURAL FORMULA	BOILING TEMPERATURE / K
pentane	H H H H H H—C—C—C—C—C—H H H H H H	309
2-methylbutane	H CH_3 H H H—C—C—C—C—H H H H H	301
2,2-dimethylpropane	CH_3 CH_3—C—CH_3 CH_3	283

table A

The more branching in the molecule, the fewer points of contact between adjacent molecules; i.e. they do not pack together as well. This leads to a decrease in the overall intermolecular force of attraction between molecules and a decrease in boiling temperature.

ALCOHOLS

Alcohols are a homologous series of compounds with the general formula $C_nH_{2n+1}OH$ (see **Topic 10**). They contain an –O—H group and can therefore form intermolecular hydrogen bonds in addition to London forces. This additional bonding affects the boiling temperature of alcohols when compared with the equivalent alkane. This is shown in **table B**.

FORMULA OF ALCOHOL	NUMBER OF ELECTRONS IN MOLECULE	BOILING TEMPERATURE / K	FORMULA OF ALKANE	NUMBER OF ELECTRONS IN MOLECULE	BOILING TEMPERATURE / K
CH_3OH	18	338	CH_3CH_3	18	184
CH_3CH_2OH	26	352	$CH_3CH_2CH_3$	26	231
$CH_3CH_2CH_2OH$	34	370	$CH_3CH_2CH_2CH_3$	34	267
$CH_3CH_2CH_2CH_2OH$	42	390	$CH_3CH_2CH_2CH_2CH_3$	42	309

table B Boiling temperatures of alcohols and alkanes.

Consider the case of methanol (CH_3OH) and ethane (CH_3CH_3). Both molecules have a similar chain length. They also have the same number of electrons. If the intermolecular interactions in each were only London forces, then their boiling points would be almost identical. However, the boiling point of the alcohol is higher than that of the alkane. This is because of the hydrogen bonding that exists between methanol molecules. This does not exist between molecules of ethane. The additional force of attraction increases the energy required to separate the molecules.

You will notice the same trend in the other compounds in **table B**.

It is sometimes stated that the predominant, or main, bonding in alcohols is hydrogen bonding. We have already mentioned that this is not always the case (**Topic 7A.1**). **Table C** provides evidence that for the first few members of the alcohol series hydrogen bonding is predominant. However, London forces eventually predominate as the chain length increases.

ALCOHOL	ENTHALPY CHANGE OF VAPORISATION / $kJ\,mol^{-1}$
CH_3CH_2OH	38.6
$CH_3CH_2CH_2OH$	47.5
$CH_3CH_2CH_2CH_2OH$	52.4
$CH_3CH_2CH_2CH_2CH_2OH$	57.0
$CH_3CH_2CH_2CH_2CH_2CH_2OH$	61.6

ALKANE	ENTHALPY CHANGE OF VAPORISATION / $kJ\,mol^{-1}$
$CH_3CH_2CH_3$	15.7
$CH_3CH_2CH_2CH_3$	21.0
$CH_3CH_2CH_2CH_2CH_3$	26.4
$CH_3CH_2CH_2CH_2CH_2CH_3$	31.6
$CH_3CH_2CH_2CH_2CH_2CH_2CH_3$	36.6

table C Enthalpy changes of vaporisation of alcohols and alkanes.

The enthalpy change of vaporisation is a measure of the amount of energy that is required to *completely* separate the molecules of a liquid and convert it into a gas at the same temperature. It is, therefore, a direct measure of the strength of the intermolecular interactions. The greater the enthalpy change of vaporisation, the greater the forces of attraction between the molecules.

In the case of ethanol, the total energy required to separate one mole of molecules is 38.6 kJ.

Of this, approximately 15.7 $kJ\,mol^{-1}$ can be attributed to London forces. For this reason, the hydrogen bonding is the predominant bonding, providing approximately 59% of the total.

[(38.6 – 15.7) = 22.9; 22.9 is 59% of 38.6]

You can calculate the percentage contribution for the other alcohols in a similar way. **Table D** shows the results of these calculations.

ALCOHOL	APPROXIMATE PERCENTAGE CONTRIBUTION OF HYDROGEN BONDING
$CH_3CH_2CH_2OH$	56
$CH_3CH_2CH_2CH_2OH$	50
$CH_3CH_2CH_2CH_2CH_2OH$	45
$CH_3CH_2CH_2CH_2CH_2CH_2OH$	41

table D Percentage contribution of hydrogen bonding in alcohols.

The calculations demonstrate the danger of making generalised statements such as, 'The predominant bonding in alcohols is hydrogen bonding'.

BOILING TEMPERATURES OF THE HYDROGEN HALIDES

The graph in **fig C** shows the boiling temperatures of the hydrogen halides HF to HI.

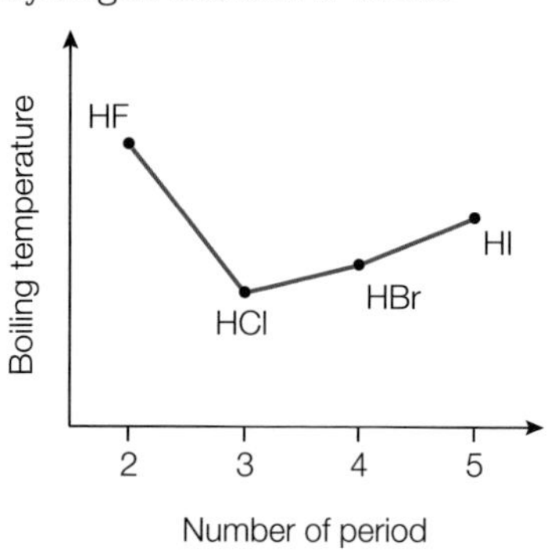

fig C Boiling temperatures of the hydrogen halides.

The steady increase in boiling temperature from HCl to HI is the result of the increasing number of electrons per molecule, which in turn results in an increase in London forces.

EXAM HINT

You may be asked to sketch a graph or plot a graph. If you are asked to sketch, you only need to show the pattern or trend. If you are asked to plot, you need to provide clear, even scales on each axis and points plotted to within an accuracy of 1 mm^2.

DID YOU KNOW?

There are three intermolecular interactions within each compound. They are:

1 London forces
2 permanent dipole–dipole interactions
3 hydrogen bonding.

However, with HCl, HBr and HI, the electronegativity of the halogen is low enough for the permanent dipole–dipole interactions and the hydrogen bonding to be relatively weak. The predominant interaction in each compound is the London force.

The boiling temperature of HF is significantly higher than that of the other three hydrogen halides, despite having fewer electrons per molecule. The London force is weaker in HF, but the hydrogen bonding is significantly greater because of the high electronegativity of fluorine.

ANOMALOUS PROPERTIES OF WATER

Water has some anomalous properties. The following two are particularly important:

1 It has a relatively high melting and boiling temperature for a molecule with so few electrons.
2 The density of ice at 0 °C is less than that of water at 0 °C.

MELTING AND BOILING TEMPERATURES

The hydrogen bonds between water molecules are relatively strong (see **Topic 3E.2**). As a result, the overall intermolecular forces of attraction in water are greater than would be expected from the number of electrons (10) in the molecule.

Water, therefore, has an abnormally high melting temperature (0 °C, 273 K) and boiling temperature (100 °C, 373 K) at 100 kPa pressure.

It is interesting to compare the boiling temperatures of water, ammonia and hydrogen fluoride (**table E**).

	BOILING TEMPERATURE	NUMBER OF ELECTRONS PER MOLECULE	STRENGTH OF HYDROGEN BONDING / $kJ mol^{-1}$
H_2O	373 K (100°C)	10	22
NH_3	240 K (−33°C)	10	17
HF	293 K (20°C)	10	29

table E Strength of hydrogen bonding in water, ammonia and hydrogen fluoride.

The number of electrons per molecule is identical, so we would expect the London forces to be similar between each set of molecules. The differences are a result of different extents of hydrogen bonding.

The hydrogen bond strength for HF is greater than the hydrogen bond strength for H_2O, but surprisingly, its boiling temperature is lower. This is because of two factors.

1 HF forms two hydrogen bonds per molecule, whereas water molecules can form up to four hydrogen bonds per molecule. This means that the hydrogen bonding is much more extensive in water.
2 Not all of the hydrogen bonds in HF are broken on vaporisation, since HF is substantially polymerised, even in the gas phase.

Ammonia has the lowest boiling temperature of the three compounds. The amount of hydrogen bonding in ammonia is limited by the fact that each nitrogen atom only has one lone pair. In a group of ammonia molecules, there are not enough lone pairs to go around to satisfy all the hydrogen atoms. This means that, on average, each ammonia molecule can form one hydrogen bond using its lone pair and one involving one of its $\delta+$ hydrogen atoms. The other hydrogen atoms are 'wasted'. The reason for the relatively low boiling temperature of ammonia is presumably the low strength of the hydrogen bonding between its molecules.

DENSITY OF ICE

As we mentioned above, water has another unusual property: the density of the solid (ice) is less than the density of the liquid at 0 °C.

The molecules in ice are arranged in rings of six, held together by hydrogen bonds (**fig D**).

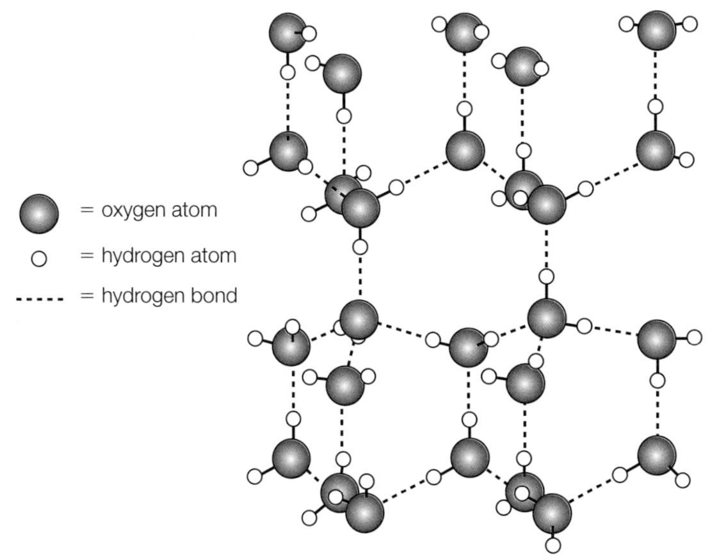

▲ **fig D** Hydrogen bonding in ice.

The structure creates large areas of open space inside the rings. When ice melts, the ring structure is destroyed and the average distance between the molecules decreases, causing an increase in density.

CHOOSING SUITABLE SOLVENTS

For a substance to dissolve, the following two conditions must be met.

1 The solute particles must be separated from each other and then become surrounded by solvent particles.

2 The forces of attraction between the solute and solvent particles must be strong enough to overcome the solvent–solvent forces and the solute–solute forces.

DISSOLVING IONIC SOLIDS

Many ionic solids dissolve in water. The energy required to separate the ions in the solid is either completely, or partially, supplied by the hydration of the ions.

Fig E shows the process of dissolving for sodium chloride.

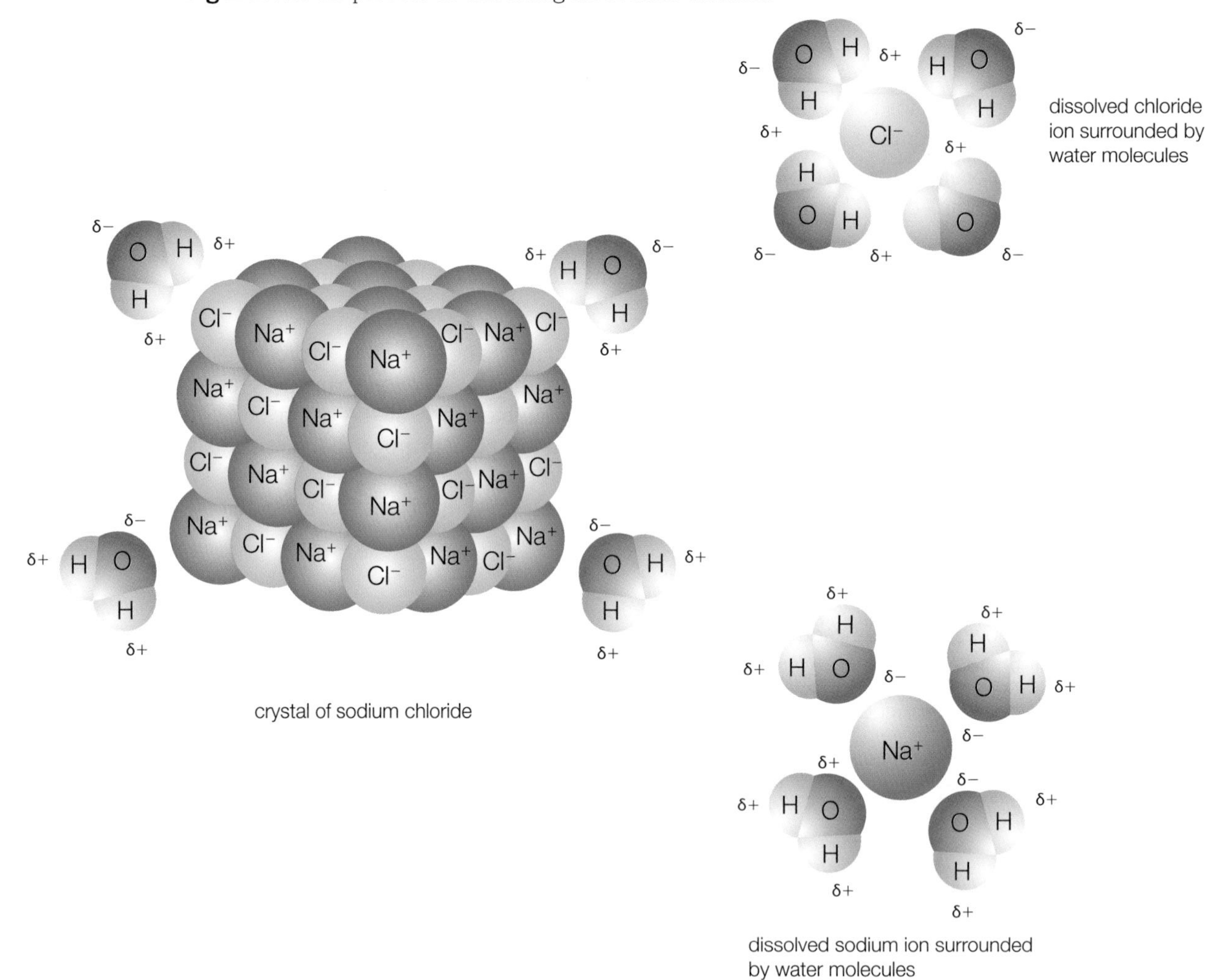

▲ **fig E** Dissolving of sodium chloride.

The $\delta-$ ends of the water molecules attract the sodium ions sufficiently to remove them from the lattice. The sodium ions then become surrounded by water molecules, as shown. The interaction between the sodium ions and the water molecules is called an ion–dipole interaction.

The $\delta+$ end of the water molecules attracts the chloride ions. Once they are in solution, the chloride ions become surrounded by water molecules, as shown. The chloride ions are hydrogen bonded to the water molecules.

The above processes are known as 'hydration' and the energy released is known as the 'hydration energy'.

DID YOU KNOW?

This explanation is a simplification of the processes that occur. A more detailed explanation of the dissolving of ionic solids in water will be discussed in Topic 12 (Book 2: IAL).

COMPOUNDS THAT CAN FORM HYDROGEN BONDS WITH WATER

Alcohols contain an –O—H group and can therefore form hydrogen bonds with water.

Fig F shows a molecule of ethanol forming a hydrogen bond to water:

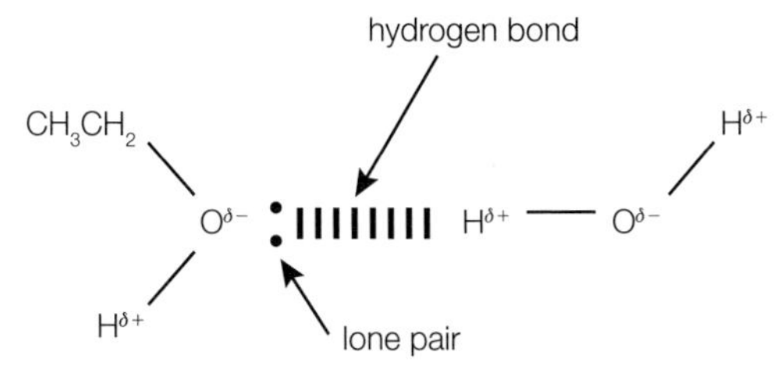

▲ **fig F** Hydrogen bonding between ethanol and water.

Ethanol and water mix in all proportions. The hydrogen bonding between the ethanol and water molecules is similar in strength to the hydrogen bonding in pure ethanol and in pure water.

The solubility of alcohols in water decreases with increasing hydrocarbon chain length as London forces predominate between the alcohol molecules.

COMPOUNDS THAT CANNOT FORM HYDROGEN BONDS WITH WATER

Non-polar molecules such as the alkanes do not dissolve in water. The attraction between the alkane molecules and water molecules is not sufficiently strong to disrupt the hydrogen bonded system between the water molecules.

Many polar molecules also have limited solubility in water. This is because they either do not form hydrogen bonds with water, or the hydrogen bonds they do form are weak compared with the hydrogen bonds in water.

Ethoxyethane, $CH_3CH_2OCH_2CH_3$, is polar (dipole moment = 1.15 D) and yet is almost totally immiscible with water. The forces of attraction between ethoxyethane and water molecules are not large enough to replace the relatively strong hydrogen bonding between the water molecules.

Halogenoalkanes (see **Topic 10**) are also not very soluble in water for similar reasons. They are much more soluble in ethanol, and this is why some reactions of halogenoalkanes are carried out in a medium of aqueous ethanol.

NON-AQUEOUS SOLVENTS

A general rule is that 'like dissolves like'. In other words, substances that are very similar dissolve each other. If you are searching for a solvent for a non-polar substance, or for a substance that has a substantial non-polar part to its molecule, then liquids that contain similar molecules are often the answer.

For example, alkanes are soluble in one another. Crude oil is a complex mixture of alkanes dissolved in each other.

Non-polar bromine dissolves readily in non-polar hexane (C_6H_{14}), and this solution is sometimes used to test for unsaturation in molecules. It is more convenient to use than bromine water, since the molecules being tested will also be soluble in hexane, whereas they are likely to be insoluble in water.

LEARNING TIP

The situation is complicated by the fact that molecules such as bromine both dissolve in, *and* react with, water. This type of reaction will be considered further in **Topic 8**.

CHECKPOINT

SKILLS REASONING

1. The hydrides of Group 4 exist as tetrahedral molecules. Explain the trend in their boiling temperatures:

   CH_4 (112 K) SiH_4 (161 K) GeH_4 (185 K) SnH_4 (221 K)

2. Suggest why water has a relatively high surface tension for a molecule of such low molecular mass.
3. A propanone molecule (CH_3COCH_3) has 32 electrons. A butane molecule ($CH_3CH_2CH_2CH_3$) has 34 electrons. Explain why propanone has a much higher boiling temperature than butane.
4. Suggest why magnesium chloride is soluble in water even though the energy required to break up the lattice is 2494 kJ mol^{-1}.

LIFE WITHOUT WATER

SKILLS CURIOSITY, CONTINUOUS LEARNING

The article below describes how the antifreeze proteins produced by some animals help them to survive in freezing conditions.

HOW FISH KEEP ONE STEP AHEAD OF ICE

Until now, no one has quite figured out how antifreeze proteins help insects, fish and plants to survive at subzero temperatures. But a detailed computer model has lent strong support to a theory that the antifreeze proteins stop the growth of ice crystals by attaching to the crystals and forcing them to change their shape – much like a pillow is deformed when stones are placed on it.

Several organisms, including icefish and Antarctic cod, produce antifreeze proteins (AFPs) to help them survive in icy waters. Normally, the formation of ice crystals in their body fluids should rupture delicate membranes and cellular structures, but the AFPs bind to the surface of ice crystals and prevent them growing.

Now, Leonard Sander and Alexei Tkachenko of the University of Michigan in Ann Arbor have shown just how this could happen. When the proteins attach to ice crystals, the ice is forced to grow in a bulge between them. The computer model shows that if the bulge gets large enough, then the ice will engulf the proteins rather than keep on growing. This slows down the expansion of the ice crystals. 'If the ice grows slowly, more proteins will [attach] than can be absorbed, so the ice can't "eat" them fast enough,' says Sander. 'The process collapses and the ice stops growing.'

As the ice stops forming, the surrounding fluid can become supercooled to below freezing point. Some animals can survive even if their body fluids are chilled to −2 °C. The researchers tested the model with proteins of different shapes to predict how far the water could be supercooled before ice started to form again, and their predictions matched experimental data collected by other researchers almost exactly (*Physical Review Letters*, vol 93, 128102). The model also shows that spherical proteins are most effective because they have no region for the ice to preferentially attack. Rod-shaped proteins are not as good because the ice can grow over their pointed ends quite easily.

Not everyone is convinced that the theory is accurate, however. One of its requirements is that the proteins must irreversibly attach to the ice. But biophysicist Yin Yeh of the University of California at Davis says this may not happen.

'We are starting to see phenomena where proteins are not as firmly anchored as postulated in the model,' he says. 'There are a number of experiments that show these molecules may be coming on and off the surface.' Yeh has detected proteins moving around the surface of ice crystals, even at −15 °C. And if the proteins do not bind to the ice permanently, they cannot prevent the ice crystals from forming.

Ice is forced to bulge as it grows between protein molecules.

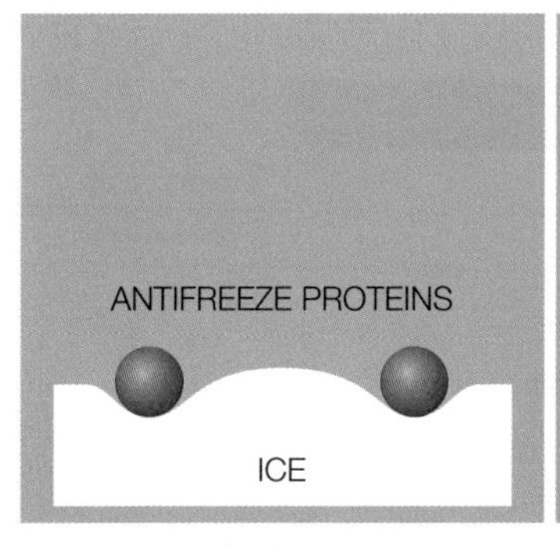

Eventually, the bulge gets so big that it takes less energy for the ice to engulf the proteins than to grow further.

If proteins attach faster to the ice than they are engulfed, ice growth is halted.

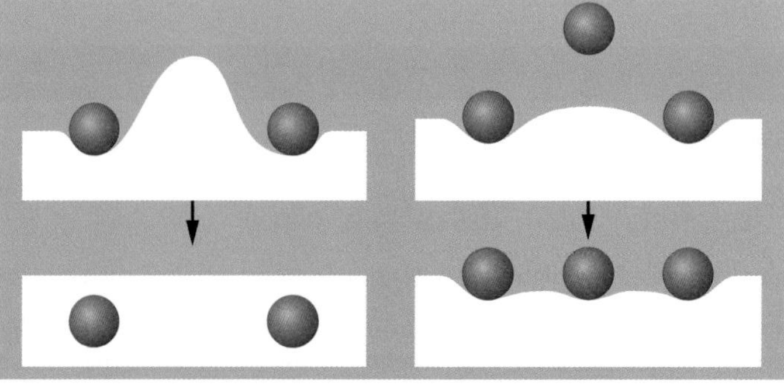

SCIENCE COMMUNICATION

1. Re-read the article and summarise the key issues using bullet points.

CHEMISTRY IN DETAIL

2. (a) Describe and explain the intermolecular forces between water molecules, using a diagram to illustrate your answer.
   (b) Using your answer to part (a), explain why ice freezing is an exothermic process.
3. The diagram below shows a representation of an amino acid called serine.

   O, HO, OH, NH_2

   Would you expect serine to be soluble in water? Explain your answer.
4. In the second paragraph of the article, the author states that 'the formation of ice crystals in their body fluids should rupture delicate membranes and cellular structures'. What unusual property does this indicate about water?

ACTIVITY

Water has been described as the 'matrix of life'. Make a list of all of water's unusual physical and chemical properties. Do you think any other substance could substitute for water as the matrix of life? Discuss your ideas in groups.

DID YOU KNOW?

Some aquatic organisms have a copper-based blue blood that is adapted for oxygen transport at temperatures near freezing.

7 EXAM PRACTICE

1 Which of the following solids consists of atoms or molecules held together by London forces?

A I_2

B Cu

C NaCl

D SiO_2 [1]

(Total for Question 1 = 1 mark)

2 Which of the following statements describes a phenomenon that can be explained by intermolecular hydrogen bonding?

A The boiling points of the alkanes increase with increasing relative molecular mass.

B CH_3COCH_3 ($M_r = 58$) has a higher boiling temperature than $CH_3CH_2CH_3$ ($M_r = 44$).

C Hydrogen chloride forms an acidic solution when dissolved in water.

D Ice has a lower density than water at 0 °C. [1]

(Total for Question 2 = 1 mark)

3 Which of the following isomers is likely to have the highest boiling temperature?

A $(CH_3)_2CHCH_2CH_2CH_3$

B $CH_3CH_2CH_2CH_2CH_2CH_3$

C $CH_3CH_2CH(CH_3)CH_2CH_3$

D $(CH_3)_2CHCH(CH_3)_2$ [1]

(Total for Question 3 = 1 mark)

4 Which quantity would best indicate the relative strengths of the hydrogen bond between the molecules in liquid hydrogen halides, HF, HCl, HBr and HI?

A bond enthalpies

B enthalpy changes of formation

C enthalpy changes of solution

D enthalpy changes of vaporisation [1]

(Total for Question 4 = 1 mark)

5 The surface of Triton, a moon of the planet Neptune, contains condensed methane that flows rapidly.

Which statement explains why condensed methane flows rapidly?

A Condensed methane has a metallic structure.

B Methane molecules contain strong covalent bonds.

C The intermolecular forces between methane molecules are weak.

D Methane molecules have a tetrahedral shape. [1]

(Total for Question 5 = 1 mark)

6 Which type of interaction is responsible for the intermolecular forces in liquid tetrachloromethane, CCl_4?

A covalent bonding

B hydrogen bonding

C instantaneous dipole-induced dipole attractions

D permanent dipole-permanent dipole attractions [1]

(Total for Question 6 = 1 mark)

7 Hydrogen bonding occurs in water and its presence has an effect on the properties of water.

(a) State what is meant by a hydrogen bond. [2]

(b) Draw a diagram to show the formation of a hydrogen bond between two water molecules. [2]

(c) State two ways in which the properties of water are affected by hydrogen bonding. [2]

(Total for Question 7 = 6 marks)

8 Ethanol (CH_3CH_2OH) and methoxymethane (CH_3OCH_3) are structural isomers.

Ethanol is soluble in water whereas methoxymethane is not.

Their boiling temperatures are: ethanol, 78.5 °C; methoxymethane, –24.8 °C.

The structural formula and shape of each molecule is:

O
CH_3CH_2 H
ethanol

O
H_3C CH_3
methoxymethane

(a) Name the types of intermolecular interactions (forces) that exist in

(i) ethanol

(ii) methoxymethane. [5]

(b) Use your answers in part (a) to explain the difference in solubility and boiling temperatures of the two substances. [4]

(Total for Question 8 = 9 marks)

9 The table shows the relative molecular masses (M_r) and boiling temperatures of water and the first eight unbranched members of the homologous series of alkanes, methane to octane.

Formula of compound	H_2O	CH_4	C_2H_6	C_3H_8	C_4H_{10}	C_5H_{12}	C_6H_{14}	C_7H_{16}	C_8H_{18}
M_r	18	16	30	44	58	72	86	100	114
Boiling temperature / °C	100	-164	-89	-42	-0.5	36	69	98	125

(a) Comment on the difference in boiling temperature between water and methane. [3]

(b) Explain the trend in boiling temperatures of the alkanes. [3]

(c) Butane has an isomer called methylpropane, $(CH_3)_3CH$. Explain whether the boiling temperature of methylpropane is the same as, lower than or higher than that of butane. [3]

(Total for Question 9 = 9 marks)

TOPIC 8 REDOX CHEMISTRY AND GROUPS 1, 2 AND 7

A REDOX CHEMISTRY | B THE ELEMENTS OF GROUPS 1 AND 2 | C INORGANIC CHEMISTRY OF GROUP 7 | D QUANTITATIVE CHEMISTRY

Redox reactions are among the most common and most important chemical reactions in everyday life. Here are two examples of redox reactions that occur in our daily lives:

1. One of the most common chemical reactions, *combustion*, is a classic redox reaction. For example, when petrol burns within the internal combustion engine of a car, carbon atoms in the fuel are oxidised to carbon dioxide and carbon monoxide. As this is happening, oxygen in the air is reduced when it combines with the hydrogen atoms of the fuel to form water.
2. *Photosynthesis* is probably the single most important process on the planet. It uses the energy of sunlight to grow plants and so provide food for all higher organisms. Photosynthesis is accomplished through a series of redox reactions in which energy from light is finally converted into carbohydrates on which animals and humans can feed. Oxygen is also a vital by-product of photosynthesis.

The Periodic Table of elements is a familiar feature in any study of chemistry. It helps us to make sense of the variety of chemical elements and their different reactions, some of which are redox reactions. It also helps us to make sense of the even greater variety of chemical compounds of the elements. In this topic, we will focus mostly on three groups (columns) of the Periodic Table to illustrate the trends within the groups and several features of the elements and compounds in them.

In these groups, you will meet familiar elements such as magnesium and chlorine, but also less familiar elements such as barium and bromine. For example, you will learn how the limewater test for carbon dioxide works, how milk of magnesia relieves indigestion and why farmers spread lime on their fields. You will learn that although barium compounds are very poisonous, you can safely eat a 'barium meal' before having an X-ray.

By learning how flame tests work, you will also understand how fireworks can have flames of different colours. The use of chlorine and its compounds in disinfectants and bleaches has saved many lives, and you will learn something of the chemistry involved in how these products work.

You will also learn about titrations, especially how to do them accurately, and carry out calculations from results.

MATHS SKILLS FOR THIS TOPIC

- Carry out calculations using numbers in ordinary form
- Use appropriate units in calculations including measurement of ionic radius and ionisation energy
- Use ratios to construct and balance equations
- Understand the connection between frequency and wavelength, and their use in representing the colour of light

What prior knowledge do I need?

Topic 1

- Calculations involving moles and concentrations

Topic 1B

- Balancing equations for charge using electrons

Topic 2B

- The structure of the Periodic Table

Topic 7

- Intermolecular forces

What will I study in this topic?

- Oxidation and reduction in terms of electron loss and gain
- Using oxidation numbers
- Reaction trends in Groups 1, 2 and 7
- Tests for anions and cations
- Practical techniques and calculations in volumetric analysis
- Accuracy, precision, errors and uncertainties

What will I study later?

Topic 16 (Book 2: IAL)

- Redox titrations
- Standard electrode (redox) potentials and their uses to calculate the electromotive force of a cell and the feasibility of a reaction

Topic 17 (Book 2: IAL)

- Redox reactions of certain transition metals

8A 1 ELECTRON LOSS AND GAIN

SPECIFICATION REFERENCE 8.5 8.6 8.7

LEARNING OBJECTIVES

- Understand oxidation and reduction in terms of electron transfer and changes in oxidation number.
- Understand the application of these ideas to reactions of s-block and p-block elements.
- Know that oxidising agents gain electrons and that reducing agents lose electrons.
- Understand that a disproportionation reaction involves an element in a single species being simultaneously oxidised and reduced.

BACKGROUND TO OXIDATION AND REDUCTION

You originally knew 'oxidation' as the addition of oxygen and 'reduction' as the removal of oxygen.

The reaction between iron(III) oxide and carbon monoxide that takes place in the blast furnace provides an example of each (**fig A**).

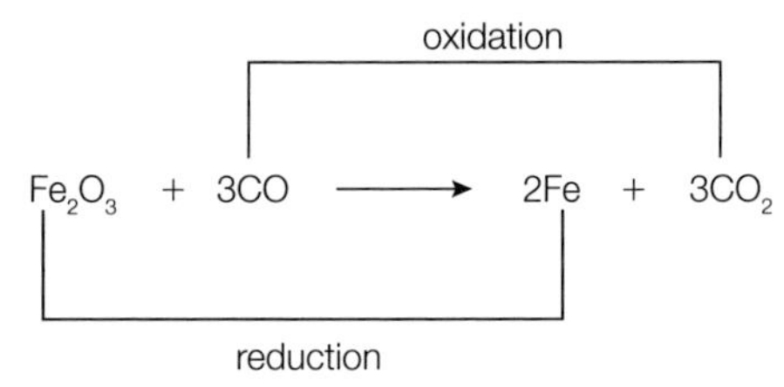

▲ **fig A** Example of oxidation and reduction during the reaction between iron(III) oxide and carbon monoxide.

We then expanded these definitions so that oxidation is described as the removal of hydrogen and reduction is described as the addition of hydrogen. The reaction between chlorine and hydrogen sulfide provides an example (**fig B**).

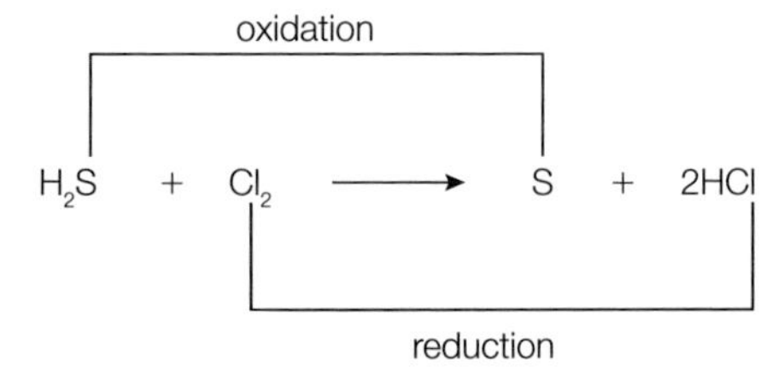

▲ **fig B** Example of oxidation and reduction during the reaction between chlorine and hydrogen sulfide.

A reaction that involves both reduction and oxidation is called a **redox reaction**.

OXIDATION AND REDUCTION IN TERMS OF ELECTRON LOSS AND GAIN

ELECTRON TRANSFER IN REDOX REACTIONS

The approach preferred by chemists today to describe oxidation and reduction involves the transfer of electrons.

When magnesium burns in oxygen it forms magnesium oxide:

$$2Mg + O_2 \rightarrow 2MgO$$

The magnesium has been oxidised because it has gained oxygen. The oxygen must have been reduced, but by what definition?

The reaction results in the formation of Mg^{2+} and O^{2-} ions. Each of the two magnesium atoms has lost two electrons:

$$Mg \rightarrow Mg^{2+} + 2e^-$$

The oxygen molecule has gained four electrons to become oxide ions:

$$O_2 + 4e^- \rightarrow 2O^{2-}$$

The magnesium has been *oxidised* because it has lost electrons.

Oxygen has been *reduced* because it has gained electrons.

DEFINITIONS OF OXIDATION AND REDUCTION

We now have new definitions of oxidation and reduction:

Oxidation is the *loss* of electrons.

Reduction is the *gain* of electrons.

These new definitions are easily remembered using the mnemonic **OIL RIG**.

Oxidation

Is

Loss (of electrons)

Reduction

Is

Gain (of electrons)

We can now describe the reduction of iron(III) oxide in the blast furnace as the gain of three electrons by the iron(III) ion in the oxide:

$$Fe^{3+} + 3e^- \rightarrow Fe$$

The carbon monoxide must have been oxidised (it gains oxygen), but it is not easy to see how it has lost electrons because carbon monoxide and carbon dioxide are both covalent compounds.

OXIDISING AND REDUCING AGENTS

In the reaction between magnesium and oxygen to form magnesium oxide, the oxygen has oxidised the magnesium.

Therefore, oxygen is an **oxidising agent**.

The magnesium has reduced the oxygen and is therefore a **reducing agent**.

EXAM HINT

An oxidising agent is reduced.

DISPROPORTIONATION

Consider the following reaction, which occurs if copper(I) oxide is added to dilute sulfuric acid:

$$Cu_2O(s) + H_2SO_4 \rightarrow CuSO_4(aq) + Cu(s) + H_2O(l)$$

One of the Cu^+ ions in copper(I) oxide has lost an electron to become Cu^{2+}. At the same time, the other Cu^+ ion has gained an electron to become Cu. Both oxidation and reduction have occurred but the *same element*, copper, is involved in both changes.

This is an example of a **disproportionation reaction**.

CHECKPOINT

1. In the reaction:

$Zn(s) + CuSO_4(aq) \rightarrow ZnSO_4(aq) + Cu(s)$

(a) Which species has been oxidised and which species has been reduced?

(b) Write ionic half-equations to represent both the oxidation and the reduction reactions.

2. In each of the following reactions identify whether the underlined species has been oxidised, reduced or neither.

(a) $\underline{Al}(s) + 1\frac{1}{2}Cl_2(g) \rightarrow AlCl_3(s)$

(b) $4\underline{Na}(s) + TiO_2(s) \rightarrow 2Na_2O(s) + Ti(s)$

(c) $\underline{Ag}^+(aq) + Cl^-(aq) \rightarrow AgCl(s)$

(d) $Cl_2(aq) + 2\underline{Br}^-(aq) \rightarrow 2Cl^-(aq) + Br_2(aq)$

(e) $\underline{Cu}O(s) + 2H^+(aq) \rightarrow Cu^{2+}(aq) + H_2O(l)$

(f) $\underline{MnO_2}(s) + 4HCl(aq) \rightarrow MnCl_2(aq) + Cl_2(aq) + 2H_2O(l)$

3. In each of the following reactions, identify the species that has been oxidised. In each case, justify your answer.

(a) $Fe(s) + H_2SO_4(aq) \rightarrow FeSO_4(aq) + H_2(g)$

(b) $Na(s) + \frac{1}{2}H_2(g) \rightarrow NaH(s)$

(c) $CuO(s) + Cu(s) \rightarrow Cu_2O(s)$

(d) $2Fe(OH)_2(s) + \frac{1}{2}O_2(aq) + H_2O(l) \rightarrow 2Fe(OH)_3(s)$

(e) $2V^{3+}(aq) + Zn(s) \rightarrow 2V^{2+}(aq) + Zn^{2+}(aq)$

SUBJECT VOCABULARY

redox reaction a reaction that involves both reduction and oxidation
oxidation when the oxidation number of an element increases; the loss of electrons
reduction when the oxidation number of an element decreases; the gain of electrons
oxidising agent a species (atom, molecule or ion) that oxidises another species by removing one or more electrons; when an oxidising agent reacts it gains electrons and is, therefore, reduced
reducing agent a species that reduces another species by adding one or more electrons; when a reducing agent reacts it loses electrons and is, therefore, oxidised
disproportionation reaction a reaction involving the simultaneous oxidation and reduction of an element in a single species

8A 2 ASSIGNING OXIDATION NUMBERS

SPECIFICATION REFERENCE 8.1 8.2 8.9

LEARNING OBJECTIVES

- Know what is meant by the term oxidation number and understand the rules for assigning oxidation numbers.
- Be able to calculate the oxidation number of elements in compounds and ions, including in peroxides and metal hydrides.
- Understand that metals, in general, form positive ions by loss of electrons with an increase in oxidation number whereas non-metals, in general, form negative ions by gain of electrons with a decrease in oxidation number.

OXIDATION NUMBER

So far we have restricted our discussion of oxidation and reduction to atoms and ions. For redox reactions involving these species it is easy to see which species are losing and which are gaining electrons. However, many compounds are covalent and for them a simple treatment involving ions is not appropriate.

To get around this difficulty, the concept of **oxidation number** has been developed.

For example, in MgO the oxidation number of magnesium is +2, since the charge on the magnesium ion is 2^+. Similarly, the oxidation number of oxygen is −2.

In SO_2, the oxidation number of the sulfur is +4 because if the compound were fully ionic the sulfur ion would have a charge of 4^+. The oxidation number of oxygen is once again −2.

RULES FOR DETERMINING THE OXIDATION NUMBER

Here are some rules to help you calculate the oxidation number.

- The oxidation number of an uncombined element is zero.
- The sum of the oxidation numbers of all the elements in a neutral compound is zero.
- The sum of the oxidation numbers of all the elements in an ion is equal to the charge on the ion.
- The more electronegative element in a substance is given a negative oxidation number.
- The oxidation number of fluorine is always −1.
- The oxidation number of hydrogen is +1, except when combined with a less electronegative element. Then it becomes −1.
- The oxidation number of oxygen is −2, except in peroxides where it is −1 and when combined with fluorine when it is +2.

The best way to get used to these rules is to put them into practice, which is what we shall now do.

WORKED EXAMPLE 1

Deduce the oxidation number of chlorine in:

(a) NaCl

(b) NaClO

(c) $NaClO_3$.

Answers

(a) The oxidation number of Na is +1. The two oxidation numbers must add up to zero, so the oxidation number of Cl must be −1.

(b) The oxidation number of Na is +1 and the oxidation number of O is −2. The oxidation numbers must add up to zero, so the oxidation number of Cl must be +1.

(c) The oxidation number of Na is +1 and the oxidation number of O is −2. The oxidation numbers must add up to zero, so the oxidation number of Cl must be +5.

$$+1 + (3 \times -2) + x = 0$$

$$x = 0 - 1 - (-6) = +5$$

WORKED EXAMPLE 2

Deduce the oxidation number of nitrogen in:

(a) NH_3

(b) NO_2^-

(c) NO_3^-.

Answers

(a) N is more electronegative than H and must therefore have a negative oxidation number. The oxidation number of H is therefore +1. The oxidation numbers must add up to zero, so the oxidation number of N is −3.

$x + (3 \times 1) = 0$

$x = -3$

(b) O is more electronegative than N, so the oxidation number of O is −2. The oxidation numbers must add up to −1 (the charge on the ion), so the oxidation number of N is +3.

$x + (2 \times -2) = -1$

$x = +3$

(c) O is more electronegative than N, so the oxidation number of O is −2. The oxidation numbers must add up to −1 (the charge on the ion), so the oxidation number of N is +5.

$x + (3 \times -2) = -1$

$x = +5$

CHECKPOINT

1. Calculate the oxidation number of the underlined element in each of the following species.

$\underline{S}O_2$	$\underline{S}O_3$	$H_2\underline{S}$	$\underline{S}O_3^{2-}$	$\underline{S}O_4^{2-}$	$\underline{S}_2O_3^{2-}$
$\underline{Cr}O_4^{2-}$	$\underline{Cr}_2O_7^{2-}$	$\underline{Mn}O_4^{2-}$	$\underline{Mn}O_4^-$	$\underline{V}O^{2+}$	$\underline{V}O_2^+$
$\underline{Cl}O^-$	$\underline{Cl}O_2^-$	$\underline{Cl}O_3^-$	$\underline{Cl}O_4^-$	$[\underline{Cu}Cl_4]^{2-}$	$[\underline{V}(H_2O)_6]^{2+}$
$H_2\underline{O}_2$	$\underline{O}F_2$	$Na\underline{H}$	$Ba\underline{O}_2$		

SUBJECT VOCABULARY

oxidation number the charge that an ion has, or the charge that it would have if the species were fully ionic

8A 3 RECOGNISING REACTIONS USING OXIDATION NUMBERS

SPECIFICATION REFERENCE 8.5 8.8

LEARNING OBJECTIVES

- Understand oxidation and reduction in terms of electron transfer and changes in oxidation number, and the application of these ideas to reactions of s-block and p-block elements.
- Know that the oxidation number is a useful concept in terms of the classification of reactions as redox and as disproportionation.

USING OXIDATION NUMBERS TO CLASSIFY REACTIONS

Consider the following ionic half-equations:

$$Zn(s) \rightarrow Zn^{2+}(aq) + 2e^-$$
$$Fe^{2+}(aq) \rightarrow Fe^{3+}(aq) + e^-$$
$$2I^-(aq) \rightarrow I_2(aq) + 2e^-$$

In all three cases, electrons have been lost so the reactions are oxidations. But also notice that the oxidation number of the element has increased in each case.

- Zn has increased from 0 to +2.
- Fe has increased from +2 to +3.
- I has increased from −1 to 0.

This leads to another definition of **oxidation** and **reduction**.

Consider the following ionic half-equations:

$Cl_2(aq) + 2e^- \rightarrow 2Cl^-(aq)$	Cl changes from 0 to −1
$MnO_4^-(aq) + 8H^+(aq) + 5e^- \rightarrow Mn^{2+}(aq) + 4H_2O(l)$	Mn changes from +7 to +2
$2H_2O(l) + 2e^- \rightarrow 2OH^-(aq) + H_2(g)$	H changes from +1 to 0

EXAM HINT

Notice that the higher oxidation numbers for metal ions in solution are always in the form of oxyions, e.g., in CrO_4^{2-} Cr has an oxidation number of +6.

In each case, the oxidation number of one of the elements involved has decreased, so reduction has taken place.

This concept can now be applied to full equations.

WORKED EXAMPLE

Use oxidation numbers to show whether the following reactions are examples of redox reactions.

(a) $H_2S(g) + Cl_2(g) \rightarrow 2HCl(g) + S(s)$

Oxidation numbers: +1 −2 0 +1 −1 0

Answer

The oxidation number of S has increased (−2 to 0), so it has been oxidised.

The oxidation number of Cl has decreased (0 to −1), so it has been reduced.

Since both oxidation and reduction have taken place, the reaction is classified as a redox reaction.

(b) $NaOH(aq) + HCl(aq) \rightarrow NaCl(aq) + H_2O(l)$

Oxidation numbers: +1 −2 +1 +1 −1 +1 −1 +1 −2

Answer

There is no change in oxidation number for any of the elements involved in the reaction, so this is not an example of a redox reaction.

(c) $2NaOH(aq) + Cl_2(aq) \rightarrow NaCl(aq) + NaClO(aq) + H_2O(l)$

Oxidation numbers: +1 −2 +1 0 +1 −1 +1 +1 −2 +1 −2

Answer

The oxidation number of Cl has both increased (0 in Cl_2 to +1 in NaClO) and decreased (0 in Cl_2 to −1 in NaCl), so this is an example of a redox reaction.

It is also an example of a **disproportionation reaction**.

REACTIONS OF S-BLOCK ELEMENTS

We can make some predictions about the reactions of s-block elements in terms of redox.

Reactive metals in Groups 1 and 2 lose electrons when they react. For example, the Group 1 element sodium loses one electron, which can be represented in this ionic half-equation:

$$Na \rightarrow Na^+ + e^-$$

At the same time, you can also recognise that this involves an increase in oxidation number from 0 to +1.

Magnesium, a Group 2 metal, reacts in a similar way:

$$Mg \rightarrow Mg^{2+} + 2e^-$$

This involves an increase in oxidation number from 0 to +2.

You will learn more about these reactions in **Topic 8B**.

REACTIONS OF P-BLOCK ELEMENTS

We can also make some predictions about the reactions of p-block elements in terms of redox. Unfortunately, it is not as straightforward as with s-block elements. This is because the p-block contains elements with different characteristics. For example, the Group 3 element aluminium is a metal with a reactivity nearly as high as magnesium, so it reacts in a similar way:

$$Al \rightarrow Al^{3+} + 3e^-$$

This involves an increase in oxidation number from 0 to +3.

Now consider the Group 7 element fluorine, which is the most reactive halogen. The half-equation for its reaction is:

$$F_2 + 2e^- \rightarrow 2F^-$$

This involves a decrease in oxidation number from 0 to −1.

Nitrogen is a p-block element in Group 5, with an electronegativity higher than aluminium, but lower than fluorine. What predictions can we make about the change in oxidation number in its reactions? This depends on which element it reacts with.

For example, nitrogen reacts with the Group 1 metal sodium as follows:

$$6Na + N_2 \rightarrow 2Na_3N$$

In this redox reaction, the oxidation number of sodium increases from 0 to +1, and that of nitrogen decreases from 0 to −3.

Now consider nitrogen's reaction with fluorine:

$$N_2 + 3F_2 \rightarrow 2NF_3$$

As always, the oxidation number of fluorine decreases from 0 to −1, but this time the oxidation number of nitrogen increases from 0 to +3. You will learn more about these reactions in **Topic 8C**.

CHECKPOINT

1. Copy the following table. Use oxidation numbers to complete it. The first example has been done for you.

EQUATION	REDOX REACTION (✓ OR ✗)	DISPROPORTIONATION (✓ OR ✗)	ELEMENT OXIDISED	ELEMENT REDUCED
$Mg + 2HCl \rightarrow MgCl_2 + H_2$ 0 +1 −1 +2 −1 0	✔	✘	Mg	H
$CaO + H_2O \rightarrow Ca(OH)_2$				
$2H_2O_2 \rightarrow 2H_2O + O_2$				
$KOH + HNO_3 \rightarrow KNO_3 + H_2O$				
$Cl_2 + H_2O \rightarrow HCl + HClO$				

SUBJECT VOCABULARY

oxidation when the oxidation number of an element *increases*; the loss of electrons

reduction when the oxidation number of an element *decreases*; the gain of electrons

disproportionation reaction a reaction involving the simultaneous oxidation and reduction of an element in a single species

8A 4 OXIDATION NUMBERS AND NOMENCLATURE

SPECIFICATION REFERENCE 8.3 8.4

LEARNING OBJECTIVES

- Be able to indicate the oxidation number of an element in a compound or ion, using a Roman numeral.
- Be able to write formulae given oxidation numbers.

SYSTEMATIC NAMES

When an element can have more than one oxidation state, the names of its compounds and its ions often include the oxidation number of the element, written as a Roman numeral in brackets. This name is often referred to as the 'systematic name'.

Table A shows some examples.

FORMULA OF COMPOUND OR ION	RELEVANT OXIDATION NUMBER	SYSTEMATIC NAME OF COMPOUND OR ION
$FeCl_2$	Fe +2	iron(II) chloride
$FeCl_3$	Fe +3	iron(III) chloride
$KMnO_4$	Mn +7	potassium manganate(VII)
K_2MnO_4	Mn +6	potassium manganate(VI)
CrO_4^{2-}	Cr +6	chromate(VI) ion
$Cr_2O_7^{2-}$	Cr +6	chromate(VI) ion

table A Examples of how to indicate the oxidation number of an element in a compound or ion using the systematic name.

WHEN TO USE SYSTEMATIC NAMES

We often use systematic names in chemistry so that we can be specific about the compounds and ions we are referring to. However, in the wider world the numbers are often left out.

For example:

- the systematic name for Na_2SO_4 is sodium sulfate(VI), but it is often just called sodium sulfate
- Na_2SO_3 should be labelled sodium sulfate(IV), but the name sodium sulfite is still commonly used
- SO_2 and SO_3 are more commonly referred to as sulfur dioxide and sulfur trioxide, rather than sulfur(IV) oxide and sulfur(VI) oxide, respectively.

You should use systematic names as often as possible, particularly during your studies.

WRITING FORMULAE WHEN YOU HAVE THE OXIDATION NUMBER

The other skill you need to develop is to work backwards from the oxidation number to deduce the formula of the compound or ion concerned.

WORKED EXAMPLE 1

Deduce the formula for iron(II) sulfate.

Answer

The formula of the iron(II) ion is Fe^{2+}.

The formula of the sulfate ion is SO_4^{2-}.

So the two ions are present in a 1 : 1 ratio to produce a neutral compound, giving the formula $FeSO_4$.

WORKED EXAMPLE 2

Deduce the formula for iron(III) sulfate.

Answer

The formula of the iron(III) ion is Fe^{3+}.

The formula of the sulfate ion is SO_4^{2-}.

So the two ions are present in a 2 : 3 ratio to give a neutral compound, giving the formula $Fe_2(SO_4)_3$.

CHECKPOINT

1. Give the systematic name for each of the following compounds:

(a) PCl_3

(b) PCl_5

(c) V_2O_5

(d) NaClO

(e) $NaClO_3$

2. Give the systematic name for each of the following ions:

(a) NO_3^-

(b) NO_2^-

(c) ClO_4^-

(d) VO_2^+

(e) VO_2^+

3. Deduce the formula for each of the following compounds:

(a) copper(I) oxide

(b) copper(II) oxide

(c) chromium(III) sulfate(VI)

(d) lead(IV) iodide

(e) cobalt(III) nitrate(V)

4. Why is it not necessary to refer to sodium chloride as sodium(I) chloride, or magnesium oxide as magnesium(II) oxide?

8A 5 CONSTRUCTING FULL IONIC EQUATIONS

SPECIFICATION REFERENCE 8.10

LEARNING OBJECTIVES

- Be able to write ionic half-equations and use them to construct full ionic equations.

BALANCING BY COUNTING ELECTRONS

STRAIGHTFORWARD EXAMPLES

When solid zinc is added to an aqueous solution of copper(II) sulfate, the following two changes take place:

$$Zn(s) \rightarrow Zn^{2+}(aq) + 2e^- \text{ and } Cu^{2+}(aq) + 2e^- \rightarrow Cu(s)$$

Both ionic half-equations involve two electrons, so to construct the full ionic equation for this reaction you simply add together the two half-equations so that the electrons cancel out:

$$Zn(s) \rightarrow Zn^{2+}(aq) + 2e^-$$

$$Cu^{2+}(aq) + 2e^- \rightarrow Cu(s)$$

$$Zn(s) + Cu^{2+}(aq) \rightarrow Zn^{2+}(aq) + Cu(s)$$

Now let's try an example where the electrons are not the same in the two ionic half-equations.

When chlorine gas is bubbled into an aqueous solution of iron(II) chloride, the iron(II) ions are oxidised to iron(III) ions and the chlorine molecules are reduced to chloride ions.

The two ionic half-equations are:

$$Fe^{2+}(aq) \rightarrow Fe^{3+}(aq) + e^- \text{ and } Cl_2(g) + 2e^- \rightarrow 2Cl^-(aq)$$

This time, one of the half-equations contains one electron, while the other contains two electrons. Before these can be added together to produce a full equation, the equation containing Fe^{2+} must be multiplied by 2.

$$2Fe^{2+}(aq) \rightarrow 2Fe^{3+}(aq) + 2e^-$$

$$Cl_2(g) + 2e^- \rightarrow 2Cl^-(aq)$$

$$2Fe^{2+}(aq) + Cl_2(g) \rightarrow 2Fe^{3+}(aq) + 2Cl^-(aq)$$

MORE COMPLICATED EXAMPLES

Here is something a little more difficult.

When an acidified aqueous solution of potassium manganate(VII) is added to an aqueous solution of iron(II) sulfate, the following two changes occur:

$$Fe^{2+}(aq) \rightarrow Fe^{3+}(aq) + e^- \text{ and}$$
$$MnO_4^-(aq) + 8H^+(aq) + 5e^- \rightarrow Mn^{2+}(aq) + 4H_2O(l)$$

In order to balance the electrons, the first half-equation must be multiplied by 5:

$$5Fe^{2+}(aq) \rightarrow 5Fe^{3+}(aq) + 5e^-$$

$$MnO_4^-(aq) + 8H^+(aq) + 5e^- \rightarrow Mn^{2+}(aq) + 4H_2O(l)$$

$$5Fe^{2+}(aq) + MnO_4^-(aq) + 8H^+(aq) \rightarrow 5Fe^{3+}(aq) + Mn^{2+}(aq) + 4H_2O(l)$$

EXAM HINT

Acidified potassium manganate(VII) is very commonly used as an oxidising agent in analytical chemistry. It will be useful to learn this half equation by heart.

Here's another challenging example.

The reaction between aqueous acidified potassium manganate(VII) and hydrogen peroxide involves two changes represented by the following ionic half-equations:

$$MnO_4^-(aq) + 8H^+(aq) + 5e^- \rightarrow Mn^{2+}(aq) + 4H_2O(l) \text{ and}$$
$$H_2O_2(aq) \rightarrow 2H^+(aq) + O_2(g) + 2e^-$$

The lowest common multiple of 2 and 5 is 10. This means that the first half-equation should be multiplied by 2 and the second by 5 before they are added together.

$$2MnO_4^-(aq) + 16H^+(aq) + 10e^- \rightarrow 2Mn^{2+}(aq) + 8H_2O(l)$$

$$5H_2O_2(aq) \rightarrow 10H^+(aq) + 5O_2(g) + 10e^-$$

$$2MnO_4^-(aq) + 16H^+(aq) + 5H_2O_2(aq) \rightarrow 2Mn^{2+}(aq) + 8H_2O(l) + 10H^+(aq) + 5O_2(g)$$

The electrons have now been cancelled out, but we are left with an equation that has H^+ ions on both sides of the equation in unequal numbers. We now have to cancel these out so that they are present on only one side. To do this, you subtract the $10H^+$ on the right-hand side from the $16H^+$ on the left-hand side to give the final equation:

$$2MnO_4^-(aq) + 6H^+(aq) + 5H_2O_2(aq) \rightarrow 2Mn^{2+}(aq) + 8H_2O(l) + 5O_2(g)$$

EXAM HINT

This final equation shows that this reaction takes place in acid conditions. Many redox reactions are pH sensitive.

BALANCING USING OXIDATION NUMBERS

WORKED EXAMPLE 1

Use oxidation numbers to balance the following equation:

$....SO_2(g) + ...H_2O(l) + ...Ag^+(aq) \rightarrow .... SO_4^{2-}(aq) + ...H^+(aq) + ...Ag(s)$

Answer

Identify the elements whose oxidation numbers have changed.

In this case:

- S changes from +4 to +6; this is a '2 electron' change
- Ag changes from +1 to 0; this is a '1 electron' change.

So the ratio of SO_2 to Ag^+ is 1 : 2. This gives:

$$SO_2(g) + ...H_2O(l) + 2Ag^+(aq) \rightarrow SO_4^{2-}(aq) + ...H^+(aq) + 2Ag(s)$$

We now need to balance the H and O atoms.

This gives:

$$SO_2(g) + 2H_2O(l) + 2Ag^+(aq) \rightarrow SO_4^{2-}(aq) + 4H^+(aq) + 2Ag(s)$$

Lastly, check the equation for balanced charges.

- The total charge on the left-hand side is 2+.
- The total charge on the right-hand side is also 2+ (−2 + +4).

The equation is now balanced.

WORKED EXAMPLE 2

Use oxidation numbers to balance the following equation:

$$....Fe^{2+} + ...ClO_3^- + ...H^+ \rightarrow .... Fe^{3+} + ...Cl^- + ...H_2O$$

Answer

Identify the elements whose oxidation numbers have changed.

In this case:

Fe changes from +2 to +3; this is a '1 electron' change

Cl changes from +5 to −1; this is a '6 electron' change.

So the ratio of Fe to ClO_3^- is 6 : 1. This gives:

$$6Fe^{2+} + ClO_3^- + ...H^+ \rightarrow 6Fe^{3+} + Cl^- + ...H_2O$$

Once again, balance the H and O atoms.

This gives:

$$6Fe^{2+} + ClO_3^- + 6H^+ \rightarrow 6Fe^{3+} + Cl^- + 3H_2O$$

Check the charges.

- The total charge on the left-hand side is +17((6 × +2) + −1 + (6 × +1)).
- The total charge on the right-hand side is +17((6 × +3) + −1).

The equation is now balanced.

CHECKPOINT

1. Use each pair of ionic half-equations to construct a full ionic equation. Include state symbols.

(a) $Zn(s) \rightarrow Zn^{2+}(aq) + 2e^-$ and $Fe^{3+}(aq) + e^- \rightarrow Fe^{2+}(aq)$

(b) $\frac{1}{2}I_2(aq) + e^- \rightarrow I^-(aq)$ and $2S_2O_3^{2-}(aq) \rightarrow S_4O_6^{2-}(aq) + 2e^-$

(c) $MnO_4^-(aq) + 8H^+(aq) + 5e^- \rightarrow Mn^{2+}(aq) + 4H_2O(l)$ and $Ce^{3+}(aq) \rightarrow Ce^{4+}(aq) + e^-$

(d) $Cr_2O_7^{2-}(aq) + 14H^+(l) + 6e^- \rightarrow 2Cr^{3+}(aq) + 7H_2O(l)$ and $Fe^{2+}(aq) \rightarrow Fe^{3+}(aq) + e^-$

(e) $FeO_4^{2-}(aq) + 8H^+(aq) + 3e^- \rightarrow Fe^{3+}(aq) + 4H_2O(l)$ and $C_2O_4^{2-}(aq) \rightarrow 2CO_2(g) + 2e^-$

2. Use oxidation numbers to balance each of the following equations.

(a) $....Cu(s) + ...H^+(aq) + ...NO_3^-(aq) \rightarrow ....Cu^{2+}(aq) + ...H_2O(l) + ...NO(g)$

(b) $....Cu(s) + ...H^+(aq) + ...NO_3^-(aq) \rightarrow ....Cu^{2+}(aq) + ...H_2O(l) + ...NO_2(g)$

(c) $....Cl_2(g) + ...OH^-(aq) \rightarrow ....Cl^-(aq) + ...ClO_3^-(g) + ...H_2O(l)$

8B 1 TRENDS IN GROUPS 1 AND 2

SPECIFICATION REFERENCE 8.11 8.12

LEARNING OBJECTIVES

- Understand the reasons for the trend in ionisation energy down Groups 1 and 2.
- Understand the reasons for the trend in reactivity of the elements down Group 1 (Li to K) and Group 2 (Mg to Ba).

INTRODUCTION TO THE GROUP 1 AND 2 ELEMENTS

There are six elements in each of Groups 1 and 2. You are not likely to see samples of francium or radium, as all of their isotopes are radioactive. The other five elements in these groups look almost the same in appearance. You can describe all of them, when pure, as bright silvery solids. However, we have to keep them away from air to look like that. When exposed to air, they combine with oxygen to form oxides as surface layers, which makes them appear dull. **Fig A** shows the first five elements of Group 2 together.

▲ **fig A** From left to right: beryllium, magnesium, calcium, strontium and barium.

In this topic, we look at some trends in the properties of Group 1 and 2 elements.

TREND IN IONISATION ENERGY

We looked at ionisation energy in **Topic 2**. You may remember that it is a fundamental property that affects physical and chemical properties.

FIRST IONISATION ENERGIES IN GROUP 1

You should remember the definition of **first ionisation energy**. It is the energy required to remove an electron from each atom in one mole of atoms in the gaseous state. A general equation for this process, using M to represent an atom of any Group 1 element, is:

$$M(g) \rightarrow M^{+}(g) + e^{-}$$

Table A shows the metallic radius and the values of the first ionisation energies for the Group 1 elements.

ELEMENT	METALLIC RADIUS / nm	FIRST IONISATION ENERGY / kJ mol^{-1}
Lithium	0.152	519
Sodium	0.186	494
Potassium	0.231	418
Rubidium	0.244	402
Caesium	0.262	376

table A The metallic radii and the value of the first ionisation energies for the Group 1 elements.

The energy needed for ionisation is used to overcome the electrostatic attraction between the electron being removed and the protons in the nucleus.

FIRST AND SECOND IONISATION ENERGIES IN GROUP 2

Now we will consider the loss of two electrons from each atom, so we need to consider the **second ionisation energy.** This is the energy required to remove an electron from each singly charged ion in one mole of positive ions in the gaseous state. A general equation for this process is:

$$M^{+}(g) \rightarrow M^{2+}(g) + e^{-}$$

Table B shows the metallic radius and the values of the first and second ionisation energies for the Group 2 elements.

ELEMENT	METALLIC RADIUS / nm	IONISATION ENERGY / kJ mol^{-1}		
		FIRST	SECOND	1st + 2nd
Beryllium	0.112	900	1757	2657
Magnesium	0.160	738	1451	2189
Calcium	0.197	590	1145	1735
Strontium	0.215	550	1064	1614
Barium	0.224	503	965	1468

table B The metallic radii and the values of the first and second ionisation energies for the Group 2 elements.

The energy needed for ionisation is used to overcome the electrostatic attraction between the electron being removed and the protons in the nucleus.

FACTORS TO CONSIDER FOR GROUPS 1 AND 2

The factors to consider when explaining trends in ionisation energy are:

- the nuclear charge (or the number of protons in the nucleus)
- the orbital in which the electron exists
- the shielding effect (sometimes called the 'screening effect'). This is the repulsion between filled inner shells and the electron being removed.

In **Topic 2**, we considered the subshell (or sublevel) from which the electron is being removed. For the Group 1 and 2 elements this is not necessary because in their reactions the electrons are always removed from an s subshell.

You should be able to understand why the trend is a decrease down both Group 1 and Group 2.

- As the nuclear charge increases, so the force of attraction for the electron being removed also increases. This means an increase in ionisation energy down the group.

- As each quantum shell is added, energy of the outermost electrons increases. This means a decrease in ionisation energy down the group.
- As the number of filled inner shells increases, their force of repulsion on the electron being removed increases. This means a decrease in ionisation energy down the group.

You can see that the first factor causes an increase, but the second and third factors cause a decrease. The combined effect of the second and third factors outweighs the effect of the first factor. This means that there is a decrease down the group.

TREND IN REACTIVITY

We will look at the reactions of the elements of Groups 1 and 2 in **Topics 8B.2 to 8B.6**. In all these reactions, the element changes into either an M^+ ion or an M^{2+} ion, and there is a general increase in reactivity down the group. This can be explained by the decrease in energy needed to remove the electron from each atom of the element.

LEARNING TIP

When writing an equation for an ionisation, you should always include the state symbol (g) after each atom and ion.

CHECKPOINT

1. Write equations to represent the first ionisation of beryllium and the second ionisation of barium.

2. Explain fully why beryllium is less reactive than barium.

SUBJECT VOCABULARY

first ionisation energy the energy required to remove an electron from each atom in one mole of atoms in the gaseous state

second ionisation energy the energy required to remove an electron from each singly charged positive ion in one mole of positive ions in the gaseous state

8B 2 REACTIONS OF GROUP 1 ELEMENTS

SPECIFICATION REFERENCE
8.12 PART 8.13 PART

LEARNING OBJECTIVES

- Know the reactions of the elements of Group 1 (Li to K) with oxygen, chlorine and water.

REACTIONS WITH OXYGEN

You know that the Group 1 metals are only shiny when kept out of air. When they are exposed to air, they tarnish. This means that they form a dull, dark layer on the surface as they are oxidised by oxygen in the air.

Group 1 metals can burn when heated in air or oxygen, but you would not normally do this in the laboratory because the reactions are extremely vigorous.

Many people are now aware of one problem to do with batteries containing lithium. Such batteries are common in mobile phones and laptops, but there have been a small number of cases where the battery has caught fire. **Fig A** shows a mobile phone on fire, caused by the lithium in the battery reacting rapidly with the oxygen in the air.

▲ **fig A** The lithium inside the battery of this mobile phone caught fire.

The general equation for all the reactions of Group 1 metals with the oxygen in the air is:

$$4M(s) + O_2(g) \rightarrow 2M_2O(s)$$

The products are oxides containing M^+ and O^{2-} ions.

Lithium is the least reactive Group 1 metal, and the reactivity of the metals increases down Group 1.

REACTIONS WITH CHLORINE

The Group 1 elements combine with chlorine when heated in chlorine gas. Just like the reactions with oxygen, the reactions with chlorine become more vigorous down the group, although this trend is harder to see than with the oxygen reactions.

The general equation for all of these reactions is:

$$2M(s) + Cl_2(g) \rightarrow 2MCl(s)$$

The products are chlorides containing M^+ and Cl^- ions.

REACTIONS WITH WATER

The reaction between lithium and water is vigorous. When a small piece of lithium is added to a beaker or trough of water, it reacts immediately and floats on the surface of the water. The fizzing is due to the formation of hydrogen gas and after a short while the lithium can no longer be seen. This is because it has reacted with the water to form a colourless solution of lithium hydroxide.

All the other Group 1 metals react in the same way, although the reactions become more vigorous down the group. When sodium is added to water, there is sometimes a flame, caused by the heat of the reaction igniting the hydrogen (**fig B**). When potassium is added to water, the hydrogen nearly always catches fire.

The general equation for all these reactions is:

$$2M(s) + 2H_2O(l) \rightarrow 2MOH(aq) + H_2(g)$$

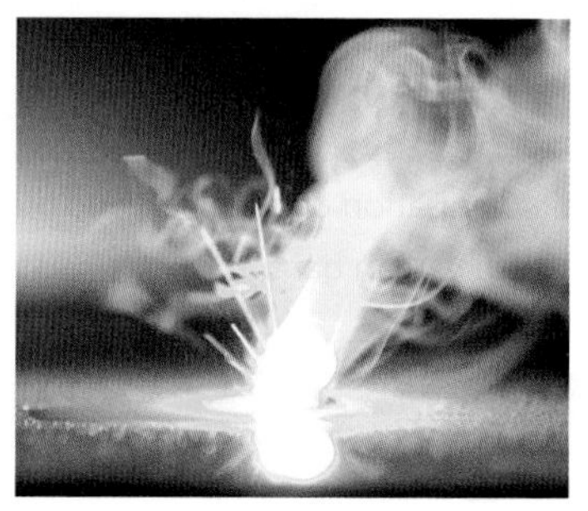

▲ **fig B** The heat energy from the reaction between these pieces of sodium and water has ignited the hydrogen formed.

The products are hydrogen gas and hydroxides containing M^+ and OH^- ions.

The equation for the reaction with lithium is:

$$2Li(s) + 2H_2O(l) \rightarrow 2LiOH(aq) + H_2(g)$$

EXAM HINT

Make sure that you remember to use the correct symbols for reactants and products if asked to do so in an exam question.

REACTIONS OF OTHER GROUP 1 METALS

You are not required to know any of the reactions of rubidium or caesium. However, because you know that the trend is increasing reactivity down the group, you can predict that both rubidium and caesium are more reactive than potassium.

LEARNING TIP

Practise writing equations for reactions of Group 1 elements with oxygen, chlorine and water.

CHECKPOINT

1. Write an equation for each of these reactions:
   (a) lithium with oxygen
   (b) sodium with chlorine
   (c) potassium with water.
2. Soon after sodium is added to water, a colourless solution is formed. Explain why it is wrong to write that sodium is soluble in water.

8B 3 REACTIONS OF GROUP 2 ELEMENTS

SPECIFICATION REFERENCE
8.12 PART 8.13 PART

LEARNING OBJECTIVES

- Know the reactions of the elements of Group 2 (Mg to Ba) with oxygen, chlorine and water.

REACTIONS WITH OXYGEN

You should be familiar with the reaction that occurs when magnesium burns in air – there is a very bright flame and the formation of a white solid.

Similar observations can be made when other Group 2 elements burn in air, but they are qualitatively different, that is, you should be able to recognise that when calcium, strontium and barium are heated in air, the reactions are more vigorous. However, this may be hard to see if the metals are not fresh samples. If the burning metal is placed in a gas jar of oxygen, then the same reaction occurs, although more vigorously.

For all these elements, the element needs to be heated for the reaction to start. However, even without heating, there is a slow reaction between the element and oxygen when the element is exposed to air. This forms a surface coating of oxide which helps to prevent the element from further reaction.

Barium is the most reactive. It is often stored under oil to keep it from reacting with oxygen and water vapour in the air.

The general equation for all these reactions is:

$$2M(s) + O_2(g) \rightarrow 2MO(s)$$

The products are oxides containing M^{2+} and O^{2-} ions.

(a)

(b)

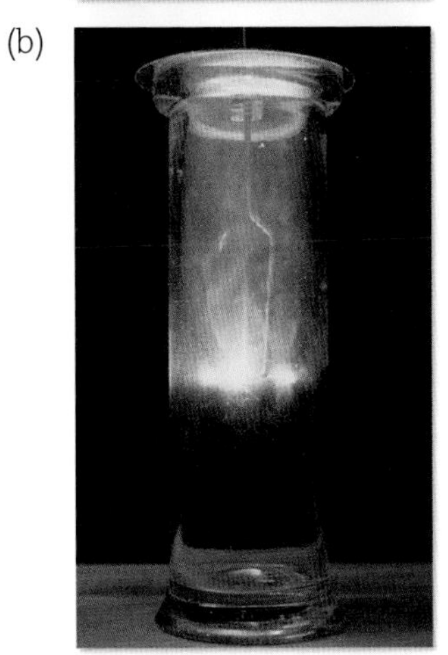

▲ **fig A** (a) Burning magnesium in air. (b) Burning calcium in oxygen.

REACTIONS WITH CHLORINE

The Group 2 elements combine with chlorine when heated in the gas. Just like the reactions with oxygen, the reactions with chlorine become more vigorous down the group. However, this trend is harder to see than with the oxygen reactions. **Fig B** shows what magnesium burning in a flask of chlorine gas looks like.

The general equation for all of these reactions is:

$$M(s) + Cl_2(g) \rightarrow MCl_2(s)$$

The products are chlorides containing M^{2+} and Cl^- ions.

▲ **fig B** Magnesium burning in chlorine.

REACTIONS WITH WATER

The reaction between magnesium and water is very slow and does not proceed completely. Calcium, strontium and barium react with increasing vigour (i.e. reactivity increases down the group), which can be seen by the increase in effervescence.

In **fig C**, you can see that a piece of magnesium in water is covered with bubbles of hydrogen gas, but that the reaction is not very vigorous.

Fig D shows piece of calcium in water is also covered with bubbles of hydrogen gas, but the reaction is more vigorous.

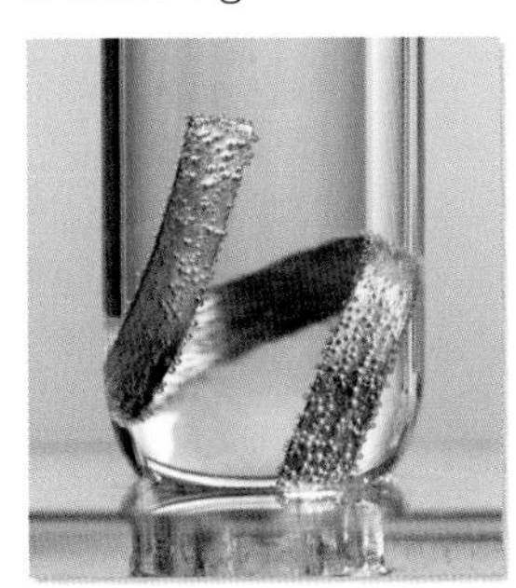

▲ **fig C** Magnesium reacting very slowly with water.

▲ **fig D** Calcium reacting vigorously with water.

The general equation for all of these reactions is:

$$M(s) + 2H_2O(l) \rightarrow M(OH)_2(aq) + H_2(g)$$

The products are hydrogen gas and hydroxides containing M^{2+} and OH^- ions.

The equation for the reaction with calcium is:

$$Ca(s) + 2H_2O(l) \rightarrow Ca(OH)_2(s) + H_2(g)$$

Calcium hydroxide is only slightly soluble in water, so the liquid in this experiment goes cloudy as a precipitate of calcium hydroxide forms.

The equation for the reaction with barium is:

$$Ba(s) + 2H_2O(l) \rightarrow Ba(OH)_2(aq) + H_2(g)$$

Note the difference in the state symbol for the hydroxide in these equations, as barium hydroxide is soluble in water. You will learn more about the solubility of Group 2 hydroxides in the next topic.

MAGNESIUM AND STEAM

Magnesium reacts differently when heated in steam – it rapidly forms magnesium oxide (a white solid) and hydrogen gas in a vigorous reaction. The equation for this reaction is:

$$Mg(s) + H_2O(g) \rightarrow MgO(s) + H_2(g)$$

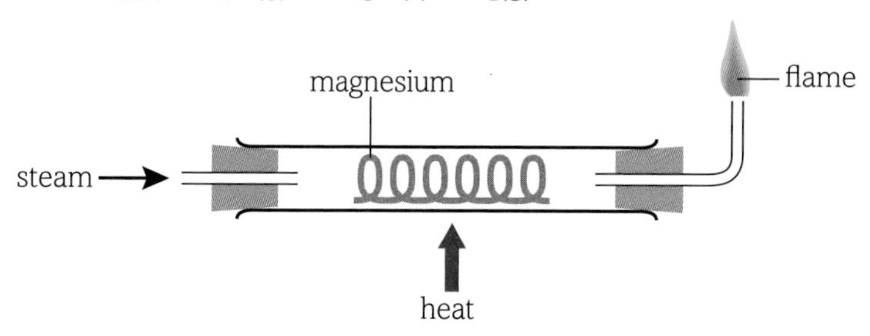

The hydrogen formed is burned as it leaves the tube. This is for safety reasons, to prevent the escape of a highly flammable gas into the laboratory.

REACTIONS OF BERYLLIUM AND RADIUM

You are not required to know any of the reactions of beryllium and radium.

In a similar way to Group 1, the trend is increasing reactivity down the group, so you should be able to predict that beryllium is less reactive than magnesium, and that radium is more reactive than barium.

LEARNING TIP

Practise writing equations for reactions of Group 2 elements with oxygen, chlorine and water.

CHECKPOINT

1. Write an equation for each of these reactions:

(a) calcium with oxygen

(b) strontium with chlorine

(c) barium with water.

2. Suggest why it is not a good idea to use water to put out a fire involving burning magnesium.

8B 4 OXIDES AND HYDROXIDES IN GROUPS 1 AND 2

SPECIFICATION REFERENCE 8.14 8.15 8.19(i,ii)

LEARNING OBJECTIVES

- Know the reactions of the oxides of Group 1 and 2 elements with water and dilute acid, and their hydroxides with dilute acid.
- Know the trends in solubility of the hydroxides and sulfates of Group 2 elements.
- Know the reactions, including ionic equations where appropriate, for identifying carbonate and sulfate ions.

REACTIONS OF THE OXIDES WITH WATER

The Group 1 and 2 oxides are classed as **basic oxides**, which means that they can react with water to form alkalis. These reactions occur when the oxides are added to water. The only observation we can make is that the solids react to form colourless solutions. The general equations for these reactions are:

for Group 1 oxides: $M_2O(s) + H_2O(l) \rightarrow 2MOH(aq)$

for Group 2 oxides: $MO(s) + H_2O(l) \rightarrow M(OH)_2(aq)$

These equations can be simplified because there is no change to the M^+ or M^{2+} ion during the reactions.

$$O^{2-} + H_2O \rightarrow 2OH^-$$

This equation shows the formation of hydroxide ions, which is why the resulting solutions are alkaline.

TRENDS IN SOLUBILITY OF THE GROUP 2 HYDROXIDES

The pH value of the alkaline solution formed depends partly on the relative amounts of oxide and water, but is also affected by differences in the solubility of the hydroxides.

For example, when magnesium oxide reacts with water, the magnesium hydroxide formed has a very low solubility in water. The solubility of the Group 2 hydroxides increases down the group. Therefore, the maximum alkalinity (pH value) of the solutions formed also increases down the group.

TESTING FOR CARBON DIOXIDE

You may remember from your previous study of chemistry that limewater is used to test for carbon dioxide. It goes cloudy (or milky) as a white precipitate forms. Limewater is the name used for a saturated aqueous solution of calcium hydroxide. Carbon dioxide reacts to form calcium carbonate, the white precipitate, which is insoluble in water. The equation for the reaction is:

$$CO_2(g) + Ca(OH)_2(aq) \rightarrow CaCO_3(s) + H_2O(l)$$

As carbon dioxide is bubbled through limewater, the amount of precipitate increases (**fig A**).

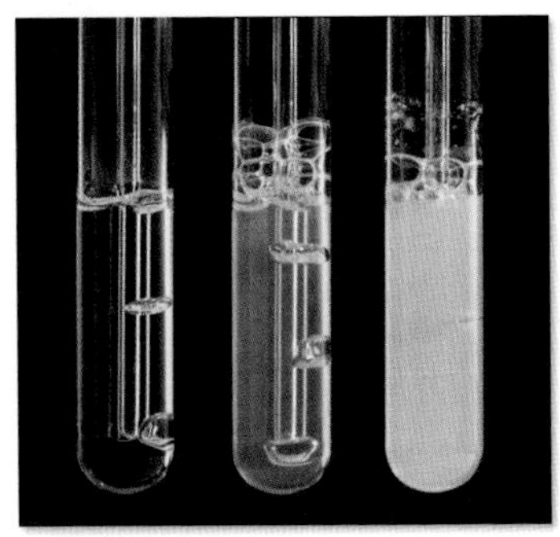

▲ **fig A** As carbon dioxide is bubbled through limewater, the amount of precipitate increases.

MILK OF MAGNESIA

For over a century, a suspension of magnesium hydroxide in water has been sold as an indigestion remedy called milk of magnesia. A bottle of this contains a saturated solution of magnesium hydroxide mixed with extra solid magnesium hydroxide, which acts as an antacid.

The human stomach contains hydrochloric acid that is needed to digest food, but sometimes there is too much acid and the person develops symptoms of indigestion. Taking milk of magnesia neutralises some of the hydrochloric acid and relieves the symptoms. The equation for the reaction is:

$$Mg(OH)_2 + 2HCl \rightarrow MgCl_2 + 2H_2O$$

Although hydroxide ions attack human tissue, the very low solubility of magnesium hydroxide means that the concentration of OH^- ions in the medicine is also very low and is not a risk to health.

▲ **fig B** A bottle of milk of magnesia.

REACTIONS OF THE OXIDES AND HYDROXIDES WITH ACIDS

All of the Group 1 and 2 oxides and hydroxides react with acids to form salts and water. These reactions can be described as neutralisation reactions.

During the reactions, the only observations to be made are that a white solid reacts to form a colourless solution. The reactions are exothermic, so you may use some of them in experiments to measure energy changes.

Here are some sample equations:

$$Na_2O + H_2SO_4 \rightarrow Na_2SO_4 + H_2O$$

$$CaO + 2HNO_3 \rightarrow Ca(NO_3)_2 + H_2O$$

$$KOH + HCl \rightarrow KCl + H_2O$$

$$Ba(OH)_2 + 2HCl \rightarrow BaCl_2 + 2H_2O$$

USE IN AGRICULTURE

For centuries, farmers have used lime to control soil acidity so that a greater yield of crops can be obtained.

▲ **fig C** Lime being spread on a field.

Lime is mostly calcium hydroxide (obtained from limestone, which is calcium carbonate), and neutralises excess acidity in the soil. Using nitric acid to represent the acid in soil, the equation for this reaction is:

$$Ca(OH)_2 + 2HNO_3 \rightarrow Ca(NO_3)_2 + 2H_2O$$

TRENDS IN SOLUBILITY OF THE GROUP 2 SULFATES

All Group 2 nitrates and chlorides are soluble, but the solubility of Group 2 sulfates decreases down the group.

- Magnesium sulfate is classed as soluble.
- Calcium sulfate is slightly soluble.
- Strontium sulfate and barium sulfate are insoluble.

You do not have to understand the reasons for this trend, but you do need to know how the very low solubility of barium sulfate is used in a test for sulfate ions in solution.

TESTING FOR SULFATE IONS

The presence of sulfate ions in an aqueous solution can be shown by adding a solution containing barium ions (usually barium chloride or barium nitrate). Any sulfate ions in the solution will react with the added barium ions to form a white precipitate of barium sulfate. The ionic equation for this reaction is:

$$Ba^{2+}(aq) + SO_4^{2-}(aq) \rightarrow BaSO_4(s)$$

There are other anions that could also form a white precipitate with barium ions, especially carbonate ions, so in the test there must be H^+ ions present to prevent barium carbonate from forming as a white precipitate. Dilute nitric acid or dilute hydrochloric acid is therefore added as part of the test.

As an example, here is how to test for the presence of sulfate ions in a solution of sodium sulfate.

- Add dilute nitric acid and barium nitrate solution to the sodium sulfate solution.
- A white precipitate forms.

The equation for the reaction is

$$Ba(NO_3)_2(aq) + Na_2SO_4(aq) \rightarrow BaSO_4(s) + 2NaNO_3(aq)$$

BARIUM MEALS

Solutions containing barium ions are poisonous to humans, however, barium sulfate is used in hospitals, where patients are sometimes given a barium 'meal'. This 'meal' contains barium sulfate, which is not poisonous because it is insoluble – although it contains barium ions, these ions are not free to move. Although bones can be seen clearly on X-rays, soft tissues cannot. If the patient has a barium meal before an X-ray, these soft tissues will show up more clearly because of the dense white solid.

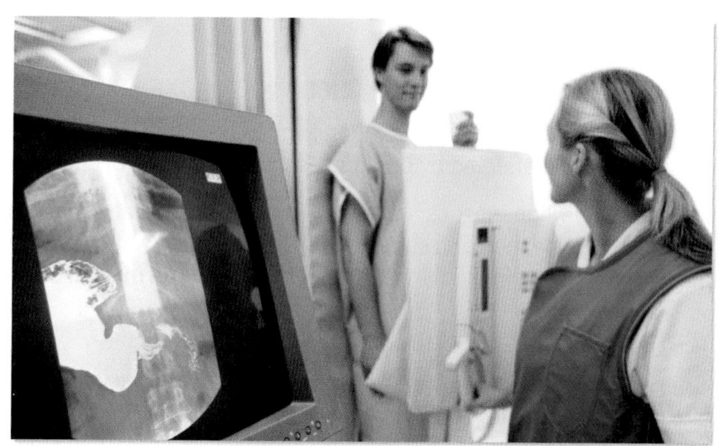

▲ **fig D** How a barium meal can help show up soft tissues on an X-ray.

LEARNING TIP

There is no trend in reactivity with water for the Group 1 and 2 oxides, because they already contain metal ions, not metal atoms.

CHECKPOINT

1. Limewater is used to test for carbon dioxide. Why should limewater not be left exposed to air before using it in this test?
2. Why is an acid added when using barium chloride or barium nitrate solution to test for sulfate ions?

SUBJECT VOCABULARY

basic oxides oxides of metals that react with water to form metal hydroxides, and with acids to form salts and water

SPECIFICATION REFERENCE 8.16 8.18(i)

8B 5 THERMAL STABILITY OF COMPOUNDS IN GROUPS 1 AND 2

LEARNING OBJECTIVES

- Understand the reasons for the trends in thermal stability of the nitrates and the carbonates of the elements in Groups 1 and 2 in terms of the size and charge of the cations involved.
- Know experimental procedures used to show patterns in the thermal decomposition of Group 1 and 2 nitrates and carbonates.

FACTORS AFFECTING THERMAL STABILITY

Thermal stability is a term that indicates how stable a compound is when it is heated. Does it not decompose at all (very thermally stable), or does it decompose as much as possible (not at all thermally stable), or somewhere in between?

You are familiar with ionic bonding in compounds and you know that the bond strength is responsible for physical properties such as the melting temperature. For example, the high melting temperature of sodium chloride (NaCl) can be explained in terms of the strong electrostatic forces of attraction between the large numbers of oppositely charged Na^+ and Cl^- ions. When sodium chloride melts, the ions change from being regularly arranged in a giant lattice to moving freely in a liquid. There is no decomposition occurring in this change of state.

The situation with Group 2 nitrates and carbonates is very different compared with a Group 1 chloride such as sodium chloride. There are three reasons for this.

1 The charge on a Group 2 cation is double that on a Group 1 cation (e.g. Ca^{2+} compared with Na^+).
2 The size (ionic radius) of a Group 2 cation is smaller than that of the Group 1 cation in the same period.
3 The nitrate (NO_3^-) and carbonate (CO_3^{2-}) anions are more complex than the Cl^- ion.

These differences mean that when Group 2 nitrates and carbonates are heated, they do not melt. Instead, they decompose. We need to look carefully at these factors to understand why this is.

- The larger, more complex nitrate ion can change into the smaller, more stable nitrite ion (NO_2^-) or oxide ion (O^{2-}) by decomposing and releasing oxygen gas and/or nitrogen dioxide gas.
- The larger, more complex carbonate ion can change into the smaller, more stable oxide ion (O^{2-}) by decomposing and releasing carbon dioxide gas, CO_2.
- The stabilities of the nitrate and carbonate anions are influenced by the charge and size of the cations present. Smaller and more highly charged cations affect these anions more.

Table A shows the charge and radius for each of the ions in Groups 1 and 2.

GROUP 1			GROUP 2		
ELEMENT	CHARGE ON ION	IONIC RADIUS / nm	ELEMENT	CHARGE ON ION	IONIC RADIUS / nm
Lithium	+1	0.074	Beryllium	+2	0.027
Sodium	+1	0.102	Magnesium	+2	0.072
Potassium	+1	0.138	Calcium	+2	0.100
Rubidium	+1	0.149	Strontium	+2	0.113
Caesium	+1	0.170	Barium	+2	0.136

table A

You can see that the cation with the greatest influence (biggest charge and smallest size) on an anion is Be^{2+}, and the one with the least influence (smallest charge and largest size) is Cs^+.

THERMAL STABILITY OF NITRATES

All of the nitrates of the Group 1 and Group 2 elements are white solids. When they are heated, they all decompose to nitrites or oxides, and give off nitrogen dioxide (brown fumes) and/or oxygen. If the nitrate contains water of crystallisation, then steam will also be observed.

If no brown fumes are observed, this indicates a lesser decomposition. This decomposition can be represented by this word equation:

metal nitrate → metal nitrite + oxygen

Oxygen gas cannot be observed as it is colourless. It can be detected using a glowing spill or splint, which relights if oxygen is present.

Nitrates and nitrites are sometimes differentiated by using oxidation numbers – nitrate(V) for nitrate, and nitrate(III) for nitrite.

If brown fumes are observed, this indicates a greater decomposition that can be represented by this word equation:

metal nitrate → metal oxide + nitrogen dioxide + oxygen

Table B shows typical observations obtained by heating samples of nitrates in test tubes over a Bunsen flame.

GROUP 1 NITRATE		GROUP 2 NITRATE	
NAME	RESULT	NAME	RESULT
Lithium nitrate	brown fumes	Beryllium nitrate	brown fumes
Sodium nitrate	no brown fumes	Magnesium nitrate	brown fumes
Potassium nitrate	no brown fumes	Calcium nitrate	brown fumes
Rubidium nitrate	no brown fumes	Strontium nitrate	brown fumes
Caesium nitrate	no brown fumes	Barium nitrate	brown fumes

table B

Table C shows what happens, in terms of decomposition, when samples of nitrates are heated in test tubes over a Bunsen flame.

GROUP 1 NITRATE		GROUP 2 NITRATE	
NAME	RESULT	NAME	RESULT
Lithium nitrate	greater decomposition	Beryllium nitrate	greater decomposition
Sodium nitrate	lesser decomposition	Magnesium nitrate	greater decomposition
Potassium nitrate	lesser decomposition	Calcium nitrate	greater decomposition
Rubidium nitrate	lesser decomposition	Strontium nitrate	greater decomposition
Caesium nitrate	lesser decomposition	Barium nitrate	greater decomposition

table C

You can see that the greater decomposition occurs when:

- the cation has a 2+ charge (all of the Group 2 nitrates)
- the cation has a 1+ charge and is also the smallest Group 1 cation.

Here are some sample equations for the reactions that occur:

$4LiNO_3 \rightarrow 2Li_2O + 4NO_2 + O_2$
(lithium nitrate – the only Group 1 nitrate that decomposes in this way)

$2NaNO_3 \rightarrow 2NaNO_2 + O_2$
(all other Group 1 nitrates decompose in this way)

$2Be(NO_3)_2 \rightarrow 2BeO + 4NO_2 + O_2$
(all Group 2 nitrates decompose in this way)

EXAM HINT

Previous exam questions suggest that many students are poor at reproducing these equations under exam conditions so be sure to learn them.

EXAM HINT

Note that these reactions involving decompositions of nitrates are redox reactions. Check by assigning oxidation numbers to nitrogen and oxygen in each reaction.

THERMAL STABILITY OF CARBONATES

All of the carbonates of the Group 1 and Group 2 elements are white solids. When they are heated, they either do not decompose, or decompose to oxides and give off carbon dioxide.

As the gas given off is colourless and the carbonate and oxide are both white solids, there are no observations that can be made.

Table D shows what happens when samples of carbonates are heated in test tubes over a Bunsen flame.

GROUP 1 CARBONATE		GROUP 2 CARBONATE	
NAME	RESULT	NAME	RESULT
Lithium carbonate	decomposition	Beryllium carbonate	decomposition
Sodium carbonate	no decomposition	Magnesium carbonate	decomposition
Potassium carbonate	no decomposition	Calcium carbonate	decomposition
Rubidium carbonate	no decomposition	Strontium carbonate	decomposition
Caesium carbonate	no decomposition	Barium carbonate	decomposition

table D Results, in terms of decomposition, when samples of carbonates are heated.

You can see that the pattern is similar to that of the nitrates. The lithium compound and all of the Group 2 compounds behave differently from the other Group 1 compounds.

Lithium carbonate decomposes at lower temperatures than the other Group 1 carbonates:

$$Li_2CO_3 \rightarrow Li_2O + CO_2$$

Other Group 1 carbonates do not decompose on heating, except at very high temperatures.

All Group 2 carbonates decompose in the same way, but with increasing difficulty down the group. A typical equation for one of these decompositions is:

$$CaCO_3 \rightarrow CaO + CO_2$$

You can see a similar pattern as with the nitrates. Decomposition occurs when:

- the cation has a 2+ charge (all of the Group 2 carbonates)
- the cation has a 1+ charge and is also the smallest Group 1 cation (only lithium carbonate).

fig A Calcium oxide forms when calcium carbonate (limestone) is heated strongly. If the strong heating is continued, there is no further chemical change, but a very bright glow is seen. This is the origin of 'limelight' – it was formerly used in theatre lighting.

LEARNING TIP

Look for the similar patterns in the thermal stability of Group 1 and Group 2 nitrates and carbonates. Although the reactions are different, the lithium compound behaves like the Group 2 compounds and not like the other Group 1 compounds.

CHECKPOINT

1. What observations would be made when these compounds are heated in a test tube over a Bunsen flame?

(a) calcium nitrate

(b) sodium carbonate.

2. Write a chemical equation for each of these reactions.

(a) the decomposition of potassium nitrate

(b) the decomposition of strontium carbonate.

SUBJECT VOCABULARY

thermal stability a measure of the extent to which a compound decomposes when heated

8B 6 FLAME TESTS AND THE TEST FOR AMMONIUM IONS

SPECIFICATION REFERENCE 8.17 8.18(ii) 8.19(iii)

LEARNING OBJECTIVES

- Know experimental procedures used to show flame colours in compounds of Group 1 and 2 elements.
- Understand the formation of characteristic flame colours by Group 1 and 2 compounds in terms of electron transitions.
- Know how to identify ammonium ions.

INTRODUCTION TO FLAME TESTS

A flame test seems a very simple chemical test to identify the presence of a cation in a compound. However, it is important to know how to do one carefully, to know how to interpret the results, and to understand how the test works.

A flame test result can indicate the presence of some metals (in the form of cations) in Groups 1 and 2 of the Periodic Table. It does not work for all of them.

HOW TO DO A FLAME TEST

We will assume that a small quantity of a solid is available in a glass dish. This is how to do a flame test.

- Wear safety glasses and a lab coat. Within a fume cupboard, light a Bunsen burner.
- Using a dropper, add a few drops of concentrated hydrochloric acid to the solid and mix together so that the metal compound begins to dissolve. (One reason for using hydrochloric acid is to convert any metal compound to a chloride – chlorides are more volatile than other salts so are more likely to give better results.)
- Dip a clean metal wire (platinum or nichrome) or silica rod into the mixture to obtain a sample of the compound.
- Hold the end of the wire or rod in the flame and observe the colour.

fig A A platinum wire loop with a sample of a strontium compound being held in a Bunsen flame.

PROBLEMS WITH A FLAME TEST

There are two main problems with a flame test.

- Many compounds contain small amounts of sodium compounds as impurities, so the intense colour of sodium can hide other colours.
- Describing colours with words is subjective – people have different levels of colour vision, and a word description of a colour may mean different colours to different people.

COLOUR DESCRIPTIONS

Some traditional descriptions of colours are problematic. For example, what is the colour 'brick red'? It may depend on the bricks used in a particular location, and these are not the same throughout the world.

Another example is the traditional colour description used for potassium. It is lilac, but can you tell the difference between lilac and lavender, magenta, mauve, pink, plum, puce, purple and violet?

Table A shows some traditional colour descriptions obtained when Group 1 and 2 compounds are tested in this way.

METAL CATION	FORMULA	COLOUR
Lithium	Li^{+}	red
Sodium	Na^{+}	yellow/orange
Potassium	K^{+}	lilac
Rubidium	Rb^{+}	red/purple
Caesium	Cs^{+}	blue/violet
Beryllium	Be^{2+}	(no colour)
Magnesium	Mg^{2+}	(no colour)
Calcium	Ca^{2+}	(brick) red
Strontium	Sr^{2+}	(crimson) red
Barium	Ba^{2+}	(apple) green

table A Colour descriptions obtained when Group 1 and 2 compounds undergo a flame test.

There are also some metal cations with characteristic flame colours that are not in Groups 1 or 2. For example, copper compounds produce a blue-green colour in a flame test.

WHAT CAUSES THE COLOURS IN FLAME TESTS?

The simple answer is electron transitions. However, we need a more complete explanation than this.

You know from **Topic 2** that electrons occupy orbitals in specific energy levels in an atom. These are often represented using electron configurations such as 2.8.1 (for sodium).

Electrons can absorb energy and move to higher energy levels. Sometimes the term 'ground state' is used to describe an atom with all its electrons in their lowest possible energy levels. If an electron moves to a higher energy level, then the new situation can be described as an 'excited state'. This movement of an electron to an excited state occurs during a flame test.

However, this movement is immediately followed by the return of the electron to its ground state, which releases energy. If this energy corresponds to radiation in the visible light spectrum, then a characteristic colour appears – this is the flame test colour.

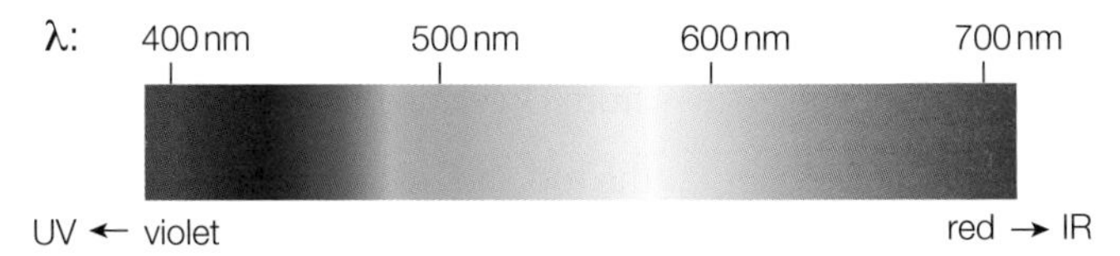

▲ **fig B** The wavelengths of the spectrum of visible light.

For example, the visible spectrum covers the part of the electromagnetic spectrum in the wavelength range 400–700 nm. The electron transition in sodium corresponds to a wavelength of about 590 nm, which is in the yellow-orange part of the spectrum, so this is the colour of a sodium compound in a flame test. The electron transition in magnesium corresponds to a wavelength outside the visible spectrum, so there is no flame colour for magnesium.

▲ **fig C** The yellow-orange colour of sodium street lights is the same as the colour of sodium in a flame test.

LEARNING TIP

Remember that doing a flame test on a magnesium compound has nothing to do with burning magnesium metal in a Bunsen flame. When magnesium metal burns, it changes from Mg atoms to Mg^{2+} ions. Magnesium compounds already contain Mg^{2+} ions.

THE TEST FOR AMMONIUM IONS

One cation that does not give a colour in a flame test is the ammonium ion. The usual test for ammonium ions in a solid or solution is to add sodium hydroxide solution and warm the mixture. The addition of sodium hydroxide causes this reaction:

$$NH_4^+ + OH^- \rightarrow NH_3 + H_2O$$

and the warming releases ammonia gas. Ammonia can be recognised by its smell, but you can use damp litmus paper as a simple chemical test, which turns blue (ammonia is the only common alkaline gas). Alternatively, hydrogen chloride gas (from concentrated hydrochloric acid) reacts with ammonia to form white smoke of ammonium chloride:

$$NH_3 + HCl \rightarrow NH_4Cl$$

EXAM HINT

If you are asked for a safety precaution for these reactions, it is worth saying that they should be carried out in a fume cupboard.

CHECKPOINT

1. Why is concentrated hydrochloric acid used in flame tests?
2. Explain why barium compounds give a characteristic flame colour but magnesium compounds do not.

8C 1 TRENDS IN GROUP 7

SPECIFICATION REFERENCE
8.24 8.28

LEARNING OBJECTIVES

- Understand reasons for the trends in melting and boiling temperatures, physical state at room temperature, electronegativity and reactivity down the group.
- Be able to make predictions about fluorine and astatine and their compounds, in terms of knowledge of trends in halogen chemistry.

INTRODUCTION TO THE GROUP 7 ELEMENTS

Group 7 of the Periodic Table contains five elements. These elements are often known as halogens and they all form salts called halides. The term 'halogen' comes from Greek and means 'salt producer'.

When considering the properties of the group, the elements at the top and bottom of the group (fluorine and astatine) are often ignored. Fluorine is ignored because it sometimes behaves differently from chlorine, bromine and iodine, and astatine is ignored because (like radium in Group 2) it only exists as radioactive isotopes.

Table A shows some information about the Group 7 elements.

ELEMENT	STATE AT ROOM TEMPERATURE	MELTING TEMPERATURE / °C	BOILING TEMPERATURE / °C	ELECTRONEGATIVITY
Fluorine	gas	−220	−188	4.0
Chlorine	gas	−101	−35	3.0
Bromine	liquid	−7	59	2.8
Iodine	solid	114	184	2.5
Astatine	solid	302	337	2.2

table A Physical properties of Group 7 elements.

TRENDS IN MELTING AND BOILING TEMPERATURE

All of the halogens exist as diatomic molecules, so their melting and boiling temperatures depend on the strengths of the intermolecular forces of attraction between these molecules. We looked at these forces, known as London forces, in **Topic 7**. Here is a reminder of how they happen.

As the two atoms in the diatomic molecule are identical, the pair of electrons forming the covalent bond between them is shared equally between the two atoms. This means that the halogen molecules are non-polar, at least on average. However, as the positive charges of the protons in the two nuclei are in fixed positions, but the electron density in a halogen molecule continuously fluctuates, sometimes the centres of positive and negative charge do not coincide. This situation results in a temporary dipole, which is often referred to as an instantaneous dipole.

We will use a rectangular shape to represent a halogen molecule. The two dots represent the nuclei of the two halogen atoms. When two molecules are close together, you would expect no interaction between the two because they are both non-polar.

If the molecule on the left becomes an instantaneous dipole, then it will cause an induced dipole in the molecule on the right. This results in a force of attraction between the two molecules:

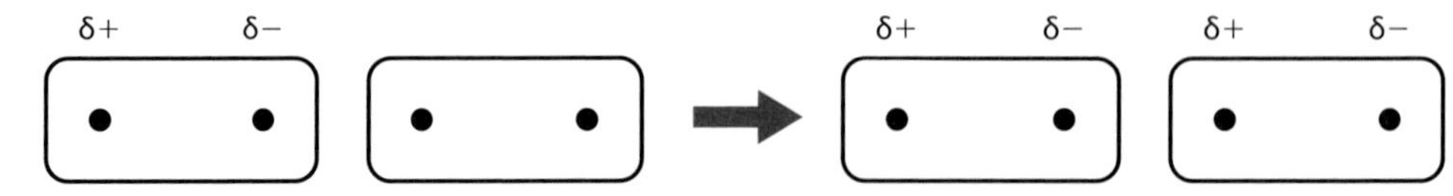

This force of attraction is described as an instantaneous dipole–induced dipole attraction, and these are the intermolecular forces of attraction that exist between halogen molecules. These weak forces increase as the number of electrons and therefore the size of the electron cloud increases. The forces increase in strength down Group 7 as the number of electrons in the molecules increases. This explains the increase in both melting temperature and boiling temperature down Group 7.

EQUATIONS FOR CHANGES OF STATE

You are probably used to writing equations for chemical changes, but we use equations for physical changes less often. As well as using the correct state symbols, it is important to remember to write the formulae (not the symbols) of the halogens.

When bromine is left at room temperature, it gives off brown vapour, as its boiling point (59 °C) is not much higher than room temperature. The equation for this change is:

$$Br_2(l) \rightarrow Br_2(g)$$

When iodine is warmed, most of it changes directly into a vapour without melting. This change is called **sublimation**. The equation for this change is:

$$I_2(s) \rightarrow I_2(g)$$

TREND IN ELECTRONEGATIVITY

We have already seen in **Topic 3B** that electronegativity is the ability of an atom to attract the pair of electrons in a covalent bond. The 0–4 scale devised by Linus Pauling is still used. The electronegativity of an atom depends on:

- its nuclear charge – the bigger the nuclear charge, the higher the electronegativity
- the distance between the nucleus and the bonding pair of electrons – the shorter the distance, the higher the electronegativity
- the shielding effect of electrons in inner energy levels – the fewer energy levels, the higher the electronegativity.

The electronegativity of the Group 7 elements is the highest of any group in the Periodic Table. The electronegativity of fluorine is the highest of all elements.

TREND IN REACTIVITY

Fluorine is an extremely reactive element, and reactivity decreases down Group 7. Because of their high electronegativity, most reactions of the halogens involve them acting as oxidising agents and gaining electrons to form negative ions or becoming the slightly negative ($\delta-$) part of a polar molecule. The decreasing reactivity down the group can therefore be explained by reference to the same factors used to explain the decreasing electronegativity down the group.

We will see examples of reactions in **Topics 8C.2 to 8C.4**.

LEARNING TIP

Write a brief summary to help you understand the difference between intermolecular forces involving permanent dipoles and temporary dipoles.

CHECKPOINT

1. Why does bromine have a higher boiling temperature than chlorine?

2. Why is fluorine the most electronegative element of all?

SUBJECT VOCABULARY

sublimation the process of a solid changing directly into a vapour without melting

8C 2 REDOX REACTIONS IN GROUP 7

SPECIFICATION REFERENCE 8.25 8.26

LEARNING OBJECTIVES

- Understand the trend in reactivity of Group 7 elements in terms of the redox reactions of chlorine, bromine and iodine with halide ions in aqueous solution, followed by the addition of an organic solvent.
- Understand, in terms of changes in oxidation number, oxidation reactions of the halogens with Group 1 and 2 metals and disproportionation reactions of chlorine with water, and with cold and hot alkali.

REACTIONS WITH METALS IN GROUPS 1 AND 2

There are 12 elements in Groups 1 and 2, and 5 elements in Group 7, so there are 60 possible reactions to consider. You do not need to know details of all of these reactions, but here are some useful generalisations that you should know.

- Reactions are most vigorous between elements at the bottom of Groups 1 and 2, and elements at the top of Group 7. The most vigorous reaction should be between caesium (or francium) and fluorine, and the least vigorous between beryllium and iodine (or astatine).
- The products of these reactions are salts, ionic solids that are usually white.
- All of these reactions involve electron transfer to the halogen, so they are redox reactions in which the halogen acts as an oxidising agent.
- The oxidation number of the halogen decreases from 0 to −1, and the oxidation number of the metal increases from 0 to +1 or +2, depending on the group.

Here are a couple of sample equations:

- lithium reacting with chlorine $2Li + Cl_2 \rightarrow 2LiCl$
- barium reacting with bromine $Ba + Br_2 \rightarrow BaBr_2$

HALOGEN/HALIDE DISPLACEMENT REACTIONS

A more reactive halogen can displace a less reactive halogen from one of its compounds. So:

- chlorine displaces bromine and iodine
- bromine displaces iodine but not chlorine
- iodine does not displace either chlorine or bromine.

These reactions occur in aqueous solution, so any reaction that occurs is indicated by a colour change. One problem in interpreting colour changes in these reactions is the similarity of some colours and the variation in colour with concentration. For example, bromine in its liquid state is red-brown, but bromine dissolved in water might be orange or yellow, depending on the concentration. Iodine dissolved in water may also appear brown at some concentrations.

When doing these reactions, it is a good idea to add an organic solvent (such as cyclohexane) after the reaction, and then shake the tube. Halogens are more soluble in cyclohexane than in water, so the halogen dissolves in the organic upper layer, where its colour can more easily be seen. **Fig A** shows the colours of the halogens dissolved in cyclohexane.

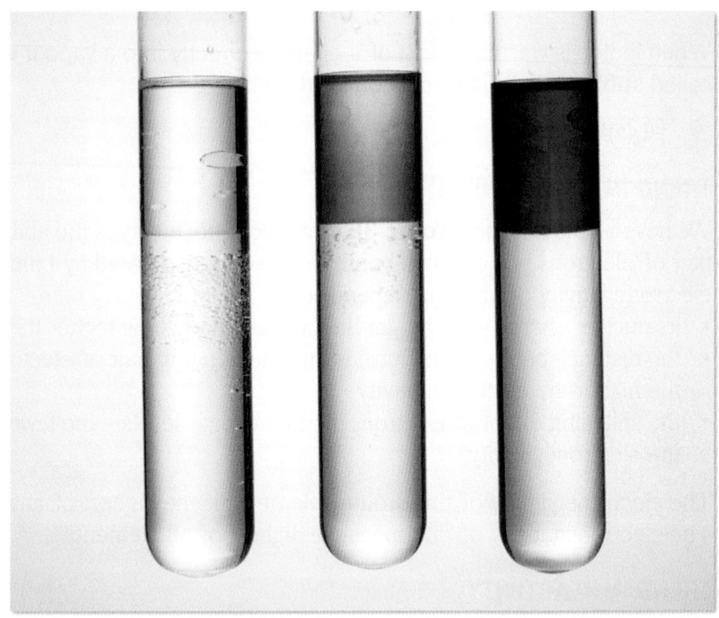

▲ **fig A** Look at the upper layer in each tube. The pale green colour of chlorine does not change much, the orange colour of bromine looks a bit darker, but the colour of iodine changes to purple or violet in cyclohexane.

Here are some sample equations:

- chlorine displacing bromine $Cl_2 + 2NaBr \rightarrow 2NaCl + Br_2$
- bromine displacing iodine $Br_2 + 2I^- \rightarrow 2Br^- + I_2$

These equations are examples of redox reactions. The reacting halogen decreases its oxidation number from 0 to −1, and the reacting halide increases its oxidation number from −1 to 0.

The decreasing reactivity of chlorine, bromine and iodine in the reactions above can be explained using the same factors as in **Topic 8C.1**. Chlorine is the most reactive of the three because:

- it is the smallest atom, so the incoming electron gets closer to, and is more attracted by, the protons in the nucleus
- it has the smallest number of complete inner energy levels of electrons, so the incoming electron experiences the least repulsion.

DISPROPORTIONATION REACTIONS OF CHLORINE

Disproportionation is a more unusual type of reaction. In this reaction, one element undergoes both oxidation and reduction at the same time. We will look at three examples of this type of reaction, all involving chlorine.

CHLORINE WITH WATER

When chlorine is added to water, it dissolves to form a solution that is sometimes called 'chlorine water' (just as 'bromine water' is used to refer to bromine dissolved in water). Some of the dissolved chlorine also reacts to form a mixture of two acids.

You are familiar with one of the acids, hydrochloric acid, but the other acid is chloric(I) acid (its old name is hypochlorous acid). Its formula is shown in the equation below as HClO, but HOCl is also commonly used. Both acids are colourless solutions, so there is no visible change during the reaction.

The disproportionation that occurs can be shown using oxidation numbers:

$Cl_2 + H_2O \rightarrow HCl + HClO$

0   −1   chlorine is reduced

0   +1   chlorine is oxidised

The addition of chlorine to disinfect water for drinking purposes has saved countless lives, and continues to do so today. It kills the pathogens responsible for water-borne diseases such as cholera.

▲ **fig B** Chlorine can reduce the risk of transmitting infections in public swimming pools.

CHLORINE WITH COLD ALKALI

When chlorine is added to cold dilute aqueous sodium hydroxide, it reacts to form the salts of the acids in the equation above. These salts are sodium chloride and sodium chlorate(I), which is also known as sodium hypochlorite.

Again, the disproportionation that occurs can be shown using oxidation numbers:

$Cl_2 + 2NaOH \rightarrow NaCl + NaClO + H_2O$

0   −1   chlorine is reduced

0   +1   chlorine is oxidised

The sodium chlorate(I) formed is also a disinfectant, but it is mainly known for its bleaching action. It is used extensively in industry and is the active ingredient in household bleach.

CHLORINE WITH HOT ALKALI

When chlorine is added to hot concentrated sodium hydroxide solution, it reacts to form sodium chloride and a different product, sodium chlorate(V).

Again, the disproportionation that occurs can be shown using oxidation numbers:

$3Cl_2 + 6NaOH \rightarrow 5NaCl + NaClO_3 + 3H_2O$

0 −1 chlorine is reduced

0 +5 chlorine is oxidised

The sodium chlorate(V) formed is also used in bleaching, and as a weed killer.

Bromine and iodine react in similar ways.

REACTIONS OF FLUORINE AND ASTATINE

If you are asked to predict reactions of fluorine and astatine that you are not familiar with, then you can use information in this topic to help you. For example, you could write an equation to represent the reaction between sodium and astatine, based on your knowledge of the reaction between sodium and iodine.

LEARNING TIP

Read the explanation for chlorine being more reactive than the other halogens (except for fluorine) at the end of **Topic 8C.1**. Now try to explain why astatine is the least reactive halogen.

CHECKPOINT

1. (a) Write a chemical equation for the reaction between chlorine and potassium iodide.

(b) Write an ionic equation for the reaction between bromine and sodium astatide.

2. Write equations and name the products of the reactions between bromine and:

(a) cold dilute aqueous sodium hydroxide

(b) hot concentrated sodium hydroxide solution.

SUBJECT VOCABULARY

disproportionation reaction a reaction involving the simultaneous oxidation and reduction of an element in a single species

3 REACTIONS OF HALIDES WITH SULFURIC ACID

SPECIFICATION REFERENCE 8.27(i)

LEARNING OBJECTIVES

- Understand the reactions of solid Group 1 halides with concentrated sulfuric acid, to illustrate the trend in reducing ability of the hydrogen halides.

REDOX REACTIONS AGAIN

In **Topics 8A.3**, **8C.1** and **8C.2**, you have learnt about several examples of the halogens acting as oxidising agents, and you know that this oxidising power decreases down the group.

In this topic, we will look at reactions of halide *ions*, not halogen *molecules*. It is important to realise that in these reactions, the halides act as reducing agents, and that the trend is different. This is shown in **table A**.

OXIDISING POWER	HALOGEN		HALIDE		REDUCING POWER
High ↑	fluorine	F_2	fluoride	F^-	Low ↓
	chlorine	Cl_2	chloride	Cl^-	
	bromine	Br_2	bromide	Br^-	
	iodine	I_2	iodide	I^-	
Low	astatine	At_2	astatide	At^-	High

table A Notice that the decreasing trend down the group in oxidising power of halogens goes with the increasing trend in reducing power of halides.

The reducing action of halide ions can be represented by this general half-equation:

$$2X^- \rightarrow X_2 + 2e^-$$

Sulfuric acid is of course an acid, but when concentrated it contains very few ions. We can write an equation for its partial ionisation:

$$H_2SO_4 \rightleftharpoons H^+ + HSO_4^-$$

Note the reversible arrow. The position of this equilibrium lies well to the left in the concentrated acid, and this first ionisation is far from complete. The second ionisation, to produce sulfate ions, occurs only to a small extent in the concentrated acid:

$$HSO_4^- \rightleftharpoons H^+ + SO_4^{2-}$$

Sulfuric acid, especially when concentrated, can act as an oxidising agent as well as an acid. When it acts as an oxidising agent, it is reduced, but the extent of its reduction and the products formed depend on the species being oxidised.

The three possible reduction products are:
- sulfur dioxide
- sulfur
- hydrogen sulfide.

Three different half-equations can be written to represent its oxidising action. Note the change in oxidation number of the sulfur in each case:

Half-equation 1. $\underset{+6}{H_2SO_4} + 2H^+ + 2e^- \rightarrow 2H_2O + \underset{+4}{SO_2}$

Half-equation 2. $\underset{+6}{H_2SO_4} + 6H^+ + 6e^- \rightarrow 4H_2O + \underset{0}{S}$

Half-equation 3. $\underset{+6}{H_2SO_4} + 8H^+ + 8e^- \rightarrow 4H_2O + \underset{-2}{H_2S}$

These half-equations may look complicated at first, but you should be able to see the pattern.

- In half-equation 1, the decrease in oxidation number (+6 to +4) is 2, which is the same as the numbers of H^+ ions and electrons in the equation.
- The pattern is similar in the other two equations. The decrease in oxidation number of the sulfur is the same as the numbers of H^+ ions and electrons in the half-equation (6 in reaction 2, 8 in reaction 3).

OBSERVATIONS AND PRODUCTS

Table B shows typical observations made, and products formed, when concentrated sulfuric acid is added to three sodium halides.

HALIDE	OBSERVATIONS	PRODUCTS	
NaCl	misty fumes	hydrogen chloride	HCl
NaBr	misty fumes brown fumes colourless gas with choking smell	hydrogen bromide bromine sulfur dioxide	HBr Br_2 SO_2
NaI	misty fumes purple fumes or black solid colourless gas with choking smell yellow solid colourless gas with rotten egg smell	hydrogen iodide iodine sulfur dioxide sulfur hydrogen sulfide	HI I_2 SO_2 S H_2S

table B Observations and products formed when concentrated sulfuric acid is added to three sodium halides.

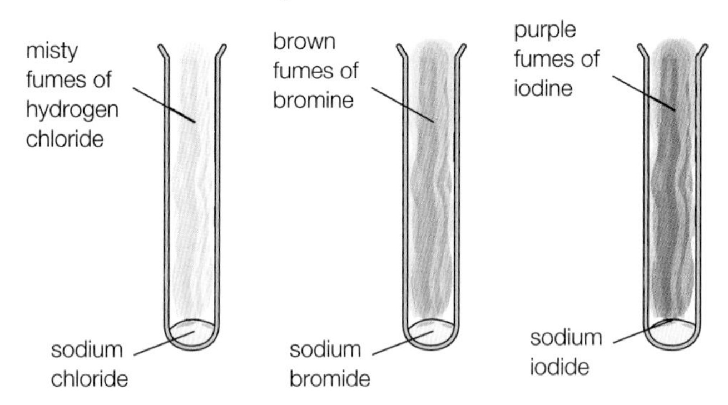

▲ **fig A** From left to right: tubes containing hydrogen chloride (formed from sodium chloride), bromine (formed from sodium bromide) and iodine (formed from sodium iodide).

The tube on the left in **fig A** shows no colour. These are the misty fumes of hydrogen chloride, formed from sodium chloride. The tube in the middle contains brown fumes. This is bromine, formed from sodium bromide. The tube on the right contains purple fumes. This is iodine, formed from sodium iodide.

With sodium chloride, the sulfuric acid behaves only as an acid, and not as an oxidising agent. This is because chloride ions have low reducing power.

With sodium bromide, the greater reducing power of bromide ions causes the sulfuric acid to be reduced, as in half-equation 1 above.

With sodium iodide, the much greater reducing power of iodide ions causes the sulfuric acid to be reduced, as in half-equations 1, 2 and 3.

CONSTRUCTING EQUATIONS

Some of the equations you see, or are asked to write, may look very complicated and be hard to remember. It is better not to try to remember them, but to work them out by the addition of half-equations. This method is recommended for the redox equations of sodium bromide and sodium iodide.

SODIUM CHLORIDE

The reaction between sodium chloride and concentrated sulfuric acid can be represented by one equation, because no redox reactions are occurring:

$$NaCl + H_2SO_4 \rightarrow NaHSO_4 + HCl$$

EXAM HINT

The chloride ion is behaving as a base in this reaction.

SODIUM BROMIDE

The formation of the misty fumes of hydrogen bromide can be represented by an equation analogous to the one for sodium chloride:

$NaBr + H_2SO_4 \rightarrow NaHSO_4 + HBr$

The table of observations shows that only one redox reaction occurs: the formation of sulfur dioxide in half-equation 1. The two relevant half-equations are:

$2Br^- \rightarrow Br_2 + 2e^-$

and

$H_2SO_4 + 2H^+ + 2e^- \rightarrow 2H_2O + SO_2$

Adding these together, then cancelling the $2e^-$ on each side, gives:

$2Br^- + H_2SO_4 + 2H^+ \rightarrow 2H_2O + SO_2 + Br_2$

You could combine the ions on the left to give:

$2HBr + H_2SO_4 \rightarrow 2H_2O + SO_2 + Br_2$

This equation represents the oxidation of the misty fumes of hydrogen bromide.

SODIUM IODIDE

The formation of the misty fumes of hydrogen iodide can be represented by an equation analogous to the one for sodium chloride:

$NaI + H_2SO_4 \rightarrow NaHSO_4 + HI$

The table of observations shows that three redox reactions occur, so the situation is more complicated. You could construct an equation showing the formation of sulfur dioxide in the same way as for sodium bromide.

Here is the result of applying the same method to the formation of sulfur in half-equation 2. The two relevant half-equations are:

$2I^- \rightarrow I_2 + 2e^-$

and

$H_2SO_4 + 6H^+ + 6e^- \rightarrow 4H_2O + S$

Before you add these together, you need to multiply the first one by three, so that the $6e^-$ on each side will cancel, giving:

$6I^- + H_2SO_4 + 6H^+ \rightarrow 4H_2O + S + 3I_2$

You could combine the ions on the left to give:

$6HI + H_2SO_4 \rightarrow 4H_2O + S + 3I_2$

This equation represents the oxidation of the misty fumes of hydrogen iodide.

You should now be able to use the same method to construct an equation to represent the oxidation of the misty fumes of hydrogen iodide to form hydrogen sulfide.

LEARNING TIP

Look carefully at all of the equations in this topic. Make sure you can identify whether they are redox reactions and, if so, what the extent of the reaction is.

CHECKPOINT

1. Explain what kind of reaction occurs when concentrated sulfuric acid is added to sodium fluoride.
2. Use half-equations to construct an overall equation for the reaction between iodide ions and concentrated sulfuric acid that results in the formation of hydrogen sulfide.

8C 4 OTHER REACTIONS OF HALIDES

SPECIFICATION REFERENCE 8.27(ii, iii)

LEARNING OBJECTIVES

- Understand precipitation reactions of the aqueous anions Cl^-, Br^- and I^- with aqueous silver nitrate solution and nitric acid, and the solubility of the precipitates in aqueous ammonia solutions.
- Understand reactions of hydrogen halides with ammonia and with water.

EXAM HINT

Make sure that you can write an ionic equation for the reaction of acid with carbonate ions to produce carbon dioxide and water.

TESTING FOR HALIDE IONS IN SOLUTION

These tests depend on the very low solubility of silver halides in water and their different solubility in aqueous ammonia.

The reagent is silver nitrate solution, but dilute nitric acid is added first to make sure that any other anions (especially carbonate ions) are removed, as they would also form precipitates.

If a precipitate is obtained, it is then usual to add some ammonia solution. This solution can be dilute or concentrated.

Table A and **fig A** show the results obtained. This test cannot be used to detect fluoride ions in aqueous solution, because silver fluoride is soluble.

	CHLORIDE IONS	BROMIDE IONS	IODIDE IONS
add silver nitrate solution	white precipitate	cream precipitate	yellow precipitate
add dilute aqueous ammonia	soluble	insoluble	insoluble
add concentrated aqueous ammonia	soluble	soluble	insoluble

table A Results obtained from precipitation reactions of the aqueous anions Cl^-, Br^- and I^- with silver nitrate solution, followed by aqueous ammonia.

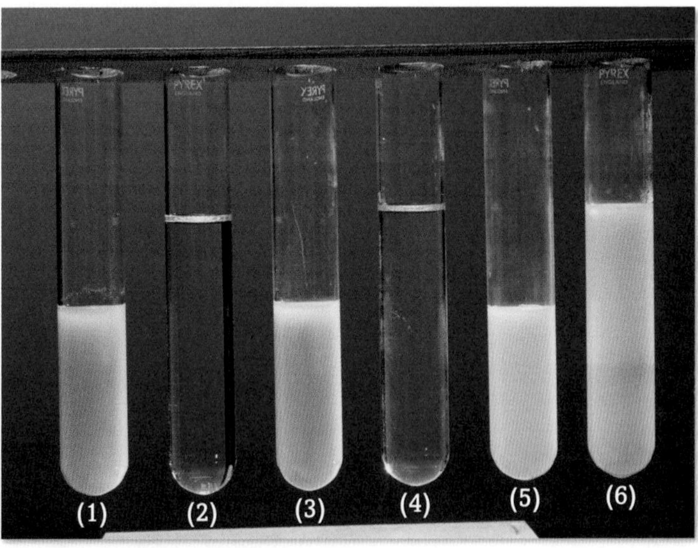

▲ **fig A** Testing for halide ions in solution.

From left to right in **fig A**:

- tube 1 shows the white precipitate formed from a chloride
- tube 2 shows the result of adding dilute aqueous ammonia to the white precipitate
- tube 3 shows the cream precipitate formed from a bromide
- tube 4 shows the result of adding concentrated aqueous ammonia to the cream precipitate
- tube 5 shows the yellow precipitate formed from an iodide
- tube 6 shows the result of adding concentrated aqueous ammonia to the yellow precipitate – it has not dissolved.

The general ionic equation for the formation of the precipitates is:

$$Ag^+(aq) + X^-(aq) \rightarrow AgX(s)$$

You can write a specific equation, such as:

$$AgNO_3(aq) + NaCl(aq) \rightarrow AgCl(s) + NaNO_3(aq)$$

The results in **table A** suggest that a halide ion could be identified without using aqueous ammonia. However, the colours of the three precipitates are similar. Even when all three are seen together, it is not easy to be sure which is which. In a single test, where only one precipitate is seen, this would be even more difficult.

Aqueous ammonia is a useful solvent because the precipitates have different solubilities in it.

- Silver chloride dissolves readily in both dilute and concentrated aqueous ammonia.
- Silver bromide dissolves readily in concentrated aqueous ammonia, but not in dilute aqueous ammonia.
- Silver iodide dissolves in neither dilute nor concentrated aqueous ammonia.

Dissolving of the precipitates occurs because of the formation of a complex ion. In the case of silver chloride, the equation for the reaction is:

$$AgCl(s) + 2NH_3(aq) \rightarrow [Ag(NH_3)_2]^+(aq) + Cl^-(aq)$$

HYDROGEN HALIDES ACTING AS ACIDS

All of the hydrogen halides are colourless gases and exist as polar diatomic molecules.

REACTIONS WITH WATER

The hydrogen halides readily react with water to form acidic solutions, all of which are colourless.

Table B shows these reactions.

HYDROGEN HALIDE	ACID FORMED	EQUATION
Hydrogen fluoride	hydrofluoric acid	$HF + H_2O \rightleftharpoons H_3O^+ + F^-$
Hydrogen chloride	hydrochloric acid	$HCl + H_2O \rightarrow H_3O^+ + Cl^-$
Hydrogen bromide	hydrobromic acid	$HBr + H_2O \rightarrow H_3O^+ + Br^-$
Hydrogen iodide	hydroiodic acid	$HI + H_2O \rightarrow H_3O^+ + I^-$

table B Reactions of hydrogen halides with water to form acidic solutions.

LEARNING TIP

Note the reversible arrow in the first equation. Although hydrogen fluoride is very soluble in water, it is only a weak acid, whereas the other acids formed are strong.

REACTIONS WITH AMMONIA

Hydrogen halides all react with ammonia gas to form salts, all of which are white ionic solids.

For example, when ammonia and hydrogen chloride gases are mixed together, they react to form ammonium chloride:

$$NH_3(g) + HCl(g) \rightarrow NH_4Cl(s)$$

You may be familiar with the use of the reaction in **fig B** to illustrate diffusion and the different rates of diffusion of ammonia and hydrogen chloride.

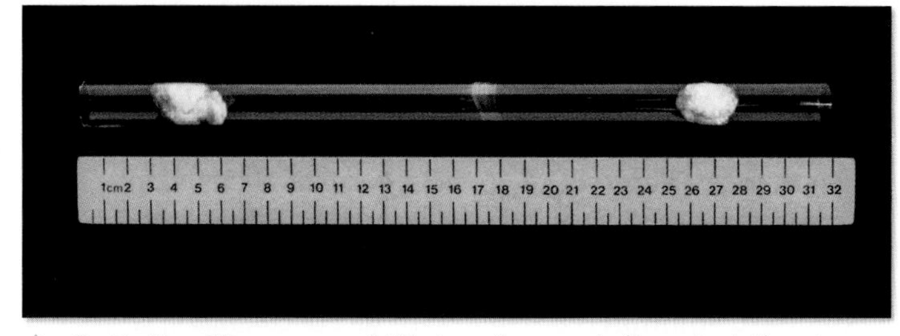

fig B The different rates of diffusion of ammonia (from the left) and hydrogen chloride (from the right).

Ammonia and hydrogen chloride gases are given off from the cotton wool pieces soaked in concentrated aqueous ammonia (on the left) and concentrated hydrochloric acid (on the right). These colourless gases move through the tube until they meet and react to form ammonium chloride. Ammonia gas molecules (M_r = 17.0) move more quickly and therefore travel further in the same time as hydrogen chloride molecules (M_r = 36.5). This means that the ammonium chloride does not form exactly halfway down the tube, it forms closer to the hydrochloric acid side.

FLUORIDES AND ASTATIDES

Fluorides and astatides have been mentioned in some of the reactions in this chapter, especially when it is not possible to make predictions about their reactions using trends in the chlorides, bromides and iodides. For example, you might predict that silver nitrate would form a white precipitate with sodium fluoride because sodium chloride does, but this is not the case.

If you are asked to predict reactions of fluorides and astatides that you are not familiar with, you can use information in this topic to help you. For example, you could write an equation to show the formation of an acid when hydrogen astatide is added to water, based on the equation for hydrogen iodide.

DID YOU KNOW?

None of the reactions in this topic involve redox. This is because the oxidation number of every halogen in a halogen compound in this topic remains unchanged at −1.

CHECKPOINT

1. Write equations to show:
   (a) the formation of the cream precipitate in the test for bromide ions
   (b) the precipitate dissolving in concentrated aqueous ammonia.

2. (a) Describe what happens when ammonia and hydrogen bromide gases are mixed together.
   (b) Name the product of the reaction and write an equation for its formation.

8D 1 MAKING STANDARD SOLUTIONS

SPECIFICATION REFERENCE

8.20 PART | CP4

LEARNING OBJECTIVES

- Be able to calculate solution concentrations, including simple acid-base titrations.

WHAT ARE STANDARD SOLUTIONS AND PRIMARY STANDARDS?

The next topic, **Topic 8D.2**, is about doing titrations. One substance needed in a titration is a standard solution, which we will look at in this topic.

A **standard solution** is a solution whose concentration is accurately known. One obvious way to prepare a standard solution is to take a known mass of a substance and dissolve it in water to make a known volume of solution.

These substances are known as **primary standards**. Ideally, primary standards should:

- be solids with high molar masses
- be available in a high degree of purity
- be chemically stable (neither decompose nor react with substances in the air)
- not absorb water from the atmosphere
- be soluble in water
- react rapidly and completely with other substances when used in titrations.

Unfortunately, several substances that are often used in titrations are not suitable as primary standards. For example, hydrochloric acid does not exist as a solid, but only as HCl(g) and HCl(aq). Sodium hydroxide is a solid, but it readily absorbs water vapour and reacts with carbon dioxide from the atmosphere. A sample of sodium hydroxide can be accurately weighed, but it is not certain what is being weighed, as there will be unknown masses of water and sodium carbonate mixed with it. You may remember seeing a white powder around the neck of a bottle of sodium hydroxide solution. This white powder is sodium carbonate.

MAKING A STANDARD SOLUTION OF SULFAMIC ACID

Sulfamic acid is probably unfamiliar to you. It is a readily available primary standard for use in acid–base titrations that has the necessary characteristics. Its formula can be shown in more than one way, including NH_2SO_3H, and it has a molar mass of 97.1 g mol^{-1}.

We will look in detail at a method used to obtain an accurately known value for the concentration of this solution. The method used is known as 'weighing by difference'. In your practical work, you may use a different method.

CALCULATING ROUGHLY HOW MUCH TO WEIGH

You need some idea of the approximate concentration and volume of the solution to be made. Typical values are 0.1 mol dm^{-3} and 250 cm^3.

Using the calculation method given in **Topic 1E.2**, the approximate amount of sulfamic acid, $n = c \times V = 0.1 \times 0.25 = 0.025$ mol, so the approximate mass needed is $m = n \times M = 0.025 \times 97.1 = 2$ g.

Note that even though we are going to do a very accurate weighing, we only need to know an approximate mass at this stage.

APPARATUS

The apparatus you need is:

- safety glasses and a lab coat
- an accurate balance (we will assume one reading to 3 decimal places)
- a weighing bottle (or weighing boat)
- a spatula
- a 250 cm^3 beaker
- a 250 cm^3 volumetric flask
- a wash bottle containing deionised water (or distilled water)
- a small funnel
- a glass stirring rod.

METHOD

1. Add between 2.3 and 2.5 g of sulfamic acid to the weighing bottle and weigh accurately.
2. Transfer as much as possible of the acid to a clean beaker and reweigh the weighing bottle.
3. Add about 100 cm^3 of deionised water to the beaker and stir until all of the sulfamic acid has dissolved.
4. Remove the stirring rod, washing traces of solution from the rod into the beaker using the wash bottle.
5. Place a funnel in the neck of the volumetric flask and pour the solution from the beaker into the flask.
6. Rinse the inside of the beaker several times using the wash bottle and transfer the rinsings to the flask.
7. Add deionised water to the flask and fill up exactly to the graduation mark.
8. Stopper the flask and invert it several times to make a uniform solution.

You can now calculate an accurate value for the concentration of the solution, using these example values.

mass of weighing bottle + sulfamic acid = 19.542 g

mass of weighing bottle + any traces of sulfamic acid = 17.151 g

mass of sulfamic acid added = 2.391 g

$$n(NH_2SO_3H) = \frac{2.391}{97.1} = 0.02462 \text{ mol}$$

$$c = \frac{0.02462}{0.250} = 0.0985 \text{ mol dm}^{-3}$$

LEARNING TIP

Look carefully at each step of the method. How does each step contribute to the accuracy of the final value of the concentration?

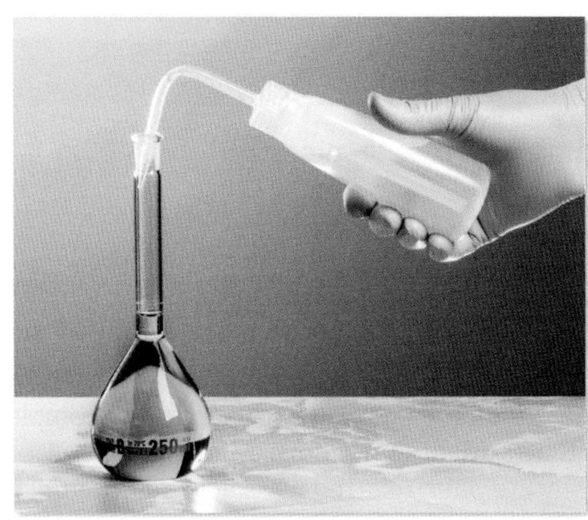

When making a standard solution in a volumetric flask, it is important to ensure that the lowest part of the liquid (the meniscus) is aligned with the graduation mark.

CHECKPOINT

SKILLS CRITICAL THINKING

1. A student used the method described but made two mistakes.
   (a) In Step 6, he poured the rinsings down the sink instead of transferring them to the flask.
   (b) In Step 7, he added water above the graduation mark.
   Explain how each mistake affected the calculated concentration.

SKILLS PROBLEM SOLVING

2. A student used the method to make 500 cm^3 of a solution of sodium carbonate ($M = 106.0\,g\,mol^{-1}$). These are her weighings:
   mass of weighing bottle + sodium carbonate = 23.382 g
   mass of weighing bottle = 18.218 g
   Calculate the concentration of the solution made.

SUBJECT VOCABULARY

standard solution a solution whose concentration is accurately known
primary standards substances used to make a standard solution by weighing

8D 2 DOING TITRATIONS

LEARNING OBJECTIVES

- Be able to calculate solution concentrations, including simple acid-base titrations.

WHAT IS A TITRATION?

A titration is a practical method with the aim of measuring the volumes of two solutions that react together, and using the results to calculate the concentration of one of the solutions. Because the method involves measuring volumes, it is sometimes known as volumetric analysis.

You may come across different titration types (redox titrations and complexometric titrations), but in this book we will consider only acid–base titrations. Many, but not all, bases are soluble in water, which makes them alkalis. They are often referred to as acid–alkali titrations.

All of the acids and bases you will use in titrations are colourless, as are the products, so there is nothing to see when the reaction is complete. This problem is solved by using an indicator, a substance which has different colours in acids and bases/alkalis.

In the method described in this topic, we will assume that a solution of sodium hydroxide of approximate concentration 0.1 $mol\,dm^{-3}$ is available and that the standard solution of sulfamic acid referred to in **Topic 8D.1** will be used. The aim is to calculate an accurate value for the concentration of the sodium hydroxide solution. The method will be described in some detail, and we will focus on avoiding common errors in technique.

In some titrations, it is important to choose which way round to do the titration (which solution goes in the pipette and which solution in the burette). In other titrations, it does not matter.

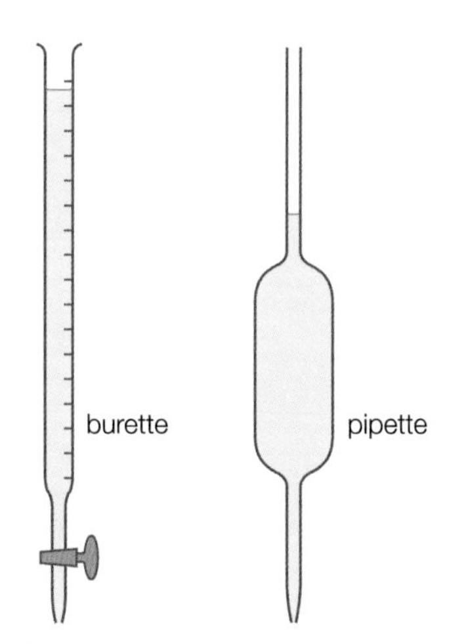

fig A Note that the space between the tap and the tip of the burette is full of solution.

OUTLINE OF THE TITRATION METHOD

Here is an outline of the titration method that introduces some key terms.

- Add the acid to the alkali until the **equivalence point** of the titration and the **end point** of the indicator is reached.
- Record the burette readings using the lowest part of the **meniscus**.
- Calculate the **titre**.
- Repeat the titration until **concordant titres** are obtained.

APPARATUS

The apparatus likely to be used is:

- a conical flask (usually 250 cm^3)
- a burette (usually 50 cm^3) and stand
- a pipette (usually 25 cm^3) and pipette filler
- a wash bottle containing deionised water (or distilled water)
- a small funnel
- a white tile.

METHOD

1 Rinse the conical flask with deionised water and place it on a white tile.
2 Using a pipette filler, rinse the pipette with deionised water and then with some of the sodium hydroxide solution.
3 Use the pipette to transfer 25.0 cm^3 of the sodium hydroxide solution to the conical flask.
4 Add about 3 drops of methyl orange indicator.
5 Rinse the burette with deionised water and then with some of the sulfamic acid solution.
6 Fill the burette with the sulfamic acid solution and set it up in the stand above the conical flask.
7 Record the burette reading.
8 Add the sulfamic acid solution to the conical flask until the indicator just changes colour, and again record the burette reading.
9 Empty and rinse the conical flask with deionised water, and repeat the titration until concordant titres have been obtained.

TITRATION TECHNIQUES

To obtain accurate results in a titration, it is important to work carefully. The diagram shows some important techniques and the reasons for them.

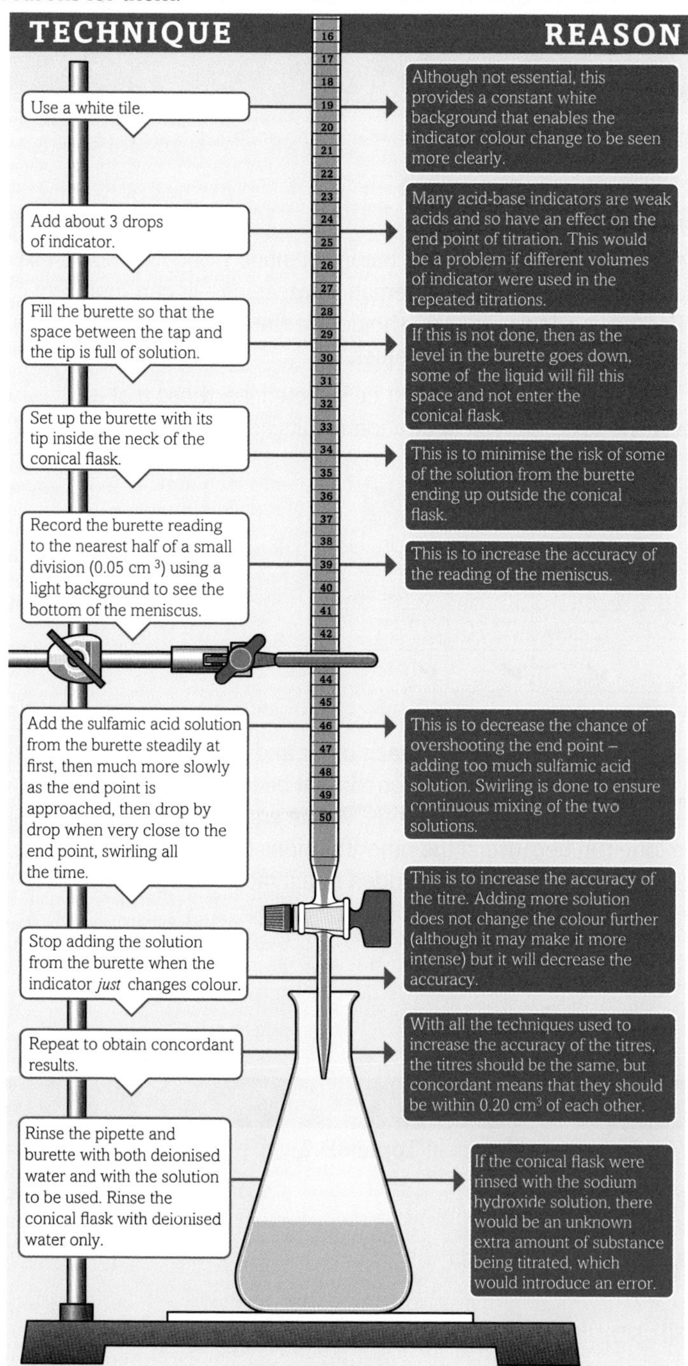

fig B Titration techniques and the reasons for them.

CHOOSING AN INDICATOR

Two common indicators are:

- methyl orange
- phenolphthalein.

Sometimes it is important to use one of these and not the other, but in other titrations it does not matter which one you use.

Table A shows the colours of these indicators and which combination of acid and base they should be used with.

INDICATOR	COLOUR IN ACID	COLOUR IN ALKALI	ACID–BASE COMBINATION
methyl orange	red	yellow	strong acid – weak base and strong acid – strong base
phenolphthalein	colourless	pink	weak acid – strong base and strong acid – strong base

table A Information about methyl orange and phenolphthalein.

Examples of strong acids are:

- hydrochloric acid
- nitric acid.

Examples of strong bases are:

- sodium hydroxide
- potassium hydroxide.

The commonest weak base is ammonia.

The commonest weak acid is ethanoic acid.

fig C This shows the burette after the titration has been done. Note the white tile and the colour of the phenolphthalein indicator. The pink colour shows that an acid has been neutralised and there is an excess of an alkali.

LEARNING TIP

The best way to understand all the features of successful titrations is to do several titrations yourself.

CHECKPOINT

SKILLS CRITICAL THINKING

1. A student does not fill the burette space between the tap and the tip in a titration. Explain the effect of this mistake on the value of the titre.

2. A student rinses out the conical flask with deionised water, then with the solution used in the pipette. Explain the effect of this mistake on the value of the titre.

SUBJECT VOCABULARY

equivalence point the point at which there are exactly the right amounts of substances to complete the reaction

end point the point at which the indicator just changes colour; ideally, the end point should coincide with the equivalence point

meniscus the curving of the upper surface of a liquid in a container; the lowest (horizontal) part of the meniscus should be read

titre the volume added from the burette during a titration

concordant titres titres that are close together (usually within $0.20\,cm^3$ of each other)

8D 3 CALCULATIONS FROM TITRATIONS

SPECIFICATION REFERENCE

8.20 PART | CP3 | CP4

LEARNING OBJECTIVES

- Be able to calculate solution concentrations, including simple acid-base titrations.

CALCULATING THE MEAN (AVERAGE) TITRE

The terms 'average' and 'mean' are often used interchangeably, but in scientific work you should take care to use the proper word. 'Average' has more than one mathematical meaning – it can represent the mean, the median or the mode. For these calculations you should use the mean: a set of values added to give a total, which is then divided by the number of values.

Before calculating a mean titre, only the concordant values must be selected, i.e. those that are within 0.20 cm^3 of each other. **Table A** shows some typical titration results and a student's choice of concordant titres.

TITRATION NUMBER	1	2	3	4
final burette reading / cm^3	24.15	25.30	24.60	23.25
initial burette reading / cm^3	1.20	2.70	1.90	0.60
titre / cm^3	22.95	22.60	22.70	22.65
concordant titres	✘	✔	✔	✔

table A A set of titration results and a student's choice of concordant titres.

In this example, titrations 2, 3 and 4 are all within 0.20 cm^3 of each other and so have been correctly ticked as concordant. It is not surprising that titration 1 is not concordant because it is normal practice to do the first titration more quickly to obtain a rough titre, so the end point is more likely to be overshot. This saves time in the long run because in the other titrations the liquid in the burette can be added quickly at first, until the end point is close, then added much more slowly.

This is how to calculate the mean:

$$\text{Mean} = \frac{22.60 + 22.70 + 22.65}{3} = 22.65\ cm^3$$

CALCULATING A CONCENTRATION

WORKED EXAMPLE 1

Here is a typical set of titration results from the method given in **Topic 8D.2**.

volume of sodium hydroxide solution used = 25.0 cm^3

volume of sulfamic acid solution used = 22.65 cm^3

concentration of sulfamic acid solution = 0.0985 $mol\ dm^{-3}$

The equation for the reaction is:

$$NaOH(aq) + NH_2SO_3H(aq) \rightarrow NH_2SO_3Na(aq) + H_2O(l)$$

You may be shown a shorter method than the following one, but this method, using moles, is more versatile and can be used in more difficult examples.

Step 1: calculate the amount of one substance, in this case the sulfamic acid.

$$n(NH_2SO_3H) = c \times V = 0.0985 \times \frac{22.60}{1000} = 0.00223\ mol$$

Step 2: calculate the amount of the other substance using the reacting ratio in the equation.

As the reacting ratio is 1 : 1, $n(NaOH) = 0.00223$ mol

Step 3: calculate the concentration of sodium hydroxide solution.

$$c = \frac{n}{V} = \frac{0.00223}{0.0250} = 0.0892\ mol\ dm^{-3}$$

WORKED EXAMPLE 2

A titration is done to calculate the concentration of a solution of nitric acid, using a standard solution of sodium carbonate. The equation for the reaction is:

$Na_2CO_3 + 2HNO_3 \rightarrow 2NaNO_3 + H_2O + CO_2$

The titration results are:

volume of sodium carbonate solution used = $25.0\,cm^3$

volume of nitric acid solution used = $27.25\,cm^3$

concentration of sodium carbonate solution = $0.108\,mol\,dm^{-3}$

Step 1: calculate the amount of sodium carbonate.

$$n = 0.108 \times \frac{25}{1000} = 0.00270\,mol$$

Step 2: calculate the amount of the other substance using the reacting ratio in the equation (1 : 2 in this example).

$$n = 0.00270 \times 2 = 0.00540\,mol$$

Step 3: calculate the concentration of the nitric acid.

$$c = \frac{0.00540}{0.02725} = 0.198\,mol\,dm^{-3}$$

WORKED EXAMPLE 3

A titration is done to calculate the concentration of a solution of hydrochloric acid, using the sodium hydroxide solution from Worked example 1. The equation for the reaction is:

$NaOH + HCl \rightarrow NaCl + H_2O$

The titration results are:

volume of sodium hydroxide solution used = $25.0\,cm^3$

volume of hydrochloric acid used = $22.68\,cm^3$

concentration of sodium hydroxide solution = $0.0892\,mol\,dm^{-3}$

This time, we will use the shorter method of calculation, using the expression

$$V_1 \times \frac{M_1}{n_1} = V_2 \times \frac{M_2}{n_2}$$

where V is the volume, M is the molar concentration and n is the coefficient in the equation. The subscript 1 refers to the first substance in the equation, and the subscript 2 to the second substance.

In this example, the unknown is M_2, so rearranging the expression and substituting the values gives the answer.

$$M_2 = \frac{V_1 \times M_1 \times n_2}{V_2 \times n_1} = \frac{25.0 \times 0.0892 \times 1}{22.68 \times 1} = 0.0983\,mol\,dm^{-3}$$

LEARNING TIP

Practise the calculation method. Make sure that you use the reacting ratio the right way round.

CHECKPOINT

SKILLS CRITICAL THINKING

1. A student records these readings during a titration:

final burette reading / cm^3	28.0	27.8	27.4	27.0
initial burette reading / cm^3	2.7	2.8	2.8	2.8
titre / cm^3	25.3	25.0	24.6	24.2

mean titre = $24.4\,cm^3$

What mistakes has the student made in recording these results?

SKILLS PROBLEM SOLVING

2. A student does a titration using this reaction:

$2KOH + H_2SO_4 \rightarrow K_2SO_4 + 2H_2O$

She records these results:

volume of KOH solution = $25.0\,cm^3$

volume of H_2SO_4 = $19.83\,cm^3$

concentration of H_2SO_4 = $0.0618\,mol\,dm^{-3}$

What is the concentration of the KOH solution, in $mol\,dm^{-3}$?

8D 4 MISTAKES, ERRORS, ACCURACY AND PRECISION

SPECIFICATION REFERENCE **8.22**

LEARNING OBJECTIVES

- Understand the difference between a mistake and an error.
- Understand the difference between accuracy and precision.
- Comment on sources of error in experimental procedures.

USING THE CORRECT TERMINOLOGY

In science, it is sometimes difficult to find the correct words to use when considering the results of experiments and the calculations based on these results. This is because in the non-scientific world, words are often used with less care. You have already seen the idea that 'amount' has a specific meaning in chemistry (amount of substance, in moles) and should not be used to refer to mass or volume.

We will look at terminology in this topic.
- Mistakes and errors are not the same thing.
- Accuracy and precision have different meanings.
- Systematic errors and random errors have different causes.

Here, we look at the differences in meaning between some of the more important terms.

To do this, we will refer back to previous sections in **Topic 8D**, which involve measuring masses and volumes using different methods and apparatus.

MISTAKES ARE NOT ERRORS

Put very simply, an error is something that even a skilled operator would find difficult to avoid, and is a consequence of the way the apparatus has been constructed and how readings can be made using it.

A mistake is something that a skilled operator can avoid by being careful.

Here are some examples of mistakes.

1. A chemist weighs a beaker on a balance without making sure the balance is tared (set to zero) beforehand. The reading on the balance could be very different from the actual mass of the beaker, so the reading should not be used, although this careless chemist may not realise this.
2. A student sees a burette reading of 27.35 cm^3 but writes it down as 23.75 cm^3. This is the student's mistake, and has nothing to do with the apparatus.
3. A student fills a burette using a funnel and forgets to remove the funnel before adding the liquid to the flask. During the addition, some of the liquid in the funnel drips into the burette, and this causes an incorrect burette reading to be recorded. This is due to the student's faulty technique. Again, it has nothing to do with the actual apparatus, only his careless use of it.

ACCURACY AND PRECISION, AND SYSTEMATIC AND RANDOM ERRORS

The most important terms to consider are:
- **error**
- **accuracy**
- **precision**.

The last two terms are often confused. We will try to understand the difference between them by considering some burette readings and an archery competition.

A teacher and some students do a titration using the same solutions. The teacher works carefully and obtains a mean titre of 24.27 cm^3, which we will assume is the 'correct' value.

Consider the titres in **table A**, which were recorded by students doing the same titration as the teacher.

STUDENT	TITRES / CM³				MEAN OF ALL TITRES / CM³
	1	2	3	4	
A	24.80	24.85	24.90	24.80	24.84
B	24.95	24.80	23.25	23.80	24.20
C	24.20	24.30	24.25	24.20	24.14

table A Titres recorded by Students A, B and C doing the same titration as their teacher.

All the mean titres have been calculated correctly.

Now for the archery competition. The aim of an archery competition is to win by managing to land all the arrows as close as possible to the centre of the target. These diagrams show how well the three students have done at archery using their titre values.

student A

student B

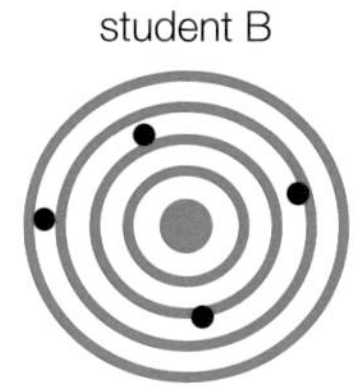

student C

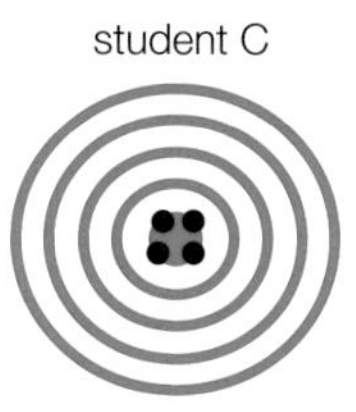

Table B provides commentary on these values.

STUDENT	COMMENTS
A	Student A has titres that are all concordant (within 0.10 cm³ of each other), but the mean is 0.57 cm³ higher than the correct value. This suggests that the titrations have been carefully carried out, but that there is probably something about the apparatus that is responsible for the large difference from the correct titre value. This is called a systematic error. The titre values are precise but not accurate.
B	Student B has no concordant titres (they are very different from each other) but the mean is within 0.07 cm³ of the correct value. This suggests that the titrations have been carelessly done, but the student has been lucky because the mean happens to be close to the correct value. This is called a random error. Even though each individual titre is not accurate, and all four of them are not precise, the mean titre is accurate.
C	Student C has titres that are all concordant, and the mean is within 0.03 cm³ of the correct value. This suggests that the titrations have been carefully done, and that the apparatus used is of the same standard as that used by the teacher to obtain the correct value. The titre values are both accurate and precise.

table B Comments explaining how Students A, B and C have done with their titre values.

LEARNING TIP

Comparing the archery competition and the titrations should help you to understand the difference between accuracy and precision.

CHECKPOINT

SKILLS CRITICAL THINKING

1. An experimental method requires a 25.0 cm³ pipette to be used to measure a volume of liquid in different experiments. A student uses a 25 cm³ measuring cylinder instead in each case. Explain whether the student has made a mistake or has introduced a random error or a systematic error.

2. Consider these titres recorded in cm³.

Student 1	27.60	27.70	27.70	27.65
Student 2	26.15	26.82	26.60	26.30
Student 3	24.40	26.50	26.50	26.40

The teacher's mean titre was 26.50 cm³, which can be assumed to be correct.

Explain whether the students' titres indicate accuracy, precision, both or neither.

SUBJECT VOCABULARY

error the difference between an experimental value and the accepted or correct value

accuracy a measure of how close values are to the accepted or correct value

precision a measure of how close values are to each other

8D 5 MEASUREMENT ERRORS AND MEASUREMENT UNCERTAINTIES

SPECIFICATION REFERENCE 8.22

LEARNING OBJECTIVES

- Understand how to minimise the sources of measurement uncertainty in volumetric analysis.

RANDOM AND SYSTEMATIC ERRORS

Consider the accepted meanings of two terms briefly mentioned in **Topic 8D.4** (**table A**).

TERM	MEANING
random error	This is an error caused by unpredictable changes in conditions such as temperature or pressure, or by a difference in recording that is difficult to get exactly right.
systematic error	This is an error caused by the apparatus, and leads to the recorded value being either too low or too high.

table A

RANDOM ERRORS

Here are examples of random errors.

- The volume of a gas collected in a syringe is measured in different experiments done on the same day. The atmospheric pressure and the temperature of the laboratory may vary during the day, and this will cause an unpredictable change in the value recorded.
- The mass of an object is measured using the same balance but at different times during the day. The values may differ slightly, as a result of changes in temperature, draughts or condensation of water vapour on the balance pan.
- An experiment is done with a flask in an electric water bath. The thermostat is set to 50 °C, but as the heater automatically switches on, the actual temperature rises slightly above 50 °C, and then falls to slightly less than 50 °C after the thermostat switches off.

Repeating an experiment should lead to a more accurate final value being recorded because these random fluctuations become less important when values are averaged.

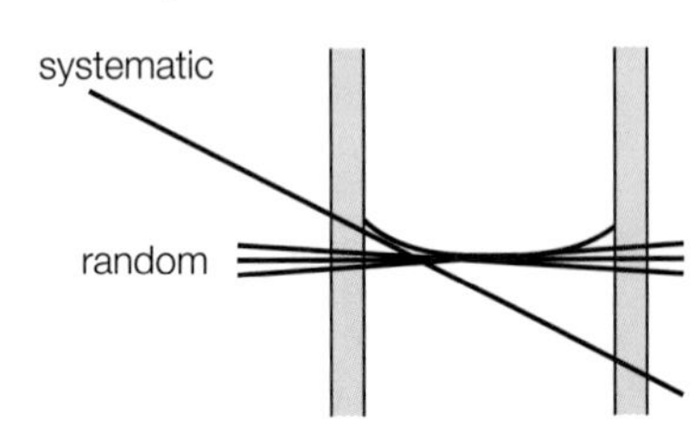

▲ **fig A** When recording a liquid level, the operator should view the meniscus at eye level (horizontally). This diagram shows that minor variations from the horizontal can be considered as random errors. If the meniscus is consistently viewed from above, then the error can be described as systematic.

SYSTEMATIC ERRORS

Here are examples of systematic errors.

- A 25.0 cm^3 pipette has been wrongly calibrated during its manufacture, so that the graduation mark is lower than it should be. This means that no matter how carefully it is used, the volume of solution added from this pipette will always be less than 25.0 cm^3.
- The amount of liquid in a thermometer is more than it should be, so that the height of the liquid at all temperatures is higher than it should be. This means that the temperature recorded will always be greater than the correct temperature.
- A measuring cylinder has markings on it from 0 to 10 cm^3, but the diameter of the cylinder is smaller than the manufacturer intended. This means that when the liquid level shows 10 cm^3, the volume is less than 10 cm^3.

Repeating an experiment using the same apparatus will not lead to a more accurate value being recorded.

LEARNING TIP

Think about how using a burette could involve both random and systematic errors.

MEASUREMENT UNCERTAINTY

When using apparatus, there is always a potential error, known as **measurement uncertainty**. The size of the measurement uncertainty is determined by the precision of the apparatus.

BALANCES

Digital balances can have various degrees of precision. A balance that measures to three decimal places is more precise that one that measures to only one decimal place. This means that the measurement uncertainty involved in using the three-decimal place balance is lower than in using the one-decimal place balance. This is illustrated in **table B**. The greater the degree of precision of the balance, the smaller the measurement uncertainty in the recorded mass.

NUMBER OF DECIMAL PLACES	MEASUREMENT UNCERTAINTY	EXAMPLE
1	±0.05 g	A reading of 17.1 g could be between 17.05 and 17.15 g
2	±0.005 g	A reading of 17.10 g could be between 17.095 g and 17.105 g
3	±0.0005 g	A reading of 17.100 g could be between 17.0995 g and 17.1005 g

table B Examples of measurement uncertainties when using a balance.

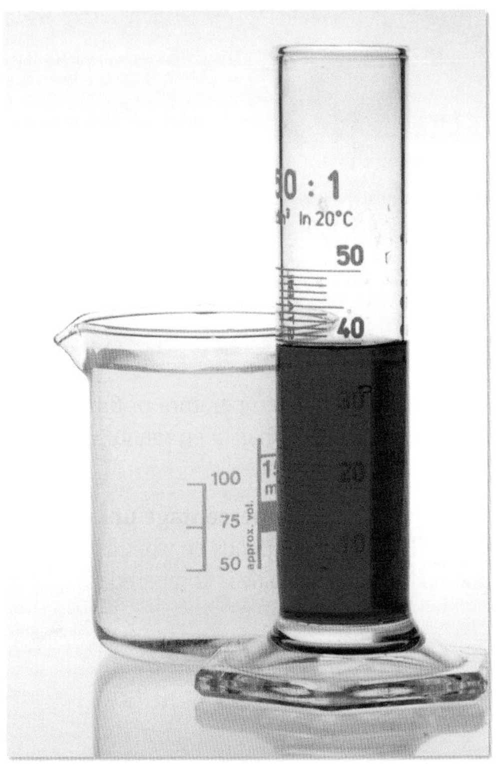

▲ **fig B** The scale on a beaker is there only as a guide and should not be used to record values to be used in calculations. A measuring cylinder is better for this purpose.

GLASSWARE

Most laboratory glassware is manufactured to Class A and Class B (or Grade A and Grade B) standards. Class A apparatus is much more expensive than Class B apparatus, and the extra cost reflects the way in which the apparatus is calibrated, rather than the quality of the actual apparatus.

You will almost certainly use Class B apparatus, which is still capable of giving measurements to a high degree of accuracy. If you look carefully, you should be able to see the manufacturer's statement about level of precision marked on the apparatus. You should see a temperature quoted (usually 20 °C). At higher temperatures, the glass and the solution it contains will expand to different extents, and the value of the volume will no longer be accurate within the given tolerances.

Table C shows information about typical pieces of Class B glassware that are used to measure volumes.

APPARATUS	CAPACITY	MEASUREMENT UNCERTAINTY
burette	50 cm^3	±0.05 cm^3 (but the burette is read twice in a titration, so the total measurement uncertainty is ±0.10 cm^3)
pipette	25 cm^3	±0.06 cm^3
volumetric flask	250 cm^3	±0.3 cm^3

table C Examples of measurement uncertainties in glassware.

LEARNING TIP

Remember that, assuming you have used the piece of apparatus correctly, you still cannot say for certain that any measurement made with the apparatus is accurate. It is only possible to know if there has been an error if you happen to know the true value of the measurement made. The best you can say is that there is a potential error involved, which is given by the *measurement uncertainty* of using the apparatus.

CHECKPOINT

SKILLS CRITICAL THINKING

1. A student carries out a titration and obtains a titre of 23.40 cm^3. She repeats the titration the next day using different apparatus and obtains a titre of 24.50 cm^3.
   (a) What is the total measurement uncertainty of the first titre?
   (b) Identify a potential source of random error that may have affected her results.
   (c) Identify a potential source of systematic error that may have affected her results.
2. A student uses a pipette and a burette to add 50.0 cm^3 of a liquid to a flask in different experiments. Use the table of measurement uncertainties (**table C** above) to calculate which piece of apparatus is the better one to use in this experiment.

SUBJECT VOCABULARY

random error an error caused by unpredictable variations in conditions
systematic error an error that is constant or predictable, usually because of the apparatus used
measurement uncertainty the potential error involved when using a piece of apparatus to make a measurement

8D 6 OVERALL MEASUREMENT UNCERTAINTY

SPECIFICATION REFERENCE
8.22
PART

LEARNING OBJECTIVES

- Understand how to estimate the overall uncertainty of the calculated result in volumetric analysis.

PERCENTAGE MEASUREMENT UNCERTAINTIES

Each piece of apparatus you use to record a value (such as mass, volume, temperature or time) has a measurement uncertainty associated with it, which depends on the way it has been manufactured and calibrated.

The actual measurement uncertainty may be fixed. However, in many cases the **percentage uncertainty** when you use the apparatus depends on the value you measure. This mostly depends on whether you use the apparatus to record only one value or two values, and on how big the value is compared to the capacity of the apparatus.

GLASSWARE

Table A shows typical percentage uncertainties for common items of glassware.

APPARATUS	CAPACITY	UNCERTAINTY	PERCENTAGE UNCERTAINTY
burette	50 cm^3	±0.05 cm^3	Note that two burette values are read in a titration, so the total measurement uncertainty is ±0.10 cm^3. If the titre is 22.50 cm^3, then the percentage uncertainty is: ±0.10 × 100/22.50 = ±0.44%
pipette	25 cm^3	±0.06 cm^3	The reading is taken only once, and for the same volume each time, so the percentage uncertainty is always: ±0.06 × 100/25 = ±0.24%
volumetric flask	250 cm^3	±0.3 cm^3	The reading is taken only once, and for the same volume each time, so the percentage uncertainty is always: ±0.3 × 100/250 = ±0.12%

table A Typical percentage uncertainties for a burette, pipette and volumetric flask.

BALANCES

The percentage uncertainty in using a balance depends on:

- the precision of the balance, i.e. the number of decimal places to which the balance can be read
- the mass being weighed, as the percentage uncertainty will be greater for a smaller mass.

WORKED EXAMPLE 1

- mass of a marble chip = 3.57 g

The measurement uncertainty in a two-decimal place balance is ±0.005 g. We have to count this uncertainty twice. This is because there is a ±0.005 g uncertainty when calibrating the balance to zero, as well as a ±0.005 g uncertainty when measuring the mass.

So the percentage uncertainty is:

$$\pm\frac{2 \times 0.005 \times 100}{3.57} = \pm 0.28\%$$

WORKED EXAMPLE 2

- mass of weighing bottle + solid = 20.354 g
- mass of weighing bottle = 19.816 g
- mass of solid = 0.538 g

The measurement uncertainty in a three-decimal place balance is ±0.0005 g. We have to count this uncertainty twice for each measurement and we have two measurements, so we need to count it four times. The percentage uncertainty is:

$$\pm\frac{4 \times 0.0005 \times 100}{0.538} = \pm 0.37\%$$

The percentage uncertainty is greater in Worked example 2, even though the balance reads to one more decimal place. This is because the balance is used twice and also because the mass being weighed is much smaller.

ADDING MEASUREMENT UNCERTAINTIES

If a final answer has been obtained using more than one piece of apparatus, then the approximate total measurement uncertainty is obtained by adding together the individual uncertainties.

For example, if a concentration of 0.118 mol dm^{-3} has been calculated from a titration that has involved balance and glassware uncertainties, then the uncertainties might be:

balance	±0.09%
volumetric flask	±0.12%
pipette	±0.24%
burette	±0.47%
overall percentage uncertainty	±0.92%

This means that there is an overall uncertainty in the concentration of 0.92% of 0.118, which is about ±0.001, so the final value can be quoted like this:

concentration = 0.118 ± 0.001 mol dm^{-3}

which means that the exact value is in the range 0.117–0.119 mol dm^{-3}.

MINIMISING ERROR AND UNCERTAINTY

How can errors and uncertainties be minimised in an experiment? This depends on a number of factors, some of which are easier to control than others.

For example, in a thermochemistry experiment carried out in the laboratory using standard equipment, there will always be transfer of heat energy to the surroundings. This heat transfer creates a random error in the measurement of the temperature change. In this type of experiment, using a balance that reads to three decimal places (instead of a balance that reads to two decimal places) will have no significant effect on the overall uncertainty of the final value. Minimising heat energy losses will have a much greater effect.

Where a measuring instrument can be used for a range of values, the percentage uncertainty can be minimised by using a higher value rather than a lower one. For example, when using a balance, weighing a sample of 5 g will lead to a much lower percentage uncertainty than weighing a 0.5 g sample. In a titration, it is not a good idea to have a titre value of 10 cm^3 instead of 30 cm^3, as the larger titre has the lower percentage uncertainty.

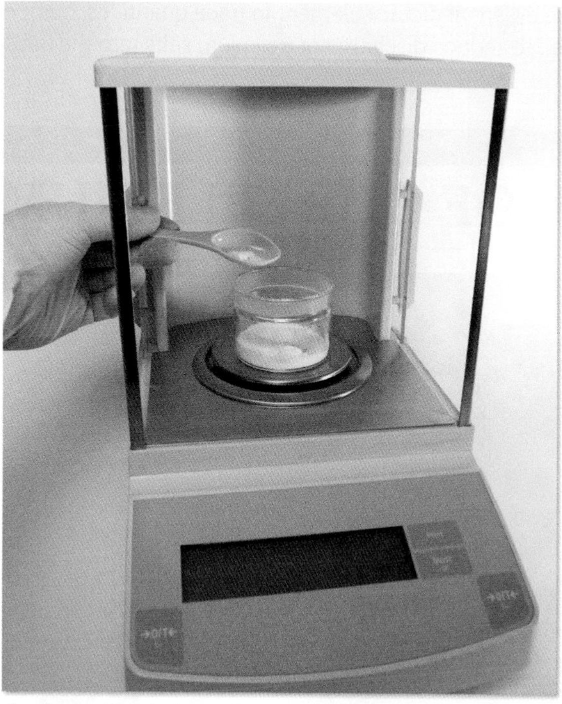

▲ **fig A** This balance is tared (set to zero) and ready to use. The pan is enclosed; this is necessary on a four-decimal place balance because of fluctuations in the reading caused by draughts of air.

LEARNING TIP

Practise calculating measurement uncertainties and percentage uncertainties for different pieces of apparatus.

CHECKPOINT

1. A student uses a one-decimal place balance to weigh a piece of zinc. He records a mass of 2.8 g. What is the percentage uncertainty?
2. A thermometer has a measurement uncertainty of 1 °C. A student uses it to measure a temperature rise and records these values:
   start temperature = 15 °C
   final temperature = 28 °C
   What is the percentage uncertainty in the temperature rise?

SUBJECT VOCABULARY

percentage uncertainty the actual measurement uncertainty in an experiment multiplied by 100 and divided by the value recorded

8 THINKING BIGGER

NUTS ABOUT SELENIUM

SKILLS INITIATIVE, SELF DIRECTION

Selenium is an essential dietary element in trace quantities, but it can be toxic in higher concentrations. The extract below describes one technique used to reduce discharge of selenium salts into the San Francisco Bay, USA.

SOME OF OUR SELENIUM IS MISSING

fig A Wetlands are an important habitat for birds and other wildlife.

In 1989, the multinational oil company Chevron discovered that it had reduced its discharge of toxic selenium salts into the San Francisco Bay by almost three-quarters. The company should have been very pleased with itself. After all, the six oil companies in the area flush up to 3000 kilograms of selenium into the bay each year, and it seemed that by simply planting a 35-hectare wetland between its outfall and the bay, Chevron had found a practical answer to the problem.

But officials from the State Regional Water Quality Control Board were not so sanguine. They wanted to know where all the selenium was going and told Chevron to find out. The company searched in the sediment at the bottom of the wetland, dug up the wetland plants, and checked for a build-up of selenium compounds in the water. But despite all the effort, half of the selenium was still missing.

Alarm bells started to ring. Perhaps the local wildlife was eating the missing selenium. But there were no telltale signs – no dead or maimed animals. Then a team of biologists from the University of California at Berkeley made an educated guess. Norman Terry, Adel Zayed and their colleagues had been studying the way that some plants take toxic selenium salts from soil and water and turn them into the volatile gas dimethyl selenide. They suggested that Chevron's selenium could literally be vanishing into thin air.

From an article in *New Scientist* magazine

SCIENCE COMMUNICATION

1. The article was written for *New Scientist* and was based on research presented in a scientific paper.
   (a) Read the article and comment on the type of writing being used. Think about whether this is a scientist reporting the results of their experiments, a scientific review of a paper or a newspaper or magazine article for a specific audience.
   (b) Is there any bias present in the report? What type of words would make you think that was the case?
   (c) Criticise how the use of language has been adapted for the audience. Would the wording be different if this was aimed at an audience of 14–16 year olds? How would it be different if written by the press officer of the company Chevron?

WRITING SCIENTIFICALLY

Note that when the word criticise is given in this context it does not mean that you should be negative. Instead, your comments should show reasoned judgement. Try to be constructive.

CHEMISTRY IN DETAIL

2. Calculate the number of moles of selenium present in 3000 kg of selenium.
3. Give the electronic configuration, using s, p, d notation, of the Se atom and the Se^{2-} ion.
4. Give the oxidation number of selenium in the following compounds:
   (a) Na_2SeO_3 (b) Ag_2Se (c) H_2SeO_4
5. Selenium is in Group 6 of the Periodic Table. Give some reasons why selenium chemistry is likely to be similar to sulfur chemistry.
6. Suggest a shape for the molecule SeF_6.
7. Give the formula for the molecule dimethyl selenide. Use your knowledge of Group 6 chemistry to suggest a shape for the molecule.
8. Dimethyl sulfide has a boiling temperature of 38 °C and dimethyl telluride has a boiling temperature of 82 °C. By drawing a suitable plot, suggest a boiling temperature for dimethyl selenide and explain your choice. What assumptions have you made?

THINKING BIGGER TIP

Consider the position of selenium relative to sulfur and tellurium in Group 6 when you are drawing your plot. Make sure you use suitable axes and scale in order to obtain a reasonable suggestion for the boiling point.

ACTIVITY

In our world today there is an increasing need to communicate scientific ideas and concepts to people who consider themselves to be 'non-scientists'. Researchers may not always be the best people to communicate science (even their own!), but there are many occupations where this is an essential skill.

Imagine you need to give a 5–10 minute presentation to the chief executive officer (CEO) of an oil company to argue the case for the use of genetically engineered plants to deal with selenium waste. How might you convince the CEO that the benefits are outweighed by any perceived concerns about genetic engineering?

INTERPRETATION NOTE

Refer back to the article's final paragraph. You might also like to do some research. If you do, make sure you carefully consider the nature of the source.

DID YOU KNOW?

Brazil nuts are a particularly rich source of selenium, but don't go overboard! Too much selenium in the diet can cause the body to break down and excrete excess selenium in the form of gaseous hydrogen selenide (H_2Se). This can leave the unwitting consumer of too many brazil nuts breathing out an odour not unlike garlic.

8 EXAM PRACTICE

1 The equation for a reaction is:

$Ca(OH)_2 + 2HNO_3 \rightarrow Ca(NO_3)_2 + 2H_2O$

Which description is correct for this reaction?

A oxidation only

B reduction only

C oxidation and reduction

D neither oxidation nor reduction [1]

(Total for Question 1 = 1 mark)

2 What is the oxidation number of chlorine in the compound $KClO_4$?

A −1

B +1

C +4

D +7 [1]

(Total for Question 2 = 1 mark)

3 Which equation represents a disproportionation reaction?

A $2LiOH + H_2SO_4 \rightarrow Li_2SO_4 + 2H_2O$

B $2H_2S + SO_2 \rightarrow 3S + 2H_2O$

C $2NO_2 + H_2O \rightarrow HNO_2 + HNO_3$

D $KF + H_2SO_4 \rightarrow KHSO_4 + HF^-$ [1]

(Total for Question 3 = 1 mark)

4 Which equation represents the process that occurs when the second ionisation energy of strontium is measured?

A $Sr(s) \rightarrow Sr^+(g) + e^-$

B $Sr^+(s) \rightarrow Sr^{2+}(g) + e^-$

C $Sr(g) \rightarrow Sr^+(g) + e^-$

D $Sr^+(g) \rightarrow Sr^{2+}(g) + e^-$ [1]

(Total for Question 3 = 1 mark)

5 A combination of reagents that can be used to test for sulfate ions in solution is $Ba(NO_3)_2$ and HNO_3.

Which other combination of reagents can also be used in this test?

A $BaO + NaOH$

B $BaCO_3 + NaOH$

C $BaSO_4 + HCl$

D $BaCl_2 + HCl$ [1]

(Total for Question 5 = 1 mark)

6 A mass of 830 mg of KOH is dissolved in water to make 150 cm^3 of solution.

What is the concentration of hydroxide ions in this solution?

A 0.0148 g dm^{-3}

B 0.0986 mol dm^{-3}

C 0.325 mol dm^{-3}

D 5.53 mg dm^{-3} [1]

(Total for Question 6 = 1 mark)

7 Most chlorine-based bleaches contain the ClO^- ion.

(a) (i) State the oxidation number of the chlorine in the ClO^- ion. [1]

(ii) Give the systematic name of the ClO^- ion. [1]

(b) The ClO^- ion is produced when chlorine gas is reacted with water.

Write an ionic equation for this reaction and explain why chlorine is said to undergo 'disproportionation'. [3]

(c) Chlorine forms another ion with the formula ClO_3^-. This ion exists in the compound $KClO_3$.

$KClO_3$ decomposes on heating according to the equation:

$4KClO_3(s) \rightarrow 3KClO_4(s) + KCl(s)$

Comment on the change in oxidation number of the chlorine in this reaction. [3]

(Total for Question 7 = 8 marks)

8 Bromine is present in sea water as aqueous bromide ions, $Br^-(aq)$. Bromine can be extracted from sea water by adding chlorine. The ionic equation for the reaction is:

$Cl_2(aq) + 2Br^-(aq) \rightarrow 2Cl^-(aq) + Br_2(aq)$

(a) (i) State what is meant by oxidation and reduction in terms of electron transfer. [1]

(ii) State which species is acting as an oxidising agent. Justify your answer. [2]

(b) The chlorine required for extraction of bromine is manufactured from a concentrated solution of sodium chloride using electrolysis.

The ionic half-equations for the reactions at the electrodes are:

At the anode: $2Cl^-(aq) \rightarrow Cl_2(aq) + 2e^-$

At the cathode: $2H_2O(l) + 2e^- \rightarrow 2OH^-(aq) + H_2(g)$

Use these two half-equations to produce an overall ionic equation for the reaction taking place during the electrolysis. [2]

(Total for Question 8 = 5 marks)

9 This question is about the chemistry of the elements in Groups 1 and 2 of the Periodic Table.

(a) When magnesium burns in air it forms a mixture of magnesium oxide and magnesium nitride.

Give the formula of both magnesium oxide and magnesium nitride. [2]

(b) (i) Write an equation for the reaction between calcium and water. Include state symbols. [2]

(ii) Explain how the reactivity of the Group 2 metals with water changes down the group. [3]

(iii) State how the solubility of the hydroxides of the Group 2 metals changes down the group. [1]

(c) Explain the trend in thermal stability of the carbonates of the Group 2 metals. [3]

(d) The nitrates of most of the Group 1 metals decompose on heating as shown in the equation:

$$MNO_3 \rightarrow MNO_2 + \frac{1}{2}O_2$$

where M represents a Group 1 metal.

However, lithium nitrate decomposes further.

Write an equation for the thermal decomposition of lithium nitrate. State symbols are not required. [2]

(Total for Question 9 = 13 marks)

10 This question is about halides.

(a) Group 1 halides react with concentrated sulfuric acid in different ways.

Sodium chloride reacts to give hydrogen chloride as the only gaseous product.

Sodium bromide reacts to give both hydrogen bromide and bromine as gaseous products.

(i) Write an equation for the reaction of both sodium chloride and sodium bromide with concentrated sulfuric acid. [2]

(ii) Explain why the two reactions produce different gaseous products. [3]

(iii) Describe how ammonia gas can be used to show that hydrogen chloride has been given off in the reaction between sodium chloride and concentrated sulfuric acid. [2]

(b) Hydrogen halides react with water.

(i) Write an equation for the reaction between hydrogen bromide and water. [1]

(ii) Explain the effect that the solution formed has on methyl orange indicator. [3]

(c) Describe how you would show that an aqueous solution contains both chloride and iodide ions. [4]

(Total for Question 10 = 15 marks)

11 Radium (Ra) and astatine (At) are radioactive elements in Groups 2 and 7 respectively in the Periodic Table.

(a) Predict the appearance at room temperature of

(i) radium [1]

(ii) astatine. [2]

(b) Predict the reactivity of astatine compared to that of iodine. Justify your answer. [3]

(c) Write an equation for the reaction of astatine with hydrogen. State symbols are not required. [1]

(d) Write an equation for the reaction between radium and chlorine. State symbols are not required. [1]

(e) The product of the reaction between radium and water, and the pH of the resulting solution, are most likely to be:

A radium oxide and 1

B radium oxide and 14

C radium hydroxide and 1

D radium hydroxide and 12 [1]

(Total for Question 11 = 9 marks)

12 This question is about some compounds of the elements in Group 2 of the Periodic Table.

(a) Water was slowly added to a solid lump of calcium oxide. The lump got very hot and broke apart to form another white solid.

Excess water was added to this white solid and the mixture was filtered to produce a colourless solution. The solution turned milky when carbon dioxide was bubbled through it.

(i) State why the lump of calcium oxide got very hot when water was added. [1]

(ii) Name the colourless solution formed when an excess of water was added to the white solid, and predict its pH. [2]

(iii) Write an equation for the reaction between this colourless solution and carbon dioxide. Include state symbols. [2]

(b) Calcium oxide can be formed in several ways.

One way is to heat calcium in oxygen. Another way is to heat calcium nitrate.

(i) Write an equation for each method of forming calcium oxide. State symbols are not required. [2]

(ii) Explain the trend in decomposition temperatures of the Group 2 nitrates. [4]

(Total for Question 11 = 11 marks)

TOPIC 9 INTRODUCTION TO KINETICS AND EQUILIBRIA

A KINETICS | B EQUILIBRIA

Reaction kinetics is the study of rates of reactions. Some reactions in everyday life take place very quickly, while other reactions are very slow. The combustion of petrol in the engine of a racing car is very rapid and allows the car to travel at very fast speeds. The formation of stalactites and stalagmites by the decomposition of dissolved calcium hydrogencarbonate into solid calcium carbonate is very slow. It has taken hundreds of years for these to form in limestone caves, such as the one shown from central Saudi Arabia. Some types of food are kept in a refrigerator or a freezer to slow down the rate at which they go bad.

Haemoglobin is the substance in red blood cells responsible for transporting oxygen around the body. Each haemoglobin molecule attaches to four oxygen molecules in a reversible reaction:

$$Hb(aq) + 4O_2(g) \rightleftharpoons Hb(O_2)_4\,(aq)$$

where 'Hb' stands for haemoglobin.

As long as there is sufficient oxygen in the air, a healthy equilibrium is maintained. However, at high altitudes, changes occur. The concentration of oxygen is lowered and this produces a shift in equilibrium to the left.

Without an adequate oxygen supply to the body's cells and tissues, you may feel light-headed. If you are not physically prepared for the change, you may need to breathe pressurised oxygen from an oxygen tank. This shifts the equilibrium to the right. For people born and raised at high altitudes, however, the body's chemistry performs the equilibrium shift to the right by producing more haemoglobin.

If you are exposed to carbon monoxide, it bonds to haemoglobin in preference to oxygen and sets up the following reversible reaction:

$$Hb(aq) + 4CO(g) \rightleftharpoons Hb(CO)_4(aq)$$

Carboxyhaemoglobin is formed, which is even redder than haemoglobin, so one sign of carbon monoxide poisoning is a flushed face.

Carbon monoxide in small quantities can cause headaches and dizziness, but larger concentrations can be fatal. To reverse the effects of the carbon monoxide, pure oxygen must be introduced to the body. It will react with the carboxyhaemoglobin to produce oxygenated haemoglobin, along with carbon monoxide:

$$Hb(CO)_4(aq) + 4O_2(g) \rightleftharpoons Hb(O_2)_4(aq) + 4CO(g)$$

The gaseous carbon monoxide thus produced is removed from the body when the person exhales.

MATHS SKILLS FOR THIS TOPIC

- Recognise and make use of appropriate units in calculations
- Recognise and use expressions in decimal and ordinary form
- Use an appropriate number of significant figures
- Plot two variables from experimental or other data
- Construct and/or balance equations using ratios

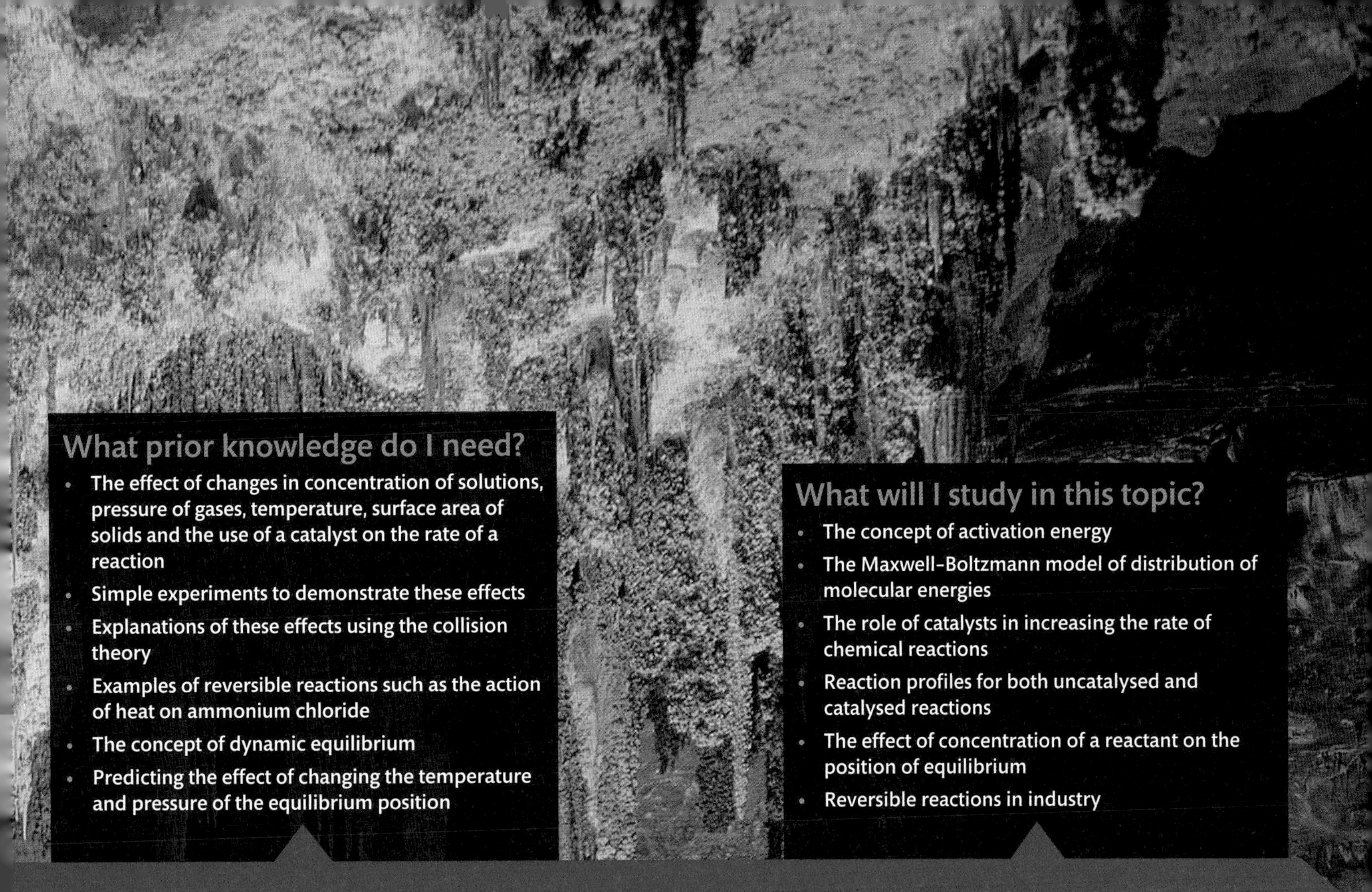

What prior knowledge do I need?

- The effect of changes in concentration of solutions, pressure of gases, temperature, surface area of solids and the use of a catalyst on the rate of a reaction
- Simple experiments to demonstrate these effects
- Explanations of these effects using the collision theory
- Examples of reversible reactions such as the action of heat on ammonium chloride
- The concept of dynamic equilibrium
- Predicting the effect of changing the temperature and pressure of the equilibrium position

What will I study in this topic?

- The concept of activation energy
- The Maxwell-Boltzmann model of distribution of molecular energies
- The role of catalysts in increasing the rate of chemical reactions
- Reaction profiles for both uncatalysed and catalysed reactions
- The effect of concentration of a reactant on the position of equilibrium
- Reversible reactions in industry

What will I study later?

Topic 11 (Book 2: IAL)

- Order of reaction and rate equations
- Experimental methods of determining rate of reaction
- Experimental method of determining activation energy
- The importance of reaction rate data in determining mechanisms for organic reactions

Topics 11 and 17 (Book 2: IAL)

- Homogeneous and heterogeneous catalysis

Topic 12 (Book 2: IAL)

- The relationship between total entropy change and equilibrium constant

Topic 13 (Book 2: IAL)

- The equilibrium constant in terms of concentrations, K_c and in terms of partial pressures, K_p
- Calculating values for K_c and K_p
- The effect of changing the temperature on the value of K_c and K_p

Topic 14 (Book 2: IAL)

- Acid-base equilibria

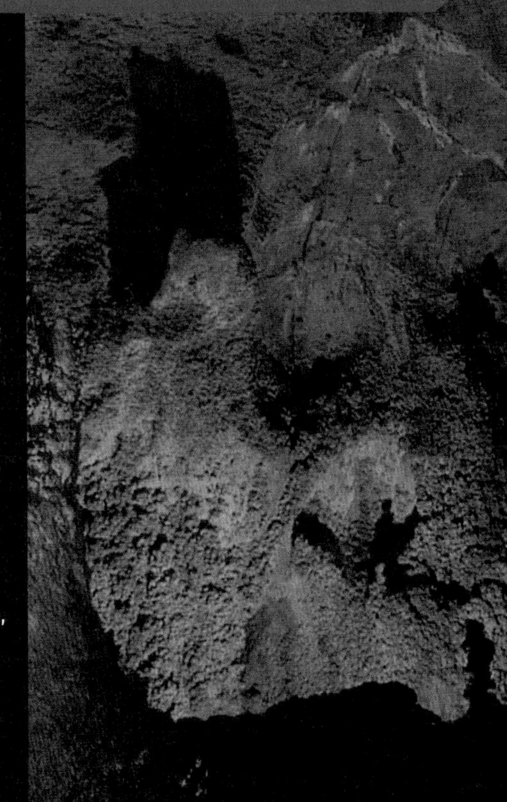

9A 1 REACTION RATE, COLLISION THEORY AND ACTIVATION ENERGY

SPECIFICATION REFERENCE 9.2 9.3

LEARNING OBJECTIVES

- Know what is meant by the term 'rate of reaction'.
- Calculate the rate of a reaction from:
  (i) the gradient of a suitable graph, by drawing a tangent, either for initial rate or at a time t
  (ii) data showing the time taken for a reaction, using
  $$\text{rate} = \frac{1}{\text{time}}.$$
- Understand that reactions take place only when collisions have sufficient energy, known as the activation energy.

RATE OF REACTION

We can determine the rate of a chemical reaction by the change in concentration of a reactant or a product per unit time.

$$\text{rate of reaction} = \frac{\text{change in concentration}}{\text{time for change to happen}}$$

To measure the rate of a reaction, we need to find out:

1 how fast one of the reactants is being used up, or

2 how fast one of the products is being formed.

The graph in **fig A** shows the concentration of a reactant against time, where the gradient (slope) of the graph indicates the rate of the reaction.

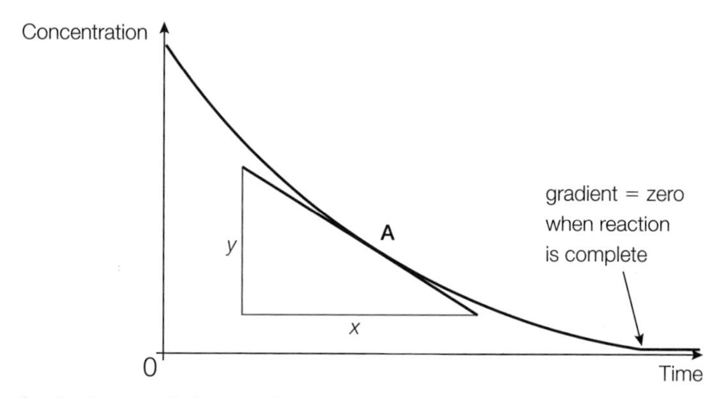

▲ **fig A** A graph showing the concentration of a reactant against time.

The gradient decreases as the rate decreases, and becomes zero when all of the reactant is used up.

In the graph in **fig A**, we determine the rate of reaction at point A by drawing a tangent to the curve at point A and measuring its gradient.

$$\text{gradient} = \frac{y}{x}$$

where y = the change in concentration and x = the change in time.

If the unit of concentration is moles per cubic decimetre ($mol\,dm^{-3}$) and the unit of time is seconds (s), then the unit of rate will be moles per cubic decimetre per second ($mol\,dm^{-3}\,s^{-1}$).

Sometimes it is more convenient to measure the concentration of the product formed over a period of time.

The graph in **fig B** shows a plot of concentration against time, with tangents drawn to obtain the initial rate of reaction and the rate at time, t.

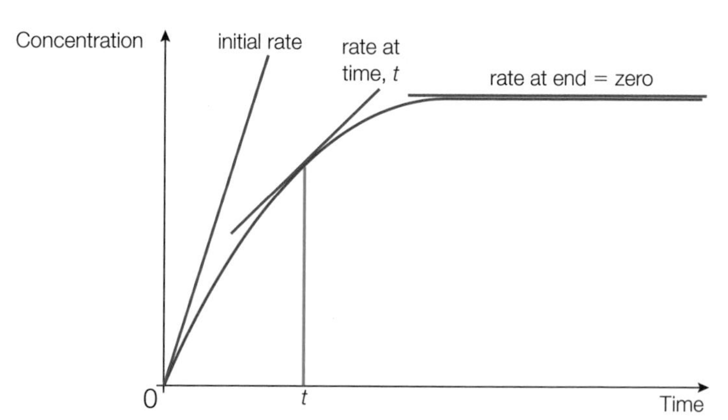

▲ **fig B** A graph showing the concentration of a product against time.

We can also calculate the rate of reaction from the time taken for a known amount of reactant to be used up, or a known amount of product to be formed. For example, if you measure, in a series of separate experiments, the time taken to collect a given volume of gas, then you can calculate the rate for each experiment using the expression:

$$\text{rate} = \frac{\text{volume of gas collected}}{\text{time taken}}$$

This is illustrated in **table A**.

TIME TAKEN TO COLLECT 20.0 cm^3 OF GAS / s	RATE OF REACTION / $cm^3\,s^{-1}$
10.0	2.00
20.0	1.00
40.00	0.50
80.00	0.25

table A The rate of reaction can be calculated from the time taken for a known amount of product to be formed.

COLLISION THEORY

Consider this reaction:

$$A + B \rightarrow C + D$$

In order for molecule A to react with molecule B, the two molecules must first of all collide with each other. If they collide they *may* react.

Why is there a possibility that the molecules may *not* react? This is because not all collisions between reactant molecules will result in a reaction. There are two requirements for a reaction to occur.

- The two molecules must collide with sufficient energy to cause a reaction – the **activation energy** E_a.
- The two molecules must collide in the correct orientation.

ACTIVATION ENERGY, E_a

The activation energy is the minimum energy that colliding particles must possess for a reaction to occur. If the particles collide with less energy than the activation energy, they simply bounce apart and no reaction occurs. Think of the activation energy as a barrier to the reaction. Only those collisions that have energies equal to or greater than the activation energy result in a reaction.

Any chemical reaction results in the breaking of some bonds (needing energy) and the making of new ones (releasing energy). Obviously, some bonds have to be broken before new ones can be made. Activation energy is involved in breaking some of the original bonds.

Where collisions are relatively gentle, there is not enough energy available to start the bond-breaking process, and so the particles do not react.

ORIENTATION

Consider the reaction between ethene and hydrogen bromide, which you met in **Topic 5**:

ethene + HBr ⟶ bromoethane

ethene + hydrogen bromide ⟶ bromoethane

The reaction can only happen if the hydrogen end of the H—Br molecule approaches the C=C of the ethene molecule. Any other collision between the two molecules will result in the molecules simply bouncing off each other.

Of the collisions shown in the figure below, only collision 1 may possibly lead to a reaction.

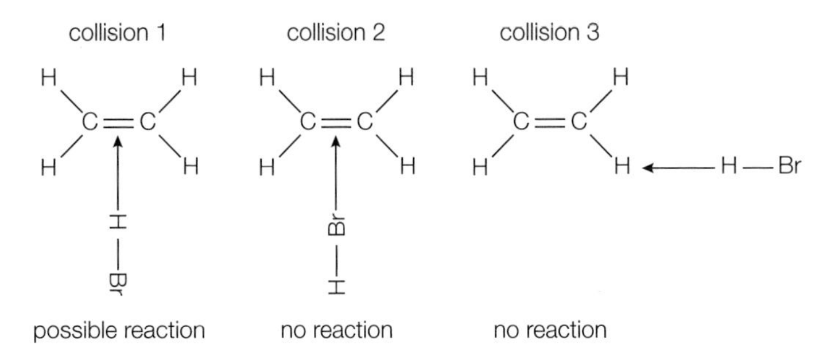

DID YOU KNOW?

When the shapes of molecules influence reactions, we say that there is a 'steric factor' involved in the reaction. In some cases, the atoms (or groups of atoms) in a molecule can hinder (interfere with) the course of a reaction. If an atom or group of atoms is particularly large, then it can get in the way of an attacking species. If this happens, we say that the reaction suffers from 'steric hindrance'.

For example, the halogenoalkane 2-bromomethylpropane, $(CH_3)_3CBr$, is hydrolysed rapidly when added to water. However, the mechanism of the reaction does not involve attack by the water molecule on the $\delta+$ carbon atom attached to the bromine, as it does with bromomethane, CH_3Br.

This is because the three methyl groups are so large that they prevent the water molecule approaching the carbon atom sufficiently close to interact with it. The mechanism for the reaction is totally different.

fig C The three methyl groups do not allow the water molecule to approach the carbon atom of the C–Br bond.

ADDITIONAL READING

If you would like to find out more, you can research S_N1 and S_N2 mechanisms for the hydrolysis of halogenoalkanes on the internet.

CHECKPOINT

1. At room temperature and pressure, in each cubic decimetre of gas there are about 1×10^{32} collisions every second. Why, therefore, are reactions between gases not completed in a fraction of a second?

2. Why do many reactions involving organic compounds have to be heated or refluxed for long periods of time?

3. Chloroalkanes, such as chloromethane (CH_3Cl), can be hydrolysed to alcohols by heating with aqueous sodium hydroxide. The hydroxide ion acts as a nucleophile attacking the $\delta+$ carbon atom and replacing the chlorine:

$$CH_3Cl + OH^- \rightarrow CH_3OH + Cl^-$$

When tetrachloromethane (CCl_4) is heated under reflux with aqueous sodium hydroxide, no reaction takes place. Suggest a reason why.

4. What will happen if a hydrogen atom and a chlorine atom collide and their combined energy of collision is 1.0×10^{-18} J? [E(H–Cl) = 431 kJ mol^{-1}].

SUBJECT VOCABULARY

activation energy, E_a the minimum energy that colliding particles must possess for a reaction to occur

steric hindrance the slowing of a chemical reaction due to large groups within a molecule getting in the way of the attacking species

9A 2 EFFECT OF CONCENTRATION, PRESSURE AND SURFACE AREA ON RATE OF REACTION

SPECIFICATION REFERENCE 9.1 PART

LEARNING OBJECTIVES

- Understand, in terms of the collision theory, the effect of changes in concentration of solution, pressure of a gas and surface area of a solid on the rate of a chemical reaction.

According to the collision theory, reactant particles have to collide with sufficient energy before they can react. It is sensible, therefore, to suggest that we can increase the rate of a reaction by increasing the frequency of collisions with sufficient energy between reactant particles. This is often true, but as in so many situations in chemistry, there are exceptions. You will learn about these in **Topic 11 (Book 2: IAL)**.

Collisions that result in a reaction are called *successful* collisions.

THE EFFECT OF CONCENTRATION

For reactions in solution, an increase in concentration often causes an increase in reaction rate. For many reactions, if the concentration of a solution is increased, then the frequency of collisions between reacting solute particles also increases. This is because they are closer together as there are more of them in a volume of solution. The frequency of successful collisions increases (i.e. there are more successful collisions per second), which in turn produces an increase in the rate of reaction.

The graph in **fig A** shows the effect of the change in volume of carbon dioxide given off with time for the reaction between calcium carbonate and excess dilute hydrochloric acid.

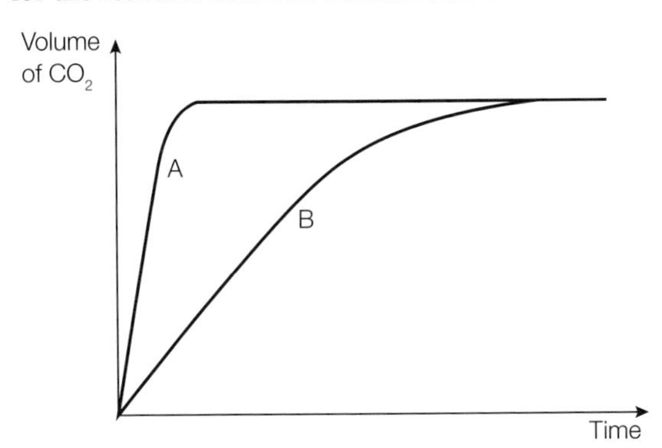

fig A A concentration-time graph showing the effect of the change in volume of carbon dioxide given off with time for the reaction between calcium carbonate and excess dilute hydrochloric acid.

Curve A represents the higher concentration of acid. You will notice that the gradient of curve A is always greater than the gradient of curve B. You will also notice that curve A levels off before curve B. Both factors indicate that the rate of reaction is greater for the higher concentration of acid.

THE EFFECT OF PRESSURE

For a reaction in which molecules collide and react in the gas phase, an increase in pressure will cause an increase in the rate of reaction.

The explanation is similar to that for concentration of solution. If the pressure of the gaseous mixture is increased, there will be more reactant molecules in a given volume of mixture. The frequency of collisions will increase. This will result in an increase in the rate of reaction.

Changing the pressure has almost no effect on reactions in the solid or liquid phase. This is because the volume of solids and liquids changes very little when they are put under pressure, so their particles do not move closer together.

ADDITIONAL READING

THE HABER PROCESS

It is interesting to look at the situation that exists when a reaction between two gases takes place on the surface of a solid that is in contact with the reaction mixture. This is the case with the catalytic conversion of hydrogen gas and nitrogen gas to make ammonia in the Haber process.

$$3H_2(g) + N_2(g) \overset{Fe(s)}{\rightleftharpoons} 2NH_3(g)$$

The reaction between the hydrogen and nitrogen molecules takes place at the surface of the iron catalyst, but only at selected sites called 'active' sites. At the high pressures used in the Haber process, there are always more reactant molecules than there are active sites. For this particular reaction, an increase in pressure will not increase the rate of reaction, since all of the active sites are already occupied.

LEARNING TIP

A heterogeneous reaction is one in which the reactants are in more than one phase; for example, a solid and a gas, or a solid and a solution.

THE EFFECT OF SURFACE AREA

For heterogeneous reactions involving a solid, a larger surface area of the solid will result in a faster reaction.

The reaction between magnesium and dilute hydrochloric acid is represented by this ionic equation:

$$Mg(s) + 2H^+(aq) \rightarrow Mg^{2+}(aq) + H_2(g)$$

Only collisions between the hydrogen ions and magnesium atoms on the *surface* of the magnesium can result in reaction. If the magnesium is powdered, the surface area is increased and hydrogen is given off more quickly.

The effectiveness of solid catalysts is also improved if they are finely divided. For example, the rate of the catalysed decomposition of hydrogen peroxide by manganese(IV) oxide is increased significantly if the catalyst is a powder rather than lumps. Here is the equation for the reaction:

$$H_2O_2(aq) \xrightarrow{MnO_2(s)\text{ catalyst}} H_2O(l) + \tfrac{1}{2}O_2(g)$$

EXAM HINT

It is important to remember that for a reaction at a position of equilibrium between reactants and products, a catalyst does not affect the position of equilibrium. It only speeds up the rate at which the system achieves equilibrium.

CHECKPOINT

SKILLS REASONING

1. Ammonia and hydrogen chloride react in the gas phase to form ammonium chloride.

   $NH_3(g) + HCl(g) \rightarrow NH_4Cl(s)$

   State and explain the effect, if any, on the rate of reaction of:

   (a) halving the volume of the container at constant temperature

   (b) increasing the pressure by adding more ammonia at constant volume and temperature.

2. (a) Draw a concentration–time graph for the reaction between marble chips (calcium carbonate) and excess dilute hydrochloric acid. Place the concentration of the hydrochloric acid on the vertical axis.

   (b) On the same axes, draw the curve that would be obtained if the reaction was repeated at the same temperature, with the same mass of powdered marble.

   (c) Explain the reason for the different-shaped curves obtained.

9A 3 EFFECT OF TEMPERATURE ON RATE OF REACTION

SPECIFICATION REFERENCE

9.4 PART

LEARNING OBJECTIVES

- Understand qualitatively, in terms of the Maxwell-Boltzmann distribution of molecular energies, how changes in temperature can affect the rate of a reaction.

MAXWELL–BOLTZMANN DISTRIBUTION CURVES

The molecules in a sample of gas have a wide range of energies. To estimate what fraction of collisions will have the required activation energy, we need to know the energy distribution of the molecules.

This was first calculated in 1860 by James Clerk Maxwell and verified in 1872 by Ludwig Boltzmann. **Fig A** shows the distribution of molecular energies at two temperatures: T_1 (in blue) and T_2 (in red). T_2 is a higher temperature than T_1.

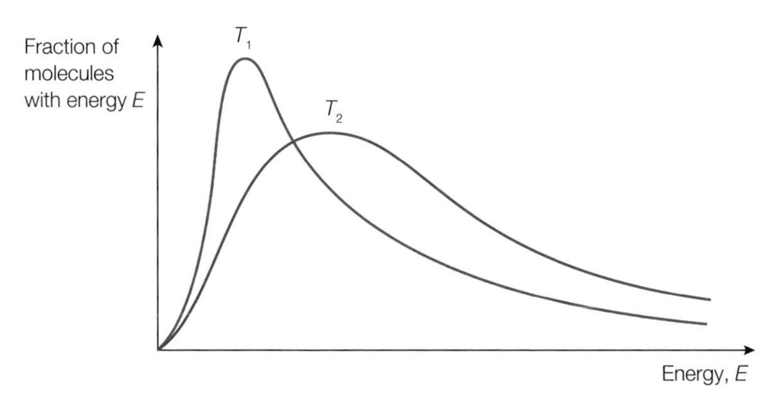

fig A A graph showing the distribution of molecular energies at two temperatures. T_2 is a higher temperature than T_1.

EXAM HINT

Make sure that you label the axes correctly if you are asked to reproduce this graph in an exam.

There are four important points to note about the curves:

- Neither curve is symmetrical.
- Both curves start at the origin and finish by approaching the *x*-axis asymptotically.
- The area under each curve is the same, since the number of molecules has not changed.
- The peak of T_2 is displaced to the right and is lower than the peak of T_1.

The curves in the graph in **fig B**, show, once again, the molecular distribution of energies at two temperatures, T_1 and T_2, where $T_2 > T_1$. E_a is the activation energy for the reaction.

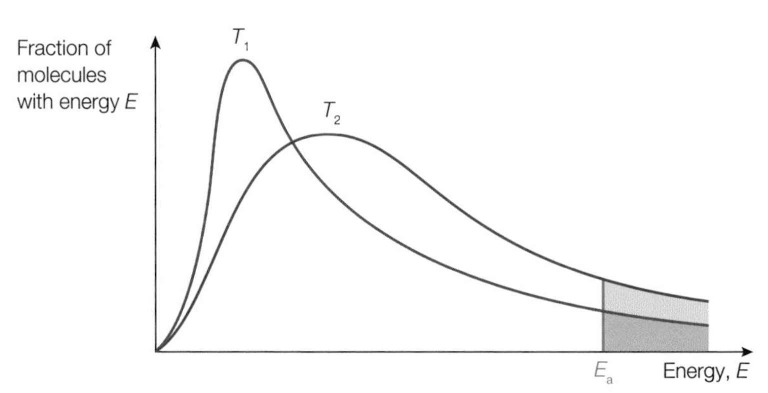

fig B A graph showing the distribution of molecular energies at two temperatures, $T_2 > T_1$.

The area shaded dark green represents the fraction of molecules that have the required energy to react at T_1.

The combined area shaded in dark green and light green represents the fraction of molecules that have the required energy to react at T_2.

It is easy to see that the fraction of molecules that can react at the higher temperature is greater.

LEARNING TIP

The Maxwell-Boltzmann diagram showing the effect of temperature consists of *two* curves and *one* activation energy line.

If you are asked to use the Maxwell-Boltzmann distribution curves as part of your explanation, you must make it clear that the area under the curve to the right of the activation energy line represents the fraction of molecules that has enough energy for the collisions to be successful.

HOW CHANGES IN TEMPERATURE CAN AFFECT THE RATE OF A REACTION

We can now explain the effect of an increase in temperature on the rate of a reaction.

An increase in temperature increases the fraction of molecules that possess the required activation energy. The rate of the reaction increases because the number of successful collisions per second increases.

The argument used above assumes that the fraction of collisions with energy greater than or equal to E_a is the same as the fraction of molecules with this energy. This is not exactly true. However, the difference is very small at high energies when the fraction is very small. Under these circumstances, it is reasonable to draw and use the molecular energy distribution curve instead of the collision distribution curve.

LEARNING TIP

When explaining the increase in reaction rate owing to an increase in temperature, it is important to mention that the number of *successful* collisions per second increases, and not just that the *overall* number of collisions per second increases.

Although there is an increase in the overall collision frequency, the effect of this is negligible (very small) compared to the increase in frequency of collisions that have energy equal to or greater than the activation energy.

COLLISIONS IN SOLUTION

In the gas phase, molecules are moving around at high speeds so they frequently collide with one another. At a pressure of 100 kPa and a temperature of 298 K, a single molecule might have somewhere between 10^9 and 10^{10} collisions per second, depending on its size.

In solution, the situation is different. The molecules are much more closely packed together so there is not a lot of 'space' between them. In low-to-medium concentrations, most of the solution is solvent, so solute molecules tend to be entirely surrounded by solvent molecules. These solute molecules are said to be trapped in a 'solvent cage'.

You may think that collisions between solute molecules would be far less frequent than similar collisions in the gas phase because the solvent molecules will get in the way. However, there are situations in which a number of solute molecules become trapped in the same solvent cage. This increases the collision rate between the solute molecules and, if the collisions are sufficiently energetic, they may react, just as in the gas phase.

There are, therefore, two distinct stages to a reaction in solution:

- Firstly, the molecules have to come together, by a process of diffusion, in the same solvent cage.
- Secondly, they have to react.

There may, of course, be some reactions between solute molecules that have 'jumped out' of their cage and just happen to meet each other.

Although what is happening in a solution phase reaction is different from what is happening in the gas phase, the resulting kinetics are the same. For simple reactions involving two species, the chance of them meeting each other in solution is proportional to their concentrations, just as in the gas phase.

CHECKPOINT

1. State the effect of increasing the temperature on the rate of:
   (a) an exothermic reaction
   (b) an endothermic reaction.

2. (a) Draw Maxwell-Boltzmann distribution curves for a gas at two temperatures, T_c and T_h, where $T_c < T_h$.
   (b) Indicate a suitable value for the activation energy of the reaction and use the curves to explain the effect of lowering the temperature on the rate of reaction.

4 EFFECT OF CATALYSTS ON RATE OF REACTION

SPECIFICATION REFERENCE

9.5 9.6 9.7 9.8

LEARNING OBJECTIVES

- Be able to interpret the action of a catalyst in terms of a qualitative understanding of the Maxwell–Boltzmann distribution of molecular energies.
- Understand the role of catalysts in providing alternative reaction routes of lower activation energy.
- Be able to draw the reaction profiles of both uncatalysed and catalysed reactions including the energy level of the intermediate formed with the catalyst.
- Understand the use of catalysts in industry to make processes more sustainable by using less energy and/or higher atom economy.
- Understand the use of a solid (heterogeneous) catalyst for industrial reactions involving gases, in terms of providing a surface for the reaction.

THE EFFECT OF CATALYSTS

A **catalyst** works by providing an alternative route for the reaction. This alternative route has a lower activation energy than the original route.

EXAM HINT

It is wrong to say that a catalyst lowers the activation energy for a particular **reaction pathway**. It provides a *different* pathway with a lower activation energy.

Fig A shows the effect on the fraction of molecules that have the required energy to react when a route of lower activation energy is available.

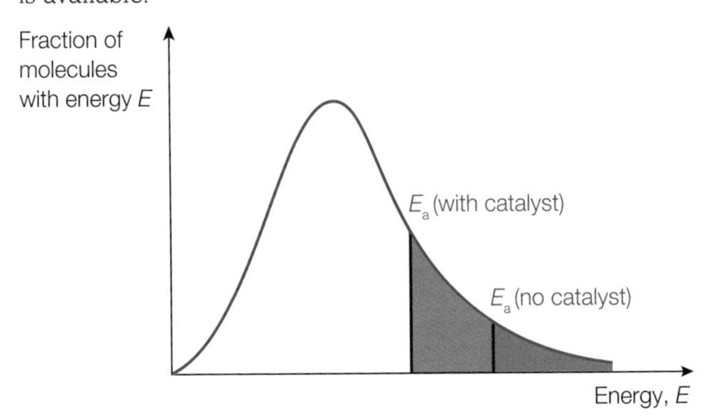

fig A A graph showing the effect on the fraction of molecules that have the required energy to react when a catalyst is present.

The blue shaded area represents the fraction of molecules that have $E > E_a$ when no catalyst is present.

The combined blue and red shaded areas represent the fraction of molecules that have $E > E_a$ when a catalyst is present.

LEARNING TIP

The Maxwell-Boltzmann diagram showing the effect of a catalyst consists of *one* curve and *two* activation energy lines. Do not confuse this with the diagram used to show the effect of temperature.

REACTION PROFILE DIAGRAMS

A reaction profile diagram is an extension of an enthalpy level diagram (see **Topic 6B**). In addition to showing the relative enthalpy levels of reactants and products, it includes the activation energy for the reaction.

A typical reaction profile, not to scale, for the combustion of methane is shown in **fig B**.

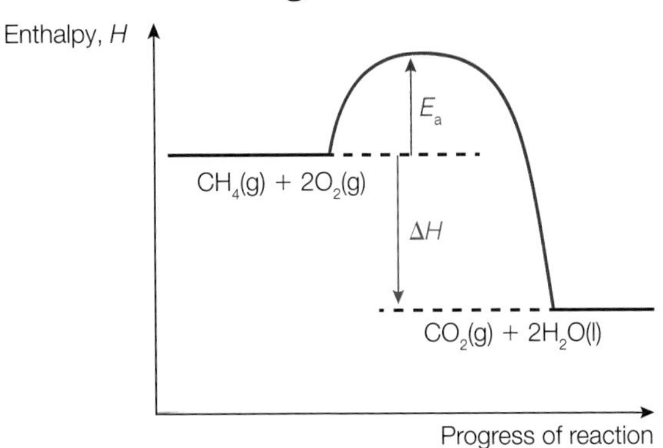

fig B A reaction profile for the combustion of methane.

For an endothermic reaction, the enthalpy level of the products is above the enthalpy level of the reactants, but otherwise the profile is the same.

Fig C shows the simplified reaction profiles for an uncatalysed reaction and a catalysed reaction.

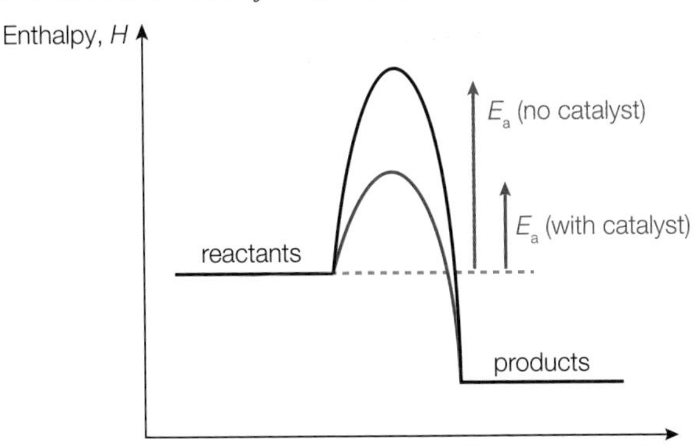

fig C Simplified reaction profiles for an uncatalysed and a catalysed reaction.

If the catalysed reaction involves the formation of an intermediate, then the reaction profile is more complicated, as shown by the example in **fig D**.

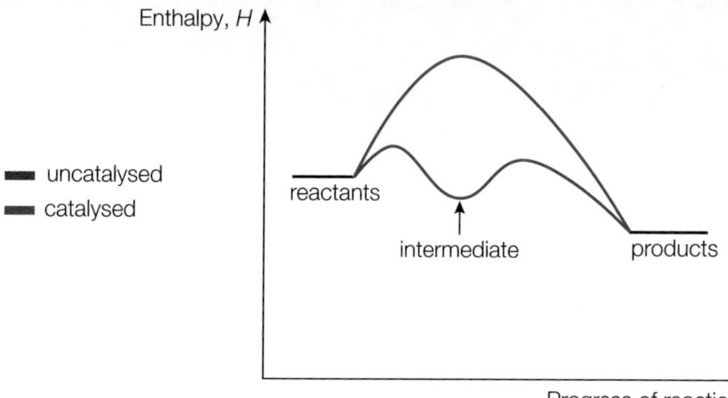

▲ **fig D** Enthalpy profile diagram for a catalysed reaction involving the formation of an intermediate.

An example of a reaction that falls into this category is the reaction between peroxydisulfate ions and iodide ions, catalysed by Fe^{2+} ions.

The equation for the uncatalysed reaction is

$$S_2O_8^{2-}(aq) + 2I^-(aq) \rightarrow 2SO_4^{2-}(aq) + I_2(aq)$$

The catalysed reaction occurs in two steps:

Step 1: $S_2O_8^{2-}(aq) + 2Fe^{2+}(aq) \rightarrow 2SO_4^{2-}(aq) + 2Fe^{3+}(aq)$

Step 2: $2Fe^{3+}(aq) + 2I^-(aq) \rightarrow 2Fe^{2+}(aq) + I_2(aq)$

The species present at the intermediate stage are $SO_4^{2-}(aq)$ and $Fe^{3+}(aq)$

N.B. The reaction can also be catalysed by Fe^{3+} ions. In this case the two steps will occur in reverse order.

CATALYSTS IN INDUSTRY

The first recorded use of a catalyst in industry was in 1746, when John Roebuck developed the lead chamber process for manufacturing sulfuric acid. Since then, catalysts have been increasingly used in the chemical industry. The development of catalysts has been mostly for economic reasons. However, more recently the development of catalysts has been for political and environmental reasons as well.

EXAM HINT

Longer exam questions may ask you to compare different industrial processes. Remember to include ideas about yield and atom economy as well as issues of energy and broader environmental issues.

The two major economic advantages of the use of catalysts are:

- they increase the rate of a chemical reaction, meaning that more of the desired product can be made in a given time period
- reactions can take place at lower temperatures, resulting in a decrease in the energy costs to the manufacturer.

Most catalysts used in industry are **heterogeneous catalysts**. A heterogeneous catalyst is one that is in a different **phase** to that of the reactants. Solids are commonly used as heterogeneous catalysts for reactions involving gases. The most well-known examples are the use of iron in the Haber process and of vanadium(V) oxide in the Contact process (see **Topic 17 (Book 2: IAL)**).

A solid catalyst provides a surface on which the gas molecules can adsorb and then react. The product molecules then desorb from the surface and more reactant molecules take their place. This process is described in more detail in **Topic 11 (Book 2: IAL)**.

DID YOU KNOW?

Environmental catalysts have also been developed to remove toxic products from industrial waste materials. Do some research to identify these catalysts.

CATALYSTS AND ATOM ECONOMY IN INDUSTRY

In the manufacture of chemicals, waste is produced when products are formed that are of little or no commercial value.

Using catalysts can lead to reactions being developed that make the desired product with little or no co-product. They may also reduce side reactions producing other unwanted products, making product separation easier.

A good example is the manufacture of phenol. The development of zeolite catalysts has meant that benzene can be oxidised to phenol using dinitrogen monoxide (N_2O), with nitrogen as the only other product. The alternative manufacture from benzene via cumene has a much lower atom economy.

The manufacture of ibuprofen is another example. The original method for making it used six steps with aluminium chloride as a catalyst, and several other reagents. Although behaving as a catalyst, the aluminium chloride cannot be recovered.

A newer method uses two catalysts, hydrogen fluoride and Raney nickel (an alloy of nickel and aluminium). These can be recovered and reused many times. The effect of using better catalysts has been to improve the atom economy of the process and to reduce the number of steps needed from six to three. You can see this in **fig E**, but you don't need to know these steps.

Some other considerations are:

- With fewer steps the energy costs are likely to be less.
- The percentage yield for each step may be higher (and fewer steps).
- There may be less product lost in separation after each step or separation may not be required until the end.
- The whole process may be quicker so that you get more product in a given time.

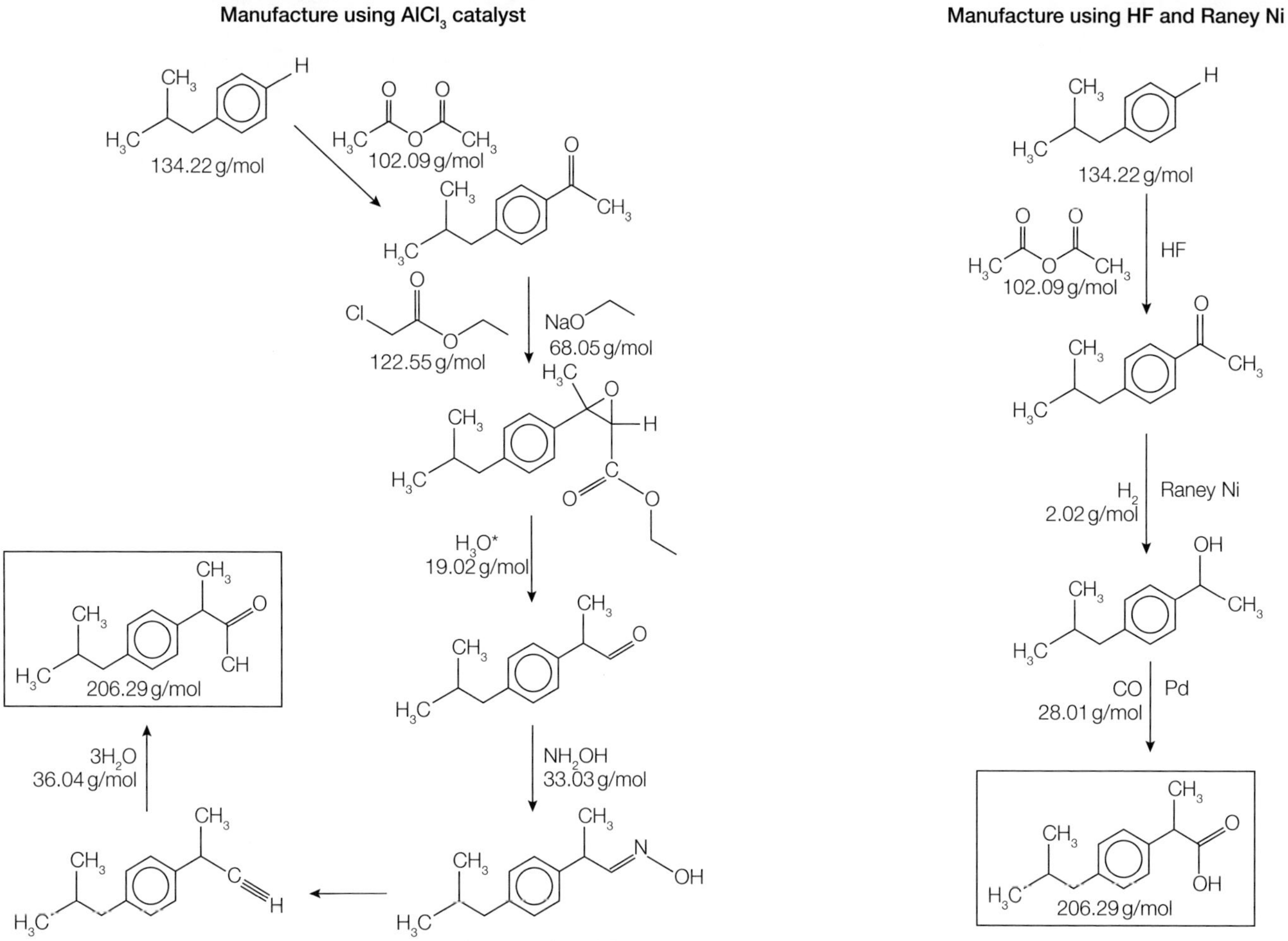

Manufacture using $AlCl_3$ catalyst

Manufacture using HF and Raney Ni

▲ **fig E** Reaction pathways for two methods of manufacturing ibuprofen.

CHECKPOINT

1. An aqueous solution of hydrogen peroxide decomposes very slowly at room temperature into water and oxygen. The decomposition is catalysed by solid manganese(IV) oxide.

$$H_2O_2(aq) \rightarrow H_2O(l) + \frac{1}{2}O_2(g) \qquad \Delta_rH^{\ominus} = -196\,kJ\,mol^{-1}$$

(a) Using the same axes, draw labelled reaction profiles for the reaction both with and without manganese(IV) oxide.

(b) Explain the change in rate of reaction that occurs when manganese(IV) oxide is added to the hydrogen peroxide solution.

SUBJECT VOCABULARY

reaction pathway the reaction, or series of reactions, that the reactants undergo in order to change into the products

catalyst a substance that increases the rate of a chemical reaction but is chemically unchanged at the end of the reaction

heterogeneous catalyst a catalyst that is in a different phase to that of the reactants

phase a physically distinct form of matter, such as a solid, liquid, gas or plasma; a phase of matter is characterised by having relatively uniform chemical and physical properties

9B 1 REVERSIBLE REACTIONS AND DYNAMIC EQUILIBRIUM

SPECIFICATION REFERENCE 9.9

LEARNING OBJECTIVES

- Know that many reactions are readily reversible.
- Know that reversible reactions can reach a state of dynamic equilibrium in which:
  (i) both forward and backward reactions are still occurring
  (ii) the rate of the forward reaction is equal to the rate of the backward reaction
  (iii) the concentrations of reactants and products remain constant.

IRREVERSIBLE AND REVERSIBLE REACTIONS

When a mixture of hydrogen and oxygen in a 2:1 molar ratio is ignited, water is produced. There is very little, if any, uncombined hydrogen or oxygen remaining at the end of the reaction. We often describe such reactions as 'irreversible'.

Most combustion reactions are in this category since they are highly exothermic. That is, ΔH is large and negative. However, ΔH is small for many reactions, especially in organic chemistry. These reactions may not go to completion. At the end of the reaction, detectable amounts of the reactants remain, mixed with the product. These reactions are called 'reversible' reactions.

HOW TO DECIDE WHETHER A REACTION IS REVERSIBLE

Deciding whether a reaction is reversible or not depends on how carefully we measure the concentrations of reactants and products. For example, the reaction between dilute hydrochloric acid and aqueous sodium hydroxide appears to go to completion. Both acid and alkali are almost completely ionised in water, so the equation for the reaction is:

$$H^+(aq) + OH^-(aq) \rightarrow H_2O(l)$$

Pure water has a slight electrical conductivity. This results from the ionisation of water molecules:

$$H_2O(l) \rightarrow H^+(aq) + OH^-(aq)$$

This indicates that the reverse reaction is taking place to a small extent. Since only one molecule in approximately 550 million is ionised, we usually ignore this small extent of ionisation. However, it becomes important when we study the pH scale of acidity.

In practical terms, if a reaction is more than 99% complete, we usually consider it to have gone to completion.

THE REACTION BETWEEN HYDROGEN AND IODINE

If a mixture of hydrogen and iodine vapour in a 1:1 molar ratio is heated to 573 K in a closed container, about 90% of the hydrogen and iodine react to form hydrogen iodide. Provided the reaction mixture remains in the closed container at 573 K, 10% of the hydrogen and iodine will remain unreacted even if you leave the reaction mixture for a long time.

If a sample of hydrogen iodide is heated to 573 K in a closed container, it partially decomposes. The mixture formed is identical to that produced when starting with an equimolar mixture of hydrogen and iodine. The reaction is clearly reversible, and when there is no further change in the concentrations of the reactants and products, the system is said to be in 'equilibrium'.

The symbol $\rightleftharpoons$ is used in an equation to represent a reversible reaction. The equation for the reaction between hydrogen and iodine is therefore written as follows:

$$H_2(g) + I_2(g) \rightleftharpoons 2HI(g)$$

When the equation is written in this way, the reaction between hydrogen and iodine is called the *forward* reaction. The decomposition of hydrogen iodide into hydrogen and iodine is called the *backward* reaction.

HOW IS EQUILIBRIUM ESTABLISHED?

When the mixture of hydrogen and iodine is heated, the two gases start to react and form hydrogen iodide. With increasing time, the concentrations of hydrogen and iodine decrease, so the rate of the forward reaction decreases.

As soon as some hydrogen iodide is formed, it starts to slowly decompose. With increasing time, however, the concentration of hydrogen iodide increases, so the rate of the backward reaction increases.

Eventually, the rates of the forward and the backward reactions become equal. After this point there is no further change in concentrations of reactants and products. The system is now in equilibrium. It is referred to as a 'dynamic' equilibrium since both forward and backward reactions are taking place at the same time, and also at the same rate.

DYNAMIC EQUILIBRIUM

Fig A shows dynamic equilibrium.

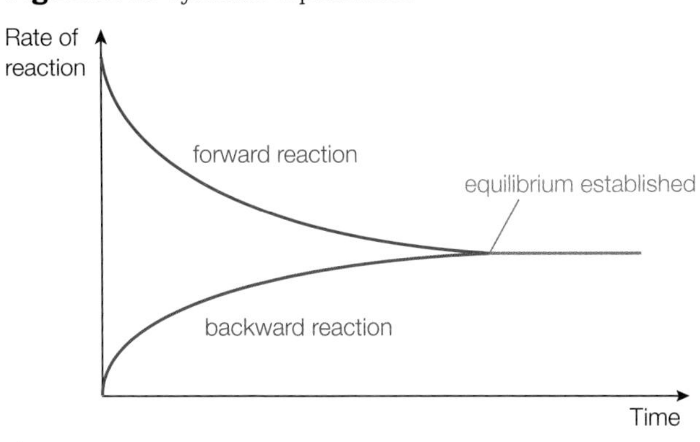

▲ **fig A** Dynamic equilibrium.

Two conditions must be met for dynamic equilibrium to be established:

1 the reaction must be reversible
2 the reaction mixture must be in a closed container.

Three important features define a system that is in dynamic equilibrium:

- Both forward and backward reactions are continuously occurring.
- The rate of the forward reaction is equal to the rate of the backward reaction.
- The concentrations of reactants and products remain constant.

LEARNING TIP

Remember these features!

CHECKPOINT

1. Explain what is meant by the term 'dynamic equilibrium' with reference to the following reaction:

$$H_2(g) + I_2(g) \rightleftharpoons 2HI(g)$$

2. Nitrogen and hydrogen react reversibly to form ammonia:

$$N_2(g) + 3H_2(g) \rightleftharpoons 2NH_3(g)$$

1 mol of nitrogen and 3 mol of hydrogen were mixed in a closed container and allowed to reach equilibrium. Twenty per cent of the nitrogen and hydrogen were converted into ammonia.

Draw three graphs, using a single set of axes, to show how the number of moles of nitrogen, hydrogen and ammonia vary with time.

9B 2 EFFECT OF CHANGES IN CONDITIONS ON EQUILIBRIUM COMPOSITION

SPECIFICATION REFERENCE 9.10

LEARNING OBJECTIVES

- Predict and justify the qualitative effect of a change in concentration, pressure or temperature, or the addition of a catalyst, on the composition of an equilibrium mixture.

CHANGING THE COMPOSITION OF AN EQUILIBRIUM MIXTURE

When a reaction mixture reaches a position of equilibrium, the composition of the equilibrium mixture (i.e. the concentration of each component) will not alter as long as the conditions remain the same.

However, if we change a condition (i.e. add some more of, or remove one of, the components, or change the temperature of the system) then the composition may change. This is often referred to as 'changing the position of equilibrium', and we refer to the position being moved to the right, to the left or not changed.

For example, if acid is added to a yellow solution containing chromate(VI) ions, CrO_4^{2-}, the solution turns orange. This is because of an increase in the amount of dichromate(VI) ions, $Cr_2O_7^{2-}$.

EXAM HINT

Note that there is no change in the oxidation number of the chromium in this reaction.

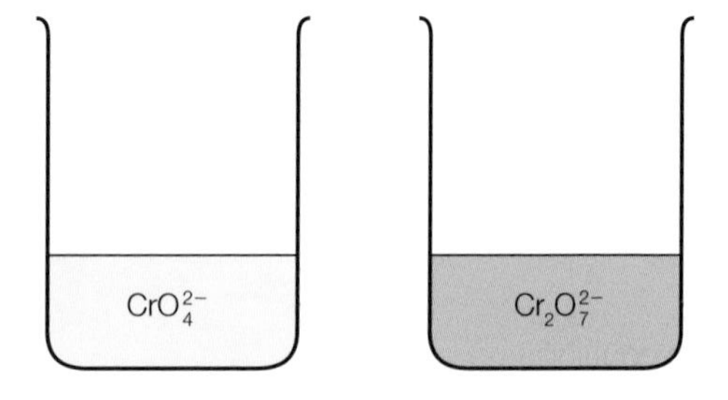

$$2CrO_4^{2-}(aq) + 2H^+(aq) \rightleftharpoons Cr_2O_7^{2-}(aq) + H_2O(l)$$

yellow — orange

▲ **fig A** Beakers containing solutions of chromate(VI) ions and dichromate(VI) ions.

The equilibrium position moves to the right when the acid (H^+) is added. If sufficient alkali is added to the orange solution, it will turn yellow as the amount of CrO_4^{2-} ions increases and exceeds the amount of $Cr_2O_7^{2-}$ ions. The equilibrium position moves to the left when alkali (OH^-) is added.

We will consider four factors that may affect the position of equilibrium of a reaction mixture. These are:

1 concentration of a component
2 pressure of the system
3 temperature of the system
4 addition of a catalyst.

EFFECT OF A CHANGE IN CONCENTRATION

If we *increase* the concentration of one of the reactants in a system in equilibrium, the rate of the forward reaction will increase and more products will form. As the concentration of the products increases, the rate of the backward reaction increases and eventually a new equilibrium is established. The equilibrium position has moved to the right, with slightly more product being present than at the original position of equilibrium.

If the concentration of one of the reactants is *decreased*, the position of equilibrium moves to the left. Similar changes occur if the concentration of the product is increased or decreased.

The changes that occur are summarised in the **table A**.

CONCENTRATION OF REACTANTS	CONCENTRATION OF PRODUCTS	CHANGE IN POSITION OF EQUILIBRIUM
increased		to the right
decreased		to the left
	increased	to the left
	decreased	to the right

table A The change in the position of equilibrium when the concentration of reactants or products is increased or decreased.

EFFECT OF A CHANGE IN PRESSURE

The effect of pressure only applies to reversible reactions involving gases. At a given temperature, the pressure of a gaseous mixture depends only on the number of gas molecules in a given volume. So, the pressure of a gaseous mixture may be increased by reducing the volume and reduced by increasing the volume.

Alternatively, the pressure at which the reaction is carried out can be:

- increased by initially using more moles of the reactants in the same volume
- decreased by using fewer moles of the reactants in the same volume.

The effect of a change in pressure (at constant temperature) caused by changing the volume of the reaction mixture can be studied using a gas syringe and pushing in, or pulling out, the plunger. The effect depends on the total number of moles of gas on each side of the balanced equation, and is summarised in **table B**.

NUMBER OF MOLES OF REACTANTS	NUMBER OF MOLES OF PRODUCTS	CHANGE IN POSITION OF EQUILIBRIUM WHEN THE PRESSURE IS INCREASED
more	fewer	to the right
fewer	more	to the left
same	same	no change

table B The change in the position of equilibrium when the pressure is increased and the number of moles of reactants or products is increased, decreased or not changed.

The reverse changes are true for a decrease in pressure.

The effect of changes in pressure as a result of using different amounts of gaseous reactants in a fixed volume container is shown in **fig B**. The reaction is between nitrogen and hydrogen, forming ammonia.

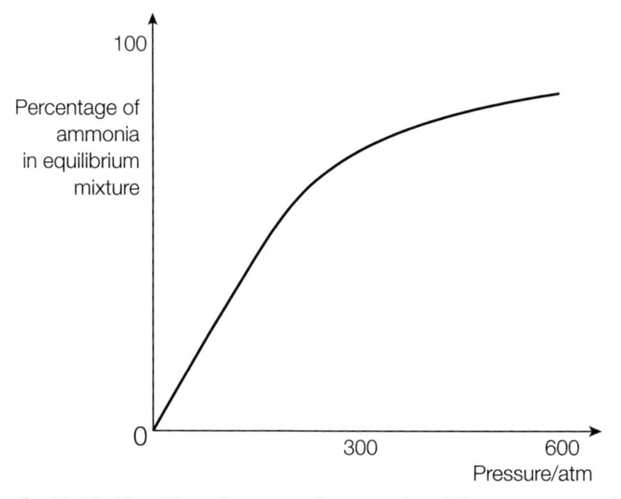

fig B The effect of pressure changes using different amounts of nitrogen and hydrogen to form ammonia.

The equation for the reaction is:

$N_2(g) + 3H_2(g) \rightleftharpoons 2NH_3(g)$

In the balanced equation, there are 4 moles of gas on the left-hand side and 2 moles of gas on the right-hand side.

The graph in **fig B** shows that the higher the pressure, the more ammonia there is in the equilibrium mixture formed. This agrees with the prediction in **table B** which states that an increase in pressure shifts the position of equilibrium to the side that has fewer moles of gas.

EXAM HINT

When you answer questions dealing with the effect of temperature change, avoid phrases such as 'an increase in temperature favours the endothermic direction'.
An increase in temperature 'favours' (i.e. increases the rate of) *both* the endothermic *and* the exothermic reactions.

The important point to remember is that it increases the rate of the endothermic reaction *more than* it increases the rate of the exothermic reaction.

EFFECT OF A CHANGE IN TEMPERATURE

If the temperature of an equilibrium mixture is raised, the rates of both the forward and the backward reactions will increase. However, the increase in the rate of the endothermic reaction will be greater than the increase in the rate of the exothermic reaction. Therefore, an increase in temperature will shift the position of equilibrium in the direction of the endothermic reaction.

So, the change in position of equilibrium will depend on whether the forward reaction is exothermic or endothermic.

The effects of temperature change are summarised in **table C**.

TEMPERATURE CHANGE	THERMICITY OF FORWARD REACTION	CHANGE IN POSITION OF EQUILIBRIUM
increased	exothermic (ΔH –ve)	to the left
decreased	exothermic (ΔH –ve)	to the right
increased	endothermic (ΔH +ve)	to the right
decreased	endothermic (ΔH +ve)	to the left

table C The change in the position of equilibrium and the thermicity of the forward reaction when the temperature changes.

EFFECT OF THE ADDITION OF A CATALYST

If a catalyst is added to a reaction mixture that is in equilibrium, the rate of both the forward and the backward reactions will increase. However, unlike the effect of increasing the temperature, the increase in rate will be the same for both reactions. So, the position of equilibrium is not altered.

The advantage of adding a catalyst at the beginning of the reaction is that it will reduce the time required to establish equilibrium.

LIMITATIONS OF MAKING QUALITATIVE PREDICTIONS

The first thing to recognise is that the qualitative predictions we have made about the effect of concentration, pressure and temperature on the position of equilibrium are just that – predictions.

The arguments we have used are *not* explanations of why changes sometimes occur. In fact, there are occasions when it is impossible to predict the direction of change, or indeed when the prediction turns out to be incorrect.

For example, if an equilibrium mixture of $NO_2(g)$ (brown) and $N_2O_4(g)$ (colourless) in a closed container is placed into a beaker of hot water at room temperature, both the temperature and the pressure of the gaseous mixture will rise.

The equation for the reaction is:

$$2NO_2(g) \rightleftharpoons N_2O_4(g) \qquad \Delta H = -57.2\,\text{kJ}\,\text{mol}^{-1}$$

Because the forward reaction is exothermic, we would predict that an increase in temperature would shift the equilibrium to the left.

However, since there are fewer moles of gas on the right-hand side of the equation, we would predict that an increase in pressure would shift the equilibrium to the right.

We do not know which effect is greater, so we cannot make a prediction of which way the equilibrium will shift. In practice, the mixture becomes darker in colour, so the temperature effect must be greater than the pressure effect.This is because the equilibrium must have shifted to the left to produce more brown $NO_2(g)$.

It might be interesting for you to do further research into the limitations of qualitative predictions. For example, what would you expect to happen if you add more nitrogen at constant pressure and temperature to an equilibrium mixture of nitrogen, hydrogen and ammonia? What would you expect to happen if you added an inert gas at constant volume and temperature to an equilibrium mixture of sulfur dioxide, oxygen and sulfur trioxide? The results may surprise you!

DID YOU KNOW?

A principle known as Le Chatelier's Principle is sometimes recommended as a useful way of working out the possible change in position of equilibrium when a condition is altered. There is no requirement to learn or use this principle in this course.

Le Chatelier's Principle does not offer an explanation as to why a position of equilibrium alters. Explanations are only possible through the use of the equilibrium constant, *K*. This concept will be introduced in Topic 13 (Book 2: IAL).

CHECKPOINT

SKILLS REASONING

1. Ethanoic acid and ethanol react reversibly to produce ethyl ethanoate and water:

$$CH_3COOH(l) + CH_3CH_2OH(l) \rightleftharpoons CH_3COOCH_2CH_3(l) + H_2O(l)$$

Assume $\Delta H = 0\,kJ\,mol^{-1}$

(a) Predict the effect on the position of equilibrium of increasing the temperature. Justify your answer.

(b) Suggest why the reaction mixture is heated when used to prepare ethyl ethanoate.

2. When carbon dioxide dissolves in water, it forms a solution containing some carbonic acid:

$$CO_2(g) + H_2O(l) \rightleftharpoons H_2CO_3(aq)$$

Carbon dioxide is less soluble in hot water than in cold water. Is ΔH for the forward reaction negative or positive? Justify your answer.

3. A sample of insoluble solid lead(II) chloride, $PbCl_2$, is shaken with some dilute hydrochloric acid and left until an equilibrium mixture containing some undissolved lead(II) chloride is established:

$$PbCl_2(s) + 2Cl^-(aq) \rightleftharpoons PbCl_4^{2-}(aq)$$

State what would be observed if concentrated hydrochloric acid was added to the mixture. Justify your answer.

4. In each case, predict whether the equilibrium position shifts to the right or the left, or is unaltered when the pressure is increased at constant temperature.

(a) $H_2(g) + I_2(g) \rightleftharpoons 2HI(g)$

(b) $2SO_2(g) + O_2(g) \rightleftharpoons 2SO_3(g)$

(c) $2O_3(g) \rightleftharpoons 3O_2(g)$

(d) $4NH_3(g) + 5O_2(g) \rightleftharpoons 4NO(g) + 6H_2O(g)$

3 REVERSIBLE REACTIONS IN INDUSTRY

LEARNING OBJECTIVES

- Evaluate data to explain the necessity, for many industrial processes, of reaching a compromise between the yield and the rate of reaction.

APPLYING THE PRINCIPLES OF REACTION RATES AND REVERSIBILITY TO INDUSTRIAL PROCESSES

The principles of reaction rates and reversibility play an important role in the design and conditions for many industrial processes. In order to maximise profits, the major problems chemists face are to convert the reactants into the products:

- as quickly as possible
- as completely as possible.

The first problem relates to kinetics (rate of reaction) and the second problem relates to reversibility.

The solution to each of these problems requires a careful choice of reaction conditions. This is easily demonstrated by considering the Haber process for the manufacture of ammonia.

THE HABER PROCESS

Ammonia is manufactured in industry by direct **synthesis** from nitrogen and hydrogen:

$$N_2(g) + 3H_2(g) \rightleftharpoons 2NH_3(g) \quad \Delta H = -92\,kJ\,mol^{-1}$$

If the reaction mixture were to reach equilibrium, the maximum yield of ammonia would be obtained by using a low temperature (forward reaction is exothermic) and a high pressure (4 mol of gas on the left; 2 mol of gas on the right).

However, the reaction mixture does not reach a position of equilibrium in the reaction chamber. You can see this by analysing the graphs in **fig A**.

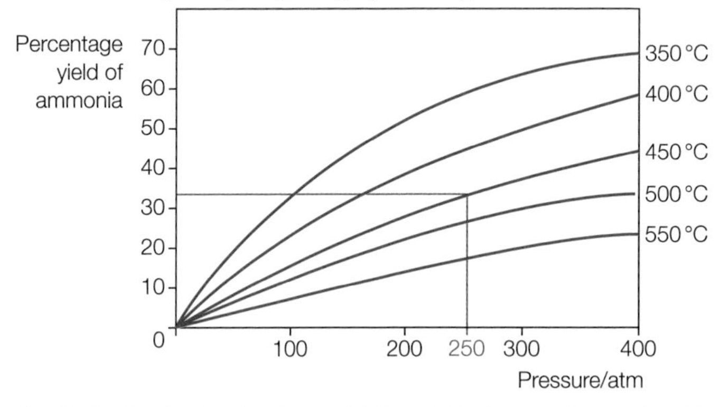

fig A Graphs showing the equilibrium yield of ammonia at different temperatures and pressures.

The conditions used in the Haber process are typically 450 °C and 250 atmospheres (atm) pressure. If the reaction mixture were to reach equilibrium, then the yield of ammonia would be just over 30%. In practice, the yield is approximately 15%. This is because the reaction mixture does not remain in the reaction chamber long enough for equilibrium to be established.

WHY ARE CONDITIONS OF 450 °C AND 250 ATM USED?

The reaction between nitrogen and hydrogen is extremely slow at room temperature. This is mainly because of the very strong nitrogen to nitrogen triple bond [$E(N{\equiv}N) = 945\,kJ\,mol^{-1}$] producing a high activation energy for the reaction. Even at high temperatures, the rate of reaction is slow in the absence of a suitable catalyst. Many metals will catalyse this reaction, including tungsten and platinum. However, these metals are very expensive, so iron is used in the industrial process.

The catalyst does not function very efficiently at low temperatures, so a relatively high temperature is necessary. However, a very high temperature would be uneconomical, because of the extra energy costs involved. It might also result in a decreased yield, since the increase in rate of the backward reaction will be greater than the increase in rate of the forward reaction. This is debatable, because the reaction mixture may not be in the reaction chamber for long enough for this to make a significant difference. For these reasons, a compromise temperature of 450 °C is used.

Under these conditions, the actual yield is about 50% of the equilibrium yield. If we assume this to be true for other pressures at a temperature of 450 °C, then a pressure of 100 atm would give a yield of around 12–13%, while a pressure of 400 atm would give a yield of around 27–28%. The higher the pressure, the larger the energy costs of compressing the gases: the lower the pressure, the lower the yield. Once again, a compromise is reached between yield and cost and a pressure of 250 atm is used.

In order to increase the efficiency of the process, the unreacted nitrogen and hydrogen, after separation from the ammonia, are mixed with fresh nitrogen and hydrogen and fed into the reaction chamber.

THE CONTACT PROCESS

Sulfuric acid is manufactured by the Contact process. The name of this process comes from the stage of the process that involves the reaction between sulfur dioxide and oxygen at the surface of a solid vanadium(V) oxide, V_2O_5, catalyst, to form sulfur trioxide.

$$SO_2(g) + \frac{1}{2}O_2(g) \rightleftharpoons SO_3(g) \qquad \Delta H = -96\,kJ\,mol^{-1}$$

The forward reaction is exothermic, so a low temperature would favour a high yield of $SO_3(g)$. You can see this in the graph in **fig B**, which shows the percentage yield of $SO_3(g)$ against temperature at a pressure of 1 atm.

Once again, the catalyst would not be very effective at low temperatures, so a moderately high temperature of 450 °C is used.

At 1 atm pressure, the yield of $SO_3(g)$ is already very high at around 97%. Higher pressures would increase the yield since there are fewer moles of gas on the right-hand side of the equation.

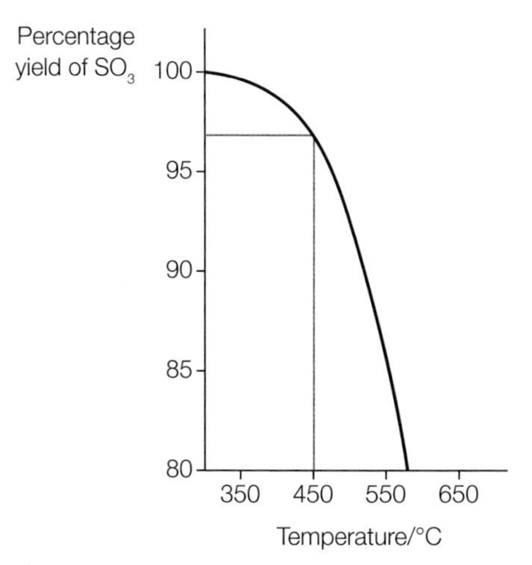

fig B A graph showing the percentage yield of sulfur trioxide against temperature at a pressure of 1 atm.

The pressure employed in the Contact process is about 2 atm. This is high enough to maintain a constant flow of gases through the reaction chamber. Pressures higher than this are unnecessary since the yield is already very high.

It is interesting to look at the mechanism of action of the vanadium(V) oxide. Unlike the iron in the Haber process, the vanadium changes its oxidation state during the reaction, but it then converts back to its original oxidation state at the end.

The mechanism for the reaction is:

$$SO_2(g) + V_2O_5(s) \rightleftharpoons SO_3(g) + V_2O_4(s)$$

$$V_2O_4(s) + \frac{1}{2}O_2(g) \rightleftharpoons V_2O_5(s)$$

CHECKPOINT

SKILLS REASONING

1. The first step in the manufacture of nitric acid, $HNO_3(l)$, from ammonia involves the exothermic reaction of ammonia with oxygen gas to form nitrogen monoxide NO(g) and steam. This is a reversible reaction that can reach a position of equilibrium.
   (a) Write an equation for the reaction between ammonia and oxygen to form nitrogen monoxide and steam.
   (b) Assuming the reaction mixture reaches a position of equilibrium, make a qualitative prediction of the conditions of temperature and pressure that would produce the maximum yield of NO(g).
   (c) In industry, the conditions used are high temperature and a pressure of around 7 atm. Suggest why these conditions are different to those you predicted in (b) to obtain a high yield of NO(g) in the equilibrium mixture. Again, assume that the reaction reaches a position of equilibrium.
2. The densities of diamond and graphite are 3.5 and 2.3 $g\,cm^{-3}$, respectively.

   The change from graphite to diamond can be represented by the equation:

   $$C(graphite) \rightleftharpoons C(diamond) \qquad \Delta H = +2\,kJ\,mol^{-1}$$

   Suggest the conditions of temperature and pressure that would favour the formation of diamond from graphite. Justify your answers.

SUBJECT VOCABULARY

synthesis the production of chemical compounds by reaction from simpler substances

IN AT THE DEEP END

SKILLS CREATIVITY

In this spread we will consider the role of chlorine in the treatment of water; specifically in swimming pools. Water-borne pathogens are still responsible for millions of deaths across the world and so effective treatment is important both for drinking water and for recreation. However, this article raises some additional issues about the use of chlorine.

SOMETHING IN THE WATER...

fig A 'A better way to cut the chloroform levels would be to reduce the amount of organic matter in the water.'

Public swimming pools contain high levels of a chemical linked to miscarriage.

A team led by Mark Nieuwenhuijsen at Imperial College, London, found that the chloroform* content of eight pools in the city was on average 20 times as high as in drinking water. The chloroform is formed when chlorine disinfectants react with organic compounds in the water.

Some studies in the US have suggested there's a correlation between the amount of chlorinated tap water drunk daily by pregnant women and their risk of miscarriage. Chloroform and related chemicals in the water have been blamed. Nieuwenhuijsen accepts that the studies linking chlorination and miscarriage are inconsistent. 'But pregnant women are advised to go swimming', he says. 'And there is a much higher level of chloroform in pools than in drinking water, so it could be a bigger pathway for exposure.'

Chloroform is produced when chlorine in the water reacts with flecks of skin, body-care products and other organic materials. Swimmers absorb the chemical through their skin, by swallowing water or by inhaling the gas.

Nieuwenhuijsen found an average of 113.3 micrograms of chloroform per litre in 40 samples from eight different pools. He also found that the more people using the pool and the warmer the water, the higher the concentration of chloroform.

But stopping chlorination is not the answer, says Nieuwenhuijsen, as alternative disinfectants such as ozone or ultraviolet light don't work so well. 'Chlorine is very effective', he says. 'A better way to cut the chloroform levels would be to reduce the amount of organic matter in the water by making sure people shower before a swim, and by improving filtration.'

*Chloroform = trichloromethane

From an article in *New Scientist* magazine

DID YOU KNOW?

Chlorine gas is an irritant that causes serious damage to the eyes and respiratory system by reacting with water to form hydrochloric acid. For this reason, it was used as a chemical weapon in the First World War. The strong odour and the green colour of the gas made it easy to detect so more effective alternatives were soon developed. One of these was phosgene – a gas with the chemical formula $COCl_2$. The gas was formed by the reaction of chlorine and carbon monoxide in the presence of light so it was named by combining the Greek 'phos' (meaning light) and 'genesis' (meaning birth). Since 1925, over 190 states have signed up to the Geneva Protocol, which prohibits the use of chemical and biological weapons.

SCIENCE COMMUNICATION

1. You can have some fun with this question! Imagine you are the editor of a local newspaper. It is a slow news day and you have to lead with this article and give the story maximum impact. Using some illustrations and the key parts of this article, write the most sensational article you can, using only the information given above.

CHEMISTRY IN DETAIL

2. (a) The IUPAC name for chloroform is trichloromethane. Draw out the displayed formula for trichloromethane.
   (b) Describe the shape of the trichloromethane molecule and, using diagrams, explain why the molecule has a permanent dipole.
3. Convert 113.3 μg/l into a micromolar concentration. (1 $\mu g = 1 \times 10^{-6}$ g)
4. When chlorine is added to water the following equilibrium is set up:

   $Cl_2(aq) + H_2O(l) \rightleftharpoons HOCl(aq) + HCl$

   (a) The reaction shown is an example of a disproportionation reaction. State what 'disproportionation' means.
   (b) There are two acids formed in the reaction above: HCl is a strong acid, and HOCl is a weak acid. Explain the difference between a strong and a weak acid.
5. When HOCl is dissolved in water, the following equilibria is set up:

   $HOCl(aq) + H_2O(l) \rightleftharpoons H_3O^+(aq) + OCl^-(aq)$

   (a) How would an increase in the pH value of this solution affect the above equilibrium? Explain your answer.
   (b) How would a greater dilution of HOCl(aq) affect the above equilibrium? Explain your answer.
6. A second chloro-organic molecule detected in the analysis in the study had the molecular formula $C_2H_3Cl_3$. Draw both structural isomers of $C_2H_3Cl_3$ and name them.

WRITING SCIENTIFICALLY

When you are asked to state the meaning of a word or phrase, you'll need to describe the term clearly and concisely in your own words. There might be more than one way it can be described.

ACTIVITY

Design an experiment to investigate how a change of pH in a sample of swimming pool water might influence bacterial growth. Your plan should include:

- all the important variables that need to be considered
- how bacterial growth might be quantified experimentally
- how you would ensure experimental reliability.

It might be worth looking at other experimental protocols that you can find either in science books or on the internet. This may help you appreciate the variables that are important and how they can be controlled.

SKILLS CREATIVITY

THINKING BIGGER TIP

Think of everything you have learned so far in the course and how it fits together to inform your understanding as a scientist. By now, you should be able to use correct terminology, analyse sources, write scientifically, and think like a scientist.

9 EXAM PRACTICE

1 Look at the following statements. Which statement explains correctly why a small increase in temperature leads to a significant increase in the rate of a gas phase reaction?

A The frequency of collisions between molecules is greater at a higher temperature.

B The activation energy of the reaction is less when the gases are at higher temperature.

C The frequency of collisions between molecules with kinetic energy greater than the activation energy is greater at a higher temperature.

D The average kinetic energy of the molecules is slightly greater at a higher temperature. [1]

(Total for Question 1 = 1 mark)

2 An exothermic reaction proceeds by two stages.

$$\text{reactants} \xrightarrow{\text{stage 1}} \text{intermediate} \xrightarrow{\text{stage 2}} \text{products}$$

The activation energy of stage 1 is $50\,\text{kJ}\,\text{mol}^{-1}$.

The overall enthalpy change of reaction is $-100\,\text{kJ}\,\text{mol}^{-1}$.

Which diagram could represent the energy profile diagram for the reaction?

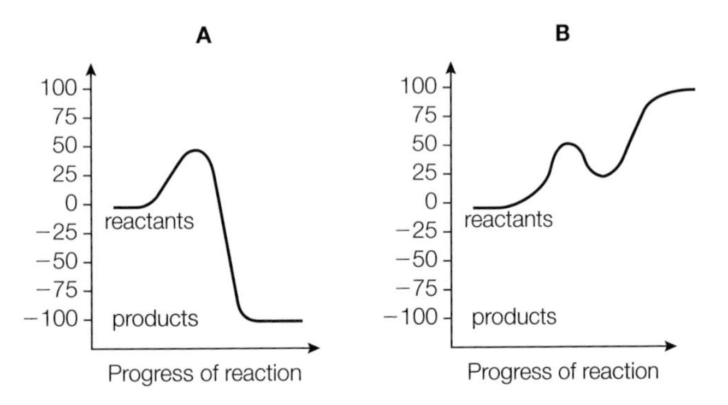

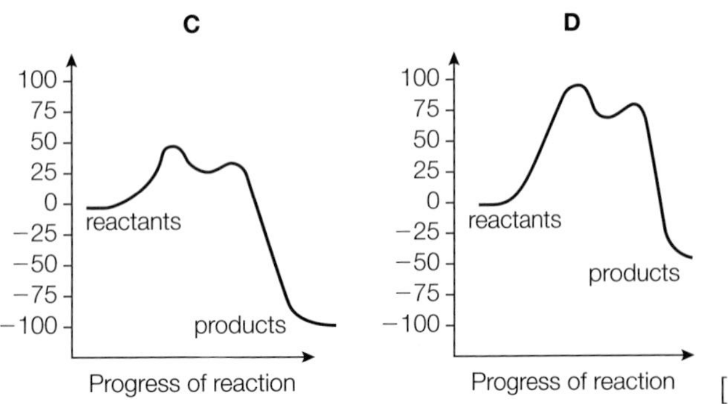

[1]

(Total for Question 2 = 1 mark)

3 The distribution of the fraction of molecules with energy E is given in the diagram for two different temperatures T_1 and T_2. T_2 is greater than T_1.

The letters P, Q and R represent separate and different shaded areas under the curves.

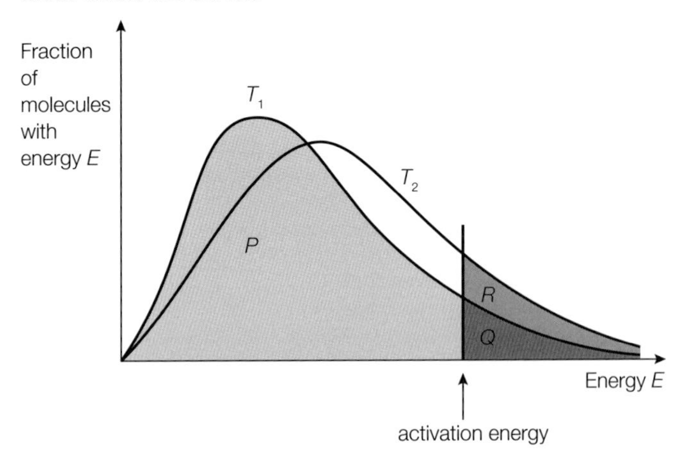

Which expression gives the fraction of molecules that have energy equal to or greater than the marked activation energy?

A $\frac{Q}{P}$ **B** $\frac{Q+P}{P}$ **C** $\frac{Q+R}{P+Q}$ **D** $\frac{Q+R}{P+Q-R}$ [1]

(Total for Question 3 = 1 mark)

4 Curve **X** in the diagram shows the volume of oxygen given off during the decomposition of $100\,\text{cm}^3$ of $1.0\,\text{mol}\,\text{dm}^{-3}$ hydrogen peroxide, catalysed by manganese(IV) oxide.

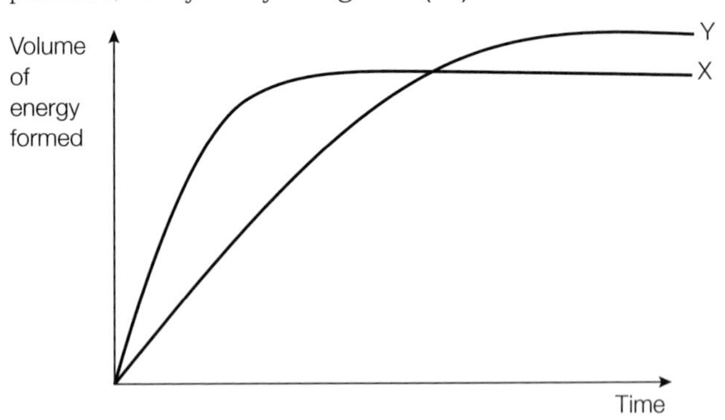

Which of the following alterations to the original experimental conditions would produce Curve **Y**?

A adding water

B adding some $0.1\,\text{mol}\,\text{dm}^{-3}$ hydrogen peroxide

C using less manganese(IV) oxide

D lowering the temperature [1]

(Total for Question 4 = 1 mark)

5 The enthalpy level diagram represents a reaction occurring with and without a catalyst.

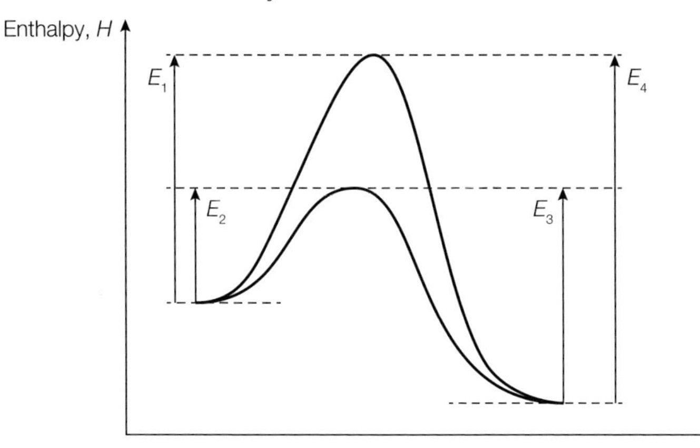

Which of the following statements is correct?

A E_4 is the activation energy for the reverses reaction.

B The forward reaction, with catalyst, is endothermic.

C The enthalpy change of reaction is $(E_2 - E_3)$.

D The enthalpy change of reaction is reduced using a catalyst. [1]

(Total for Question 5 = 1 mark)

6 (a) *R* reacts with *T* in a reversible, exothermic reaction. The reaction is allowed to reach equilibrium.

The concentration of one reactant, *R*, is plotted against time.

The graph shows the results.

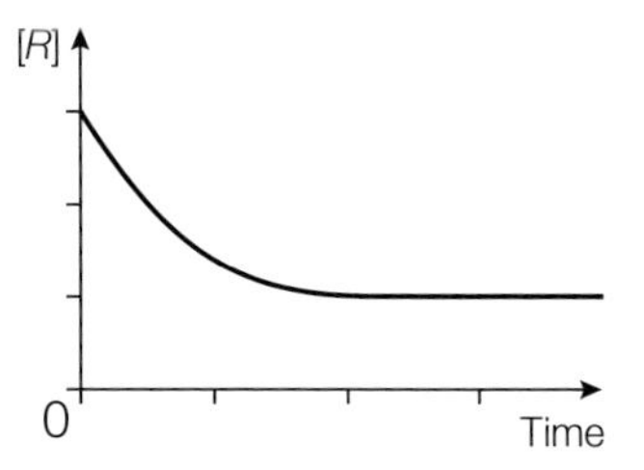

Which of the graphs below could be obtained if the reaction were repeated at a higher temperature, but with the same concentrations of *R* and *T*?

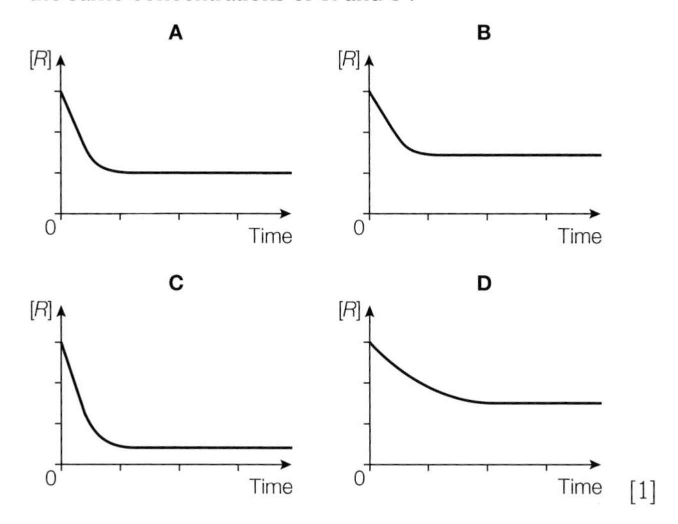

[1]

(b) Which statement about the effect of a catalyst on an endothermic, reversible reaction that is in dynamic equilibrium is correct?

A It increases decreases the yield of the product.

B It increases the rate of the forward reaction but not that of the backward reaction.

C It increases the rate of the backward reaction but not that of the forward reaction.

D It increases the rate of both the forward and the backward reactions. [1]

(Total for Question 6 = 2 marks)

7 This question is about some graphs that you may have seen during your course.

(a) The graph below is a Maxwell-Boltzmann distribution curve.

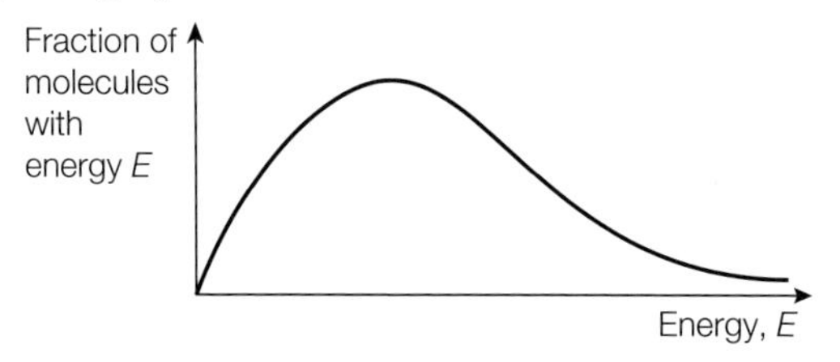

Explain, using the graph, the effect of adding a catalyst on the rate of reaction. [4]

(b) The graph below shows the change in mass observed in an experiment to investigate the rate of reaction between marble chips and dilute hydrochloric acid.

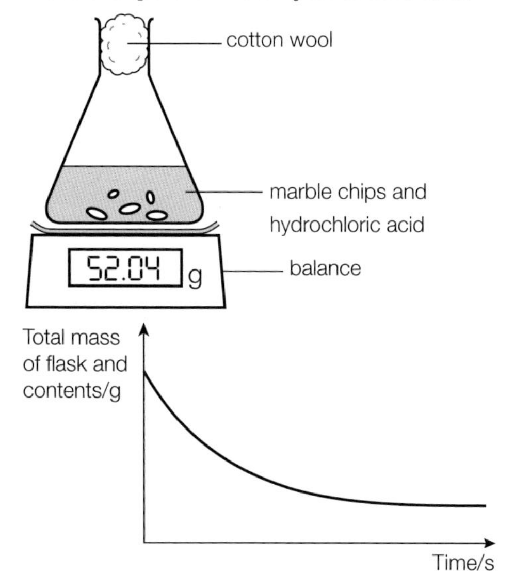

(i) State the purpose of the cotton wool in the neck of the conical flask. [1]

(ii) Explain why the mass of the flask and contents decreases during the course of the experiment. [2]

(iii) The experiment was repeated at the same temperature, with the same volume and concentration of acid and with the same mass of powdered marble.

Sketch, on the graph, the curve you would expect to obtain. [2]

(Total for Question 7 = 9 marks)

8 This question is about the effect that an increase in temperature has on the rate of a chemical reaction.

(a) A student found this statement in a text book:

'The rate of a chemical reaction increases as the temperature increases because the particles collide more often.'

Discuss the extent to which this statement is true. [5]

(b) The graph below shows a Maxwell-Boltzmann distribution curve for a gas at temperature T_1.

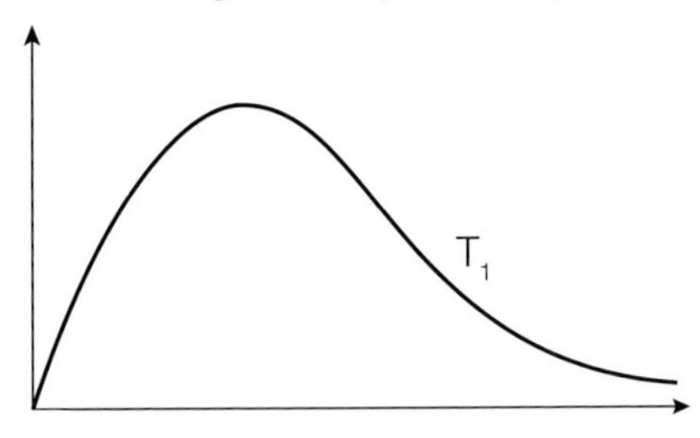

(i) Label the axes on the diagram. [2]

(ii) On the graph, sketch a curve to represent the distribution at higher temperature.

Label this curve T_2. [2]

(Total for Question 8 = 9 marks)

9 The graph below shows the reaction profile for the uncatalysed decomposition of hydrogen peroxide (H_2O_2) into water and oxygen.

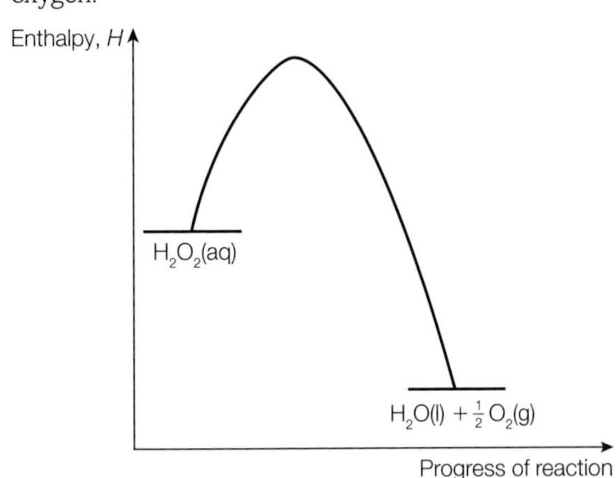

(a) On the graph, label the activation energy (E_a) and the enthalpy change of reaction (ΔH). [2]

(b) The decomposition is catalysed by the addition of solid manganese(IV) oxide (MnO_2).

(i) Draw the curve to represent the catalysed decomposition. [2]

(ii) State, in principle, how the catalyst works in this reaction. [2]

(c) State whether the reaction is exothermic or endothermic. Justify your answer. [1]

(Total for Question 9 = 7 marks)

10 Many chemical reactions are easily reversible and can form equilibrium mixtures. Hydrogen and iodine react together in the gaseous state to form hydrogen iodide:

$$H_2(g) + I_2(g) \rightleftharpoons 2HI(g) \qquad \Delta_r H^\ominus = +52\,kJ\,mol^{-1}$$

Hydrogen (colourless) and iodine vapour (purple) are placed into a sealed container at 400 °C and allowed to reach equilibrium with hydrogen iodide (colourless). The resulting mixture has a very pale purple colour.

(a) When the reaction mixture is cooled to 200 °C, the reaction mixture becomes notably darker in colour. Give a reason for this observation in terms of the change to the equilibrium composition of the mixture. [2]

(b) In another experiment, all three gases were mixed together in a sealed tube and once again allowed to establish an equilibrium. The concentrations of the gases were monitored and the results were plotted on a graph. At time *t*, a change was made to the composition of the mixture.

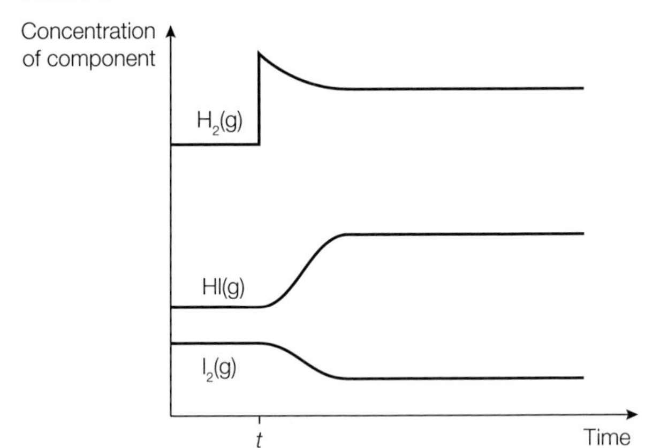

(i) State the change that was made to the reaction mixture at time *t*. [1]

(ii) Describe the changes that occur to the reaction mixture after time *t*. [3]

(Total for Question 10 = 6 marks)

11 Many industrial processes involve reversible reactions. Chemists were investigating the production of a chemical XY_2 that can be formed from X_2 and Y_2 as shown in the equation

$$X_2(g) + 2Y_2(g) \rightleftharpoons 2XY_2(g)$$

The chemists carried out a series of experiments at different temperatures, each time allowing the reaction mixture to establish equilibrium.

The chemists measured the percentage conversion of Y_2 at each temperature.

The results are shown in the graph.

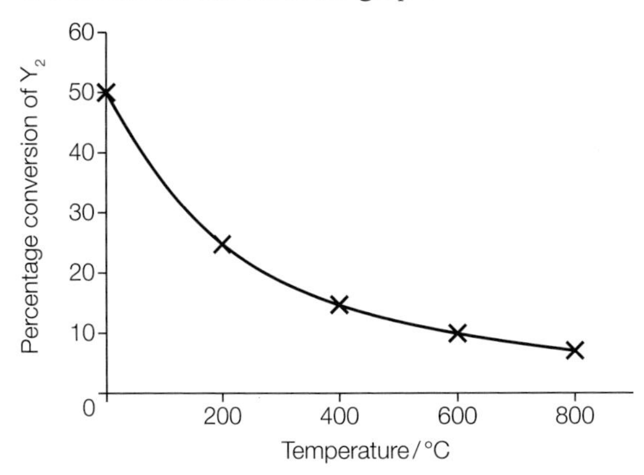

(a) Explain how the graph shows that the forward reaction is exothermic. [2]

(b) The chemists decide to use a catalyst in the process. Explain the effect of a catalyst on:

(i) the rate at which equilibrium is established

(ii) the percentage conversion of Y_2 at equilibrium. [4]

(Total for Question 11 = 6 marks)

12 Methanol (CH_3OH) can be manufactured from carbon monoxide and hydrogen.

$$CO(g) + 2H_2(g) \rightleftharpoons CH_3OH(g) \quad \Delta_r H^{\ominus} = -128\,kJ\,mol^{-1}$$

(a) Predict how the composition of the equilibrium mixture is affected by:

(i) carrying out the reaction at a higher temperature but keeping the pressure the same

(ii) carrying out the reaction at a higher pressure but keeping the temperature the same.

In each case justify your answer. [4]

(b) Assuming the reaction takes place by collision of molecules in the gas phase, explain the effect on the rate of reaction of increasing the pressure. [2]

(Total for Question 12 = 6 marks)

13 Graphite and diamond are two allotropes of carbon. Artificial diamonds can be made from graphite.

$$C(graphite) \rightleftharpoons C(diamond) \quad \Delta_r H^{\ominus} = +1.8\,kJ\,mol^{-1}$$

Substance	Density / g cm^{-3}
graphite	2.25
diamond	3.51

(a) Complete the enthalpy profile diagram below for the conversion of graphite into diamond. Label the enthalpy change, ΔH. [2]

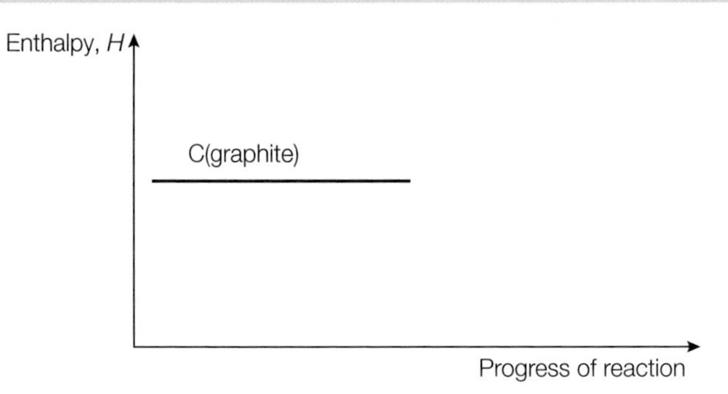

(b) State and justify which allotrope is the more thermodynamically stable. [1]

(c) State and justify in which allotrope the atoms take up less space per mole of substance. [1]

(d) Predict the conditions of temperature and pressure that are required to convert graphite into diamond. Justify your predictions. [4]

(Total for Question 13 = 8 marks)

14 An acid-base indicator, HIn, dissolves in water to form a green solution. The following equilibrium is established:

$$HIn(aq) \rightleftharpoons H^+(aq) + In^-(aq)$$
(yellow) (blue)

The indicator appears green because it contains sufficient amounts of both the yellow and blue species.

(a) Hydrochloric acid is added to the indicator solution. State what you would observe. Give a reason for your answer. [2]

(b) Aqueous sodium hydroxide is added dropwise to the resulting solution from (a) until no further colour change takes place. State all of the colours that would be observed. Give reasons for your answers. [4]

(Total for Question 14 = 6 marks)

15 The hydrogen required for the manufacture of ammonia and margarine is made by reacting methane with steam. The two gases react in a reversible reaction as shown by the equation:

$$CH_4(g) + H_2O(g) \rightleftharpoons CO(g) + 3H_2(g) \quad \Delta_r H^{\ominus} = +210\,kJ\,mol^{-1}$$

(a) State the conditions of temperature and pressure that would provide a high yield of hydrogen. Justify your answers. [4]

(b) State the conditions of temperature and pressure that would produce a fast rate of reaction. Justify your answers. Assume that the reaction takes place by molecules colliding in the gas phase. [4]

(c) Use your answers to (a) and (b) to explain why a temperature of 800 °C and a pressure of 30 atm are used for the manufacture of hydrogen from methane. [3]

(Total for Question 15 = 11 marks)

TOPIC 10 ORGANIC CHEMISTRY: HALOGENOALKANES, ALCOHOLS AND SPECTRA

A GENERAL PRINCIPLES | B HALOGENOALKANES | C ALCOHOLS | D MASS SPECTRA AND IR

In **Topics 4 and 5**, you learned about the basics of organic chemistry and about two homologous series, alkanes and alkenes. In this topic, you will learn about two more homologous series, halogenoalkanes and alcohols. There is also a new reaction mechanism to consider, nucleophilic substitution.

Although you have already learned something about mass spectrometry, we will build on this knowledge to see how the technique was developed to become a useful tool to discover the structures of organic compounds. A new technique, infrared spectroscopy, is introduced to show another way to discover different information about the structures of organic compounds.

MATHS SKILLS FOR THIS TOPIC

- Use ratios to construct and balance equations
- Represent chemical structures using angles and shapes in 2D and 3D structures

What prior knowledge do I need?

Topic 1D

- Calculation of empirical and molecular formulae

Topic 4

- The names of simple organic compounds
- Homologous series and general formula
- Representing organic compounds by structural formulae

Topics 4B and 5

- Alkanes and alkenes

What will I study in this topic?

- The hydrolysis reactions of halogenoalkanes and how to compare their rates experimentally
- Explaining the different rates of hydrolysis reactions of halogenoalkanes
- Substitution and elimination reactions of halogenoalkanes
- Oxidation and hydration reactions of alcohols
- Converting alcohols to halogenoalkanes
- Practical techniques used to prepare and purify organic compounds
- How fragmentation in mass spectrometry can give information about organic structures
- How infrared spectroscopy can give information about bonds in organic structures

What will I study later?

Topic 15A (Book 2: IAL)

- Optical isomerism

Topics 15B and 15C (Book 2: IAL)

- Reactions of carbonyl and carboxyl compounds

Topic 15D (Book 2: IAL)

- Condensation polymers

Topic 18 (Book 2: IAL)

- The type of bonding in benzene and other aromatic compounds

Topic 19 (Book 2: IAL)

- Nitrogen-containing compounds, including amino acids and proteins

Topic 20 (Book 2: IAL)

- Planning reaction schemes to prepare organic compounds
- More practical techniques for preparing and purifying organic compounds

10A GENERAL PRINCIPLES IN ORGANIC CHEMISTRY

SPECIFICATION REFERENCE

10.1 10.2 10.3 10.5

LEARNING OBJECTIVES

- Be able to classify reactions as addition, elimination, substitution, oxidation, reduction, hydrolysis or polymerisation.
- Understand the concept of a reaction mechanism.
- Understand that heterolytic bond breaking results in species that are electrophiles or nucleophiles.
- Understand the link between bond polarity and the type of reaction mechanism a compound will undergo.

REACTIONS IN ORGANIC CHEMISTRY

In **Topic 4**, you learned that there were five different types of organic reaction. In this topic you will learn about two more types. All seven types of reaction are considered here so that you don't have to refer back to **Topic 4**.

ADDITION REACTIONS

In this type of reaction, two reactant species combine together to form a single product species. Usually all the species are molecules. A general equation for an addition reaction is:

$$A + B \rightarrow C$$

One example is the reaction between ethene and bromine:

$$C_2H_4 + Br_2 \rightarrow C_2H_4Br_2$$

ELIMINATION REACTIONS

This is one of the two new types of reaction you are learning about. In this type, two atoms or groups are removed from a molecule (they are *eliminated* from the molecule). The atoms or groups are removed from neighbouring carbon atoms. This results in the formation of a second bond between the two carbon atoms. A general equation for an elimination reaction is:

```
  H   H   H   H                  H   H   H   H
  |   |   |   |                  |   |   |   |
—C—C—C—C—        ———→      —C—C=C—C—   +  XY
  |   |   |   |                  |           |
  H   X   Y   H                  H           H
```

X and Y represent atoms or groups of atoms. In this example, the organic product has a double bond between the central carbon atoms and a small molecule, XY, is also formed.

One example you will meet later is the reaction between bromoethane and ethanolic potassium hydroxide:

$$CH_3CH_2Br + KOH \rightarrow CH_2{=}CH_2 + KBr + H_2O$$

The small molecule eliminated is HBr (the H comes from the left-hand carbon, and the Br from the right-hand carbon). This is not easy to see just by looking at the equation. This is because the H goes to the H_2O molecule and the Br to KBr.

SUBSTITUTION REACTIONS

In this type of reaction, two reactant species combine together to form two product species. Usually all the species are molecules or ions. A general equation for a substitution reaction is:

$$A + B \rightarrow C + D$$

One example is the reaction between bromoethane and potassium hydroxide. Potassium hydroxide is an ionic compound, and as the potassium ion is a spectator ion, the reaction only involves the hydroxide ion. An equation for this reaction is:

$$C_2H_5Br + OH^- \rightarrow C_2H_5OH + Br^-$$

In this reaction, the OH group has taken the place of, or *substituted*, the Br atom. You might think that this type of reaction is like the displacement reactions you met in **Topic 8C.2**. The equation for one of these displacement reactions is $Mg + CuSO_4 \rightarrow MgSO_4 + Cu$. However, this equation can be simplified to $Mg + Cu^{2+} \rightarrow Mg^{2+} + Cu$, which means that electrons are being transferred from Mg to Cu^{2+}, so it is a redox reaction. The reaction between bromoethane and the hydroxide ion is not a redox reaction, so it is better not to think of substitution and displacement reactions as being the same type.

OXIDATION REACTIONS

In this type of reaction, one organic compound is oxidised, usually by an inorganic reagent. This means that the organic compound can either lose hydrogen or gain oxygen. There isn't a suitable general equation that can be used for this type of reaction, but here is one example you will see later in this topic: the oxidation of ethanol by a mixture of potassium dichromate(VI) and sulfuric acid. The equation is not written to include the inorganic reagent because it would be very complicated. Usually the oxygen atoms produced by the oxidising agent are shown using the symbol [O], so the equation then becomes:

$$C_2H_5OH + [O] \rightarrow CH_3CHO + H_2O$$

You can now see why this reaction is classified as oxidation: the ethanol molecule loses two hydrogen atoms.

EXAM HINT

It is important to remember that different oxidising agents carry out different oxidation reactions. For example, acidified potassium manganate(VII) will oxidise an alkene to a diol. However, acidified potassium dichromate is needed to oxidise a primary or secondary alcohol.

REDUCTION REACTIONS

In this type of reaction, an organic compound is reduced, sometimes by hydrogen gas and a catalyst and sometimes by an inorganic reagent. This means that the organic compound can either gain hydrogen or lose oxygen. There isn't a suitable general equation that can be used for this type of reaction, but here is one example you saw in **Topic 5A.3**: the reduction of an alkene to an alkane by hydrogen gas and a nickel catalyst. The equation for the reaction is:

$$C_2H_4 + H_2 \rightarrow C_2H_6$$

You can now see why this reaction is classified as reduction: the ethene molecule gains two hydrogen atoms. Note that this is also an example of an addition reaction.

HYDROLYSIS REACTIONS

This is another new type of reaction. In hydrolysis reactions, an organic compound reacts with water. The OH group of water replaces an atom or group in the organic compound. A general equation for a hydrolysis reaction is:

$$RX + H_2O \rightarrow ROH + HX$$

You might think that this is an example of a substitution reaction, and you would be right! Hydrolysis reactions are really substitution reactions in which the OH from a water molecule replaces an atom or group (usually a halogen) in an organic compound.

POLYMERISATION REACTIONS

At IAS, all the polymerisation reactions you see are examples of addition polymerisation. If you study chemistry in more detail at IAL, you will also meet examples of condensation polymerisation. However, we will not consider these in this topic. In addition polymerisation, very large numbers of a reactant molecule (sometimes of two different reactant molecules) react together to form one very large product molecule. A general equation for a polymerisation reaction is:

$$n\,\text{WYC}{=}\text{CXZ} \longrightarrow \left[\!-\text{C(W)(Y)}-\text{C(X)(Z)}-\!\right]_n$$

A familiar example of this type of reaction is the polymerisation of ethene to poly(ethene).

REACTION MECHANISMS

You have already learnt about reaction mechanisms in **Topics 4** and **5**. Just to remind you, a mechanism tries to explain the actual changes that occur during a reaction, especially in the bonding between the atoms. A mechanism is a sequence of two or more steps, each one represented by an equation, that shows how a reaction takes place.

So far in this book, you have met two different types of reaction mechanism:

Mechanism 1 free radical substitution in alkanes

Mechanism 2 electrophilic addition to alkenes

In this topic, you will look at one more:

Mechanism 3 nucleophilic substitution in halogenoalkanes

HOMOLYTIC AND HETEROLYTIC BOND BREAKING

In mechanism 1, the type of bond breaking is homolytic. This is when a covalent bond breaks, and each atom keeps one electron from the shared pair of electrons in the bond. Free radicals are formed.

In mechanisms 2 and 3, the type of bond breaking is heterolytic, when a covalent bond breaks, and one atom keeps both electrons from the shared pair of electrons in the bond. This atom becomes a negative ion, and can act as a nucleophile. The other atom becomes a positive ion, and can act as an electrophile.

ELECTROPHILES AND NUCLEOPHILES

In mechanism 1, the attacking species are free radicals, such as Cl•.

In mechanism 2, the attacking species are electrophiles, such as the H of an HBr molecule.

In mechanism 3, the attacking species are nucleophiles, such as the OH^- ion.

THE ROLE OF BOND POLARITY IN MECHANISMS

In mechanism 1, the reactants are alkanes and halogens, whose bonds are either non-polar or only very slightly polar. This means that the type of bond breaking is most likely to be homolytic.

In mechanisms 2 and 3, the reactants are molecules such as hydrogen halides and halogenoalkanes, whose bonds are polar. This means that the type of bond breaking is most likely to be heterolytic.

LEARNING TIP

You can often work out the type of an unfamiliar reaction by looking at its equation. Also, some reactions can be classified as more than one type, such as addition and reduction, if they both involve hydrogen reacting with an organic reactant to form a single organic product.

CHECKPOINT

1. The equations for some reactions are given below. Classify each reaction, by looking at the equations.

(a) $CH_3CHClCH_3 + LiOH \rightarrow CH_3CH(OH)CH_3 + LiCl$

(b) $CH_3CH{=}CH_2 + H_2O \rightarrow CH_3CH(OH)CH_3$

(c) $CH_3CH_2CHO + [O] \rightarrow CH_3CH_2COOH$

(d) $CH_3CHClCH_3 + NaOH \rightarrow CH_3CH{=}CH_2 + NaCl + H_2O$

2. A student wrote a mechanism for the reaction between propene and hydrogen chloride:

$CH_3-CH{=}CH_2$ + H—Cl ⟶ $CH_3-\overset{\oplus}{C}H-CH_3$ + $Cl^{\ominus}$

$CH_3-\overset{\oplus}{C}H-CH_3$ + $:Cl^{\ominus}$ ⟶ $CH_3-CHCl-CH_3$

Identify the mistakes in the student's mechanism.

10B 1 HALOGENOALKANES AND HYDROLYSIS REACTIONS

SPECIFICATION REFERENCE

10.4 10.6 10.7

LEARNING OBJECTIVES

- Understand the nomenclature of halogenoalkanes and be able to draw their structural, displayed and skeletal formulae.
- Understand the distinction between primary, secondary and tertiary halogenoalkanes.
- Know the definition of the term 'nucleophile'.

WHAT ARE HALOGENOALKANES?

The halogenoalkanes are a homologous series of compounds with the general formula $C_nH_{2n+1}X$. Think of them as the result of replacing a hydrogen atom in a hydrocarbon by a halogen atom. X represents a halogen atom – usually bromine or chlorine, but it could also be fluorine or iodine. You have already met halogenoalkanes earlier in this book – chloromethane is a halogenoalkane and is the product of the reaction between methane and chlorine.

The symbol R is often used in organic chemistry to represent any alkyl group (such as methyl or ethyl). The formula of a halogenoalkane could be simplified to RX.

The number of halogen atoms in a halogenoalkane molecule can be more than one, so the general formula is different for these compounds (for example, $C_nH_{2n}X_2$ and $C_nH_{2n-1}X_3$ are other possible general formulae).

▲ **fig A** The most problematic halogenoalkanes are those containing both chlorine and fluorine. They are known as chlorofluorocarbons (CFCs for short) and were used in aerosol cans for decades. We have known for a long time that they damage the ozone layer in the upper atmosphere and, through international agreement, they are mostly no longer used.

NAMING HALOGENOALKANES

We have already met the names of some halogenoalkanes, but **table A** shows some examples to remind you.

EXAM HINT

Practise drawing skeletal formulae for halogenoalkanes. This will help you identify the different isomers for a given molecular formula.

STRUCTURE	NAME
$CH_2Cl-CHCl-CH_3$	1,2-dichloropropane
$CH_2Br-CH_2-CH_2Cl$	1-bromo-3-chloropropane
CCl_4	tetrachloromethane
$CH_3-CHF-CH_2-CH_3$	2-fluorobutane

table A Examples of how to name halogenoalkanes.

If there are two or more different halogens, then the prefixes used to represent the halogens appear in alphabetical order. As with other homologous series, the longest carbon chain is the basis of the name, and when the prefix numbers are added together the total should be as small as possible.

CLASSIFYING HALOGENOALKANES

You have met the terms 'primary', 'secondary' and 'tertiary' in the topic referring to carbocations (see **Topic 5A.4**).

Halogenoalkanes are classified in a similar way, depending on the number of alkyl groups joined to the C atom bonded to the halogen atom. **Table B** shows some examples.

STRUCTURE	ABBREVIATED FORMULA	NUMBER OF ALKYL GROUPS	CLASSIFICATION
$CH_3{-}CH_2{-}CH_2F$	RX	1	primary
$CH_3{-}CHBr{-}CH_3$	R_2CHX	2	secondary
$(CH_3)_2CCl{-}CH_2{-}CH_3$	R_3CX	3	tertiary

table B Examples of how to classify halogenoalkanes.

WHAT MAKES HALOGENOALKANES REACTIVE?

Hydrocarbons contain only hydrogen and carbon atoms, which have similar electronegativities, so their bonds are almost non-polar.

Halogenoalkanes contain a halogen atom with an electronegativity higher than that of carbon, so the C—X bond is polar. This bond polarity can be indicated using the partial charges $\delta+$ and $\delta-$.

Down Group 7 of the Periodic Table, the electronegativities of the halogens decrease from fluorine to iodine, so the polarity of the C—X bond also decreases.

H
|
H — C$^{\delta+}$ — Br$^{\delta-}$
|
H

The carbon atom joined to the halogen is always the slightly positive ($\delta+$) or electron-deficient part of the molecule, and this is what makes halogenoalkanes react as they do.

These carbon atoms attract other species called **nucleophiles**. 'Nucleo' indicates positive and 'phile' indicates liking, so nucleophiles are species that are attracted to slightly positive or electron-deficient parts of a molecule. Nucleophiles are either negative ions or molecules with a slightly negative atom, but they always use a lone pair of electrons when attacking another species.

We will look at the mechanisms of halogenoalkane reactions in **Topic 10B.3**, but first let us take a look at a reaction that can be used to compare the reactivities of different halogenoalkanes.

HYDROLYSIS REACTIONS

When a halogenoalkane is added to water, a reaction begins, but it may take some time to complete. A water molecule contains polar bonds, and the $\delta-$ oxygen atom in water is attracted to the $\delta+$ carbon atom in the halogenoalkane. The reaction that occurs can be represented by the equation below, in which R represents any alkyl group:

$$RX + H_2O \rightarrow ROH + HX \text{ or } RX + H_2O \rightarrow ROH + H^+ + X^-$$

The product ROH is an alcohol and, as both organic substances are usually colourless liquids, no colour change can be seen. You can see from this simple equation that the C—X bond breaks, so the RX molecule breaks into two parts (R and X), although the R group then combines with the OH group of water.

This type of reaction is known as a **hydrolysis reaction**. 'Hydro' refers to water and 'lysis' refers to splitting, so it means splitting with water.

LEARNING TIP

In an organic formula, the symbol R can be used to represent any alkyl group, such as methyl or ethyl.

This should not be confused with the same symbol R used in a completely different context for the gas constant in the equation $pV = nRT$.

CHECKPOINT

1. Write the IUPAC name for each of these halogenoalkanes and classify them as primary, secondary or tertiary.

(a) $CH_3{-}C(CH_3)(Cl){-}CH_2{-}CH_3$

(b) $CH_3{-}CH(Br){-}CH_2{-}CH_3$

(c) $CH_3{-}CH(CH_3){-}CH_2{-}I$

2. Why are nucleophiles more strongly attracted to fluoroalkanes than to chloroalkanes?

SUBJECT VOCABULARY

nucleophile a species that donates a lone pair of electrons to form a covalent bond with an electron-deficient atom

hydrolysis reaction a reaction in which water or hydroxide ions replace an atom in a molecule with an –OH group

10B 2 COMPARING THE RATES OF HYDROLYSIS REACTIONS

SPECIFICATION REFERENCE

10.10 10.12 10.13 CP5

LEARNING OBJECTIVES

- Understand that experimental observations and data can be used to compare the relative rates of hydrolysis of primary, secondary and tertiary structural isomers of a halogenoalkane, and primary chloro-, bromo- and iodo-alkanes, using aqueous silver nitrate in ethanol.
- Know the trend in reactivity of primary, secondary and tertiary halogenoalkanes.
- Understand, in terms of bond enthalpy, the trend in reactivity of chloro-, bromo- and iodo-alkanes.

PRACTICAL ASPECTS

If instead of adding water in a hydrolysis reaction, silver nitrate solution is added, then the progress of the reaction can be followed. You may remember that silver nitrate can be used in a test for halide ions because the silver ions in silver nitrate react with the halide ions formed in the hydrolysis to give a precipitate:

$$Ag^+ + X^- \rightarrow AgX$$

This means that we can tell how quickly the hydrolysis reaction occurs by observing how quickly the precipitate of AgX forms.

Without going into full practical details, a comparison of the rates of these reactions involves:

- using ethanol as a solvent for the mixture (halogenoalkanes and aqueous silver nitrate do not mix, but form separate layers)
- controlling variables such as temperature and the concentration and quantity of halogenoalkanes
- timing the appearance of the precipitate (although this is difficult to do accurately because a precipitate may first appear faint, but become thicker with time).

In an experiment, three tubes containing silver nitrate dissolved in ethanol are left at the same temperature for several minutes. Each tube contains a different halogenoalkane. **Fig A** shows the tubes after several minutes.

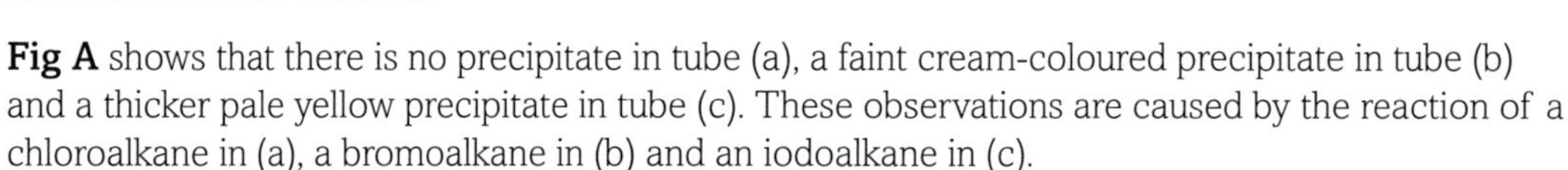

Fig A shows that there is no precipitate in tube (a), a faint cream-coloured precipitate in tube (b) and a thicker pale yellow precipitate in tube (c). These observations are caused by the reaction of a chloroalkane in (a), a bromoalkane in (b) and an iodoalkane in (c).

fig A Three tubes containing silver nitrate dissolved in ethanol, with a different halogenoalkane added to each one.

Two types of comparison can be made. You can:

- compare halogenoalkanes with the same structure but containing different halogens
- compare halogenoalkanes containing the same halogen but with different structures.

Fluoroalkanes are comparatively very unreactive, so reactions involving them are often omitted.

INTERPRETING THE RESULTS FOR DIFFERENT HALOGENS

Table A shows the trend when the halogen is different.

SAME STRUCTURE BUT DIFFERENT HALOGEN	RESULT
1-iodobutane	fastest
1-bromobutane	
1-chlorobutane	slowest

table A Examples comparing halogenoalkanes with the same structure but with different halogens.

You might suppose that the halogenoalkane with the most polar bond would be the fastest to be hydrolysed (in this case, 1-chlorobutane) because the $\delta+$ charge on the carbon atom is greatest, so the attacking nucleophile should be attracted more strongly.

This is true, but there is another, more important factor to consider. Bond breaking requires energy, and weaker bonds break more easily than stronger bonds. **Table B** shows (mean) bond enthalpies. You can see that the C—I bond is the weakest and the C—Cl bond is the strongest.

BOND	BOND ENTHALPY / kJ mol^{-1}
C–Cl	+346
C–Br	+290
C–I	+228

table B Mean bond enthalpies.

Under the same conditions, the C—I bond breaks most easily, forming I^- ions, and so a precipitate of AgI forms more quickly.

The C—F bond is much stronger (+467 kJ mol^{-1}) than any of the others, which explains why fluoroalkanes are not often used in these hydrolysis experiments.

INTERPRETING THE RESULTS FOR DIFFERENT STRUCTURES

Table C shows the trend when the structure is different.

SAME HALOGEN BUT DIFFERENT STRUCTURE	RESULT
2-bromo-2-methylpropane (tertiary)	fastest
2-bromobutane (secondary)	
1-bromobutane (primary)	slowest

table C Examples comparing halogenoalkanes with the same halogen but with different structures.

To explain why tertiary halogenoalkanes are more rapidly hydrolysed than secondary and primary compounds requires a detailed understanding of two different reaction mechanisms, and is beyond the scope of this book, but will be dealt with in **Topic 15A (Book 2: IAL)**.

LEARNING TIP

Molecules with partial charges (δ+ and δ–) are not ions. They are molecules with no overall charge but with electron-rich and electron-deficient parts.

CHECKPOINT

1. 2-bromopropane and 2-iodopropane are put into separate test tubes and warmed with water. Explain which one is hydrolysed more quickly.
2. Why are fluoroalkanes not readily hydrolysed compared with chloroalkanes?

10B 3 HALOGENOALKANE REACTIONS AND MECHANISMS

SPECIFICATION REFERENCE 10.8 10.9

LEARNING OBJECTIVES

- Understand the reactions of halogenoalkanes with aqueous potassium hydroxide, aqueous silver nitrate in ethanol, potassium cyanide, ammonia and ethanolic potassium hydroxide.
- Understand the mechanisms of the nucleophilic substitution reactions between primary halogenoalkanes and aqueous potassium hydroxide and ammonia.

SUBSTITUTION REACTIONS

The hydrolysis reactions in **Topics 10B.1** and **10B.2** involved replacing a halogen atom (X) with a hydroxyl group (OH). These are substitution reactions, and here is a summary of the four reactions you need to know.

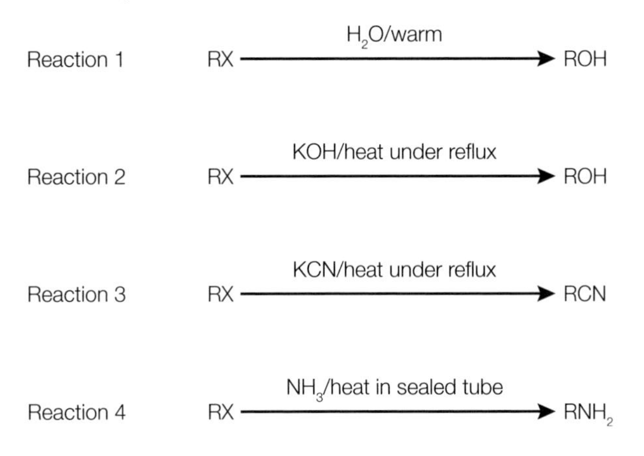

REACTION 1

This is the hydrolysis reaction you met in **Topic 10B.1**.

REACTION 2

Heating a halogenoalkane with aqueous potassium hydroxide under reflux is one way of making alcohols. The attacking nucleophile is the OH^- ion. An example of an equation for this reaction is the conversion of 1-chloropropane into propan-1-ol:

$$CH_3CH_2CH_2Cl + KOH \rightarrow CH_3CH_2CH_2OH + KCl$$

or

$$CH_3CH_2CH_2Cl + OH^- \rightarrow CH_3CH_2CH_2OH + Cl^-$$

The advantage of an ionic equation is that the nucleophile is clearly shown.

REACTION 3

Heating a halogenoalkane with potassium cyanide dissolved in ethanol under reflux is one way of making **nitriles**. The attacking nucleophile is the CN^- ion. An example of an equation for this reaction is the conversion of bromoethane into propanenitrile:

$$CH_3CH_2Br + KCN \rightarrow CH_3CH_2CN + KBr$$

or

$$CH_3CH_2Br + CN^- \rightarrow CH_3CH_2CN + Br^-$$

Note that the organic product contains one more carbon atom than the starting material. This reaction is a useful way of increasing the length of the carbon chain. It is also an important way of synthesising more complex compounds.

EXAM HINT

If you were to carry out this reaction in the lab then there would need to be a very careful risk assessment. KCN is highly toxic!

REACTION 4

Heating a halogenoalkane with ammonia solution under pressure in a sealed tube is one way of making **primary amines**. The sealed tube is needed because ammonia is a gas and would otherwise escape from the apparatus before it could react. The attacking nucleophile is the NH_3 molecule. An example of an equation for this reaction is the conversion of 1-iodobutane into butylamine:

$$CH_3CH_2CH_2CH_2I + NH_3 \rightarrow CH_3CH_2CH_2CH_2NH_2 + HI$$

This equation looks similar to those for Reactions 1–3, but it isn't quite the full story. Like NH_3, the organic product is a base, so it would react with the inorganic product, the acid HI, to form a salt. So a better equation is:

$$CH_3CH_2CH_2CH_2I + NH_3 \rightarrow CH_3CH_2CH_2CH_2NH_3^+ + I^-$$

However, this is only the first step because the product is a salt, not a primary amine. To produce a high yield of the amine, the ammonia is used in excess, and some of this excess ammonia reacts in a second step to produce the amine:

$$CH_3CH_2CH_2CH_2NH_3^+I^- + NH_3 \rightarrow CH_3CH_2CH_2CH_2NH_2 + NH_4^+I^-$$

The final products are butylamine and ammonium iodide. These two steps are often combined as:

$$CH_3CH_2CH_2CH_2I + 2NH_3 \rightarrow CH_3CH_2CH_2CH_2NH_2 + NH_4^+I^-$$

NUCLEOPHILIC SUBSTITUTION MECHANISMS

You have already learned about mechanisms for free radical substitution and electrophilic addition; here is a different mechanism to learn about.

In each reaction the attacking species is a nucleophile, so the reaction type is described as **nucleophilic substitution**. You only need to know about mechanisms for primary halogenoalkanes undergoing Reactions 2 and 4.

MECHANISM OF REACTION 2

One example is the reaction between bromoethane and aqueous potassium hydroxide. The reaction starts with the donation of a lone pair of electrons from the oxygen of a hydroxide ion to the electron-deficient carbon atom and the formation of a C—O bond. At the same time, the electrons in the C—Br bond move to the Br atom, resulting in the breaking of the C—Br bond. This type of bond breaking is known as heterolytic fission. 'Hetero' indicates different, and you already know that 'lysis' indicates breaking.

$$CH_3-\overset{\delta+}{C}H_2-Br^{\delta-} \;\; (:OH^-) \longrightarrow CH_3-CH_2-OH + Br^-$$

MECHANISM OF REACTION 4

One example is the reaction between chloroethane and ammonia. The first step of the reaction involves the donation of a lone pair of electrons from the nitrogen of an ammonia molecule to the electron-deficient carbon atom and the formation of a C—N bond. At the same time, the electrons in the C—Cl bond move to the Cl atom, resulting in the breaking of the C—Cl bond.

$$CH_3-\overset{\delta+}{C}H_2-Cl^{\delta-} \;\; (:NH_3) \longrightarrow CH_3-CH_2-N^+H_3 + Cl^-$$

The second step of the reaction involves another ammonia molecule acting as a base and removing a hydrogen ion from the ion formed in the first step.

$$CH_3-CH_2-N^+H_3 \;\; (:NH_3) \longrightarrow CH_3-CH_2-NH_2 + NH_4^+$$

ELIMINATION REACTIONS

Reaction 2 above is a nucleophilic substitution reaction, but using a different solvent (ethanol instead of water) causes a different reaction to occur. When a halogenoalkane is heated with **ethanolic** potassium hydroxide, the OH^- ion acts as a base and not as a nucleophile. You know that a base reacts with a hydrogen ion (H^+), but in the case of a halogenoalkane, the hydrogen that reacts with the OH^- ion is the one attached to a carbon atom next to the C in the C—Br bond.

For example, the equation for the reaction between 2-bromopropane and ethanolic potassium hydroxide is:

$$CH_3—CHBr—CH_3 + KOH \rightarrow CH_2{=}CH—CH_3 + H_2O + KBr$$

The organic product is propene (an alkene), and water and potassium bromide are the other products. You can see why the reaction is referred to as elimination. H and Br are removed from the halogenoalkane but they are not replaced by any other atoms.

You do not need to know the mechanism for this **elimination reaction**.

EXAM HINT

The fact that the same reagents can give different products depending on which solvent you use highlights the need for you to make sure that you know both reactants *and* conditions for an organic synthesis.

LEARNING TIP

Both the hydroxide ion and the ammonia molecule can act as nucleophiles and as bases.

CHECKPOINT

1. Chloromethane is heated with:
   (a) ammonia
   (b) potassium cyanide dissolved in ethanol.
   Give the name of the organic product formed in each case.

2. 2-chlorobutane is heated with:
   (a) aqueous potassium hydroxide
   (b) ethanolic potassium hydroxide.
   State the mechanism of each reaction and explain why there are two different organic products in (b).

SUBJECT VOCABULARY

nitrile organic compound containing the C–CN group
primary amine compound containing the C–NH_2 group
nucleophilic substitution a reaction in which an attacking nucleophile replaces an existing atom or group in a molecule
ethanolic a solution in which ethanol is the solvent
elimination reaction a reaction in which a molecule loses atoms attached to adjacent carbon atoms, forming a C=C double bond

10C 1 ALCOHOLS AND SOME OF THEIR REACTIONS

SPECIFICATION REFERENCE
10.15 10.16
10.17 CP6

LEARNING OBJECTIVES

- Understand the nomenclature of alcohols and be able to draw their structural, displayed and skeletal formulae.
- Understand the distinction between primary, secondary and tertiary alcohols.
- Understand the combustion, halogenation and dehydration reactions of alcohols.

▲ **fig A** Sanitising hand gels containing alcohols are used in hospitals to reduce infection risks to patients.

WHAT ARE ALCOHOLS?

The alcohols are a homologous series of compounds with the general formula $C_nH_{2n+1}OH$. Think of them as the result of replacing a hydrogen atom in a hydrocarbon with a hydroxyl group. You have already met alcohols in **Topic 10B.1** as the products of the hydrolysis of halogenoalkanes.

The symbol R can be used to represent an alkyl group, as with halogenoalkanes. The formula of an alcohol can be simplified to ROH.

NAMING ALCOHOLS

We have already discussed how to name alcohols in **Topic 4A.4**, but **table A** shows some examples to remind you.

STRUCTURE	NAME
$CH_3-CH(OH)-CH_3$	propan-2-ol
$CH_3-CH_2-CH_2-CH_2OH$	butan-1-ol
$(CH_3)_3C-CH_2OH$	2,2-dimethylpropan-1-ol
$CH_2(OH)-CH(OH)-CH_2OH$	propane-1,2,3-triol

table A Examples of how to name alcohols.

You will have come across the last example in foods, medicines and personal care products. Its common names are glycerol and glycerine. You can see how the IUPAC system can be adapted to name compounds with more than one OH group.

CLASSIFYING ALCOHOLS

You have met the terms primary, secondary and tertiary applied to carbocations (in **Topic 5A.4**) and to halogenoalkanes (in **Topic 10B.1**).

Alcohols are classified in a similar way, depending on the number of alkyl groups joined to the C atom bonded to the hydroxyl group. **Table B** shows some examples.

STRUCTURE	ABBREVIATED FORMULA	NUMBER OF ALKYL GROUPS	CLASSIFICATION
$CH_3-CH_2-CH_2OH$	ROH	1	primary
$CH_3-CH(OH)-CH_3$	R_2CHOH	2	secondary
$(CH_3)_2C(OH)-CH_2-CH_3$	R_3COH	3	tertiary

table B Examples of how to classify alcohols.

REACTIONS

In this topic, the reactions of alcohols we cover are:

- combustion
- conversions to halogenoalkanes
- dehydration to alkenes.

Reaction mechanisms are not required for any of these reactions.

COMBUSTION

You have already seen in **Topic 4B.3** that alcohols are used as biofuels. If combustion is complete, the products are carbon dioxide and water. This is the equation for the complete combustion of ethanol:

$$C_2H_5OH + 3O_2 \rightarrow 2CO_2 + 3H_2O$$

CONVERSIONS TO HALOGENOALKANES

These reactions involve replacing the hydroxyl group in an alcohol molecule with a halogen atom. The reaction is known as **halogenation**. However, just adding a halogen to an alcohol does not work. A different method is needed for each halogen.

Chlorination is carried out using phosphorus(V) chloride (a white solid, also known as phosphorus pentachloride). The reaction is very vigorous at room temperature, so the alcohol and phosphorus(V) chloride reaction mixture does not need heating. There are also two inorganic products: phosphoryl chloride and hydrogen chloride. This is the equation for the reaction with propan-1-ol:

$$CH_3CH_2CH_2OH + PCl_5 \rightarrow CH_3CH_2CH_2Cl + POCl_3 + HCl$$

Chlorination of tertiary alcohols can be done in a different way, using a method that does not work well for primary and secondary alcohols. The alcohol needs only to be mixed (by shaking) with concentrated hydrochloric acid at room temperature. This is the equation for the reaction with 2-methylpropan-2-ol:

$$(CH_3)_3COH + HCl \rightarrow (CH_3)_3CCl + H_2O$$

Bromination is carried out using a mixture of potassium bromide and about 50% concentrated sulfuric acid. The reaction mixture is warmed with the alcohol. It is better to write two equations, rather than one, as the inorganic reagents first react together to form hydrogen bromide.

The other inorganic product is either potassium hydrogensulfate or potassium sulfate.

$$KBr + H_2SO_4 \rightarrow KHSO_4 + HBr$$

or

$$2KBr + H_2SO_4 \rightarrow K_2SO_4 + 2HBr$$

This is the equation for the reaction with butan-1-ol:

$$CH_3CH_2CH_2CH_2OH + HBr \rightarrow CH_3CH_2CH_2CH_2Br + H_2O$$

EXAM HINT

The sulfuric acid must be no more concentrated than a 50% solution. More concentrated sulfuric acid would oxidise bromide ions to bromine and so result in different products.

Iodination is carried out using a mixture of red phosphorus and iodine. The reaction mixture, including the alcohol, is heated under reflux. As with bromination, it is better to write two equations, as the inorganic reagents first react to form phosphorus(III) iodide:

$$2P + 3I_2 \rightarrow 2PI_3$$

This is the equation for the reaction with ethanol:

$$3C_2H_5OH + PI_3 \rightarrow 3C_2H_5I + H_3PO_3$$

The inorganic product is phosphonic acid (often known as phosphorous acid).

DEHYDRATION TO ALKENES

Dehydration is done by heating the alcohol with concentrated phosphoric acid. The reaction is similar to the elimination reaction of a halogenoalkane, with the OH group and a hydrogen atom from an adjacent carbon atom being removed and a C=C double bond formed in the carbon chain.

You can see why the reaction is described as dehydration, as water is the only inorganic product. These are the equations for a reaction in which there are two possible products, starting with butan-2-ol:

$$CH_3CH(OH)CH_2CH_3 \rightarrow \underset{\text{but-1-ene}}{CH_2{=}CHCH_2CH_3} + H_2O$$

and

$$CH_3CH(OH)CH_2CH_3 \rightarrow \underset{\text{but-2-ene}}{CH_3CH{=}CHCH_3} + H_2O$$

Remember that but-2-ene actually exists as a pair of *E–Z* (*cis-trans*) isomers. It is therefore more correct to state that there are three products – but-1-ene and the two isomeric forms of but-2-ene.

The formula for phosphoric acid does not appear in the equation. This is because the water formed in the reaction mixes with the concentrated phosphoric acid to dilute the acid.

LEARNING TIP

When you balance an equation for the combustion of an alcohol, remember that there is already one atom of oxygen in an alcohol molecule.

CHECKPOINT

SKILLS REASONING

1. Write an equation for the complete combustion of butanol, C_4H_9OH.
2. When butan-2-ol is dehydrated, there is more than one organic product. Explain why there is only one pair of *E/Z* isomers when pentan-3-ol is dehydrated.

SUBJECT VOCABULARY

halogenation a reaction where the hydroxyl group in an alcohol molecule is replaced by a halogen atom

dehydration a reaction where the hydroxyl group from an alcohol molecule, and a hydrogen atom from an adjacent carbon atom, are removed, forming a C=C double bond

10C 2 OXIDATION REACTIONS OF ALCOHOLS

SPECIFICATION REFERENCE

10.18 10.19(i) CP7

LEARNING OBJECTIVES

- Understand the oxidation reactions of primary and secondary alcohols by potassium dichromate(VI) to form aldehydes, carboxylic acids and ketones.
- Understand the techniques of distillation and heating under reflux used to maximise the yields of different products.

BACKGROUND

There are different ways of considering oxidation and reduction. In this topic, the best way to consider oxidation is as the loss of hydrogen from an alcohol molecule. Unlike dehydration, covered in **Topic 10C.1**, oxidation affects only one carbon atom. The atoms removed from an alcohol molecule are the hydrogen of the OH group and a hydrogen atom from the carbon atom joined to the OH group, as shown below.

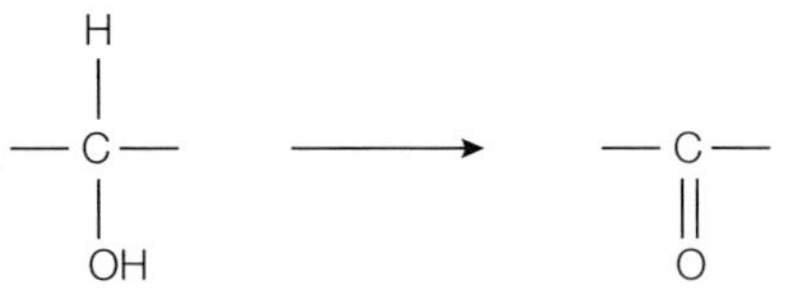

H
—C— (with OH) ⟶ —C— (with =O)
OH   O

The organic product contains a C=O group, known as a carbonyl group.

The diagrams below should help you to understand why primary and secondary alcohols, but not tertiary alcohols, can be oxidised in this way.

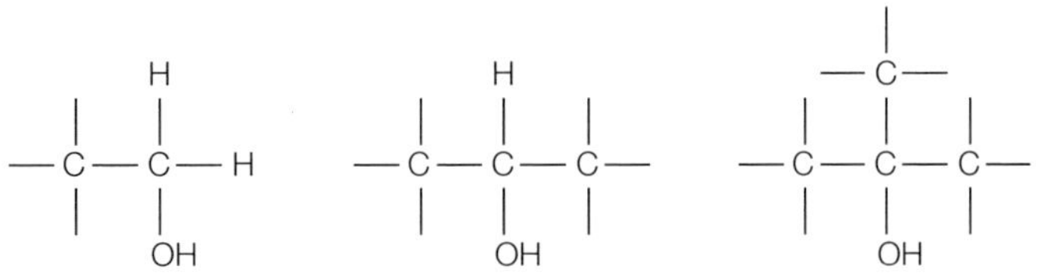

primary   secondary   tertiary

Only the primary and secondary structures have a hydrogen atom on the C of the C—OH group – the tertiary structure does not.

THE PRODUCTS OF OXIDATION

KETONES

When a secondary alcohol is oxidised, the organic product belongs to a homologous series called **ketones**. A simplified formula for a ketone is RCOR. Using the same symbol R twice suggests that the two alkyl groups are the same (for example, two ethyl groups). This is not necessarily the case, but if the two alkyl groups are meant to be different, then the R symbols can be altered slightly. For example, one could be R and the other could be R'. Another way to show that the alkyl groups are different is for one to be R_1 and the other R_2.

ALDEHYDES AND CARBOXYLIC ACIDS

When a primary alcohol is oxidised, the organic product belongs to a homologous series called **aldehydes**. A simplified formula for an aldehyde is RCHO. This should *not* be written as RCOH, which would imply that the molecule contains an OH group.

There is a complication with aldehydes that is not the case with ketones. This is that aldehydes are more easily oxidised than alcohols, so when a primary alcohol is oxidised, the aldehyde formed may be oxidised further. When an aldehyde is oxidised, the process involves the gain of an oxygen atom, not the loss of hydrogen. The oxygen atom gained goes between the C and H of the CHO group. The organic product belongs to a homologous series called **carboxylic acids**. A simplified formula for a carboxylic acid is RCOOH.

These reactions are summarised below.

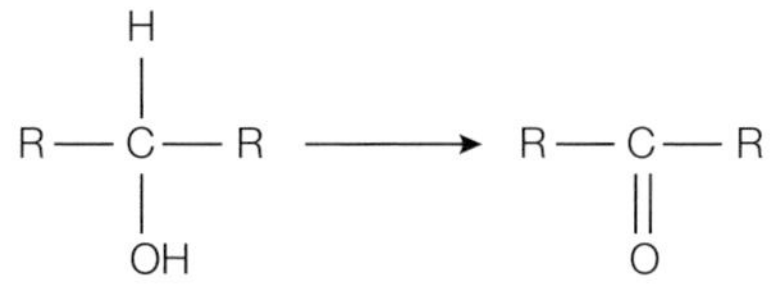

R—CH(OH)—R ⟶ R—C(=O)—R

secondary alcohol   ketone

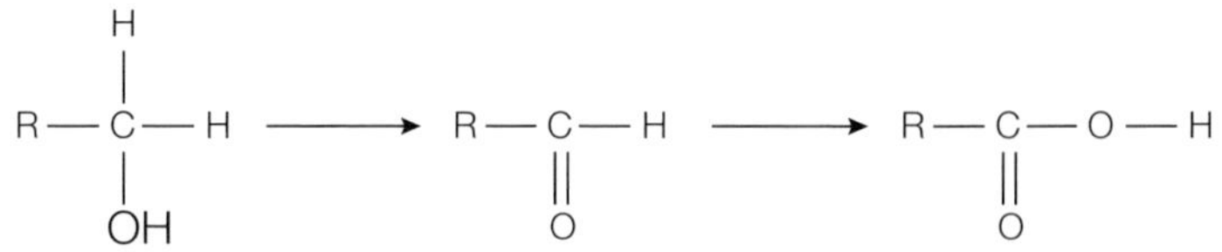

R—CH(OH)—H ⟶ R—C(=O)—H ⟶ R—C(=O)—O—H

primary alcohol   aldehyde   carboxylic acid

The usual reagent for these oxidation reactions is a mixture of potassium dichromate(VI) and dilute sulfuric acid. Unlike with the inorganic reagents used previously, equations involving this mixture do not need to be written (because they are very complicated). Instead, the oxidising agent is represented by [O]. This symbol simply represents an oxygen atom provided by the oxidising agent. Whenever the mixture of potassium dichromate(VI) and sulfuric acid is used as an oxidising agent, there is a colour change from orange to green.

So now we can write some equations for examples of oxidation reactions. The names of the organic products are given, but a fuller study of these products is outside the scope of IAS.

Propan-1-ol to propanal:

$$CH_3CH_2CH_2OH + [O] \rightarrow CH_3CH_2CHO + H_2O$$

Propanal to propanoic acid:

$$CH_3CH_2CHO + [O] \rightarrow CH_3CH_2COOH$$

Propan-2-ol to propanone:

$$CH_3CH(OH)CH_3 + [O] \rightarrow CH_3COCH_3 + H_2O$$

You do not need to know the mechanisms of any of these reactions.

DIFFERENT PRACTICAL TECHNIQUES

Because of the easier oxidation of aldehydes compared with alcohols, two different techniques are used. These are:

- heating under reflux
- distillation with addition.

HEATING UNDER REFLUX

When the oxidation is intended to be complete (to obtain a ketone or a carboxylic acid), the technique used is **heating under reflux**. You can see the apparatus for this technique in the diagram.

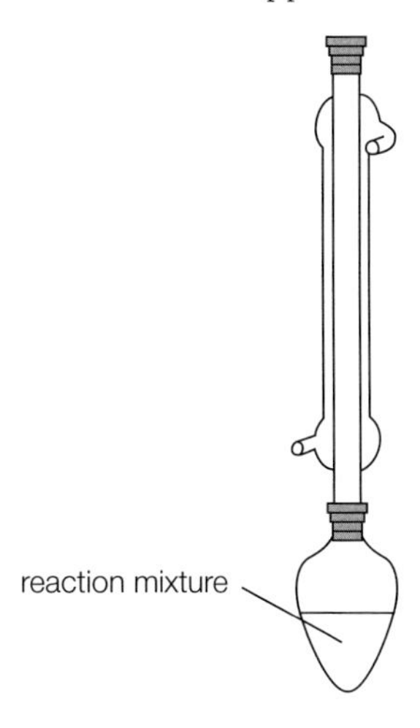

In this apparatus, the products of oxidation stay in the reaction mixture, because if they do boil off, they condense in the vertical condenser and return to the heating flask.

DISTILLATION WITH ADDITION

When the oxidation is intended to be incomplete (to obtain an aldehyde, and not a carboxylic acid), the technique used is **distillation with addition**. The apparatus for this technique is shown in the diagram.

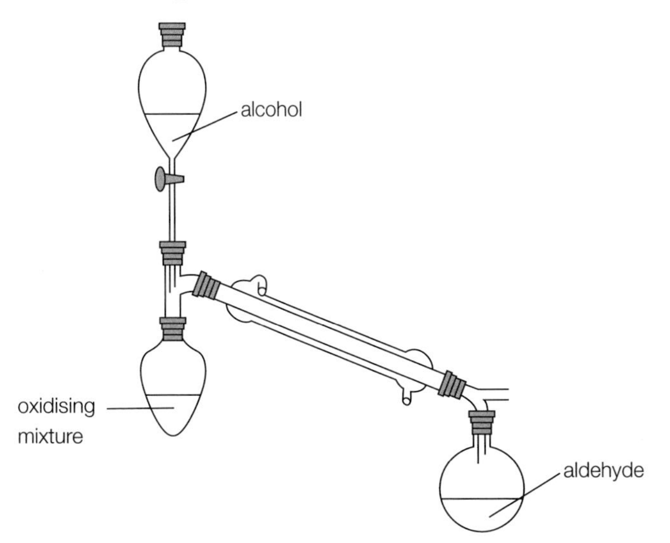

In this apparatus, only the oxidising agent is heated, and the alcohol is slowly added to the oxidising agent. When the aldehyde is formed, it immediately distils off (it has a much lower boiling temperature than the alcohol used to make it), and is collected in the receiver.

LEARNING TIP

Practise writing equations to show how a given alcohol can be separately dehydrated and oxidised. This will help you to understand the difference between the two reactions.

CHECKPOINT

1. Butan-1-ol can be dehydrated to form one organic product, but can be oxidised to form two organic products. Give the names and structures of these three products.
2. Pentan-2-ol can be dehydrated to form more than one organic product, but can be oxidised to form only one organic product. Give the names and structures of these products.
3. There are four isomeric alcohols with the formula C_4H_9OH. Draw a structural formula for each isomer and use the formulae to explain which ones can be oxidised by heating with acidified potassium dichromate(VI).

SUBJECT VOCABULARY

ketone one of a homologous series of organic compounds formed by oxidation of secondary alcohols, formula RCOR

aldehyde one of a homologous series of organic compounds formed by the partial oxidation of primary alcohols, formula RCHO

carboxylic acid one of a homologous series of organic compounds formed by the complete oxidation of primary alcohols, formula RCOOH

heating under reflux heating a reaction mixture with a condenser fitted vertically

distillation with addition heating a reaction mixture, but adding another liquid and distilling off the product as it forms

10C 3 PURIFYING AN ORGANIC LIQUID

SPECIFICATION REFERENCE

10.19 (ii-v) CP6

LEARNING OBJECTIVES

- Know when and how to use the following techniques in the preparation and purification of a liquid organic compound: simple distillation, fractional distillation, solvent extraction, drying, and boiling temperature determination.

BACKGROUND

So far, you have met many different reactions that can be used in the preparation of organic compounds. In most cases the intended organic product is not pure. It could be contaminated with:

- unreacted starting materials
- other organic products
- the inorganic reagents used, or the inorganic products formed from them
- water.

This means that organic chemists need to use several techniques to separate the intended product from a reaction mixture. These techniques will be different, depending on whether the intended product is a gas, a liquid or a solid. In this topic, we consider only the techniques used to purify an organic liquid.

The techniques we consider are:

- simple distillation
- fractional distillation
- solvent extraction
- drying
- boiling point determination.

APPARATUS

In a chemistry textbook, you will see many different diagrams of apparatus used in organic chemistry. You will have some of these pieces of apparatus in your laboratory, but you will probably not be able to set up every piece of apparatus you see in the book.

In simple laboratory experiments not involving organic compounds, you are likely to use pieces of glassware connected by tubing and corks or bungs. This kind of apparatus is relatively inexpensive to buy and easy to use. If one of the gases you are using, such as carbon dioxide, leaks from the apparatus, this is not a major problem.

Organic compounds are more of a problem because they may be flammable, or toxic, or both. They may also attack corks and bungs, and so increase the risk of leaks and contamination. One solution to these problems is to use a type of apparatus made only (or mostly) of glass, and which can be fitted together tightly using ground-glass joints.

Fig A shows a selection of this type of apparatus. Using these eight pieces of apparatus, you can create a wide variety of experimental set-ups.

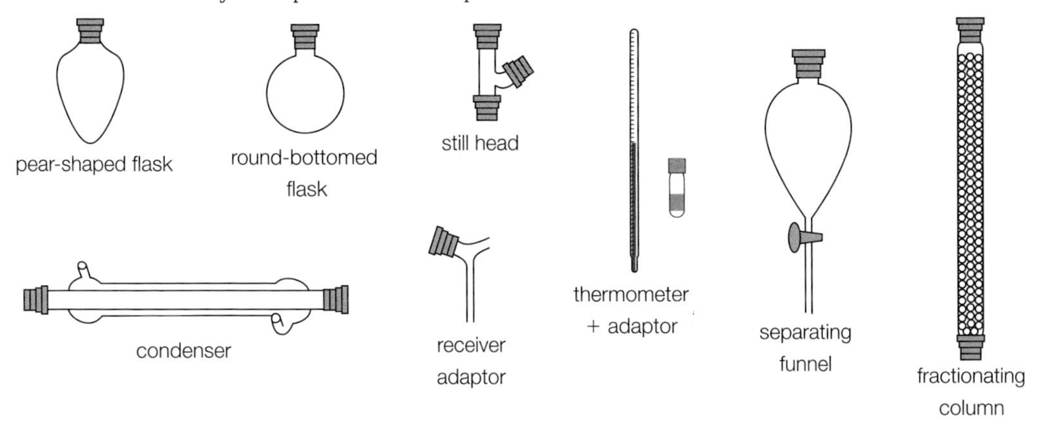

▲ **fig A** A selection of apparatus made mainly of glass, ideal for experiments using organic compounds.

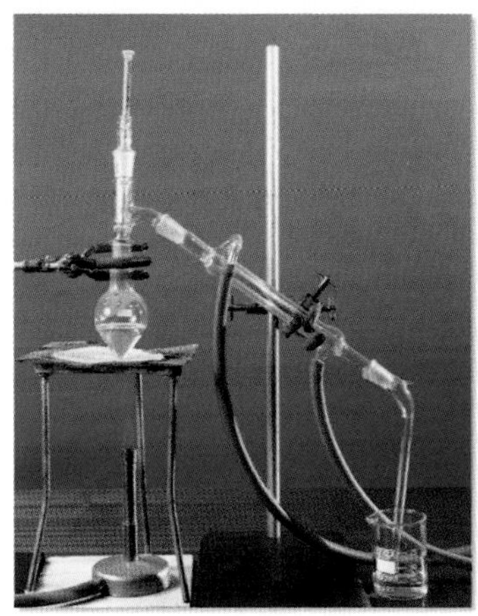

▲ **fig B** This is just one of the many apparatus set-ups using ground-glass joints.

Heating is needed in many experiments. You can do this using a Bunsen burner, but often an electric heating mantle or a hot water bath or hot oil bath is preferred for safety reasons.

SIMPLE DISTILLATION

Here is a summary of the process of **simple distillation**.

- Distillation of an impure liquid involves heating the liquid in a flask connected to a condenser.
- The liquid with the lowest boiling temperature evaporates or boils off first and passes into the condenser first. This means it can be collected in the receiver separately from any other liquid that evaporates later.
- The purpose of the thermometer is to monitor the temperature of the vapour as it passes into the condenser. If the temperature remains steady, this is an indication that one compound is distilling over. If, after a while, the temperature begins to rise, this indicates that a different compound is beginning to distil over.

Fig C shows the apparatus used for simple distillation.

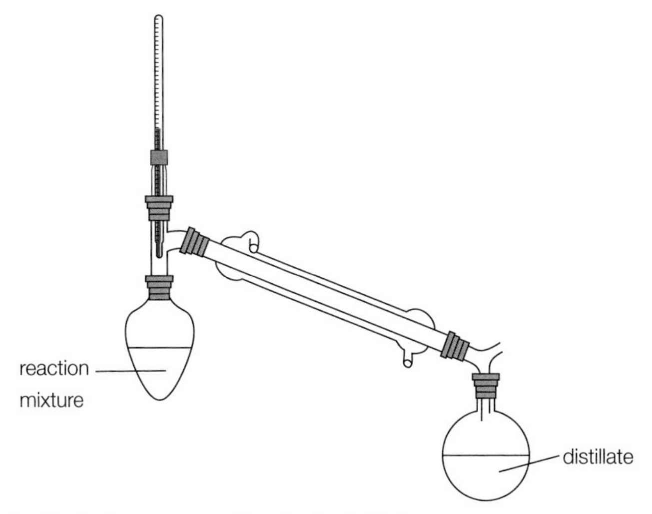

▲ **fig C** Apparatus used for simple distillation.

FRACTIONAL DISTILLATION

Here is a summary of the process of **fractional distillation**.

- Fractional distillation uses the same apparatus as simple distillation, but with a fractionating column between the heating flask and the still head (**fig D**).
- The column is usually filled with glass beads or pieces of broken glass, which act as surfaces on which the vapour leaving the column can condense. It can then be evaporated again as more hot vapour passes up the column.
- Effectively, the vapour undergoes several repeated distillations as it passes up the column. This provides a better separation.

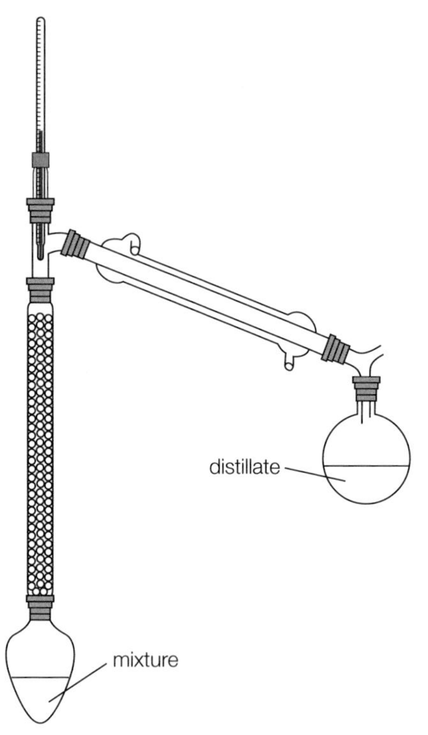

▲ **fig D** Apparatus used for fractional distillation.

Fractional distillation takes longer than simple distillation, and is best used when the difference in boiling temperatures is small, and when there are several compounds to be separated from a mixture.

ADVANTAGES AND DISADVANTAGES

The advantages of using simple distillation, rather than fractional distillation, are that it is easier to set up and it is quicker.

The disadvantage is that it does not separate the liquids as well as fractional distillation. Simple distillation should only be used if the boiling temperature of the liquid being purified is very different from the other liquids in the mixture, ideally a difference of more than 25 °C.

SOLVENT EXTRACTION

As the name suggests, this method involves using a solvent to remove the desired organic product from the other substances in the reaction mixture. There are several solvents that can be used, but the choice depends mainly on these features.

- The solvent added should be immiscible (i.e. does not mix) with the solvent containing the desired organic product.
- The desired organic product should be much more soluble in the solvent added than in the reaction mixture.

Here is a summary of the process of **solvent extraction**.

- Place the reaction mixture in a separating funnel, and then add the chosen solvent – it should form a separate layer.
- Place the stopper in the neck of the funnel and gently shake the contents of the funnel for a while.
- Allow the contents to settle into two layers.
- Remove the stopper and open the tap to allow the lower layer to drain into a flask. Then pour the upper layer into a separate flask.

▲ **fig E** Apparatus used for solvent extraction.

If a suitable solvent is used and the method is followed correctly, most of the desired organic product will have moved into the added solvent. It is better to use the solvent in small portions than in a single large volume (for example, four portions of 25 cm^3 rather than one portion of 100 cm^3) because this is more efficient. Using more portions of solvent, but with the same total volume, removes more of the desired organic product.

The desired organic product has been removed from the reaction mixture, but is now mixed with the added solvent. Simple distillation or fractional distillation now has to be used to separate the desired organic product from the solvent used.

DRYING

Many organic liquids are prepared using inorganic reagents, which are often used in aqueous solution. A liquid organic product may partially or even completely dissolve in water, so water may be an impurity that needs to be removed by a drying agent. One important feature of a drying agent is that it does not react with the organic liquid.

There are several drying agents available, but the most common ones are anhydrous metal salts, often calcium sulfate, magnesium sulfate and sodium sulfate. What these compounds have in common is that they form hydrated salts, so when they come into contact with water in an organic liquid they absorb the water as water of crystallisation. Anhydrous calcium chloride can also be used for some organic compounds, although it does react with others and is soluble in alcohols.

Here is a summary of how to dry an organic liquid.

- The drying agent is added to the organic liquid and the mixture is swirled or shaken, and then left for a period of time.
- Before use, a drying agent is powdery, but after absorbing water it looks more crystalline.

- If a bit more drying agent is added, and it remains powdery, this is an indication that the liquid is dry.
- The drying agent is removed either by decantation (pouring the organic liquid off the solid drying agent), or by filtration.

TESTING FOR PURITY

Using one or more of the purifying techniques, you have purified an organic compound. But is it really pure?

For liquids, there is a simple way to test whether it is pure – measure its boiling temperature. Impurities raise the boiling temperature.

The boiling temperatures of pure organic compounds have been carefully measured and are widely available in data books and online. If you measure the boiling temperature of your organic compound, you can compare it with an accurate value and then make your decision about how pure it is.

The apparatus used depends on the volume of liquid available, and whether it is toxic or flammable. The apparatus used for simple distillation can be used.

A word of caution – this test may not be conclusive, because you may not be able to measure the boiling temperature of your organic compound accurately enough. Your thermometer might read too low or too high, so even if your measured boiling temperature exactly matches the one in the data book or online, you may wrongly assume that your compound is pure.

Also remember that different organic compounds can, by coincidence, have the same boiling temperature. For example, both 1-chloropentane and 2-methylpropan-1-ol boil at 108 °C.

LEARNING TIP

Give examples of when you would use fractional distillation instead of simple distillation. Think back to the advantages and disadvantages of both methods listed above.

CHECKPOINT

1. List the apparatus you could use to perform the experiments below. You can refer to diagrams in this chapter to help you:
   (a) a simple distillation
   (b) fractional distillation
   (c) solvent extraction.
2. What is the purpose of using a thermometer during distillation?
3. What are the limitations of using the measurement of a boiling temperature as a way of assessing the purity of an organic liquid?

SUBJECT VOCABULARY

simple distillation a method used to separate liquids with very different boiling temperatures, by boiling and condensing

fractional distillation a method used to separate liquids with similar boiling temperatures, by boiling and condensing, using a fractionating column

solvent extraction a method used to separate a liquid from a mixture by causing it to move from the mixture to the solvent

10D 1 MASS SPECTROMETRY OF ORGANIC COMPOUNDS

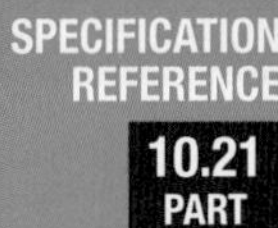

LEARNING OBJECTIVES

- Be able to interpret data from mass spectra to suggest possible structures of simple organic compounds.

BACKGROUND TO MASS SPECTROMETRY

You have already come across the principles of mass spectrometry in **Topic 2A.2**. We will now look at how this technique can be used to determine the relative molecular mass of organic compounds. We will also determine the structures of some of these compounds.

The mass spectrum of an element often appears very simple, with a very small number of vertical lines, called peaks. Each peak represents an isotope of the element.

In contrast, the mass spectrum of an organic compound often appears complex, with many peaks. If all the peaks are included, there may be more peaks than there are atoms in a molecule of the compound, so something different is happening.

THE MOLECULAR ION PEAK

The obvious thing to do when first looking at the mass spectrum of an organic compound is to find the peak furthest to the right. This is the one with the greatest m/z value (mass to charge ratio). This peak is the **molecular ion peak** – the result of the organic molecule losing an electron in the mass spectrometer. The equation for this process, using butane as an example, is:

$$C_4H_{10} + e^- \rightarrow C_4H_{10}^{+} + 2e^-$$

An electron collides with a butane molecule and knocks out an electron, so forming a positive ion from the molecule. The m/z value of the $C_4H_{10}^{+}$ peak (58) indicates the relative molecular mass of butane.

OTHER PEAKS

The spectrum may show a very small peak just to the right of the molecular ion peak (sometimes referred to as the M+1 peak). This is caused by the presence of a naturally occurring isotope of carbon (^{13}C, rather than the usual ^{12}C isotope) in the molecule. Approximately 99% of all carbon atoms are ^{12}C, with most of the remaining 1% being ^{13}C.

You may have met the ^{14}C isotope, which also occurs naturally in organic compounds. This isotope is radioactive and used in radiocarbon dating. The proportion of ^{14}C atoms in a sample of an organic compound is extremely small and can be ignored in mass spectrometry.

The peaks with smaller m/z values result from fragmentation in the mass spectrometer, rearrangement reactions and the loss of more than one electron. Rearrangement is often unpredictable and will not be considered here, but it does help to explain the large number of peaks in some spectra. The breaking of a carbon–hydrogen bond can occur, but this is not usually described as fragmentation.

FRAGMENTATION IN HYDROCARBONS

Fragmentation is very common and can often be used to work out the structure of an organic molecule.

Consider the breaking of a carbon–carbon bond in the molecular ion formed from a hydrocarbon. Two species are formed. They are:

- another positive ion
- a neutral species (usually a free radical).

EXAMPLES OF FRAGMENTATION

A very simple example of fragmentation is the molecular ion of ethane, which can fragment to form a methyl cation and a methyl radical:

$$(CH_3–CH_3)^+ \rightarrow CH_3^+ + CH_3$$

The ethane molecule is symmetrical, and there is only one carbon–carbon bond in ethane, so you can imagine that the right-hand carbon is just as likely to become the positive ion. The equation for this fragmentation would be:

$$(CH_3–CH_3)^+ \rightarrow CH_3 + CH_3^+$$

You can see that the products are identical, so the spectrum does not depend on how the bond breaks – there will be a peak at $m/z = 15$.

Sometimes the free radical formed is shown with a dot (representing the unpaired electron), and the molecular ion is shown with a dot as well as a positive charge, as in this example:

$$(CH_3–CH_3)^{+\bullet} \rightarrow CH_3\bullet + CH_3^+$$

Now consider propane. There are two carbon–carbon bonds, but they are equivalent – they can both be described as the bond between the central carbon and one of the two terminal carbons. However, there are now two possible fragment ions that can form:

$$(CH_3–CH_2–CH_3)^+ \rightarrow CH_3^+ + CH_2–CH_3$$

$$(CH_3–CH_2–CH_3)^+ \rightarrow CH_3 + (CH_2–CH_3)^+$$

You would therefore expect to see peaks at $m/z = 15$ (the methyl cation) and $m/z = 29$ (the ethyl cation) in its spectrum. These peaks are present, although there are several others that are difficult to explain and are of no help in deducing the structure.

Free radicals are not detected in a mass spectrometer, so all the peaks formed by fragmentation are caused by positive ions.

POSSIBLE INFORMATION

Here is the information you could be given in exam questions.

- A complete mass spectrum – the disadvantage of this is the possible large number of peaks that cannot be used to work out the structure and would be distracting. Another possibility is a complete mass spectrum, but with only the m/z values of the useful peaks marked on the spectrum.
- A simplified mass spectrum showing only the peaks that will help you work out a structure.
- A list of the m/z values of the useful peaks.

With practice, you will be able to work out the structure of an organic compound from this information.

A TYPICAL MASS SPECTRUM

The traditional way to present a mass spectrum is to label the vertical axis as relative intensity (%), always from 0% to 100%. The horizontal axis is labelled m/z (with no units). The horizontal axis usually, but not necessarily, starts from zero and continues to just beyond the molecular ion peak.

The tallest peak is sometimes referred to as the **base peak**.

This base peak represents the ion with the highest abundance, and is shown with a relative intensity of 100%. It represents the most stable fragment.

This is the mass spectrum of butane:

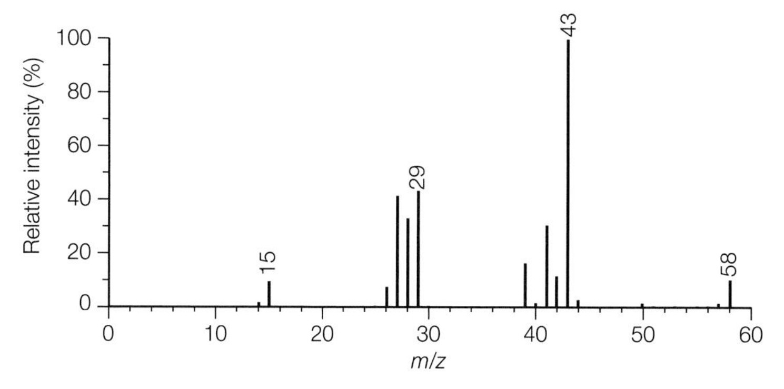

Among the visible peaks are those labelled 15, 29, 43 and 58. **Table A** shows the origin of these peaks.

m/z	ION	NOTES
15	CH_3^+	
29	$(CH_3–CH_2)^+$	
43	$(CH_3–CH_2–CH_2)^+$	This is the most abundant peak.
58	$(CH_3–CH_2–CH_2–CH_3)^+$	This is the molecular ion peak.

table A The origin of the peaks in the mass spectrum of butane.

LEARNING TIP

Be clear about the difference between a peak with a large m/z value and a peak with a large abundance.

The m/z value indicates which ion is present, but the height of the peak indicates its abundance.

CHECKPOINT

1. Explain the origin of the $M+1$ peak seen in some mass spectra.
2. Write an equation to show the formation of the base peak in the fragmentation of the molecular ion of butane.

SUBJECT VOCABULARY

molecular ion peak peak for the species formed from the molecule by the loss of one electron

fragmentation occurs when the molecular ion breaks into smaller pieces

base peak the peak with the greatest abundance

10D 2 DEDUCING STRUCTURES FROM MASS SPECTRA

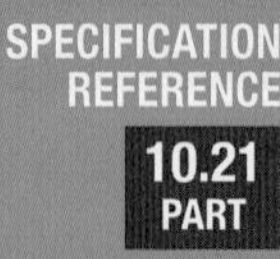

LEARNING OBJECTIVES

- Be able to interpret data from mass spectra to suggest possible structures of simple organic compounds.

FRAGMENTATION IN OTHER ORGANIC COMPOUNDS

So far, we have looked only at the mass spectra of alkanes. Now let's look at other organic compounds – those containing oxygen.

In this topic, we will look at two examples of simplified mass spectra and see how they can be used to work out the structures of the compounds responsible for them.

Oxygen is present in many organic compounds, so you need to be familiar with other *m/z* values. You may also come across compounds containing nitrogen (amines) and halogens (halogenoalkanes).

Table A shows some common *m/z* values and possible ions responsible for these peaks.

M/Z VALUE	POSSIBLE IONS
15	CH_3^+
17	OH^+
28	CO^+
29	$CH_3CH_2^+$ and CHO^+
31	CH_2OH^+
43	$CH_3CH_2CH_2^+$ and $CH_3CHCH_3^+$ and CH_3CO^+
45	$COOH^+$ and CH_3CHOH^+
57	$C_4H_9^+$ (this represents four possible structures)

table A Common *m/z* values and the possible ions responsible for these peaks.

LEARNING TIP

It is sometimes worth comparing the *m/z* values of fragment ions with that of the molecular ion. The difference between the two indicates what has been lost during fragmentation.

For example, if the molecular ion peak is at *m/z* = 60 and there is a fragment ion with *m/z* = 45, the difference is 15, which suggests the loss of a CH_3 group.

WORKED EXAMPLE 1

Two compounds, A and B, have the molecular formula C_3H_6O. Their simplified mass spectra are shown below.

Can you deduce the structure of each one?

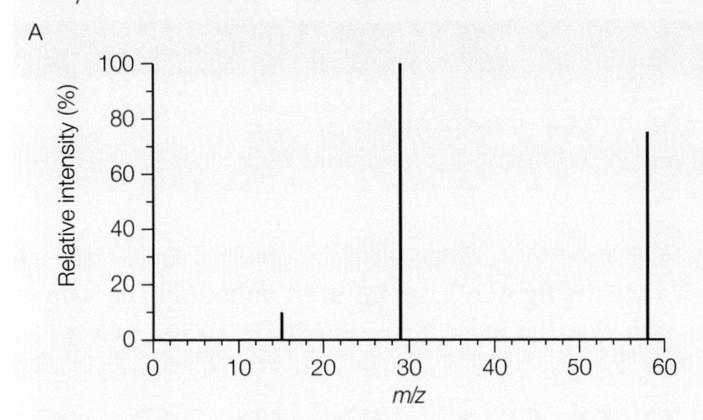

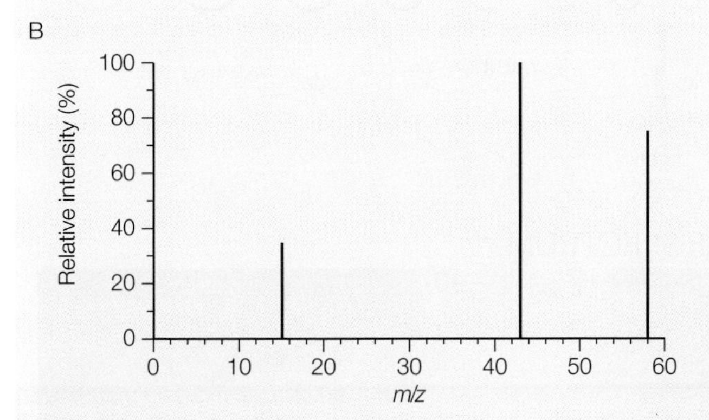

INTERPRETATION

Here are the main points that can be deduced from the simplified mass spectra.

- Both mass spectra show a peak at $m/z = 58$. This corresponds to the molecular ion $C_3H_6O^+$ (relative molecular mass of C_3H_6O is 58.0), but this does not help in deducing the structure.
- Both A and B have a peak at $m/z = 15$, which is caused by the CH_3^+ ion. As this is present in both, it does not help to distinguish between the two structures.
- One obvious difference is a major peak at $m/z = 29$ in A, which is not present in B. This could be caused by either $CH_3CH_2^+$ or CHO^+, or by both of them. The structure CH_3CH_2CHO fits perfectly – this is the aldehyde propanal.
- The other obvious difference is a major peak at $m/z = 43$ in B, which is not present in A. This could not be caused by either $CH_3CH_2CH_2^+$ or $CH_3CHCH_3^+$ because the radical produced at the same time would have a mass of 15 (they must add up to 58), and oxygen has a mass of 16. The other possibility is CH_3CO^+, which when considered with the peak at $m/z = 15$ (CH_3^+) suggests the structure CH_3COCH_3 – this is the ketone propanone.

You may be asked to write equations to show the formation of the ions you have used in your deduction.

In this example, they are:

A ($m/z = 29$) $(CH_3CH_2CHO)^+ \rightarrow CH_3CH_2^+ + CHO\cdot$ and
$(CH_3CH_2CHO)^+ \rightarrow CH_3CH_2\cdot + CHO^+$

B ($m/z = 43$) $(CH_3COCH_3)^+ \rightarrow CH_3CO^+ + CH_3\cdot$

WORKED EXAMPLE 2

Two compounds, C and D, have the molecular formula C_3H_8O. Their simplified mass spectra are shown below.

Can you deduce the structure of each one?

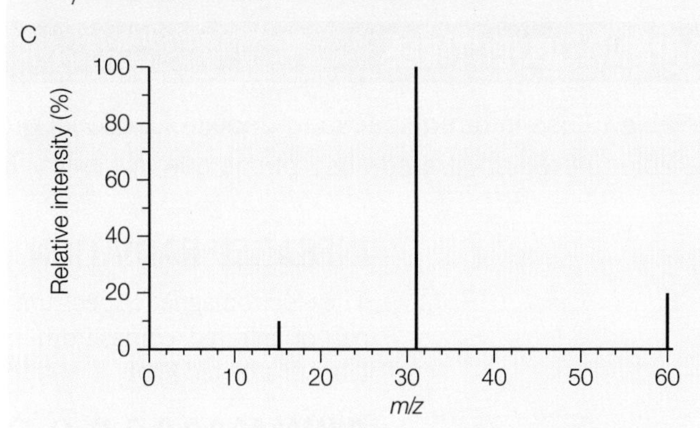

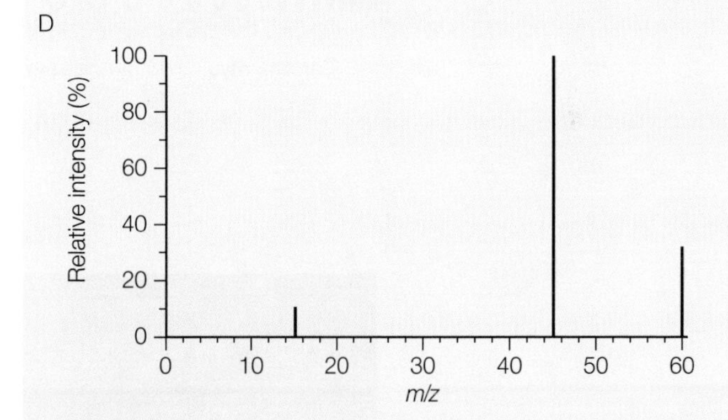

INTERPRETATION

Here are the main points that can be deduced from the simplified mass spectra.

- Both mass spectra show a peak at $m/z = 60$. This corresponds to the molecular ion $C_3H_8O^+$ (relative molecular mass of C_3H_8O is 60.0), but this information does not help in deducing the structure.
- Both C and D have a peak at $m/z = 15$, which is caused by the CH_3^+ ion. As this is present in both, this information does not help to distinguish between the two structures.
- One obvious difference is a major peak at $m/z = 31$ in C, which is not present in D. This could be caused by CH_2OH^+. When considered with the peak at $m/z = 15$ (CH_3^+) and a CH_2 group, this suggests the structure $CH_3CH_2CH_2OH$, which is propan-1-ol.
- The other obvious difference is a major peak at $m/z = 45$ in D, which is not present in C. This could not be caused by $COOH^+$ because D contains only 1 oxygen atom. The other possibility is CH_3CHOH^+, which when considered with the peak at $m/z = 15$ (CH_3^+) suggests the structure $CH_3CH(OH)CH_3$, which is propan-2-ol.

CHECKPOINT

1. Write an equation to show the formation of the ethyl cation in the fragmentation of the molecular ion of pentane.
2. Write equations to show the formation of the ions used in the deduction of the structures of C and D (from Worked example 2 above).

10D 3 INFRARED SPECTROSCOPY

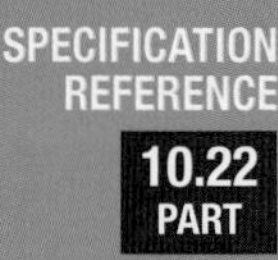

LEARNING OBJECTIVES

- Be able to use infrared spectra to deduce functional groups present in organic compounds.
- Be able to predict infrared absorptions due to familiar functional groups including wavenumber data.

INFRARED RADIATION

The electromagnetic spectrum of radiation shown in **fig A** includes **infrared radiation**. The 'infra' part of 'infrared' comes from the Latin for 'below', so this radiation has a frequency below, or less than, that of red light.

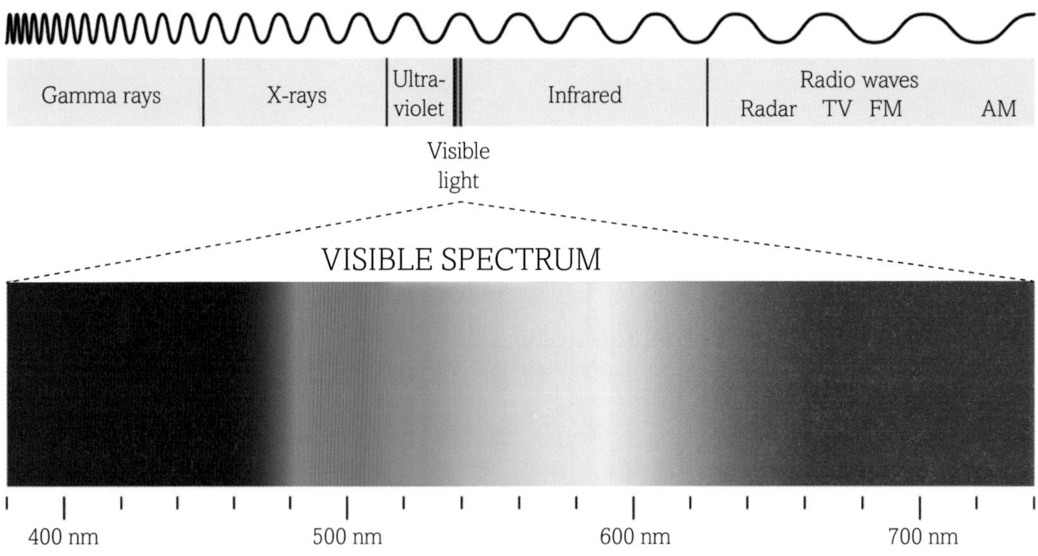

▲ **fig A** You can see how infrared radiation fits into the wider electromagnetic spectrum, next to the red part of the visible spectrum.

WHAT HAPPENS WHEN MOLECULES ABSORB INFRARED RADIATION

The importance of infrared radiation in chemistry is that it is absorbed by molecules and causes two possible effects, both described as vibrations. These effects are:

- **stretching** – where the bond length increases and decreases
- bending – where the bond angle increases and decreases.

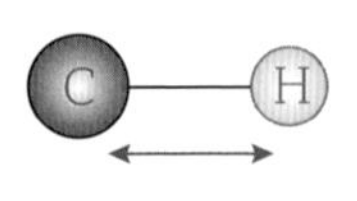

The C–H bond stretches when it absorbs infrared radiation.

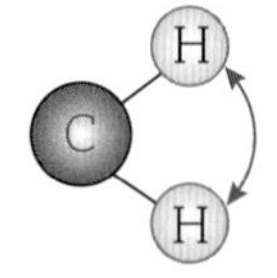

The C–H bond bends when it absorbs infrared radiation.

In this book, only stretching vibrations will be considered.

When a molecule absorbs infrared radiation, the amount of energy absorbed depends on:

- the length of the bond
- the strength of the bond
- the mass of each atom involved in the bond.

The absorption of infrared radiation is linked to changes in the polarity of the molecule, so simple non-polar molecules (such as H_2 and Cl_2) do not absorb infrared radiation.

WHAT DOES AN INFRARED SPECTRUM LOOK LIKE?

When a compound is irradiated by infrared radiation, the bonds in the molecules absorb radiation from some parts of the spectrum, but not from others.

AXES

The spectrum is normally shown with the vertical axis labelled **transmittance**, shown as a percentage from 0 to 100. A value of 100% transmittance means that 100% of the radiation is transmitted and none is absorbed.

The horizontal axis could be labelled either as frequency or wavelength, but a different unit is used: **wavenumber**. This is the reciprocal of the wavelength, and so it represents frequency. It is usually quoted in the unit cm^{-1}. The numerical scale normally starts at 4000 cm^{-1} and ends at 500 cm^{-1}. It may seem unusual for the numbers to decrease from left to right, but the left to right direction does represent increasing frequency. Another unusual feature is that the scale changes after 2000 cm^{-1}.

Fig B is an example of a typical infrared spectrum.

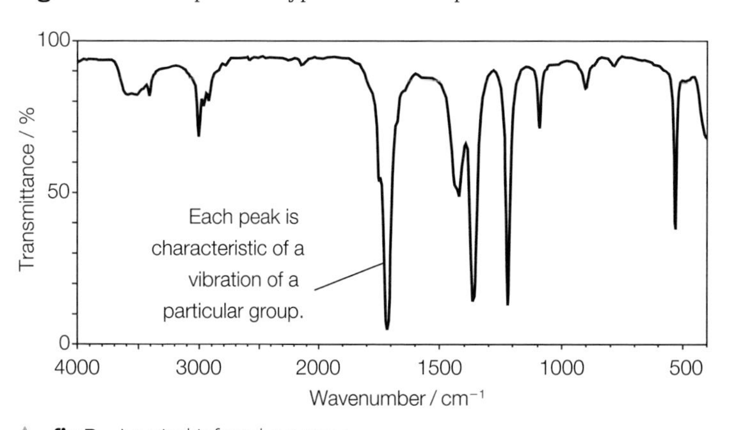

▲ **fig B** A typical infrared spectrum.

ABSORPTIONS AND THEIR INTENSITIES

You can see that some of the spectrum consists of a wavy but almost horizontal line close to 100%, but at specific wavenumbers there are dips or troughs. These are referred to as absorptions (or sometimes as peaks). The actual transmittance value of an absorption is not very important, but its **intensity** is.

Weak intensities refer to high transmittance values, and strong intensities to low transmittance values.

WAVENUMBER VALUES

The wavenumber values are also very important. The spectrum in **fig B** shows that the absorption with the lowest transmittance occurs at about 1700 cm^{-1}. Sometimes, another important feature to note is whether the absorption is sharp (i.e. a narrow wavenumber range) or broad (a wide wavenumber range).

CHARACTERISTIC ABSORPTIONS

You do not need to remember any of the information in **table A**, as it will be provided in an exam. However, it is important to be familiar with how to use it when interpreting infrared spectra.

WAVENUMBER / cm^{-1}	BOND	FUNCTIONAL GROUP
3750–3200	O–H	alcohol
3500–3300	N–H	amine
3300–2500	O–H	carboxylic acid
3095–3010	C–H	alkene
2962–2853	C–H	alkane
2900–2820 2775–2700	C–H	aldehyde
1740–1720	C=O	aldehyde
1725–1700	C=O	carboxylic acid
1720–1700	C=O	ketone
1669–1645	C=C	alkene

table A Information to help interpret an infrared spectrum.

In the example spectrum in this topic, the strong absorption at 1700 cm^{-1} is caused by C=O, but the compound responsible for this spectrum could be a carboxylic acid or a ketone.

Sometimes a missing absorption is just as useful. In the example, there is no absorption in the 3300–2500 cm^{-1} region, so the compound does not contain an OH group of a carboxylic acid. Therefore, the absorption at 1700 cm^{-1} strongly suggests that the compound is a ketone and not a carboxylic acid.

LEARNING TIP

In an infrared spectrum, remember that the horizontal scale uses wavenumber to represent frequency, and that the scale changes at 2000 cm^{-1}.

CHECKPOINT

1. Explain why hydrogen fluoride absorbs infrared radiation but fluorine does not.
2. A compound has an infrared spectrum that shows a broad absorption centred on 2850 cm^{-1} and a sharp absorption at 1710 cm^{-1}. Suggest what homologous series the compound belongs to.

SUBJECT VOCABULARY

infrared radiation the part of the electromagnetic spectrum with frequencies below that of red light

stretching when a bond absorbs infrared radiation and uses it to alter the length of the bond

transmittance the amount of radiation absorbed at a particular wavenumber

wavenumber the frequency of infrared radiation absorbed by a particular bond in a molecule

intensity the amount of infrared radiation absorbed

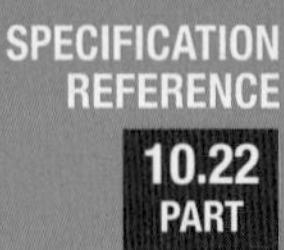

LEARNING OBJECTIVES

- Be able to use infrared spectra to deduce functional groups present in organic compounds.
- Be able to predict infrared absorptions due to familiar functional groups including wavenumber data.

DIFFERENT WAYS OF USING INFRARED SPECTRA

There are three main ways your understanding of infrared spectra could be tested, and you need to become familiar with all of them. They are:

- predicting the absorptions in the spectrum of an organic compound
- deducing the functional groups in a compound from a list of wavenumbers
- deducing the structure of a compound from wavenumbers and molecular formula.

These all require the use of wavenumber data, like the data shown in **Topic 10D.3**. This will either be provided in the question, or you can use information from the Data booklet provided in the exam. **Table A** is the table from **Topic 10D.3** reproduced here for your convenience.

WAVENUMBER / cm^{-1}	BOND	FUNCTIONAL GROUP
3750–3200	O–H	alcohol
3500–3300	N–H	amine
3300–2500	O–H	carboxylic acid
3095–3010	C–H	alkene
2962–2853	C–H	alkane
2900–2820 2775–2700	C–H	aldehyde
1740–1720	C=O	aldehyde
1725–1700	C=O	carboxylic acid
1720–1700	C=O	ketone
1669–1645	C=C	alkene

table A Information to help interpret an infrared spectrum.

FINGERPRINT REGION

You may come across the term 'fingerprint region'. This term is worth explaining. The table of wavenumbers starts at 3750 and ends at 1645 cm^{-1}, even though there are many absorptions in the infrared region between 1500 to 500 cm^{-1}. Most of the absorptions in this region (1500 and 500 cm^{-1}) result from bending vibrations (not considered in this book) or from absorptions by bonds not listed in the table of wavenumbers.

This region is sometimes referred to as the 'fingerprint region' because, although individual absorptions are not easily recognised, the whole pattern acts like a fingerprint that is slightly different for similar molecules.

PREDICTING THE SPECTRUM OF AN ORGANIC COMPOUND

Suppose you are given the identity of an organic compound. This might be a formula (displayed, skeletal or structural) or a name. You should then be able to predict the wavenumber ranges of the compound's infrared spectrum.

WORKED EXAMPLE 1

What absorptions would you expect to find in the infrared spectrum of propanal?

Once you recognise propanal as an aldehyde, you could predict:

- absorptions in the ranges 2900–2820 and 2775–2700 cm^{-1} resulting from the C–H bond
- absorption in the range 1740–1720 cm^{-1} resulting from the C=O bond.

WORKED EXAMPLE 2

What absorptions would you expect to find in the infrared spectrum of $CH_3CH(OH)CH_3$?

This compound is an alcohol, so you could predict a broad absorption in the range 3750–3200 cm^{-1} resulting from the O–H bond.

DEDUCING THE FUNCTIONAL GROUPS FROM A LIST OF WAVENUMBERS

WORKED EXAMPLE 3

An organic compound has absorptions in the infrared region at these wavenumbers: 3675, 2870 and 1735 cm^{-1}. Which functional groups does it contain?

Answer

O–H (alcohol), C–H and C–O (aldehyde).

WORKED EXAMPLE 4

An organic compound has absorptions in these wavenumber ranges in the infrared region: 3500–3300 and 3300–2500 cm^{-1}. Which functional groups does it contain?

Answer

O–H (carboxylic acid) and N–H (amine): it could be an amino acid.

DEDUCING THE STRUCTURE FROM WAVENUMBERS AND MOLECULAR FORMULA

This is a bit more complicated. Some molecular formulae could represent different combinations of functional groups, so to make the decision you need to consider the actual functional groups from the spectrum and how they could be used in conjunction with the molecular formula.

WORKED EXAMPLE 5

A compound has infrared absorptions at 1730 and 3450 cm^{-1}, and has a molecular formula of $C_2H_4O_2$. Deduce a possible structure for it.

Answer

The functional groups are O–H (alcohol) and C=O (aldehyde).
The only structure with the molecular formula $C_2H_4O_2$ that fits is $CH_2(OH)CHO$. Note that it is not CH_3COOH (ethanoic acid) because this would have an absorption in the range 3300–2500 cm^{-1}, not at 3450 cm^{-1}.

WORKED EXAMPLE 6

An organic compound has a molecular formula of $C_3H_6O_2$. Its infrared spectrum is shown below.

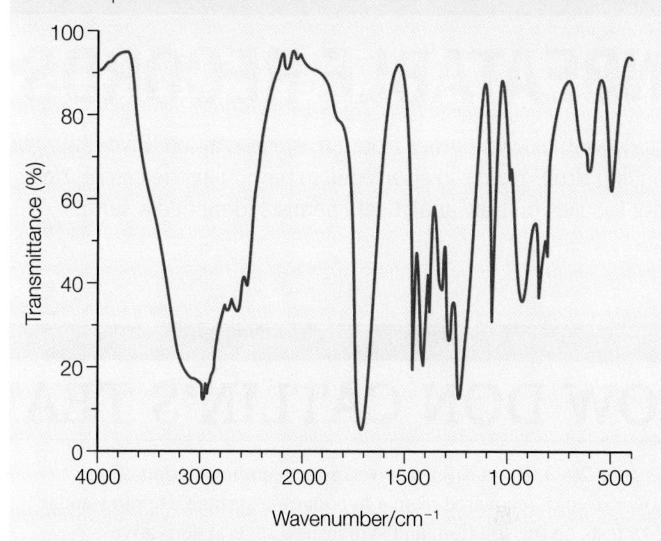

▲ **fig A** The infrared spectrum of an unknown compound.

Deduce a possible structure for the compound.

Answer

There is a broad absorption at 3000 cm^{-1} and a narrow absorption at 1700 cm^{-1}. These suggest the presence of O–H (carboxylic acid) and C=O (carboxylic acid). The compound must contain the COOH group, which leaves C_2H_5 to make up the rest of the molecular formula. The only possible structure is CH_3CH_2COOH, which is propanoic acid.

LEARNING TIP

Concentrate on the region between 4000 and 1500 cm^{-1}.
Practise making predictions of the spectra of molecules in **Topics 4**, **5** and **10**.

CHECKPOINT

1. Why is it not easy to use infrared spectra to distinguish between propan-1-ol and propan-2-ol?
2. A compound has the molecular formula C_3H_6O. How can its infrared spectrum be used to show that it has the structure $(CH_3)_2CO$ and not $CH_3C(OH)=CH_2$?

UNBEATABLE RECORDS

SKILLS ANALYSIS, INTERPRETATION, CRITICAL THINKING

The following extract comes from an article entitled 'Five rings good, four rings bad'. The article highlights the problem of performance-enhancing drug misuse in sport and, in particular, the use of 'designer' steroids that are not detectable by routine drug testing. This extract focuses on how analytical chemists identified a sample of an unknown steroid as tetrahydrogestrinone (THG).

HOW DON CATLIN'S TEAM CRACKED THG

On 13 June 2003, Don Catlin received a methanolic solution of an unknown steroid, recovered from a hypodermic syringe. He ran standard GC-MS tests on the solution, and synthesised several derivatives. Attempts to identify the steroid failed because the mass spectrum contained a large number of unidentifiable peaks. The only compound that they could identify at this stage was a small amount of another anabolic steroid, norbolethone, evidently present as an impurity.

Catlin suspected that the 'unknown' shared a common carbon skeleton with norbolethone. However, they noted a peak in its mass spectrum with $m/z = 312$, and thought this was the molecular ion. Accurate mass measurement gave 312.2080, from which they deduced the compound had the molecular formula $C_{21}H_{28}O_2$.

fig A Don Catlin

When they compared the mass spectrum of the unknown with other steroids, it became clear that it shared features with gestrinone and trenbolone.

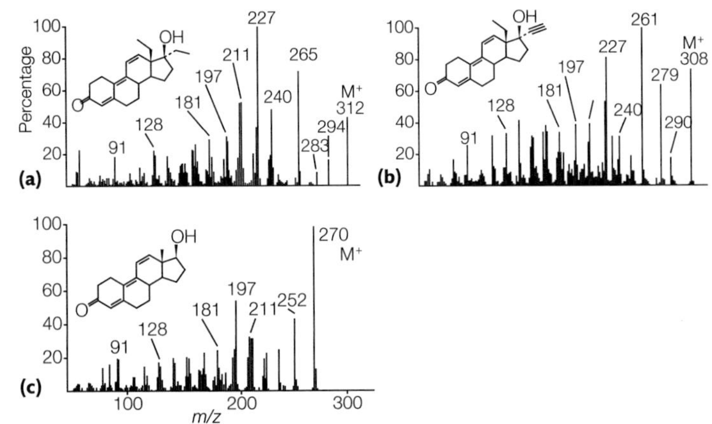

fig B The mass spectra of (a) the unknown substance, (b) gestrinone and (c) trenbolone.

All three compounds had the same fragments with m/z values at 211 and below present, so Catlin deduced that they contained the same A, B and C rings.

Furthermore, when the MS of the unknown was compared with gestrinone, the fragments with m/z above 240 occurred 4 Da higher in the unknown, suggesting that it was gestrinone with four additional hydrogen atoms. A possibility was that the terminal alkyne group in gestrinone had been reduced to an ethyl group.

Having tentatively identified the unknown steroid as tetrahydrogestrinone, the team then prepared an authentic sample of THG by catalytic hydrogenation of gestrinone. This required careful control of conditions (0 °C) to prevent hydrogenation of C=C double bonds (see equation in **fig C**). The retention time and mass spectra of the synthetic THG matched the unknown material exactly.

gestrinone $+ 2H_2$ —(0°C, Pd/C catalyst)→ tetrahydrogestrinone

fig C The hydrogenation of gestrinone.

To study the metabolism of THG in mammals, the team gave intravenous doses of THG to a baboon, and collected urine samples from the animal over several days. Detectable amounts of THG were found in urine for many hours after administration.

THG was thus directly detectable in urine samples, though it defies detection by the standard procedure involving derivatisation into the Me_3Si derivatives.

From an article in *Education in Chemistry* magazine, published by the Royal Society of Chemistry

SCIENCE COMMUNICATION

1. (a) Who do you think is the intended audience for this article? Evaluate the ways in which the author has written for them.

(b) Why do you think value judgements are avoided by the author even though the article considers a very emotive issue?

CHEMISTRY IN DETAIL

2. The analytical techniques of IR spectroscopy and mass spectrometry can be used to identify unknown molecules. Suggest which of these two techniques would be more useful in distinguishing samples of two steroids with similar structures. You should be prepared to justify your choice.

3. The mass spectrum of trenbolone is shown in **fig B**. Suggest how the fragment at $m/z = 252$ is generated.

THINKING BIGGER TIP

The dalton (Da) is the standard unit for indicating mass on an atomic or molecular scale.

ACTIVITY

SKILLS ANALYSIS

The online database at NIST Chemistry WebBook allows you to search for a range of organic compounds and related data.

Use the database to find the mass spectrum of one of the following compounds:

(a) chlorobenzene

(b) bromoethane

(c) ethylamine

(d) cyclohexane.

Prepare a 3–5 minute presentation showing the mass spectrum of your chosen molecule and identifying the most important peaks. Your presentation should include:

- a picture of the mass spectrum of your chosen molecule
- an identification of the main fragment and isotopic abundance peaks of your molecule, explaining how each peak is formed.

WRITING SCIENTIFICALLY

An evaluation should review all the information to form a conclusion. You should think about the strengths and weaknesses of the evidence and information, and come to a supported judgement.

DID YOU KNOW?

The women's 800 m record was set in 1983 by the Czech Jarmila Kratochvilova, who ran the distance in 1:53.28. This was before the test for human growth hormone was in routine use at athletics competitions. Since then, only one athlete has managed to come within a second of her record.

10 EXAM PRACTICE

1 A halogenoalkane has the structure $CH_3CHClCH(CH_3)_2$.

Which response shows its name and classification?

A 2-chloro-3-methylbutane primary

B 2-chloro-3-methylbutane secondary

C 2-methyl-3-chlorobutane secondary

D 2-methyl-3-chlorobutane tertiary [1]

(Total for Question 1 = 1 mark)

2 These are the equations for two hydrolysis reactions of halogenoalkanes:

reaction 1 $CH_3CH_2CH_2Br + H_2O \rightarrow CH_3CH_2CH_2OH + HBr$

reaction 2 $CH_3CH_2CH_2Cl + H_2O \rightarrow CH_3CH_2CH_2OH + HCl$

Which statement explains why reaction 1 is faster than reaction 2, when both reactions occur under the same conditions?

A Bromine has a higher atomic radius than chlorine.

B Chlorine is more electronegative than bromine.

C The carbon-chlorine bond has a greater bond enthalpy than the carbon-bromine bond.

D The relative atomic mass of bromine is greater than that of chlorine. [1]

(Total for Question 2 = 1 mark)

3 A mixture of 1-bromobutane, potassium cyanide and ethanol is heated under reflux to prepare a sample of a nitrile.

Which formula shows the structure of the organic product in this preparation?

A $CH_3CH_2CH_2CN$

B $CH_3CH_2CH_2CH_2CN$

C $CH_3CH(CN)CH_3$

D $CH_3CH_2CH(CN)CH_3$ [1]

(Total for Question 3 = 1 mark)

4 Alcohols can be converted into chloroalkanes in different ways.

Which reagent is the most suitable for this conversion?

$CH_3CH_2CH_2CH_2OH \rightarrow CH_3CH_2CH_2CH_2Cl$

A Cl_2

B HCl

C PCl_5

D H_3PO_4 [1]

(Total for Question 4 = 1 mark)

5 Alcohols can be oxidised to aldehydes, ketones or carboxylic acids.

Which practical method is the most suitable to use in this oxidation?

$CH_3CH_2CH_2CH_2OH \rightarrow CH_3CH_2CH_2COOH$

A distillation with addition

B fractional distillation

C heating under reflux

D simple distillation [1]

(Total for Question 5 = 1 mark)

6 An organic compound with the formula $CH_3CH_2CH_2CH_2Cl$ is heated under reflux with NaOH(aq) for some time.

What is the formula of the main organic product of this reaction?

A $CH_3CH_2CH_2OH$

B $CH_3CH_2CH_2CH_2OH$

C $CH_3CH{=}CH_2$

D $CH_3CH_2CH{=}CH_2$ [1]

(Total for Question 6 = 1 mark)

7 The following is a description of how to prepare a dry, pure sample of 2-chloro-2-methylpropane starting from methylpropan-2-ol.

The equation for the reaction is:

$$(CH_3)_3C{-}OH + HCl \rightarrow (CH_3)_3C{-}Cl + H_2O$$

methylpropan-2-ol 2-methyl-2-chloropropane

Step 1: Place about 9 cm³ of methylpropan-2-ol into a separating funnel and carefully add 20 cm³ of concentrated hydrochloric acid, about 3 cm³ at a time. After each addition, hold the stopper tap firmly in place, and invert the funnel a few times. Then, with the funnel in the upright position, loosen the stopper briefly to release any pressure.

Step 2: Leave the separating funnel and contents in a fume cupboard for about twenty minutes. Shake gently at intervals.

Step 3: Allow the layers in the separating funnel to separate and then run off, and discard, the lower aqueous layer.

Step 4: Add sodium hydrogencarbonate solution 2 cm³ at a time. Shake the funnel carefully after each addition and release the pressure of gas by loosening the stopper. Repeat until no more gas is evolved.

Step 5: Allow the layers to separate and then run off, and discard, the lower aqueous layer.

Step 6: Run the organic layer into a small, dry conical flask and add some anhydrous sodium sulfate. Swirl the flask occasionally for about five minutes.

Step 7: Carefully decant the organic liquid from the solid sodium sulfate into a pear-shaped flask. Add a few anti-bumping granules and set up the flask for distillation. Collect the liquid that distils over between 47–53 °C.

(a) Other than wearing safety spectacles and a protective coat, explain one safety precaution you should take when using concentrated hydrochloric acid. [2]

(b) State the purpose of adding sodium hydrogencarbonate in step 4. [1]

(c) State the purpose of adding anhydrous sodium sulfate in step 6. [1]

(d) Draw a labelled diagram of the distillation of the organic liquid in step 7. [4]

(e) A student started with 7.40 g of methylpropan-2-ol and obtained 7.82 g of 2-chloro-2-methylpropane. Calculate the percentage yield of the product. [3]

(Total for Question 7 = 11 marks)

8 Spectrum A and Spectrum B are the mass spectra of pentan-2-one ($CH_3COCH_2CH_2CH_3$) and pentan-3-one, ($CH_3CH_2COCH_2CH_3$), but not necessarily in that order.

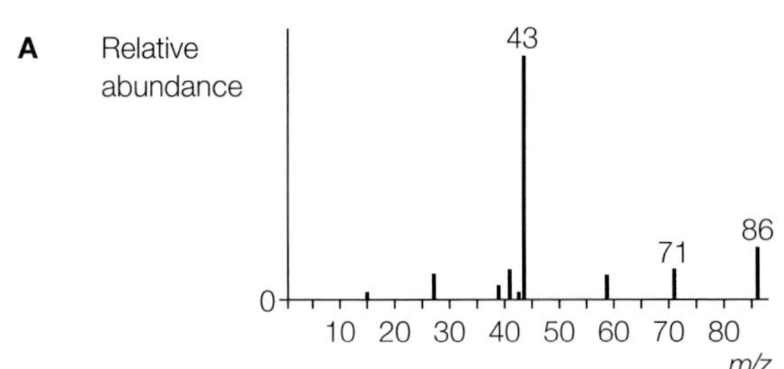

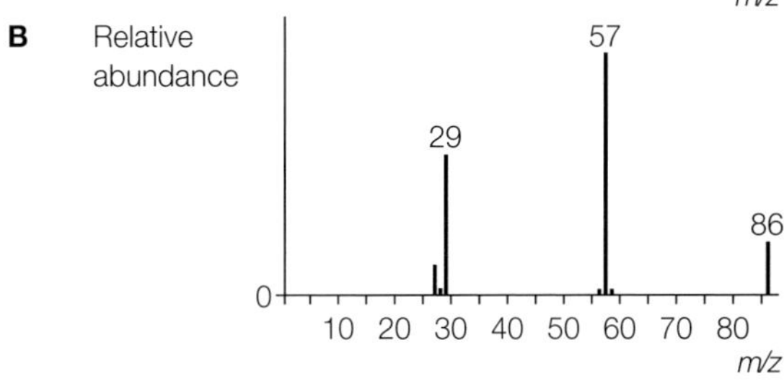

Explain which spectrum belongs to each compound. [4]

(Total for Question 8 = 4 marks)

9 A primary alcohol can be oxidised by reaction with acidified potassium dichromate(VI). The major product obtained depends on the conditions used.

If the oxidising agent is slowly added to the alcohol and then the product is distilled off as it forms, an aldehyde is collected.

If the alcohol is heated under reflux with an excess of the oxidising agent, a carboxylic acid is formed.

The infrared spectrum below is that of a product formed by the oxidation of butan-1-ol.

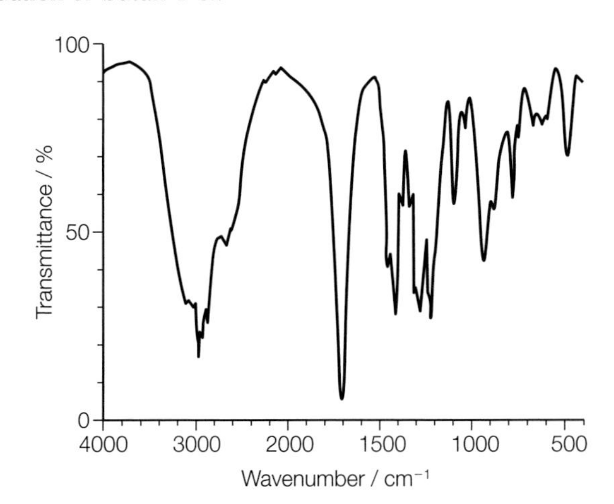

(a) Identify the product and explain your reasoning. [3]

(b) Write an equation for the oxidation of butan-1-ol to this product. Use [O] to represent the oxidising agent. [2]

(Total for Question 9 = 5 marks)

10 Compound X has the following composition by mass: C, 62.07%; H, 10.34%; O, 27.59%.

(a) Calculate the empirical formula of compound X. [3]

(b) The mass spectrum of compound X is shown below.

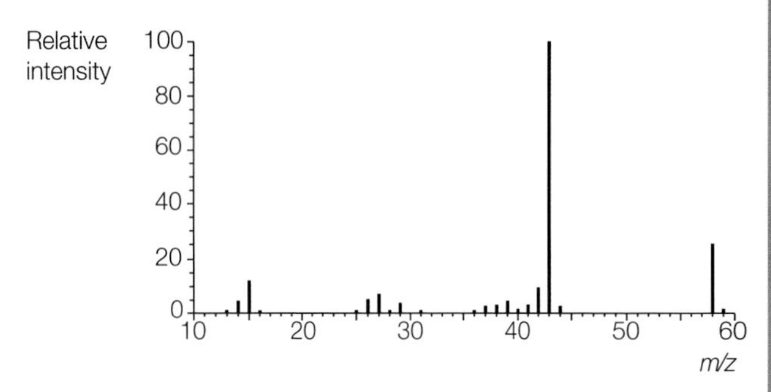

Deduce the molecular formula of compound X. Show how you obtained your answer. [3]

(c) Compound X is one of two structural isomers. One is an aldehyde, the other is a ketone.

The infrared spectrum of compound X is shown below.

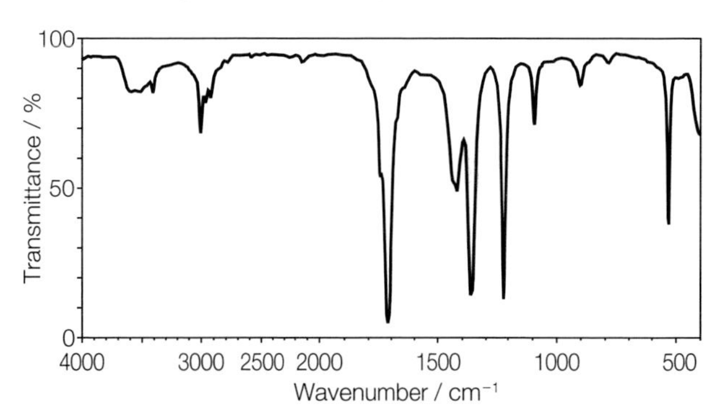

Compound X produced no observable change when heated with Fehling's solution.

(i) Identify compound X. [1]

(ii) Identify the species responsible for the peaks at 15 and 43 in the mass spectrum of compound X. [2]

(iii) Identify the functional group responsible for the absorbance at 1700 cm^{-1} in the infrared spectrum of compound X. [1]

(Total for Question 10 = 10 marks)

MATHS SKILLS

In order to be able to develop your skills, knowledge and understanding in Chemistry, you will need to have developed your mathematical skills in a number of key areas. This section gives more explanation and examples of some key mathematical concepts you need to understand. Further examples relevant to your IAS / IAL Chemistry studies are given throughout the book.

ARITHMETIC AND NUMERICAL COMPUTATION

USING STANDARD FORM

Dealing with very large or small numbers can be difficult. For example, Avogadro's constant is an important value in Chemistry which is approximately equal to 602 000 000 000 000 000 000 000 mol^{-1}. To make such numbers easier to handle, we can write them in the format $a \times 10^b$. This is called standard form. Using standard form, Avogadro's constant can be written as $6.02 \times 10^{23}\ mol^{-1}$.

To change a number from decimal form to standard form:

Count the number of positions you need to move the decimal point by until it is directly to the right of the first number that is not zero.

This number is the index number that tells you how many multiples of 10 you need. If the original number was a decimal, your index number must be negative.

Here are some examples:

DECIMAL NOTATION	STANDARD FORM NOTATION
0.000 000 012	1.2×10^{-8}
15	1.5×10^{1}
1000	1.0×10^{3}
3 700 000	3.7×10^{6}

USING RATIOS, FRACTIONS AND PERCENTAGES

Ratios, fractions and percentages help you to express one quantity in relation to another with precision. Ratios compare like quantities using the same units. Fractions and percentages are important mathematical tools for calculating proportions.

RATIOS

A ratio is used to compare quantities. You can simplify ratios by dividing each side by a common factor. For example, 12 : 4 can be simplified to 3 : 1 by dividing each side by 4.

WORKED EXAMPLE

Divide 180 into the ratio 3 : 2

Answer

Our strategy is to work out the total number of parts. Then divide 180 by the number of parts to find the value of one part.

Total number of parts = 3 + 2 = 5

Value of one part = 180 ÷ 5 = 36

Answer = 3 × 36 : 2 × 36 = 108 : 72

Check your answer by making sure the parts add up to 180
108 + 72 = 180

WORKED EXAMPLE

An excess of magnesium is added to 0.2 dm³ of 1 mol dm⁻³ dilute hydrochloric acid. The equation for the reaction is:

$$Mg + 2HCl \rightarrow MgCl_2 + H_2$$

How many moles of hydrogen are formed?

Since there is an excess of magnesium, we know that all of the hydrochloric acid will react. We can use the following equation to calculate the number of moles of hydrochloric acid:

$$\text{concentration in mol dm}^{-3} = \frac{\text{amount in moles}}{\text{volume in dm}^3}$$

$$1\ \text{mol dm}^{-3} = \frac{\text{amount in moles of HCl}}{0.2\ \text{dm}^3}$$

amount of HCl reacted = 0.2 mol

The ratio for HCl : H_2 in this reaction is 2 : 1, so when two moles of HCl react, one mole of H_2 is formed.

number of moles of H_2 formed = 0.5 × number of moles of HCl reacted
= 0.5 × 0.2 mol
= 0.1 mol

FRACTIONS

When using fractions, make sure you know the key strategies for the four operators:

To add or subtract fractions, find the lowest common multiple (LCM) and then use the golden rule of fractions. The golden rule states that a fraction remains unchanged if the numerator and denominator are multiplied or divided by the same number.

WORKED EXAMPLE

$$\frac{1}{2} + \frac{1}{5} = \frac{5}{10} + \frac{2}{10} = \frac{7}{10}$$

To multiply fractions together, simply multiply the numerators together and multiply the denominators together.

WORKED EXAMPLE

$$\frac{2}{7} \times \frac{4}{9} = \frac{8}{63}$$

To divide fractions, simply invert (flip) the second fraction and multiply.

WORKED EXAMPLE

$$\frac{2}{3} \div \frac{7}{9} = \frac{2}{3} \times \frac{9}{7} = \frac{18}{21} = \frac{6}{7}$$

PERCENTAGES

When using percentages, it is useful to recall the different types of percentage questions.

To increase a value by a given percentage, use a percentage multiplier.

WORKED EXAMPLE

Increase 30 mg by 23%.

If we increase by 23%, our new value will be 123% of the original value. We therefore multiply by 1.23.

Answer = 30 × 1.23 = 36.9 mg

To decrease a value by a given percentage, you need to focus on the part that is left over after the decrease.

WORKED EXAMPLE

Decrease 30 mg by 23%.

If we decrease by 23%, our new value will be 100 − 23 = 77% of the original value. We therefore multiply by 0.77.

Answer = 30 × 0.77 = 23.1 mg

To calculate a percentage increase, use the following equation:

$$\text{Percentage change} = \frac{\text{difference between values}}{\text{original value}} \times 100$$

To calculate percentage decrease, use the same equation but remember that your answer should be negative.

WORKED EXAMPLE

The volume of a solution increased from 40 ml to 50 ml. Calculate the percentage increase.

Change in volume = 10 ml

$$\text{Percentage increase} = \frac{10}{40} \times 100 = 25\%.$$

ALGEBRA

CHANGING THE SUBJECT OF AN EQUATION

It can be very helpful to rearrange an equation to express the variable that interests you, in terms of other variables. Always remember that any operation that you apply to one side of the equation must also be applied to the other side.

WORKED EXAMPLE

A sample of 2.5 mol of a substance has a mass of 12.5 g.

What is the molar mass of the substance?

The equation for calculating moles is $n = \frac{m}{M}$ where m = mass in grams, M = molar mass and n = amount of substance in moles.

If we wished to rearrange this equation to make M the subject, we would first multiply each side by M to obtain:

$$nM = m$$

Now to obtain the formula in terms of M, we divide each side by n to obtain:

$$M = \frac{m}{n}$$

We can now simply substitute in the values for the question:

$$m = 12.5\,\text{g}$$

$$n = 2.5\,\text{mol}$$

$$M = \frac{m}{n} = \frac{12.5\,\text{g}}{2.5\,\text{mol}} = 5\,\text{g mol}^{-1}$$

HANDLING DATA

USING SIGNIFICANT FIGURES

Often when you do a calculation, your answer will have many more figures than you need. Using an appropriate number of significant figures will help you to interpret results in a meaningful way.

Remember the 'rules' for significant figures:

1 The first significant figure is the first figure which is not zero.

2 Digits 1–9 are always significant.

3 Zeros that come after the first significant figure are significant unless the number has already been rounded.

Here are some examples:

EXACT NUMBER	TO ONE S.F.	TO TWO S.F	TO THREE S.F
45 678	50 000	46 000	45 700
45 000	50 000	45 000	45 000
0.002 755	0.003	0.002 8	0.002 76

APPLYING YOUR SKILLS

You will often find that you need to use more than one maths technique to answer a question. In this section, we will look at three example questions and consider which maths skills are required and how to apply them.

WORKED EXAMPLE

10 g of potassium was added to excess water, producing potassium hydroxide and hydrogen gas. The hydrogen was set alight because of the exothermic nature of the reaction.

(i) Calculate the number of moles of potassium used in the reaction.

(ii) Calculate the number of moles of $H_2(g)$.

(iii) Calculate the volume of $H_2(g)$ produced in dm^3.

This type of question is very common. The key is to be familiar with the three different types of moles equations and when to use them. You will also usually have to work with ratios in order to find the correct number of moles of a reactant or product. This tends to be where lots of mistakes are made.

The relevance of this question to the wider world is that, as chemists, we must know how much product we are going to make. This makes sense both for economic reasons, and in the case above where we are producing a highly flammable gas, for our safety as well!

Moles equations

(a) $\text{moles} = \frac{\text{mass}}{\text{molar mass}}$

This equation tends to be used when dealing with solids or masses.

(b) **moles = volume × concentration**

This equation can be used when dealing with solutions or titrations.

(c) $\text{moles} = \frac{\text{volume in dm}^3}{24}$ or $\frac{\text{volume in cm}^3}{24\,000}$

This equation is used for gases.

If you are confident with mathematics then you can also combine the above equations and rearrange them to solve the missing value.

$$\frac{\text{mass}}{\text{molar mass}} = \text{volume} \times \text{concentration}$$

$$\frac{\text{mass}}{\text{molar mass}} = \frac{\text{volume in dm}^3}{24}$$

Using the information above:

Step 1: Calculate the number of moles of potassium used in the reaction.

In order to answer this question, we first need to know which equation to use. We are given a mass, and know we are dealing with solids, so we should use equation (a). Knowing this, it is now a simple case of substituting in the numerical values and solving the equation.

$$\text{moles} = \frac{\text{mass}}{\text{molar mass}}$$

$$\text{moles} = \frac{10\text{ g}}{39.1\text{ g mol}^{-1}} = 0.2558\text{ mol}$$

Step 2: Calculate the number of moles of $H_2(g)$.

To find the number of moles of $H_2(g)$ requires us to work with ratios. First we write the chemical equation. Then make sure it is balanced. Sometimes this will already be done for you or it could be part of a previous question. Either way, balancing the chemical equation is key to determining the ratio of reactants to products, etc. From this, we can now determine the ratio of K to H_2, which tells us how many moles of K there are compared to H_2.

$$2K(s) + 2H_2O(l) \rightarrow 2KOH(aq) + H_2(g)$$

$K : H_2 = 2 : 1$ therefore 0.2558 mol : 0.1279 mol

Step 3: Calculate the volume of $H_2(g)$ produced in dm^3.

The final part of the question requires us to calculate the volume of H_2 produced. To do this we must decide which moles equation to use. As we are dealing with gases it leaves us with one obvious choice.

$$\text{moles} = \frac{\text{volume in dm}^3}{24} \text{ or moles} = \frac{\text{volume in cm}^3}{24\,000}$$

Always check what units are required as you could lose valuable marks otherwise. This is usually stated either in the question or next to the space provided to write the answer. This question has asked for the answer to be shown in dm^3. Rearranging this equation, substituting in the numerical values before solving provides us with the answer.

$$\text{volume in } dm^3 = 0.1279\text{ mol} \times 24\text{ mol dm}^{-3} = 3.069\text{ dm}^3$$

It is always good to check your answer following a long question like this. The best way is to work backwards, and make sure we get 10 g of K as the end point if you have time!

WORKED EXAMPLE

In the United Kingdom, there are currently 16 operational nuclear reactors located at 9 power stations. However, as the country's energy demands increase this number could rise. Recent surveys have found public opinion is not in favour of this increase, due to risk of leaks and radiation exposure following an explosion. One possible radioactive contaminant following an explosion is radioactive iodine, I, which is taken up by a person's thyroid gland. To counter this, potassium iodide, KI, is given to patients. This blocks the uptake of radioactive iodine by the thyroid gland by swamping the body with non-radioactive iodine instead.*

The current recommended dose of iodine, I_2*, is 130 mg.*
1 mg = 1 × 10^{-3} *g*

Calculate the mass of potassium iodide required to obtain the recommended dose of iodine.
Show your answer in mg.

This style of question requires you to think carefully about what the question is asking. Whilst there are only two steps to finding the correct answer, this can still prove troublesome. This question is made harder by the fact that we are using iodine, which is I_2, rather than just I^-. This means the molar mass is $2 \times 126.9\text{ g mol}^{-1}$.

We must first determine the number of moles of iodine required, using the basic moles equation to do so. We know it is this equation because we are dealing with solids and masses. Therefore the equation to solve is

$$\text{moles} = \frac{\text{mass}}{\text{molar mass}}$$

$$\frac{0.130\text{ g}}{253.8\text{ g mol}^{-1}} = 5.122 \times 10^{-4}\text{ mol}$$

However, it takes two moles of potassium iodide, KI, to release one mole of I_2. Therefore, the required moles of KI is $2 \times 5.122 \times 10^{-4} = 1.024 \times 10^{-3}$ mol. We can now work out the mass of potassium iodide by simply changing the subject of the equation and solving it using the molar mass of KI as 166.0 g mol^{-1}.

$$\text{mass} = \text{moles} \times \text{molar mass}$$

However, the question has asked for the mass to be shown in units of mg. This means we must multiply the answer by 1000. When answering any question it is better to leave the rounding to the end, in order to avoid rounding errors.

$$(1.024 \times 10^{-3}\text{ mol} \times 166.0) \times 1000 = 170\text{ mg}$$

WORKED EXAMPLE

Copper(II) chloride solution, $CuCl_2(aq)$, can be prepared by reacting excess powdered copper(II) oxide, CuO(s), with hot hydrochloric acid:

$$CuO(s) + 2HCl(aq) \rightarrow CuCl_2(aq) + H_2O(l)$$

Unreacted copper(II) is then removed by filtration, and water is evaporated from the filtrate to leave crystals of copper(II) chloride-2-water, $CuCl_2.2H_2O$.

In one such preparation, 50.0 cm^3 of 0.500 mol dm^{-3} hydrochloric acid is used.

(a) *Calculate the amount (in moles) of hydrochloric acid used.*

(b) *Calculate the minimum amount (in moles) of copper(II) oxide needed.*

(c) *Calculate the mass of copper(II) oxide needed, if an excess of 20% is necessary. Give your answer to two significant figures.*

(d) *Calculate the molar mass of copper(II) chloride-2-water, $CuCl_2.2H_2O$.*

(e) *1.81 g of copper(II) chloride-2-water is obtained. Calculate the percentage yield.*

In this type of question, you will need to perform a series of calculations, often using answers from earlier in the question. Pay particular attention to units and follow the 'NAUTE' rule – no approximation until the end! Give your final answer to an appropriate number of significant figures.

(a) 50.0 cm^3 = 0.50 dm^3

$$0.50\ dm^3 \times 0.5\ mol\ dm^{-3} = 0.025\ mol$$

(b) From the equation we can see that one mole of CuO reacts with two moles of HCl. We therefore need half the number of moles of CuO than of HCl.

$$0.025\ mol \div 2 = 0.0125\ mol$$

(c) To obtain an excess of 20%, we need 120% of the minimum number of moles we found in part (b). To calculate 120%, we can multiply by 1.2.

excess = 0.0125 × 1.2 = 0.015 mol

We can now find the mass required by multiplying the number of moles by the molar mass of CuO.

molar mass of CuO = 63.5 + 16.0 = 79.5

mass of CuO required = 0.015 × 79.5 = 1.1925 g

The data in the question was given to three significant figures, so we give our final answer to two significant figures:

mass of CuO required = 1.2 g

(d) We can calculate the molar mass using the relative atomic masses of the elements:

molar mass of $CuCl_2.2H_2O$ = 63.5 + (2 × 35.5) + (2 × 18.0) = 170.5

(e) From part (b) we know that the expected number of moles is 0.0125. We can use our answer from (d) to calculate the mass of the expected yield.

mass of expected yield = 0.0125 × 170.5 = 2.13 g

percentage yield = 1.82 ÷ 2.13 × 100 = 85%

PREPARING FOR YOUR EXAMS

IAS AND IAL OVERVIEW

The Pearson Edexcel International Advanced Subsidiary (IAS) in Chemistry and the Pearson Edexcel International Advanced Level (IAL) in Chemistry are modular qualifications. The IAS can be claimed on completion of the International Advanced Subsidiary (IAS) units. The International Advanced Level can be claimed on completion of all the units (IAS and IAL units).

- International AS students will sit three exam papers. The IAS qualification can either be standalone or contribute 50% of the marks for the International Advanced Level.
- International A Level students will sit six exam papers, the three IAS papers and three IAL papers.

The tables below give details of the exam papers for each qualification.

IAS Papers	Unit 1: Structure, Bonding and Introduction to Organic Chemistry	Unit 2: Energetics, Group Chemistry, Halogenoalkanes and Alcohols*	Unit 3: Practical Skills in Chemistry I
Topics covered	Topics 1–5	Topics 5–10	Topics 1–10
% of the IAS qualification	40%	40%	20%
Length of exam	1 hour 30 minutes	1 hour 30 minutes	1 hours 20 minutes
Marks available	80 marks	80 marks	50 marks
Question types	multiple-choice short open open response calculation	multiple-choice short open open response calculation extended writing	short open open response calculation
Mathematics	For both Unit 1 and Unit 2, a minimum of 18 marks will be awarded for mathematics at Level 2 or above. For Unit 3, a minimum of 6 marks will be awarded for mathematics at Level 2 or above.		

* This paper will contain some synoptic questions which require knowledge and understanding from Unit 1.

IAL Papers	Unit 4: Rates, Equilibria and Further Organic Chemistry**	Unit 5: Transition Metals and Organic Nitrogen Chemistry†	Unit 6: Practical Skills in Chemistry II
Topics covered	Topics 11–15	Topics 16–20	Topics 11–20
% of the IAL qualification	20%	20%	10%
Length of exam	1 hour 45 minutes	1 hour 45 minutes	1 hour 20 minutes
Marks available	90 marks	90 marks	50 marks
Question types	multiple-choice short open open response calculation extended writing	multiple-choice short open open response calculation extended writing	short open open response calculation
Mathematics	For Unit 4, a minimum of 22 marks will be awarded for mathematics at Level 2 or above. For Unit 5, a minimum of 18 marks will be awarded for mathematics at Level 2 or above. For Unit 6, a minimum of 6 marks will be awarded for mathematics at Level 2 or above.		

** This paper will contain some synoptic questions which require knowledge and understanding from Units 1 and 2.

† This paper will contain some synoptic questions which require knowledge and understanding from Units 1, 2 and 4.

EXAM STRATEGY

ARRIVE EQUIPPED

Make sure you have all of the correct equipment needed for your exam. As a minimum you should take:

- pen (black ink or ball-point pen)
- pencil (HB)
- ruler (ideally 30 cm)
- eraser (make sure it's clean and doesn't smudge the pencil marks or rip the paper)
- calculator (scientific).

ENSURE YOUR ANSWERS CAN BE READ

Your handwriting does not have to be perfect but the examiner must be able to read it! When you're in a hurry it's easy to write key words that are difficult to decipher.

PLAN YOUR TIME

Note how many marks are available on the paper and how many minutes you have to complete it. This will give you an idea of how long to spend on each question. Be sure to leave some time at the end of the exam for checking answers. A rough guide of a minute a mark is a good start, but short answers and multiple choice questions may be quicker. Longer answers might require more time.

UNDERSTAND THE QUESTION

Always read the question carefully and spend a few moments working out what you are being asked to do. The command word used will give you an indication of what is required in your answer.

Be scientific and accurate, even when writing longer answers. Use the technical terms you've been taught.

Always show your working for any calculations. Marks may be available for individual steps, not just for the final answer. Also, even if you make a calculation error, you may be awarded marks for applying the correct technique.

PLAN YOUR ANSWER

In questions marked with an *, marks will be awarded for your ability to structure your answer logically showing how the points that you make are related or follow on from each other where appropriate. Read the question fully and carefully (at least twice!) before beginning your answer.

MAKE THE MOST OF GRAPHS AND DIAGRAMS

Diagrams and sketch graphs can earn marks – often more easily and quickly than written explanations – but they will only earn marks if they are carefully drawn.

- If you are asked to read a graph, pay attention to the labels and numbers on the x and y axes. Remember that each axis is a number line.
- If asked to draw or sketch a graph, always ensure you use a sensible scale and label both axes with quantities and units. If plotting a graph, use a pencil and draw small crosses or dots for the points.
- Diagrams must always be neat, clear and fully labelled.

CHECK YOUR ANSWERS

For open-response and extended writing questions, check the number of marks that are available. If three marks are available, have you made three distinct points?

For calculations, read through each stage of your working. Substituting your final answer into the original question can be a simple way of checking that the final answer is correct. Another simple strategy is to consider whether the answer seems sensible. Pay particular attention to using the correct units.

SAMPLE EXAM ANSWERS

QUESTION TYPE: MULTIPLE CHOICE

Solutions containing chlorate (I) ions are used as household bleaches and disinfectants. These solutions decompose on heating as shown in the equation below.

$3ClO^- \rightarrow ClO_3^- + 2Cl^-$

Which oxidation state is shown by chlorine in each of these three ions?

	ClO^-	ClO_3^-	Cl^-
☐ **A**	+1	+3	–1
☐ **B**	–1	+3	+1
☐ **C**	–1	+5	+1
☐ **D**	+1	+5	–1

[1]

Three oxidation states have to be correctly calculated before an answer can be selected. However, before any answers are given, the question is very clearly stated: 'Which oxidation state is shown by chlorine in each of these three ions?' Calculate the three oxidation states before looking at any of the answers and then select the correct answer from the list based on the three oxidation states that you have already determined.

Multiple choice questions always have one mark and the answer is given! For this reason students often make the mistake of thinking that they are the easiest questions on the paper. Unfortunately, this is not the case. These questions often require several answers to be worked out and error in one of them will lead to the wrong answer being selected. The three incorrect answers supplied (distractors) will feature the answers that students arrive at if they make typical or common errors. The trick is to answer the question before you look at any of the answers.

Question analysis

Multiple choice questions look easy until you try to answer them. Very often they require some working out and thinking.

- In multiple choice questions you are given the correct answer along with three incorrect answers (called distractors). You need to select the correct answer and put a cross in the box of the letter next to it.
- If you change your mind, put a line through the box (☒) and then mark your new answer with a cross (☒).

Average student answer

	ClO^-	ClO_3^-	Cl^-
☒ **A**	+1	+3	–1

In ClO^- chlorine has an oxidation state of +1, in Cl^- it is –1 and in ClO_3^- it is +5. The correct answer is D. However, the student has selected answer A, and so does not gain any marks.

COMMENTARY

This is an incorrect answer because:

- The student has calculated the oxidation state of chlorine in ClO^- and $2Cl^-$ correctly, however, the oxidation state of chlorine in ClO_3^- is +5. The student should have selected answer D.

If you have any time left at the end of the paper go back and check your answer to each part of a multiple choice question so that a slip like this does not cost you a mark.

QUESTION TYPE: SHORT OPEN

Compounds in Group 2 show trends in their properties. State the block in the Periodic Table in which the Group 2 elements are found. [1]

State, give or name are used interchangeably when you are required to simply recall and write down a piece of information. These are usually simple short answers, often one word, requiring you to recollect the chemistry you have been taught.

Question analysis

- The command word in this question is state. It requires you to recall and write down one or more pieces of information.

- Generally one piece of information is required for each mark given in the question. There is one mark available for this question and so one piece of information is required.
- Clarity and brevity are the keys to success on short open questions. For one mark, it is not always necessary to write complete sentences.

Average student answer

S-block

The Periodic Table is split into blocks; s, p, d and f according to the orbitals that their highest energy electrons occupy. For Group 1 and 2 elements in the s-block, the highest energy electrons are in the s-block as the student has correctly stated.

COMMENTARY

This is a strong answer because:

- The student clearly understands that there are different blocks on the Periodic Table and has correctly identified Group 2 elements as s-block elements.

QUESTION TYPE: OPEN RESPONSE

This question is about Group 7 and redox chemistry.

Explain the trend in the boiling temperatures of the elements down Group 7, from fluorine to iodine. [4]

The command word in this question is explain. This requires a justification of a point, which in this case is the trend in boiling temperatures down Group 7. The question doesn't actually give the trend in boiling temperatures so it is a reasonable assumption that one mark will be given for simply stating how they change from fluorine to iodine, so start by doing that. There are four marks available for this question and so you should then aim to make three valid points as to why the boiling temperatures increase down Group 7.

Question analysis

- With any question worth three or more marks, think about your answer and the points that you need to make before you write anything down. Keep your answer concise, and the information you write down relevant to the question. You will not gain marks for writing down chemistry that is not relevant to the question (even if correct) but it will cost you time.
- Remember that you can use bullet points or diagrams in your answer.

Average student answer

The boiling temperature of the elements increases as you go down the group. This is because the intermolecular forces called London forces get bigger from fluorine to iodine and so the boiling temperature increases. In some groups such as Group 1 the boiling temperature decreases because the intermolecular bonding is different and so that has a different effect on the boiling temperature. In other groups the boiling temperature increases and then decreases.

At this level, your answers need technical terms and clarity in expression otherwise you will find yourself losing marks.

COMMENTARY

This is a weak answer because:

- Although the student has correctly stated that the boiling temperature increases down Group 7 and that this is due to increasing London forces, they have not explained why London forces increase down Group 7.
- The student has written about the trend in boiling temperatures in Group 1 and other groups. The question specifically asks about Group 7 so this information cannot gain any marks.

QUESTION TYPE: EXTENDED WRITING

An experiment was carried out involving the addition of aqueous silver nitrate, followed by aqueous ammonia, to distinguish between aqueous solutions containing chloride, bromide or iodide ions.

State the observations that would be made in this experiment and show how they can be used to deduce which ion is in each solution. [6]

It is reasonable to assume that the marks will be divided equally between the two parts of the question; three marks for describing the result you would expect to see for each ion (one for each) when silver nitrate is added with an explanation of how this is used to distinguish each ion, and three marks for describing what you would observe when aqueous ammonia is added to each of these solutions.

Question analysis

- There will be questions in your exams which assess your understanding of practical skills and draw on your experience of the core practicals. For these questions, think about:
- how apparatus is set up
- the method of how the apparatus is to be used
- how readings are to be taken
- how to make the readings reliable
- how to control any variables.
- It helps with extended writing questions to think about the number of marks available and how they might be distributed. For example, if the question asked you to give the arguments for and against a particular case, then assume that there would be equal numbers of marks available for each side of the argument and balance the viewpoints you give accordingly. However, you should also remember that marks will also be available for giving an overall conclusion so you should be careful not to omit that.
- It is vital to plan out your answer before you write it down. There is always space given on an exam paper to do this so just jot down the points that you want to make before you answer the question in the space provided. This will help to ensure that your answer is coherent and logical and that you don't end up contradicting yourself. However, once you have written your answer go back and cross these notes out so that it is clear they do not form part of the answer.

Average student answer

When the aqueous silver nitrate is added to chloride ions a white precipitate of silver chloride forms. This dissolves when ammonia is added.

When it is added to bromide ions a yellow precipitate of silver bromide forms. This dissolves when concentrated ammonia is added.

When it is added to iodide ions a yellow precipitate forms. This doesn't dissolve when ammonia is added however concentrated it is

Be very careful about making statements with the word 'it'. In this answer, it is clear that the student means 'silver nitrate' but if there is ambiguity you may not gain the mark.

COMMENTARY

This is an average answer because:

- The student has given results for all of the tests. However, in several cases the results are not given precisely enough for the marks to be scored. For example, for chloride ions the student does not specify whether the silver chloride precipitate will dissolve in dilute or concentrated ammonia and so does not score the second mark.

QUESTION TYPE: CALCULATION

But-1-ene reacts with hydrogen bromide to form two different products. Analysis of one of the products showed that it contained 35.0% carbon, 6.6% hydrogen and 58.4% bromine by mass.

Calculate the empirical formula of this product. [2]

The command word here is calculate. This means that you need to obtain a numerical answer to the question, showing relevant working. If the answer has a unit, this must be included.

Question analysis

- The important thing with calculations is that you must show your working clearly and fully. The correct answer on the line will gain all the available marks. However, an incorrect answer can gain all but one of the available marks if your working is shown and is correct.
- Show the calculation that you are performing at each stage and not just the result. When you have finished, look at your result and see if it is sensible. In an empirical formula calculation such as this one, if you do not end up with a whole number ratio for the atoms then you probably have an error in your calculation (have you divided the percentage of each element by the atomic number rather than the relative atomic mass?). Go back and check.

Average student answer

carbon	hydrogen	bromine
35	6.6	58.4
$\frac{35}{12} = 2.9$	$\frac{6.6}{1} = 6.6$	$\frac{58.4}{79.9} = 0.73$
$\frac{2.9}{0.73} = 4$	$\frac{6.6}{0.73} = 9$	$\frac{0.73}{0.73} = 1$

Ratio = 4 : 9 : 1

Empirical formula C_4H_9Br

The first mark is awarded for dividing each of the percentage masses given in the question by the relative atomic mass (A_r) of the element to obtain the number of moles. You can see that the student has clearly done this in the second line of the answer above. The second mark is awarded for calculating the ratio of each of the elements present by dividing by 0.73 and then using the result to work out the empirical formula. Again this is clearly shown in the student's answer.

COMMENTARY

This is a strong answer because:

- The student has correctly divided each of the percentage masses by the M_r of the element to obtain the number of moles and so gained the first mark.
- The student has used the moles to correctly work out the ratio of the number of atoms of each element and so has obtained the second of the available marks.
- The answer has been laid out very clearly so that even if an error had been made then a mark could have been awarded for part of the calculation.

COMMAND WORDS

The following table lists the command words used across the IAS/IAL Science qualifications in the external assessments. You should make sure you understand what is required when these words are used in questions in the exam.

COMMAND WORD	THIS TYPE OF QUESTION WILL REQUIRE STUDENTS TO:
ADD/LABEL	Requires the addition or labelling to stimulus material given in the question, for example labelling a diagram or adding units to a table.
ASSESS	Give careful consideration to all the factors or events that apply and identify which are the most important or relevant. Make a judgement on the importance of something, and come to a conclusion where needed.
CALCULATE	Obtain a numerical answer, showing relevant working. If the answer has a unit, this must be included.
COMMENT ON	Requires the synthesis of a number of factors from data/information to form a judgement. More than two factors need to be synthesised.
COMPARE AND CONTRAST	Looking for the similarities **and** differences of two (or more) things. Should not require the drawing of a conclusion. Answer must relate to both (or all) things mentioned in the question. The answer must include at least one similarity and one difference.
COMPLETE/RECORD	Requires the completion of a table/diagram/equation.
CRITICISE	Inspect a set of data, an experimental plan or a scientific statement and consider the elements. Look at the merits and/or faults of the information presented and back judgements made.
DEDUCE	Draw/reach conclusion(s) from the information provided.
DERIVE	Combine two or more equations or principles to develop a new equation.
DESCRIBE	To give an account of something. Statements in the response need to be developed as they are often linked but do not need to include a justification or reason.
DETERMINE	The answer must have an element which is quantitative from the stimulus provided, or must show how the answer can be reached quantitatively.
DEVISE	Plan or invent a procedure from existing principles/ideas.
DISCUSS	Identify the issue/situation/problem/argument that is being assessed within the question. Explore all aspects of an issue/situation/problem. Investigate the issue/situation/problem etc. by reasoning or argument.

COMMAND WORD	THIS TYPE OF QUESTION WILL REQUIRE STUDENTS TO:
DRAW	Produce a diagram either using a ruler or using freehand.
ESTIMATE	Give an approximate value for a physical quantity or measurement or uncertainty.
EVALUATE	Review information then bring it together to form a conclusion, drawing on evidence including strengths, weaknesses, alternative actions, relevant data or information. Come to a supported judgement of a subject's qualities and relation to its context.
EXPLAIN	An explanation requires a justification/exemplification of a point. The answer must contain some element of reasoning/justification, this can include mathematical explanations.
GIVE/STATE/NAME	All of these command words are really synonyms. They generally all require recall of one or more pieces of information.
GIVE A REASON/REASONS	When a statement has been made and the requirement is only to give the reasons why.
IDENTIFY	Usually requires some key information to be selected from a given stimulus/ resource.
JUSTIFY	Give evidence to support (either the statement given in the question or an earlier answer).
PLOT	Produce a graph by marking points accurately on a grid from data that is provided and then drawing a line of best fit through these points. A suitable scale and appropriately labelled axes must be included if these are not provided in the question.
PREDICT	Give an expected result or outcome.
SHOW THAT	Prove that a numerical figure is as stated in the question. The answer must be to at least 1 more significant figure than the numerical figure in the question.
SKETCH	Produce a freehand drawing. For a graph this would need a line and labelled axes with important features indicated, the axes are not scaled.
STATE WHAT IS MEANT BY	When the meaning of a term is expected but there are different ways of how these can be described.
SUGGEST	Use your knowledge and understanding in an unfamiliar context. May include material or ideas that have not been learnt directly from the specification.
WRITE	When the questions ask for an equation.

GLOSSARY

accuracy a measure of how close values are to the accepted or correct value

activation energy, E_a the minimum energy that colliding particles must possess for a reaction to occur

actual yield the actual mass of product obtained in a reaction

addition reaction a reaction in which two molecules combine to form one molecule

aldehyde (RCHO) one of a homologous series of organic compounds formed by the partial oxidation of primary alcohols

atom economy the molar mass of the desired product divided by the sum of the molar masses of all the products, expressed as a percentage

atomic number (*Z*) the number of protons in the nucleus of an atom of that element

Avogadro constant (*L*) 6.02×10^{23}, the number of particles in one mole of a substance

base peak the peak with the greatest abundance

basic oxides oxides of metals that react with water to form metal hydroxides, and with acids to form salts and water

bioalcohol fuel made from plant matter, often using enzymes or bacteria

biodegradable can be broken down by microbes

biofuel fuel obtained from living matter that has died recently

bond enthalpy the enthalpy change when one mole of a bond in the gaseous state is broken

bond length the distance between the nuclei of two atoms that are covalently bonded together

carbocation a positive ion in which the charge is shown on a carbon atom

carbon neutral a considered net zero effect on the amount of carbon dioxide in the atmosphere

carboxylic acid (RCOOH) one of a homologous series of organic compounds formed by the complete oxidation of primary alcohols

catalyst a substance that increases the rate of a chemical reaction but is chemically unchanged at the end of the reaction

coefficient the technical term for the number written in front of species when balancing an equation

complete combustion all of the atoms in the fuel are fully oxidised

concordant titres titres that are close together (usually within 0.20 cm3 of each other)

cracking the breakdown of molecules into shorter ones by heating with a catalyst

curly arrows (full ones, not half-arrows) represent the movement of electron pairs

dative covalent bond the bond formed when an empty orbital of one atom overlaps with an orbital containing a lone pair of electrons of another atom

dehydration a reaction where the hydroxyl group from an alcohol molecule, and a hydrogen atom from an adjacent carbon atom, are removed, forming a C=C double bond

delocalised electrons electrons that are not associated with any single atom or any single covalent bond

diol a compound containing two OH (alcohol) groups

dipole exists when two charges of equal magnitude but opposite signs are separated by a small distance

dipole moment the difference in magnitude between $\delta+$ and $\delta-$ multiplied by the distance of separation between the charges

discrete (simple) molecule an electrically neutral group of two or more atoms held together by covalent bonds

displacement reaction a reaction in which one element replaces another, less reactive, element in a compound

displayed (full structural) formula a formula that shows each bonding pair as a line drawn between the two atoms involved

disproportionation reaction a reaction involving the simultaneous oxidation and reduction of an element in a single species

distillation with addition heating a reaction mixture, but adding another liquid and distilling off the product as it forms

electron pair repulsion (EPR) theory the electron pairs on the central atom of a molecule or ion arrange themselves in order to create the minimum repulsion between them; lone pair-lone pair repulsion is greater than lone pair-bond repulsion, which in turn is greater than bond pair-bond pair repulsion

electronegativity the ability of an atom to attract a bonding pair of electrons in a covalent bond

electronic configuration (of an atom) the number of electrons in each sub-shell in each energy level of the atom

electron-releasing group a group that pushes electrons towards the atom it is joined to

electrophile a species that is attracted to a region of high electron density

electrophilic addition a reaction in which two molecules form one molecule and the attacking molecule is an electrophile

elimination reaction a reaction in which a molecule loses atoms attached to adjacent carbon atoms, forming a C=C double bond

empirical formula the smallest whole-number ratio of atoms of each element in a compound

end point the point at which the indicator just changes colour; ildeally, the end point should coincide with the equivalence point

endothermic a reaction where heat energy is transferred from the surroundings to the system

enthalpy level diagram a diagram that shows the relationship between the enthalpy of the reactants and the enthalpy of the products in a chemical reaction

equivalence point the point at which there are exactly the right amounts of substances to complete the reaction

error the difference between an experimental value and the accepted or correct value

ethanolic a solution in which ethanol is the solvent

exothermic a reaction where heat energy is transferred from the system to the surroundings

first ionisation energy (of an element) the energy required to remove an electron from each atom in one mole of atoms in the gaseous state

fractional distillation a process used to separate liquids with similar boiling temperatures, by boiling and condensing, using a fractionating column

fragmentation occurs when the molecular ion breaks into smaller pieces

free radical a species that contains an unpaired electron

functional group an atom or group of atoms in a molecule that is responsible for its chemical reactions

geometric isomers compounds containing a C=C bond with atoms or groups attached at different positions

groups the vertical columns in the Periodic Table

halogenation a reaction involving the addition of a halogen; a reaction where the hydroxyl group in an alcohol molecule is replaced by a halogen atom

hazard something that could cause harm to a user

heating under reflux heating a reaction mixture with a condenser fitted vertically

Hess's Law the law states that the enthalpy change of a reaction is independent of the path taken in converting reactants into products, provided the initial and final conditions are the same in each case

heterogeneous catalyst a catalyst that is in a different phase to that of the reactants

heterolytic fission the breaking of a covalent bond so that both bonding electrons are taken by one atom

homologous series a family of compounds with the same functional group, which differ in formula by CH2 from the next member

homolytic fission the breaking of a covalent bond where each of the bonding electrons leaves with one species, forming a free radical

Hund's rule electrons will occupy the orbitals singly before pairing takes place

hydrate compound containing water of crystallisation, represented by formulae such as $CuSO_4.5H_2O$

hydration a reaction involving the addition of water (or steam); water molecules are attracted to ions in solution and surround the ions; the oxygen ends of the water molecules are attracted to the positive ions (cations); the hydrogen ends of the water molecules are attracted to the negative ions (anions)

hydrocarbon a compound that contains only carbon and hydrogen atoms

hydrogen bond an intermolecular interaction (in which there is some evidence of bond formation) between a hydrogen atom of a molecule (or molecular fragment) bonded to an atom which is more electronegative than hydrogen and another atom in the same or a different molecule

hydrogenation a reaction involving the addition of hydrogen

hydrolysis reaction a reaction in which water or hydroxide ions replace an atom in a molecule with an –OH group

incineration converting polymer waste into energy by burning

incomplete combustion some of the atoms in the fuel are not fully oxidised

infrared radiation the part of the electromagnetic spectrum with frequencies below that of red light

initiation first step that starts the reaction, involving the formation of free radicals, usually as a result of bond breaking caused by ultraviolet radiation

intensity the amount of infrared radiation absorbed

ionic bonding the electrostatic attraction between oppositely charged ions

isotopes atoms of the same element that have the same atomic number but different mass number

ketone (RCOR) one of a homologous series of organic compounds formed by oxidation of secondary alcohols

locant a number used to indicate which carbon atom in the chain an atom or group is attached to

mass concentration (of a solution) the mass (in g) of the solute divided by the volume of the solution

mass number the sum of the number of protons and the number of neutrons in the nucleus of an atom

mean bond enthalpy the enthalpy change when one mole of a bond, averaged out over many different molecules, is broken

measurement uncertainty the potential error involved when using a piece of apparatus to make a measurement

mechanism the sequence of steps in an overall reaction; each step shows what happens to the electrons involved in bond breaking or bond formation

meniscus the curving of the upper surface in a liquid in a container; the lowest (horizontal) part of the meniscus should be read

metallic bonding the electrostatic force of attraction between the metal cations and delocalised electrons

molar concentration (of a solution) the amount (in mol) of the solute divided by the volume of the solution

molar mass the mass per mole of a substance; it has the symbol *M* and the units $g\,mol^{-1}$

molar volume the volume occupied by 1 mol of any gas; this is normally $24\,dm^3$ or $24\,000\,cm^3$ at r.t.p.

mole the amount of substance that contains the same number of particles as the number of carbon atoms in exactly 12 g of ^{12}C

molecular formula the actual number of atoms of each element in a molecule

molecular ion peak the peak with the highest *m/z* ratio in the mass spectrum, formed from the molecule by the loss of one electron, the
M peak

monomers the small molecules that combine together to form a polymer

nitrile organic compound containing the C–CN group

nucleophile a species that donates a lone pair of electrons to form a covalent bond with an electron-deficient atom

nucleophilic substitution a reaction in which an attacking nucleophile replaces an existing atom or group in a molecule

orbital a region within an atom that can hold up to two electrons with opposite spins

oxidation number the charge that an ion has, or the charge that it would have if the species were fully ionic

oxidation reaction reaction in which a substance gains oxygen or loses hydrogen

oxidation when the oxidation number of an element increases; the loss of electrons

oxidising agent a species (atom, molecule or ion) that oxidises another species by removing one or more electrons; when an oxidising agent reacts it gains electrons and is, therefore, reduced

parts per million (ppm) the number of parts of one substance in one million parts of another substance; a measure used to describe chemical concentration; usually, 'parts' refers to masses of both substances, or to volumes of both substances

Pauli Exclusion Principle two electrons cannot occupy the same orbital unless they have opposite spins. Electron spin is usually shown by using upward and downward arrows: ↑ and ↓

percentage uncertainty the actual measurement uncertainty in an experiment multiplied by 100 and divided by the value recorded

percentage yield the actual yield divided by the theoretical yield, expressed as a percentage

periodic properties (periodicity) regularly repeating patterns of atomic, physical and chemical properties, which can be predicted using the Periodic Table and explained using the electron configurations of the elements

periods the horizontal rows in the Periodic Table

phase a physically distinct form of matter, such as a solid, liquid, gas or plasma; a phase of matter is characterised by having relatively uniform chemical and physical properties

pi bonds covalent bonds formed when electron orbitals overlap sideways

polar covalent bond a type of covalent bond between two atoms where the bonding electrons are unequally distributed; because of this, one atom carries a slight negative charge and the other a slight positive charge

polarisation the distortion of the electron density of a negative ion (anion)

polarising power the ability of a positive ion (cation) to distort the electron density of a neighbouring negative ion (anion)

polymerisation reaction reaction in which a large number of small molecules react together to form one very large molecule

precipitation reaction reaction in which an insoluble solid is formed when two solutions are mixed

precision a measure of how close values are to each other

prefix a set of letters written at the beginning of a name

primary amine compound containing the C–NH2 group

primary standards substances used to make a standard solution by weighing

propagation the two steps that, when repeated many times, convert the starting materials into the products of a reaction

quantum shell the energy level of an electron

random error an error caused by unpredictable variations in conditions

reaction pathway the reaction, or series of reactions, that the reactants undergo in order to change into the products

recycling converting polymer waste into other materials that are useful

redox reaction a reaction that involves both reduction and oxidation

reducing agent a species that reduces another species by adding one or more electrons; when a reducing agent reacts it loses electrons and is, therefore, oxidised

reduction reaction reaction in which a substance loses oxygen or
gains hydrogen

reduction when the oxidation number of an element decreases; the gain of electrons

reforming the conversion of straight-chain hydrocarbons into branched-chain and cyclic hydrocarbons

relative atomic mass (A_r) (of an element) the weighted mean (average) mass of an atom of the element compared to 1/12 of the mass of an atom of carbon-12

relative isotopic mass the mass of an individual atom of a particular isotope relative to 1/12 of the mass of an atom of carbon-12

repeat unit the set of atoms that are joined together in large numbers to produce the polymer structure

risk assessment the identification of the hazards involved in carrying out a procedure and the control measures needed to reduce the risks from those hazards

risk the chance of a hazard causing harm

saturated a compound containing only single bonds

second ionisation energy (of an element) the energy required to remove an electron from each singly charged positive ion in one mole of positive ions in the gaseous state

sigma bonds covalent bonds formed when electron orbitals overlap axially (end-on)

simple distillation a method used to separate liquids with very different boiling temperatures, by boiling and condensing

skeletal formula shows all the bonds between carbon atoms

solute a substance that is dissolved

solution a solute dissolved in a solution

solvent a substance that dissolves a solute

solvent extraction a method used to separate a liquid from a mixture by causing it to move from the mixture to the solvent

spectator ion an ion that is there both before and after the reaction but is not involved in the reaction

standard enthalpy change of atomisation the enthalpy change measured at a stated temperature, usually 298 K, and 100 kPa when one mole of gaseous atoms is formed from an element in its standard state

standard enthalpy change of combustion ($\Delta_c H^\ominus$) the enthalpy change measured at 100 kPa and a stated temperature, usually 298 K, when one mole of a substance is completely burned in oxygen

standard enthalpy change of formation the enthalpy change measured
at 100 kPa and a specified temperature, usually 298 K, when one mole of a substance is formed from its elements in their standard states

standard enthalpy change of neutralisation the enthalpy change measured at 100 kPa and a stated temperature, usually 298 K, when one mole of water is produced by the neutralisation of an acid with an alkali

standard enthalpy change of reaction the enthalpy change which occurs when equation quantities of materials react under standard conditions

standard solution a solution whose concentration is accurately known

stereoisomers compounds with the same structural formula (and the same molecular formula), but with the atoms or groups arranged differently in three dimensions

steric hindrance the slowing of a chemical reaction due to large groups within a molecule getting in the way of the attacking species

stretching when a bond absorbs infrared radiation and uses it to alter the length of the bond

structural formula shows (unambiguously) how the atoms are joined together

structural isomers compounds with the same molecular formula but different structural formulae

sublimation when a solid changes directly into a vapour without melting

substitution reaction reaction in which an atom or group is replaced by another atom or group

suffix a set of letters written at the end of a name

synthesis the production of chemical compounds by reaction from simpler substances

systematic error an error that is constant or predictable, usually because of the apparatus used

temperature gradient the way in which the temperature changes up and down the column

termination final step that involves the formation of a molecule from two free radicals, halting the reaction

theoretical yield the maximum possible mass of a product in a reaction, assuming complete reaction and no losses

thermal stability a measure of the extent to which a compound decomposes when heated

thermometric titration a titration where the endpoint is indicated by a temperature change

third ionisation energy (of an element) the energy required to remove an electron from each doubly charged positive ion in one mole of positive ions in the gaseous state

titre the volume added from the burette during a titration

transmittance the amount of radiation absorbed at a particular wavenumber

unsaturated a compound containing one or more double bonds

wavenumber the frequency of infrared radiation absorbed by a particular bond in a molecule

Key

Atomic (proton number)
Atomic symbol
Name
Relative atomic mass

Group

Period	1 (1)	2 (2)	(3)	(4)	(5)	(6)	(7)	(8)	(9)	(10)	(11)	(12)	3 (13)	4 (14)	5 (15)	6 (16)	7 (17)	8 (18)
1								1 H Hydrogen 1.0										2 He Helium 4.0
2	3 Li Lithium 6.9	4 Be Beryllium 9.0											5 B Boron 10.8	6 C Carbon 12.0	7 N Nitrogen 14.0	8 O Oxygen 16.0	9 F Fluorine 19.0	10 Ne Neon 20.2
3	11 Na Sodium 23.0	12 Mg Magnesium 24.3											13 Al Aluminium 27.0	14 Si Silicon 28.1	15 P Phosphorus 31.0	16 S Sulfur 32.1	17 Cl Chlorine 35.5	18 Ar Argon 39.9
4	19 K Potassium 39.1	20 Ca Calcium 40.1	21 Sc Scandium 45.0	22 Ti Titanium 47.9	23 V Vanadium 50.9	24 Cr Chromium 52.0	25 Mn Manganese 54.9	26 Fe Iron 55.8	27 Co Cobalt 58.9	28 Ni Nickel 58.7	29 Cu Copper 63.5	30 Zn Zinc 65.4	31 Ga Gallium 69.7	32 Ge Germanium 72.6	33 As Arsenic 74.9	34 Se Selenium 79.0	35 Br Bromine 79.9	36 Kr Krypton 83.8
5	37 Rb Rubidium 85.5	38 Sr Strontium 87.6	39 Y Yttrium 88.9	40 Zr Zirconium 91.2	41 Nb Niobium 92.9	42 Mo Molybdenum 95.9	43 Tc Technetium (98)	44 Ru Ruthenium 101.1	45 Rh Rhodium 102.9	46 Pd Palladium 106.4	47 Ag Silver 107.9	48 Cd Cadmium 112.4	49 In Indium 114.8	50 Sn Tin 118.7	51 Sb Antimony 121.8	52 Te Tellurium 127.6	53 I Iodine 126.9	54 Xe Xenon 131.3
6	55 Cs Caesium 132.9	56 Ba Barium 137.3	57 La* Lathanum 138.9	72 Hf Hafnium 178.5	73 Ta Tantalum 180.9	74 W Tungsten 183.8	75 Re Rhenium 186.2	76 Os Osmium 190.2	77 Ir Iridium 192.2	78 Pt Platinum 195.1	79 Au Gold 197.0	80 Hg Mercury 200.6	81 Ti Thallium 204.4	82 Pb Lead 207.2	83 Bi Bismuth 209.0	84 Po Polonium (209)	85 At Astatine (210)	86 Rn Radon (222)
7	87 Fr Francium (223)	88 Ra Radium (226)	89 Ac* Actinium (227)	104 Rf Rutherfordium (261)	105 Db Dubnium (262)	106 Sg Seaborgium (266)	107 Bh Bohrium (264)	108 Hs Hassium (277)	109 Mt Meitnerium (268)	110 Ds Darmstadtium (271)	111 Rg Roentgenium (272)	Cn Copernicium 112		Fl flerovium 114		Lv livermorium 116		
				58 Ce Cerium 140.1	59 Pr Praseodymium 140.9	60 Nd Neodymium 144.2	61 Pm Promethium 144.9	62 Sm Samarium 150.4	63 Eu Europium 152.0	64 Gd Gadolinium 157.2	65 Tb Terbium 158.9	66 Dy Dysprosium 162.5	67 Ho Holium 164.9	68 Er Erbium 167.3	69 Tm Thulium 168.9	70 Yb Ytterbium 173.0	71 Lu Lutetium 175.0	
				90 Th Thorium 232.0	91 Pa Protactinium (231)	92 U Uranium 238.1	93 Np Neptunium (237)	94 Pu Plutonium (242)	95 Am Americium (243)	96 Cm Curium (247)	97 Bk Berkelium (245)	98 Cf Californium (251)	99 Es Einsteinium (254)	100 Fm Fermium (253)	101 Md Mendeleevium (256)	102 No Nobelium (254)	103 Lr Lawrencium (257)	

INDEX

P

Q

R

S